D1288318

(4) CHURCH OF ALL SAINTS, NORTH STREET. Window I. St. Anne teaching the Virgin, early 15th-century.

AN INVENTORY OF THE HISTORICAL MONUMENTS IN THE

CITY OF YORK

VOLUME III
SOUTH-WEST OF THE OUSE

ROYAL COMMISSION ON HISTORICAL MONUMENTS
ENGLAND

MCMLXXII

SBN 11 700466 9*

Printed in England for Her Majesty's Stationery Office by
Alden & Mowbray Ltd, at the Alden Press, Oxford

TABLE OF CONTENTS

LIST OF COMMISSIONERS

The Most Honourable the Marquess of Salisbury, K.G. (Chairman)
Her Majesty's Lieutenant in the City of York and the West Riding (*ex officio*)
Henry Clifford Darby, Esq., O.B.E.
Courtenay Arthur Ralegh Radford, Esq.
Sir John Newenham Summerson, C.B.E.
Howard Montagu Colvin, Esq., C.B.E.
William Abel Pantin, Esq.
Arnold Joseph Taylor, Esq., C.B.E.
William Francis Grimes, Esq., C.B.E.
Maurice Willmore Barley, Esq.
Sheppard Sunderland Frere, Esq.
Richard John Copland Atkinson, Esq.
Sir John Betjeman, C.B.E.
John Nowell Linton Myres, Esq.
Harold McCarter Taylor, Esq., C.B.E.
George Zarnecki, Esq., C.B.E.

Secretary

Arthur Richard Dufty, Esq., C.B.E.

LIST OF ILLUSTRATIONS

(The prefixed numerals in brackets refer to the Monument numbers in the text. Houses without such numbers are included in the Lists of 19th-century houses on pp. 123–131).

SECULAR

THE ROYAL WARRANTS

WHITEHALL,
2ND OCTOBER, 1963

The QUEEN has been pleased to issue a Commission under Her Majesty's Royal Sign Manual to the following effect:

ELIZABETH R.

ELIZABETH THE SECOND, by the Grace of God of the United Kingdom of Great Britain and Northern Ireland and of Our other Realms and Territories, QUEEN, Head of the Commonwealth, Defender of the Faith,

To

Our Right Trusty and Entirely-beloved Cousin and Counsellor Robert Arthur James, Marquess of Salisbury, Knight of Our Most Noble Order of the Garter;

Our Trusty and Well-beloved:

Sir Albert Edward Richardson, Knight Commander of the Royal Victorian Order;
Sir John Newenham Summerson, Knight, Commander of Our Most Excellent Order of the British Empire;
Nikolaus Pevsner, Esquire, Commander of Our Most Excellent Order of the British Empire;
Christopher Edward Clive Hussey, Esquire, Commander of Our Most Excellent Order of the British Empire;
Ian Archibald Richmond, Esquire, Commander of Our Most Excellent Order of the British Empire;
Henry Clifford Darby, Esquire, Officer of Our Most Excellent Order of the British Empire;
Donald Benjamin Harden, Esquire, Officer of Our Most Excellent Order of the British Empire;
John Grahame Douglas Clark, Esquire;
Howard Montagu Colvin, Esquire;
Vivian Hunter Galbraith, Esquire;
William Abel Pantin, Esquire;
Stuart Piggott, Esquire;
Courtenay Arthur Ralegh Radford, Esquire;
Arnold Joseph Taylor, Esquire;
Francis Wormald, Esquire.

GREETING!

Whereas We have deemed it expedient that the Commissioners appointed to the Royal Commission on the Ancient and Historical Monuments and Constructions of England shall serve for such periods as We by the hand of Our First Lord of the Treasury may specify and that the said Commissioners shall, if The National Buildings Record is liquidated, assume the control and management of such part of The National Buildings Record's collection as does not solely relate to Our Principality of Wales and to Monmouthshire, and that a new Commission should issue for these purposes:

Now Know Ye that We have revoked and determined, and do by these Presents revoke and determine, all the Warrants whereby Commissioners were appointed on the twenty-ninth day of March one thousand nine hundred and forty six and on any subsequent date:

And We do by these Presents authorize and appoint you, the said Robert Arthur James, Marquess of Salisbury (Chairman), Sir Albert Edward Richardson, Sir John Newenham Summerson, Nikolaus Pevsner, Christopher Edward Clive Hussey, Ian Archibald Richmond, Henry Clifford Darby, Donald Benjamin Harden, John Grahame Douglas Clark, Howard Montagu Colvin, Vivian Hunter Galbraith, William Abel Pantin, Stuart Piggott, Courtenay Arthur Ralegh Radford, Arnold Joseph Taylor and Francis Wormald to be Our Commissioners for such periods as We may specify in respect of each of you to make an inventory of the Ancient and Historical Monuments and Constructions connected with or illustrative of the contemporary culture, civilisation and conditions of life of the people in England, excluding Monmouthshire, from the earliest times to the year 1714, and such further Monuments and Constructions subsequent to that year as may seem in your discretion to be worthy of mention therein, and to specify those which seem most worthy of preservation.

And Whereas We have deemed it expedient that Our Lieutenants of Counties in England should be appointed ex-officio Members of the said Commission for the purposes of that part of the Commission's inquiry which relates to ancient and historical monuments and constructions within their respective counties:

Now Know Ye that We do by these Presents authorize and appoint Our Lieutenant for the time being of each and every County in England, other than Our County of Monmouth, to be a Member of the said Commission for the purposes of that part of the Commission's inquiry which relates to ancient and historical monuments and constructions within the area of his jurisdiction as Our Lieutenant of such County:

And for the better enabling you to carry out the purposes of this Our Commission, We do by these Presents authorize you to call in the aid and co-operation of owners of ancient monuments, inviting them to assist you in furthering the objects of the Commission; and to invite the possessors of such papers as you may deem it desirable to inspect to produce them before you:

And We do further authorize and empower you to confer with the Council of The National Buildings Record from time to time as may seem expedient to you in order that your deliberations may be assisted by the reports and records in the possession of the Council: and to make such arrangements for the furtherance of objectives of common interest to yourselves and the Council as may be mutually agreeable:

And We do further authorize and empower you to assume the general control and management (whether as Administering Trustees under a Scheme established under the Charities Act 1960 or otherwise) of that part of the collection of The National Buildings Record which does not solely relate to our Principality of Wales or to Monmouthshire and (subject, in relation to the said part of that collection, to the provisions of any such Scheme as may be established affecting the same) to make such arrangements for the continuance and furtherance of the work of The National Buildings Record as you may deem to be necessary both generally and for the creation of any wider record or collection containing or including architectural, archaeological and historical information concerning important sites and buildings throughout England:

And We do further give and grant unto you, or any three or more of you, full power to call before you such persons as you shall judge likely to afford you any information upon the subject of this Our Commission; and also to call for, have access to and examine all such books, documents, registers and records as may afford you the fullest information on the subject and to inquire of and concerning the premises by all other lawful ways and means whatsoever:

And We do by these Presents authorize and empower you, or any three or more of you, to visit and personally inspect such places as you may deem it expedient so to inspect for the more effectual carrying out of the purposes aforesaid:

And We do by these Presents will and ordain that this Our Commission shall continue in full force and virtue, and that you, Our said Commissioners, or any three or more of you, may from time to time proceed in the execution thereof, and of every matter and thing therein contained, although the same be not continued from time to time by adjournment:

And We do further ordain that you, or any three or more of you, have liberty to report your proceedings under this Our Commission from time to time if you shall judge it expedient so to do:

And Our further Will and Pleasure is that you do, with as little delay as possible, report to Us, under your hands and seals, or under the hands and seals of any three or more of you, your opinion upon the matters herein submitted for your consideration.

Given at Our Court at Saint James's the Twenty-eighth day of September, 1963, in The Twelfth Year of Our Reign.

By Her Majesty's Command,

HENRY BROOKE

ROYAL COMMISSION ON THE ANCIENT AND HISTORICAL MONUMENTS AND CONSTRUCTIONS OF ENGLAND

REPORT to The Queen's Most Excellent Majesty

MAY IT PLEASE YOUR MAJESTY

We, the undersigned Commissioners appointed to make an Inventory of the Ancient and Historical Monuments and Constructions connected with or illustrative of the contemporary culture, civilisation and conditions of life of the people of England, excluding Monmouthshire, from the earliest times to the year 1714, and such further Monuments and Constructions subsequent to that year as may seem in our discretion to be worthy of mention therein, and to specify those which seem most worthy of preservation, do humbly submit to Your Majesty the following Report, being the twenty-ninth Report on the Work of the Commission since its first appointment.

2. We have pleasure in reporting the completion of our recording of the Monuments in the south-western part of the City of York, an area including seven parishes containing 164 monuments.

3. Following our usual practice we have prepared a full illustrated Inventory of the monuments in south-west York, which will be issued as *York* III. As in other recent Inventories, the Commissioners have adopted the terminal date 1850 for the monuments included in the Inventory.

4. The methods adopted in previous Inventories of describing post-Roman buildings and mediaeval and later earthworks, etc., have been adhered to in general. An innovation is the method of recording the many houses built between 1800 and 1850. The development and redevelopment in this period had a considerable effect upon the life and aspect of the city; for brevity therefore, without emasculation of the record, a simple tabulated list of these houses has been given, accompanied by short, general accounts of their architectural features and components.

5. York has a particularly rich inheritance of documentary records and has numbered many devoted historians among its citizens during the past three hundred years. The wealth of primary and secondary sources of information thus available has enabled us to establish close dating for many of the monuments included in the Inventory and often to attribute works to named architects and craftsmen. As a result, the Inventory establishes chronological and stylistic criteria for the region.

6. An Inventory of the Roman remains in the city was published in 1962 under the title: *City of York* I. *Eburacum, Roman York*. Monuments of a military character are listed and described in *York* II, *The Defences* (1972), which includes, as well as the Bars and Walls, Baile Hill and the former earthwork from which The Mount takes its name.

7. Our special thanks are due to incumbents and churchwardens and to owners and occupiers who have allowed access by our staff to the monuments in their charge. We are indebted to the Directors, Curators and those in charge of many institutions for their ready assistance to us, and particularly to Mr. T. Doherty and Mr. O. S. Tomlinson, successive York City Librarians, to the late Mr. J. Biggins and Mr. L. M. Smith of the Reference Library, to the Rev. Canon R. Cant and Mr. C. B. L. Barr of the

York Minster Library, to Mrs. N. K. M. Gurney, of the Borthwick Institute, to Mr. N. Higson of the East Riding County Record Office, to Mr. C. K. C. Andrew and Mr. M. Y. Ashcroft of the North Riding County Record Office, to Sir Anthony Wagner, Garter King of Arms, and the Chapter of the College of Arms, to Dr. R. W. Hunt and the Bodleian Library, Oxford, and to Mr. J. Hopkins of the Society of Antiquaries of London; also to Mr. G. F. Willmott, formerly Keeper of the Yorkshire Museum, to the officials of the York City Art Gallery, to Miss F. E. Wright, Hon. Secretary of the Yorkshire Architectural and York Archaeological Society, to Miss A. G. Foster and Mr. D. J. H. Michelmore of the Yorkshire Archaeological Society, and also to Miss E. Brunskill, the Ven. the Archdeacon of York, C. R. Forder, the partners in the firm of Messrs. Brierley, Leckenby and Keighley, architects, and Mr. W. Jesse Green, some of whose photographs of stained glass are reproduced by permission of the Dean and Chapter. The late the Rev. Canon J. S. Purvis, the late the Rev. Canon B. A. Smith and the late the Rev. Angelo Raine all gave us valuable help for which we are grateful. We wish also to record our gratitude to Mr. T. D. Tremlett, who has continued to advise us on heraldic matters.

8. We humbly recommend to Your Majesty's notice the following Monuments in the south-west area of the City of York as most worthy of preservation:

ECCLESIASTICAL:

(4) PARISH CHURCH OF ALL SAINTS, North Street, dating from *c.* 1100, with good tower and spire; chancel and aisle roofs of *c.* 1470; remarkable 14th and 15th-century stained glass.

(5) PARISH (former PRIORY) CHURCH OF HOLY TRINITY, Micklegate, on a Saxon site, consisting of nave of alien Benedictine Priory and dating from *c.* 1100; tower of 1453.

(6) PARISH CHURCH OF ST. JOHN THE EVANGELIST, Micklegate, dating from the early 12th century but mostly of later mediaeval date; tower with timber-framed upper part of 1646.

(7) PARISH CHURCH OF ST. MARTIN-CUM-GREGORY, Micklegate, dating from the 11th century but mostly of later mediaeval date; brick tower of 1677; good stained glass and other fittings.

(8) PARISH CHURCH OF ST. MARY BISHOPHILL JUNIOR, dating from the 10th century; Saxon tower, for the rest 12th to 15th-century.

(12) PARISH CHURCH OF ST. EDWARD THE CONFESSOR, Dringhouses, 1847–9, in the Gothic style; connected with the Oxford Movement.

(13) CONVENT OF THE INSTITUTE OF ST. MARY, called THE BAR CONVENT (R.C.), a well documented group of buildings dating from 1765 and including a chapel designed by Thomas Atkinson.

SECULAR:

(20) OUSE BRIDGE, 1810–20, designed by Peter Atkinson II.

(24) MIDDLETON'S HOSPITAL, Skeldergate, 1827–9, designed by Peter Atkinson II.

(28) THE OLD RAILWAY STATION, opened 1841, designed by G. T. Andrews.

(38) BISHOPHILL HOUSE, dating from the early 18th century, for the fine rococo ceiling, probably by Francesco Cortese, and the bow window with bronze glazing bars.

(52) HOLGATE HOUSE, central block finished by 1774, with original staircase and fireplaces.

(55) Nos. 3–9 MICKLEGATE, completed by 1727; two large houses of the Thompson family, with impressive exteriors in fine brickwork, first-floor saloon and good fittings.

(65) Nos. 53–55 MICKLEGATE, finished c. 1755; house, probably by John Carr, with fine wood and plasterwork.

(66) No. 54 MICKLEGATE, finished 1757; house of Garforth family, probably by John Carr, with fine fittings; surviving more complete than perhaps any other house of the period in the area.

(72) No. 68 MICKLEGATE, mid 17th-century; for the fine oak staircase of that date. Of historical interest as Henry Gyles's house.

(79) Nos. 85, 87, 89 MICKLEGATE, late mediaeval timber-framed houses, with important jettied exterior, recently restored.

(80) No. 86 MICKLEGATE, house of the Bathurst family, dating from *c.* 1720, with good staircase.

(81) Nos. 89–90 MICKLEGATE, house of the Bourchiers, finished by 1752 and probably designed by John Carr; though largely gutted, retains fine staircase and ceiling and other fittings and is the most important town house in the area.

(87) Nos. 99, 101 and 103 MICKLEGATE, mid 14th-century timber-framed houses, with original roof structure.

(95) Nos. 118, 120 MICKLEGATE, built *c.* 1742; house and fittings.

(96) Nos. 122–126 MICKLEGATE, for the staircase and panelling.

(104) CHURCH COTTAGES, No. 31 North Street with Nos. 1, 2 All Saints Lane, timber-framed, late 15th-century.

SECULAR (continued)

(114) No. 52 Skeldergate, saw mill, 1839; industrial building.

(117) No. 56 Skeldergate, *c.* 1770; house, probably by John Carr, with noteworthy front and good fittings.

(120) No. 1 Tanner Row with No. 39 North Street; late 15th-century timber-framed house.

(122) No. 7A Tanner Row (The Old Rectory), timber-framed house of *c.* 1600, with staircase from Alne Hall.

(125) Jacob's Well, Trinity Lane; late mediaeval house, timber framed, with late mediaeval door-canopy from elsewhere.

The houses on either side of The Mount and in Mount Vale together with The Mount Villas (Nos. 304, 306 Tadcaster Road) are worthy of preservation for their scenic value; many contain good fittings.

Acomb

(137–8) Manor Farm, No. 14 Front Street, and Manor House, No. 14A Front Street; house of the late 15th or early 16th century, the oldest in Acomb.

(141) No. 21 Front Street (The Lodge), built *c.* 1700; for the staircase.

(142) No. 23 Front Street (Acomb House), of the first half of the 18th century and later, with good staircase and other fittings.

Middlethorpe

(163) Middlethorpe Hall, 1699–1701; the most important country house in the area, with fine original fittings; in a good setting.

9. In compiling the foregoing list our criteria have been architectural or archaeological importance, rarity, not only in the national but in the local field, and the degree of loss to the nation that would result from destruction, always bearing in mind the extent to which the monuments are connected with or illustrative of the contemporary culture, civilisation and conditions of life of the people of England, as required by Your Majesty's Warrant. The list thus has an entirely scholarly basis.

10. We desire to express our acknowledgement of the good work accomplished by our executive staff in the preparation and production of this Inventory, in particular by Mr. J. H. Harvey, F.S.A., the results of whose industry and careful researches before his resignation in 1970 have added so much to the interest and value of the Inventory, and by Dr. E. A. Gee, M.A., F.S.A., F.R.HIST.S., whose work has complemented that of Mr. Harvey; by Messrs. R. W. McDowall, O.B.E., M.A., F.S.A., T. W. French, M.A., F.S.A., J. E. Williams, E.R.D., A.R.C.A., F.S.A., and D. W. Black, B.A.; by Messrs. H. G. Ramm, M.A., F.S.A., D. P. Dymond, M.A., F.S.A., A. G. Chamberlain, the late J. Radley, M.A., F.S.A., and I. R. Pattison, B.A.; by our photographers Mr. W. C. Light and Mr. C. J. Bassham, by our draughtsman Mr. A. R. Whittaker, and by Mrs. J. Bryant, who helped with the editorial work throughout.

11. We desire to add that our Secretary and General Editor, Mr. A. R. Dufty, C.B.E., F.S.A., A.R.I.B.A., has afforded constant assistance to us Your Commissioners.

12. The record cards for York may be consulted by accredited persons who give notice of their intention to the Secretary of the Commission. Copies of photographs may be bought on application to the National Monuments Record.

13. The next Inventory of the Monuments in the City of York will be devoted to the monuments outside the city walls, E. of the river Ouse.

Signed:

Salisbury (*Chairman*)
Kenneth Hargreaves
H. C. Darby
C. A. Ralegh Radford
John Summerson
Francis Wormald
H. M. Colvin
D. B. Harden
W. A. Pantin
A. J. Taylor
W. F. Grimes
M. W. Barley
S. S. Frere
R. J. C. Atkinson
John Betjeman
J. N. L. Myres

A. R. Dufty (*Secretary*)

ABBREVIATIONS USED IN THE TEXT

AASRP	Associated Architectural Societies' *Reports and Papers*
Allen	T. Allen, *A New and Complete History of the County of York*, 3 vols. (1828–31)
Ant. J.	Society of Antiquaries of London, *Antiquaries Journal*
APS	Architectural Publication Society
Arch. J.	*Archaeological Journal*
Benson, I	G. Benson, *York from its Origins to the end of the 11th century* (1911)
„ II	G. Benson, *Later Mediaeval York, 1100–1603* (1919)
„ III	G. Benson, *York from the Reformation to the year 1925* (1925)
Borthwick Inst.	Documents in the Borthwick Institute, York
Browne	J. Browne, *A History of the Metropolitan Church of St. Peter, York* (1847)
Brunton Knight	*see* Knight
Cave	H. Cave, *Antiquities of York* (1813). See also *Picturesque Buildings of York* (1810)
CCR	*Calendar of Close Rolls*
CFR	*Calendar of Fine Rolls*
Colvin	H. M. Colvin, *Biographical Dictionary of English Architects 1660–1840* (1954)
Cooper	T. P. Cooper, York: miscellaneous notes in manuscript, MS. in York City Library
CPL	*Calendar of Papal Letters*
CPR	*Calendar of Patent Rolls*
CSP	*Calendar of State Papers*
CWAAS	Cumberland and Westmorland Antiquarian and Archaeological Society
Davies	R. Davies, *Walks through the City of York* (1880)
Drake	F. Drake, *Eboracum* (1736)
EPNS	English Place-Name Society
EYC	*Early Yorkshire Charters*, 12 vols.: Vols. I–III, ed. W. Farrer (1914–16); Vols. IV–XII, ed. C. T. Clay, Yorkshire Archaeological Society *Record Series*, Extra Series, I–X (1935–65)
Fallow and McCall	T. M. Fallow and H. B. McCall, *Yorkshire Church Plate* (1912)
Freemen	*Register of the Freemen of the City of York*, I, 1272–1558; II, 1559–1759. Surtees Society, XCVI (1897), CII (1900)
Gent	T. Gent, *The Antient and Modern History of the Famous City of York* (1730)
Gent's Ripon	T. Gent, *History of the Loyal Town of Rippon* (1733)
Greenwood	C. Greenwood, Map of the County of York, 1818
Gunnis	R. Gunnis, *Dictionary of British Sculptors 1660–1851* (1953)
Halfpenny	J. Halfpenny, *Fragmenta Vetusta, or the Remains of Ancient Buildings in York* (1807)
Hargrove	W. Hargrove, *History and Description of the Ancient City of York*, 3 vols. (1818)
Harrison	F. Harrison, *The Painted Glass of York* (1927)
Haverfield and Greenwell	F. Haverfield and W. Greenwell, *Catalogue of Stones in the Cathedral Library, Durham* (1899)
Knight	C. B. Knight, *A History of the City of York* (1944)
Knowles	J. W. Knowles, MSS. in York City Library: Notes and Newspaper cuttings on the Lives and Works of York Artists and Sculptors, 2 vols.; The Stained Glass of York Churches, 2 vols.
Lawton	G. Lawton, *Collectio rerum ecclesiasticarum de Diœcesi Eboracensi*, 2nd ed. (1842)
LPH	*Letters and Papers of Henry VIII*
Med. Arch.	*Medieval Archaeology*
Morrell *Monuments*	J. B. Morrell, *York Monuments* (1944)
Morrell *Woodwork*	J. B. Morrell, *Woodwork in York* (1949)
NMR	National Monuments Record
NRCRO	North Riding County Record Office
OS	Ordnance Survey
Raine	A. Raine, *Mediaeval York* (1955)
RCHM	Royal Commission on Historical Monuments (England)
RS	Rolls Series
Salzman	L. F. Salzman, *Building in England down to 1540* (1952)
Shaw	Rev. P. J. Shaw (ed.), *An Old York Church, All Hallows in North Street* (1908)
Sheahan and Whellan	J. J. Sheahan and T. Whellan, *History and Topography of the City of York, the Ainsty Wapentake and the East Riding of Yorkshire*, 2 vols. (1855–6)
Skaife	R. H. Skaife, Civic Officials and Parliamentary Representatives of York, MS. in York City Library
SS	Surtees Society
Stukeley	W. Stukeley, *Diaries and Letters*, 3 vols., Surtees Society, LXXIII (1880), LXXVI (1883), LXXX (1885)
TE	*Testamenta Eboracensia*, 6 vols., Surtees Society, IV (1836), XXX (1855), XLV (1865), LIII (1868), LXXIX (1884), CVI (1902)
VCH	The Victoria History of the Counties of England
Wellbeloved	C. Wellbeloved, *Eburacum or York under the Romans* (1842)
Wills	Except where otherwise stated, Wills in the York Registry, now in the Borthwick Institute
Wood	M. Wood, *The English Mediaeval House* (1965)
YAJ	*Yorkshire Archaeological Journal*

YAS	Yorkshire Archaeological Society
YASRS	Yorkshire Archaeological Society *Record Series*
YAYAS	Yorkshire Architectural and York Archaeological Society
YC	*York Courant*
YCA	York City Archives
YCL	York City Library
YCR	York Civic Records, 8 vols., Yorkshire Archaeological Society *Record Series*, XCVIII, CIII, CVI, CVIII, CX, CXII, CXV, CXIX (1939–52)
YFR	*Fabric Rolls of York Minister*, Surtees Society, XXXV (1858)
YG	*Yorkshire Gazette*
YGS	York Georgian Society
YH	*York Herald*
YM	Yorkshire Museum
YMH	*A Hand-Book to the Antiquities in the Grounds and Museum of the Yorkshire Philosophical Society*
York I	Royal Commission on Historical Monuments, *The City of York*: Volume 1, *Eburacum* (1962)
York Survey 1959	G. F. Willmot, J. M. Biggins, P. M. Tillot (ed.), *York, A Survey, 1959*
YPS	Yorkshire Philosophical Society *Reports* (offprints repaginated).

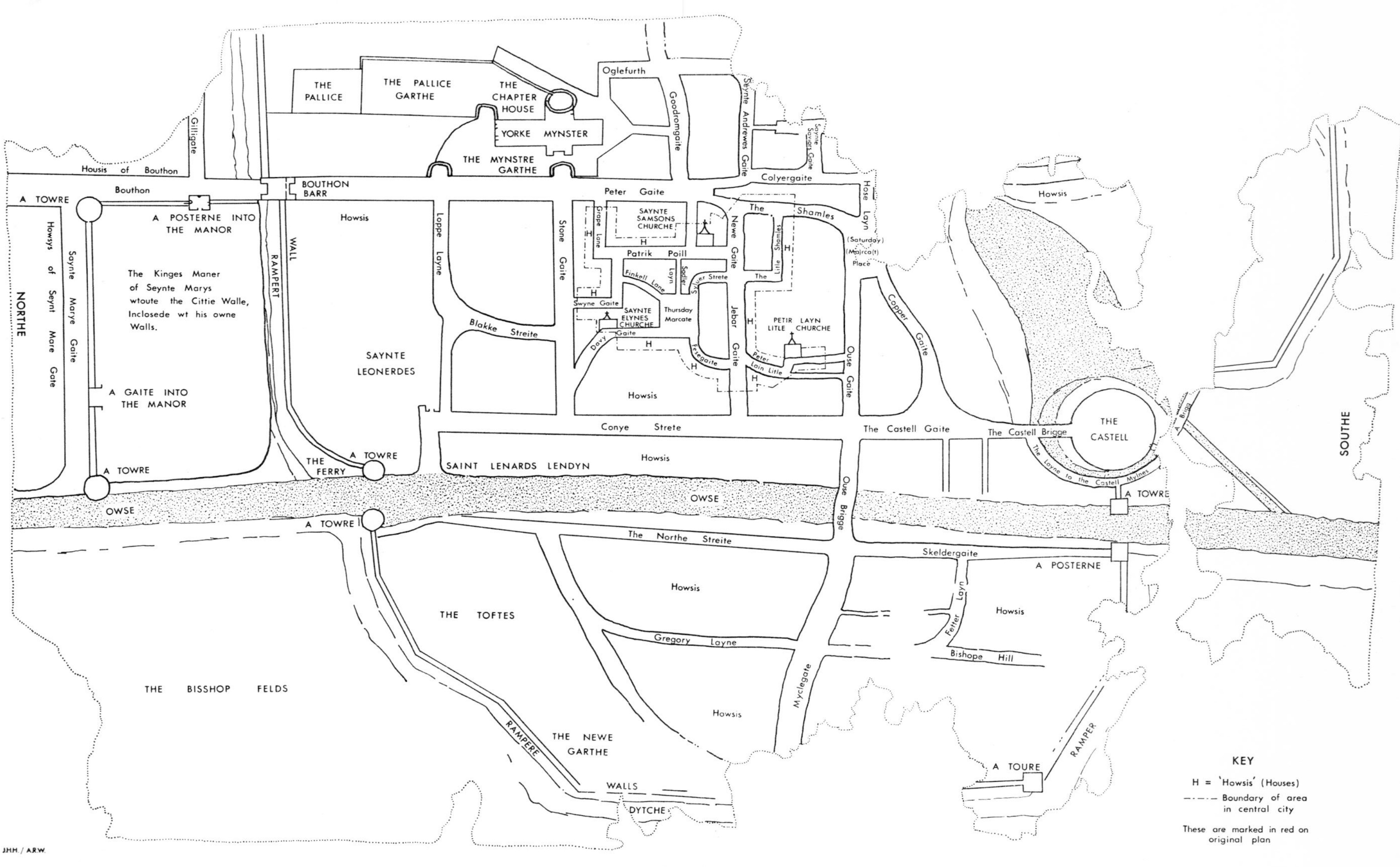

Fig. 1. Plan of York *c.* 1545.

CITY OF YORK
SOUTH-WEST OF THE OUSE

THE area described in this volume comprises the part of the walled city S.W. of the Ouse, both the ancient suburb outside Micklegate Bar and Clementhorpe, which have always been part of the liberty of the city, Holgate, which was included in the city in 1884, and a ring of outer villages, Middlethorpe, Dringhouses and Acomb, included in 1937.

The walled part of the city, of 60 acres, extends some 800 yds. along the river bank northward from the natural gully which separated the city from Clementhorpe; it has a maximum width of 500 yds. and is divided into two unequal parts by Micklegate, which links the main S.W. gate, Micklegate Bar, and Ouse Bridge. It includes part of the morainic ridge (Bishophill and Micklegate Hill) with slopes to the river.

The outer villages, now suburbs, were formally separated from the city and inner suburbs by their own fields and by the large open spaces of the Knavesmire, Hob Moor, and Bishopfields, over which the citizens had rights and of which the first two still survive.

Roman monuments and the city defences have been described in the Inventories *York* I and II respectively.

AUTHORITIES

The sources for the history and dating of York buildings are extremely diverse. For the period between the Norman Conquest and the reign of Henry VIII much information has been gleaned from the study of title-deeds, of which many are in print in published monastic cartularies and in two series produced by the Yorkshire Archaeological Society (*Early Yorkshire Charters*, and *Yorkshire Deeds*). There is unpublished material of the same kind in the York City Archives and among the muniments of the Dean and Chapter. From the 14th century onwards these are supplemented by the evidence of wills (Borthwick Institute and York Minster Library). For the churches, and particularly for the building of chantry chapels, there is much miscellaneous material in the calendared Chancery enrolments supplemented by original documents in York collections.

Since the 16th century a long series of antiquaries, local and national, has collected an immense body of material, much of it published. Short notes on the principal antiquarians and historians who have worked on York are appended (pp. xxx–xxxi). Their written work is sometimes supplemented by plans and drawings from their own hands, and also by the output of a number of artists and surveyors (*see* pp. xxxii–xxxiv). The evidence provided by the series of plans of the city is of considerable importance, though York is not so fortunate as to have any early plan showing details of individual properties, comparable to that of Cambridge by John Hamond engraved in 1592 (*see* pp. xxxiv–xxxvi).

For the history of houses the most important sources, apart from a few surviving series of early title-deeds in the possession of the present owners, are the Registers of Deeds and the Parish Rate Books. The volumes of Deeds registered before the Lord Mayor (YCA, E.93–E.98), covering the period 1719–1866 (few after 1832), include a high proportion of the conveyances and mortgages of real estate within the

City and Ainsty though the system was not, as in each of the three Ridings, compulsory. Occasionally the text of successive deeds at a short interval affords proof of a precise date of building. This is often confirmed by changes of assessment in the Parish Rates, where these survive. In the part of York S.W. of the Ouse there are continuous assessments for Holy Trinity, Micklegate, from 1774, from 1798 for St. Martin-cum-Gregory (both in the Borthwick Institute); for St. John the Evangelist only from 1837 (York City Library); for St. Mary Bishophill Senior a broken series from the mid 18th century (St. Clement's Rectory); for All Saints', North Street, fragments; for Bishophill Junior nothing.

Mention must also be made of the rich stores of contemporary information on York houses of the Georgian and Victorian periods, and on their owners and occupiers, to be found in the local newspapers. The extant files of the York journals (from 1728) are at the York City Library, where are detailed indexes extending from the earliest issues until after 1850, and in progress for later dates.

Antiquaries and Historians of York

Roger Dodsworth (1585–1654), the earliest and most diligent of the great Yorkshire antiquaries, compiled 160 volumes in MS. before the outbreak of the Civil War; these are preserved in the Bodleian Library, Oxford. Several of them are of particular importance as containing detailed accounts of church monuments in the city of York *c.* 1620, notably the complementary MSS. Dodsworth 157 and 161. The work of Dodsworth is for the most part accurate, and his errors are largely due to the illegibility of inscriptions and the decayed state of monuments and stained glass.

Sir Thomas Widdrington (*c.* 1600–64), recorder of York, compiled the first narrative history of the city, not published until the end of the 19th century (*Analecta Eboracensia*, ed. C. Caine (1897)). Widdrington made extensive use of the original records of the Corporation, but his work needs to be used with caution.

Sir William Dugdale (1605–86) took copious notes of Yorkshire heraldry in 1641 and again in 1666 and caused a fully illustrated and indexed compendium of these, together with material taken from the earlier heralds' visitations and the notes of Dodsworth, to be compiled. This is known as the 'Book of Yorkshire Arms' and is preserved at the College of Arms.

Matthew Hutton, D.D. (1639–1711), made extremely accurate church notes in various parts of England and also copied extensively from records not now extant. His extracts from the Chapter Act Books and other registers of York Minster are in the British Museum (Harleian MSS. 6950–85, 7519–21; especially 6984 of 1661), but a volume of church notes largely from York is in the Minster Library (another holograph copy is BM, Lansdowne MS. 919).

Nathaniel Johnston, M.D. (1627–1705), of Pontefract, made very extensive collections for the history of Yorkshire, now mostly in the Bodleian Library, Oxford (MSS. Top. Yorks. C.13–45) and in the Public Libraries of Leeds (Bacon Frank MSS.) and Sheffield (Bacon Frank MSS. 1–13). At least four volumes of collected plans and drawings (Bernard, *Cat. MSS. Angliae* (1697), II, p. 101, nos. 3861, 3863–5) have been lost but two, containing many drawings by Nathaniel's younger brother Henry Johnston (*c.* 1640–1723), survive (Bodleian, MSS. Top. Yorks. C.13, 14).

James Torre (1649–99), made extremely detailed collections on York Minster, with plans and sketches, and for the churches of the county and diocese. He used, and made abstracts from, many registers now lost (see York Minster Library, L.1.2 etc.) and collected 'testamentary burials' for the Minster and churches from the registers of York Wills. His work, though occasionally inaccurate, is of the utmost importance for the state of fabrics and monuments in *c.* 1687–91. Among his other work is 'The Antiquities of York City collected out of the Papers of Christopher Hildyard Esq.', with further detailed notes on churches (MS. in York City Library, Y.942.74–4259).

Henry Keepe (1652–88), the author of *Monumenta Westmonasteriensia* (1682), also made substantial collections in 1680 for a history and description of the city of York, *Monumenta Eboracensia.* What survives is preserved in the library of Trinity College, Cambridge (MS. O.4.33), and includes a detailed perambulation in which Keepe was guided by 'Mr. Andrew Davye [1630–82] an antient native here, a great lover of this his birthplace and not a little verst in Antiquities . . .'.

Thomas Gent (1693-1778), the eccentric printer, settled in York in 1724 and published his *History of the Famous City of York* in 1730; substantial addenda are included in his later *History of the Loyal Town of Rippon* (1733). Though Gent's methods were slapdash and his woodcut illustrations crude, he preserved much valuable information on York churches, and especially their stained glass, not recorded elsewhere.

Francis Drake, F.R.S., F.S.A. (1696–1771), settled in York in 1718 and became city surgeon in 1727; after several years of painstaking research he published *Eboracum* in 1736. This has ever since remained the standard work on the city, though superseded on many points of detail. Drake's notes of corrections and additions are included in his own interleaved copy (York City Library).

Thomas Beckwith, F.S.A. (*c.* 1730–86), a herald painter, made substantial collections for a revised edition of Drake's *Eboracum.* These include full transcripts of monumental inscriptions in many of the parish churches, from 1736 brought down to 1782, and corrections to Drake, by pages (YAS MSS., especially M.69). Letters by Beckwith on local archaeology are preserved in the collections of John Charles Brooke (College of Arms, *e.g.* MS. 17B).

William Hargrove (1788–1862), a York newspaper proprietor, was a keen topographer and in 1818 published a *History and Description of the Ancient City of York.* His work is usually accurate and his descriptions are from first-hand observation. Hargrove's *New Guide for Strangers and Residents in the City of York* (1838), illustrated with many woodcuts, is of great importance for the state of York immediately before the coming of the railway.

Robert Davies, F.S.A. (1793–1875), Town Clerk of York 1827–48, was a scholarly antiquary deeply versed in the records of the city. His precise knowledge of many of the individual houses in the streets, brought together for a series of lectures delivered from 1854, was published posthumously by his widow and his nephew, R. H. Skaife, as *Walks through the City of York* (1880).

Robert Hardisty Skaife (1830–1916), nephew of Robert Davies, was a scholar of scrupulous accuracy and editor of many texts of importance for Yorkshire. In 1864 he published a large *Plan of Roman, Mediaeval, & Modern York*, compiled from all available sources. He also completed but did not publish a very large biographical and genealogical compilation, 'Civic Officials of York' (York City Library, three vols.), including every freeman of York who served the office of Chamberlain. This includes much information on the builders and inhabitants of houses as well as on individual architects and craftsmen of York.

George Benson, A.R.I.B.A. (1856–1935), was a careful observer of local antiquities and incorporated many of his discoveries in his three-volume account of *York* from its origin to 1925 (1911, 1919, 1925, reprint 1968). Among many other works he published *The Bells of the Ancient Churches of York* (1885) and an article on 'The Plans of York' (*AASRP*, XXXVIII, ii (1927), 331–52).

William Arthur Evelyn, M.A., M.D. (1860–1935), settled in York in 1891 and built up a very large collection of plans, prints, drawings, watercolours and oil paintings of all periods illustrative of the city. The whole collection was acquired for York (City Art Gallery; plans kept in City Library). Evelyn also sponsored a large collection of photographs, including copies of material in public and private collections (now the property of YAYAS, but deposited in York City Library).

The Rev. Angelo Raine, M.A. (1877–1962), honorary archivist to the Corporation, published a detailed topographical survey *Mediaeval York* (1955), based upon original records and, to some extent, upon extracts from wills and other York documents made by his father, the Rev. James Raine, D.C.L. (1830–96).

John Speed (1552–1629) was apparently the first to produce a plan of York from original survey. It was engraved in 1610, and another version appeared in the sixth, and last, volume of the *Civitates Orbis Terrarum* of Braun and Hogenberg, published in 1618 (not, as often stated, in 1574, the date of the first volume). Speed's survey was accurately triangulated, and though engraved on a small scale is surprisingly precise in details which can be checked.

Samuel Parsons (*fl.* 1618–39) was a surveyor of national repute, for he worked in Middlesex, Essex and Shropshire as well as at York. In 1624 he did the fieldwork for a plan of the Manor of Dringhouses, drawn out in 1629 (York City Library). This is the earliest surviving large-scale plan of any part of York or its near neighbourhood, and is of notable accuracy (Plate 4).

Benedict Horsley (*c.* 1627–*fl.* 1702) belonged to a noted York family of herald painters. He took up the freedom as a painter stainer in 1648 and served as Chamberlain in 1688. In 1694 he surveyed the city and published the work as an engraved plan in 1697 (Fig. 2). In 1702 the Corporation ordered him to be paid £15 'for surveying the City and for drawing severall Mapps and Planns therunto relateing', but these are not known to survive.

Henry Johnston (*c.* 1640–1723), younger brother of Dr. Nathaniel Johnston, made an extensive antiquarian tour in Yorkshire in 1669–70, the results surviving as two large MS. volumes (Bodleian, MS. Top. Yorks. C.13, 14). Johnston, who was later employed by Sir William Dugdale, was mainly interested in heraldry, but he also made drawings of houses and monuments and of a few stained-glass windows of outstanding interest (Plate 101). A prospect of York from the S.W. by him was formerly in the Boyne Collection (vol. x, 12) but has been lost since 1900.

James Archer (*fl.* 1671–90), a military engineer, drew out a large and detailed plan of the city *c.* 1682. The drawing survives (now in York City Library) and is the earliest original survey of the whole city with some pretensions to accuracy. It shows the castle, walls and moats in rather precise detail, but it is inaccurate in some respects and indicates built-up areas conventionally. See also *York* II, Appendix.

Francis Place (1647–1728), a topographical draughtsman of great brilliance, produced the earliest of the large series of detailed prospects and sketches of York and its buildings. He was the close friend of William Lodge (below) and there is some evidence that the two men exchanged sketches of the city for working up into detailed engravings. Place's distant view of York from Middlethorpe (Plate 198) includes most of the area covered by this Inventory.

Gregory King (1648–1712), a heraldic artist, was a boy of 17 when he accompanied Sir William Dugdale on his Visitation of 1665–6. King's rather crude prospects of the city (bound into the MS. 'Book of Yorkshire Arms', *see* p. xxx above) are of some importance as the earliest surviving general views of York.

William Lodge (1649–89), an amateur artist and engraver and friend of Francis Place (above), drew and engraved several views in York, notably the S.W. prospect from The Mount engraved *c.* 1678 as 'The Ancient and Loyall Citty of York'. This is based in part upon Lodge's own drawing (Leeds Public Library, Graingerised Thoresby, III, opp. p. 56) but also upon a prospect attributed to Place (R. E. G. Tyler in *Leeds Art Calendar*, no. 62 (1968), 14) and upon another version now in London (Society of Antiquaries, vol. marked '1750'; Plate 2). All these original drawings by Lodge and by Place show the old spire of St. Martin, Micklegate, taken down in July 1677; Lodge's engraving shows the new tower with classical balustrade.

John Cossins (*fl.* 1725–48) produced an engraved survey plan of York, to be dated upon internal evidence to *c.* 1727, and a revised edition of it in 1748. Though showing little independence of Horsley as a survey, Cossins's plate is of interest for the sixteen surrounding views of private houses in the city, of which four

YORK FROM S.W., 1965.

YORK FROM W., 1970.

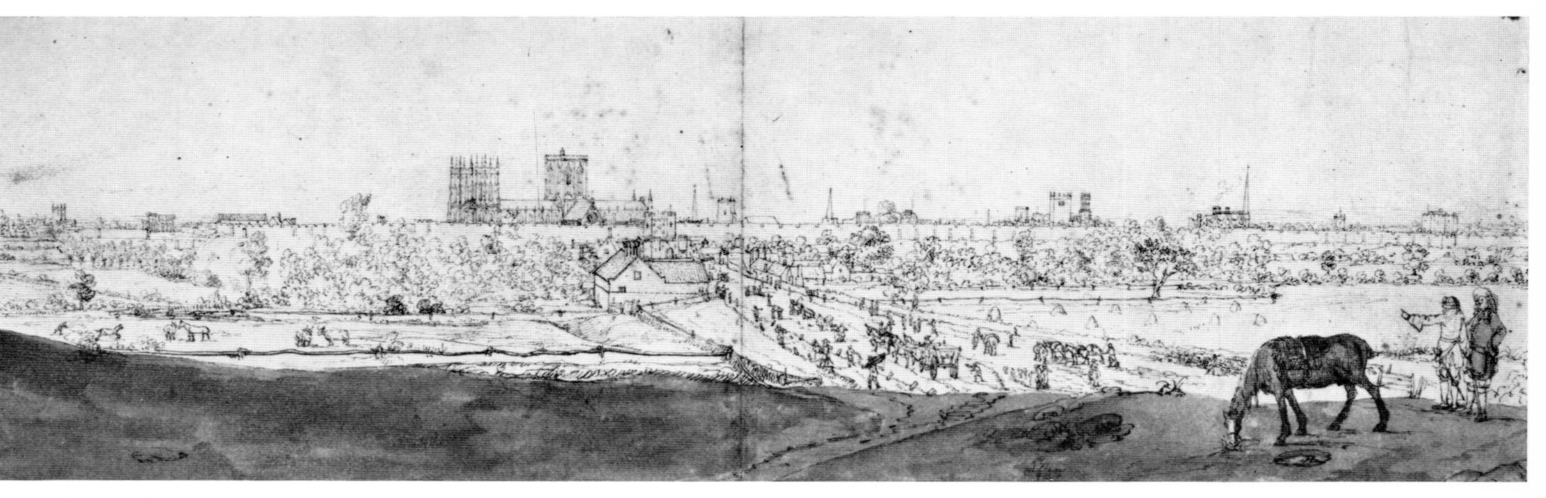

By William Lodge, *c.* 1675.

By John Haynes, 1731.

YORK FROM S.W.

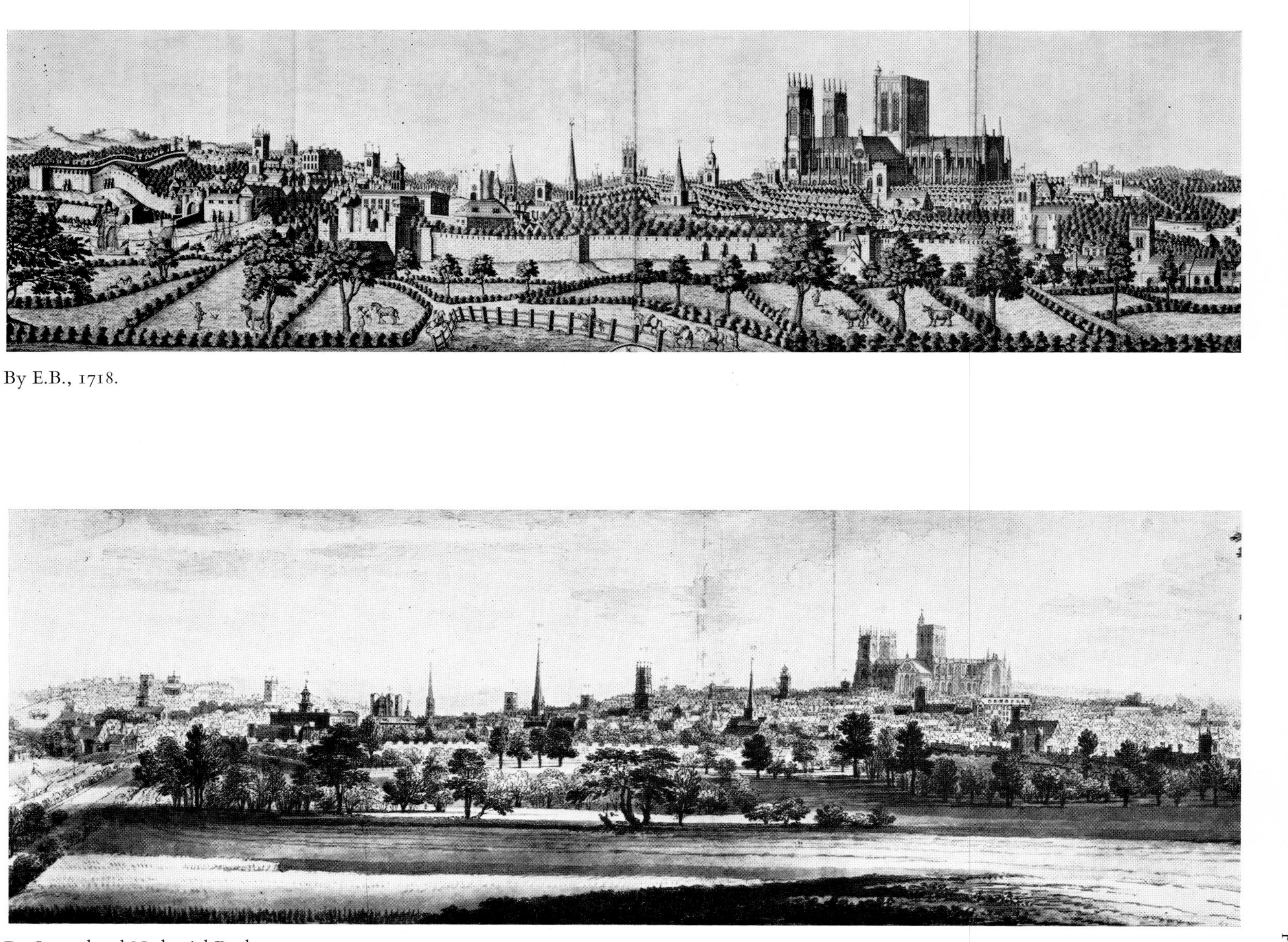

By E.B., 1718.

By Samuel and Nathaniel Buck, 1743.

YORK FROM S.E.

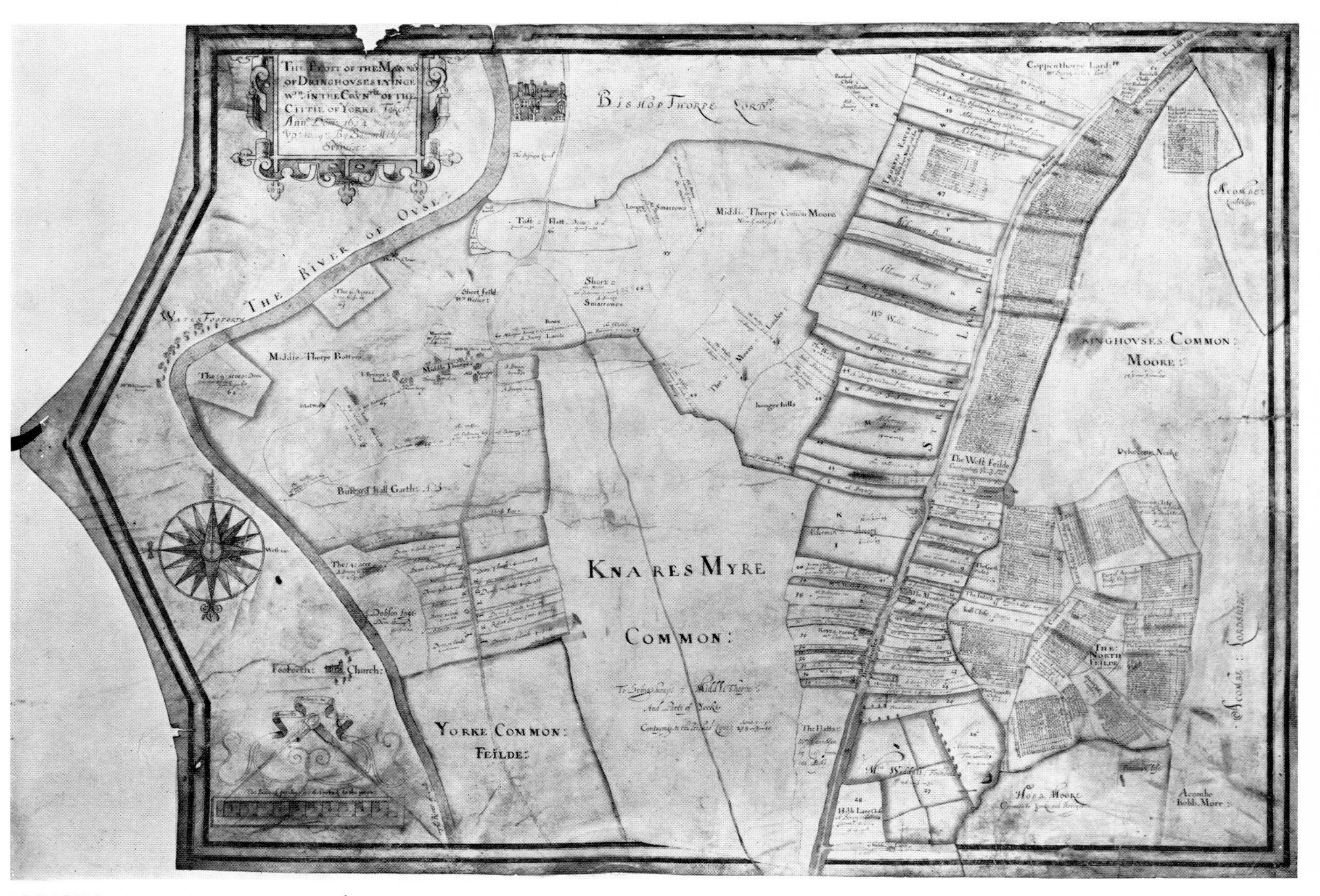

DRINGHOUSES MANOR. By Samuel Parsons, 1624.

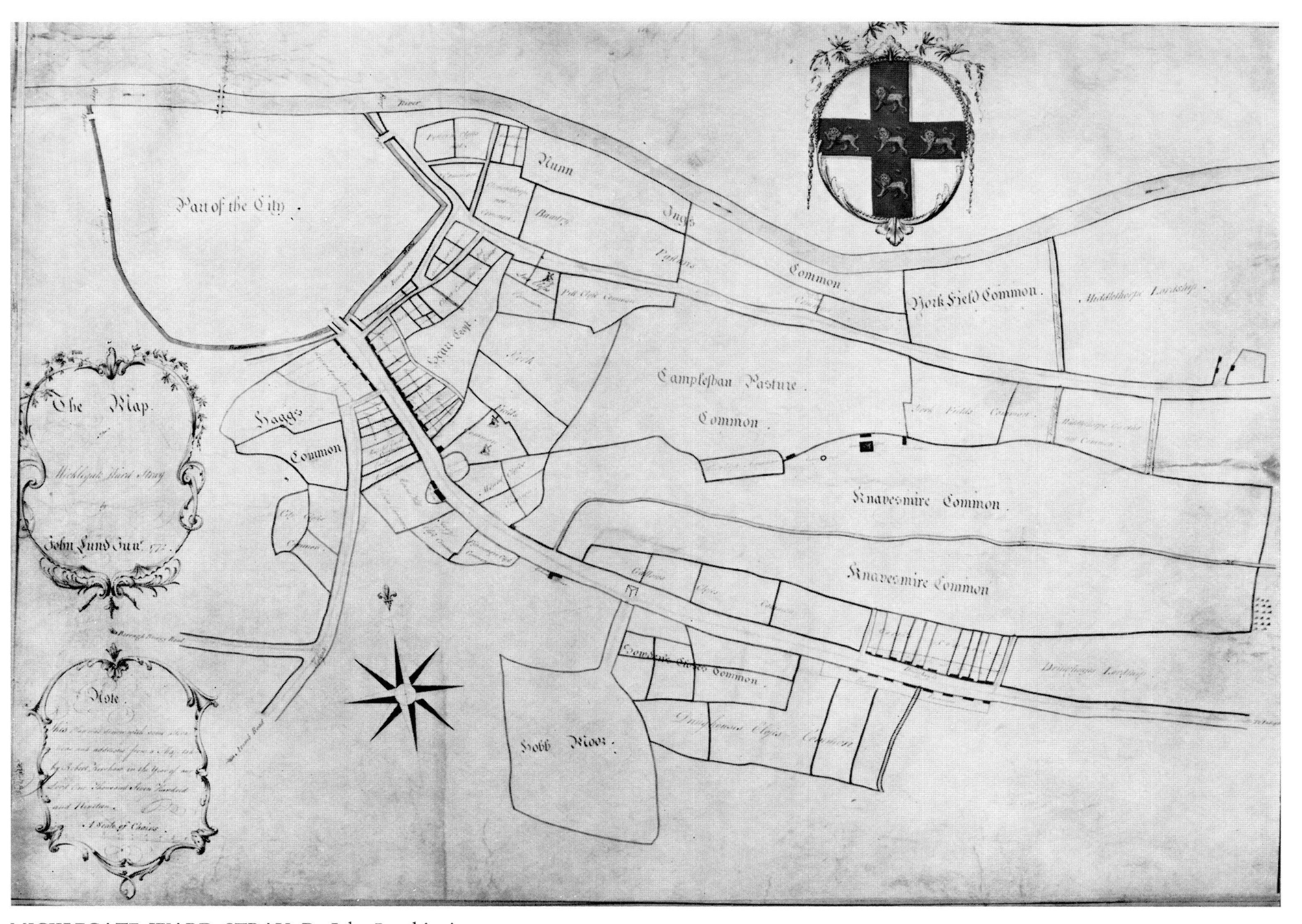

MICKLEGATE WARD STRAY. By John Lund junior, 1772.

(19) CLEMENTHORPE NUNNERY. By J. Poole, *c.* 1705.

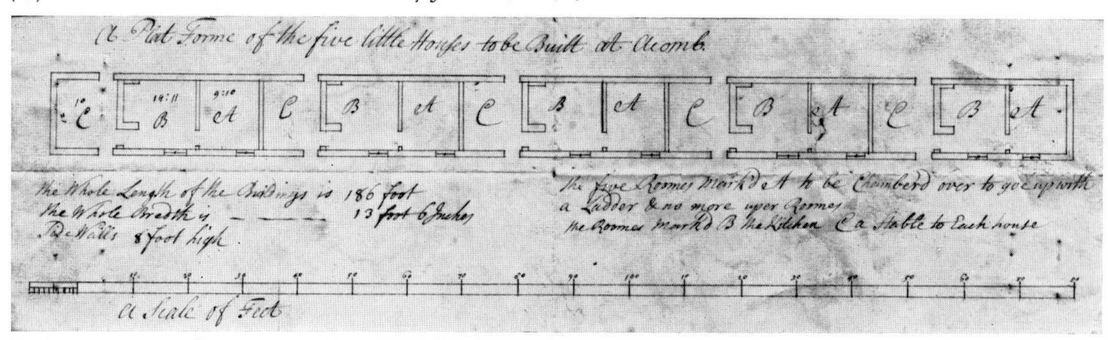

(149) ACOMB, White Row, Front Street. Plan, late 17th century.

BLOSSOM STREET FROM THE MOUNT. By Thomas White, 1802.

BLOSSOM STREET AND MICKLEGATE BAR, 1812.

YORK FROM S.W. By Nathaniel Whittock, 1858.

lay S.W. of the Ouse (Figs. 22, 23, 49, 56). These are the earliest details of individual house fronts in York, and show the reigning style of 1720–5.

John Haynes (*c.* 1705–*fl.* 1751) took up the freedom of York as a sadler in 1728 but engaged also in engraving and land surveying: in 1740 he advertised (*York Courant*, 9 Sep.) that 'Saddlers that have Occasion to have their Leather printed . . . may have it done . . . by John Haynes, Engraver, and Copper-Plate-Printer, in the Minster Yard, York. He likewise surveys Land, and draws Perspective Views of Gentlemen's Seats . . .'. He engraved for Gent and Drake, and in 1744 produced a large map of Antiquities on the Wolds in the Malton area, and also an exquisite MS. plan of Newburgh Park (NRCRO, Northallerton, ZDV (VI)). By 1751 he had probably left York, as he then published an engraved survey of the Chelsea Botanic Garden. Haynes's S.W. Prospect of York, published in 1731, though rather crude, shows many buildings in detail (Plate 2).

Thomas Jefferys (*fl.* 1735–94), a surveyor of national reputation, produced his Atlas of Yorkshire in 1772 and included a plan of York (Fig. 3). Jefferys' work is independent of earlier surveys and reasonably accurate.

Edward Abbot (*fl.* 1774–6), a painter, stayed with Thomas Beckwith (*see* above) when in 1774 he wrote a 'History of the Cathedral Church of York' (BM, Stowe MS. 884). He also, in 1774 and 1776, made many small watercolour views of churches and other buildings in and around York (Wakefield Museum, Gott Collection; there are photographs of most of the York items in the Evelyn Photographic Collection deposited in the York City Library). Abbot's work was probably meant to illustrate Beckwith's proposed enlargement of Drake's *Eboracum.*

Francis White (*fl.* 1770–1801), a surveyor of great accuracy, published in 1785 a good plan of York with a map of the whole of the Ainsty on the same sheet. The map is said to be 'from a Reduction of the Plans of the Townships', implying that White had made a large-scale cadastral survey of every township in the Ainsty. Nothing of White's original work is known to survive.

Joseph Halfpenny (1748–1811), son of Thomas Halfpenny, the Archbishop's gardener at Bishopthorpe, was apprenticed to a house-painter and later acted as clerk of works to John Carr. He was a fine draughtsman and engraver and published a series of detailed views in York as *Fragmenta Vetusta* in 1807.

John Lund junior (*c.* 1750–98), one of a family of York surveyors settled in Bootham, produced for the Corporation in 1772 the surviving plans of the four Strays (YCA, D, Vv). They were based on earlier surveys, but indicate the existence of rights of half-year common over enclosed lands around York, and show the extent of the built-up area in the suburbs (Plate 5).

Peter Atkinson junior (1776–1843), the well known architect, was also City Steward and in 1810–13 produced survey plans of all the Corporation properties in the York area (YCA, M.10 A–D).

Henry Cave (1779–1836), a prolific draughtsman and engraver, came of a York family of whitesmiths and engravers. Many of his original drawings survive (York City Art Gallery and elsewhere), as well as his copperplates for *Picturesque Buildings of York*, published in 1810 (T. P. Cooper, *The Caves of York* (1934)).

George Nicholson (1787–1878), a topographical artist of great accuracy, produced a large series of sketches of individual buildings *c.* 1825–30. Some of these are in the Evelyn Collection (York City Art Gallery), but others are now known only from photographs (Evelyn Photographic Collection).

Alfred Smith (*fl.* 1819–1834) was a land surveyor of Little Preston near Leeds, whose engraved plan of York in 1822 was published in Edward Baines, *History, Directory and Gazetteer of the County of York* (1822–3). Smith's plan was the first to show detail of property boundaries throughout the city; it is generally of remarkable accuracy for its small scale.

Nathaniel Whittock (*fl.* 1828–1860), an excellent draughtsman, produced many views in York and in 1858 a large bird's-eye view from the S.W. So far as the viewpoint permits, this gives precise details of the street elevations of all the buildings shown on the Micklegate side of the city (Plate 8). A revised edition of the view was issued *c.* 1860.

Robert Cooper (*fl.* 1830–42), a first-class land surveyor, produced in 1831 a map of York and the Ainsty, published the next year. This must have been reduced from a large-scale plan now lost, since individual buildings are marked with great accuracy. Cooper also produced plans of York and other boroughs in connection with Parliamentary Reform.

Plans of York

It has been suggested that a coloured view of a city of the early 15th century in a manuscript of the Brut (BM, Harleian MS. 1808, f. 45v.) represents York (R. A. Skelton and P. D. A. Harvey in *Journal of the Society of Archivists*, III, no. 9 (April 1969), 496), but if so no strictly topographical features other than the river and walls can be recognised.

The earliest true plan of the city seems to be that on two large skins of parchment, now separated, in the Public Record Office (MPB 49, 51), found among miscellanea of the Exchequer. Though somewhat damaged, the detail can be made out and the wording is legible (Fig. 1). The date is subsequent to the dissolution of St. Mary's Abbey and its renaming as the King's Manor, but on palaeographical grounds the map is not likely to be much later than the middle of the 16th century. It seems possible that it may be the plan referred to on 12 August 1541, when the Corporation 'paid to the armet (hermit) of the Kyngs Manour at York for drawing of the platt of the sanctuary' a sum of 9*s.* (YCR, IV, 63). In the centre of the city an area is very precisely marked out in red, including three parish churches, Thursday Market and the houses fronting on certain adjacent streets. This area could well be that designated to take the place of rights of sanctuary formerly existing within the precinct of St. Mary's Abbey. The plan is diagrammatic in that the Ouse is shown as a straight line from N. to S. and the streets are set out at right angles. The main interest of the plan historically lies in the naming of the streets and lanes and of other features.

John Speed's plan, 1610, published on his map of the W. Riding, shows the individual buildings in bird's-eye view. Apart from ribbon development along the main streets leading out from the city bars, there is little building outside the walls. Groups of windmills are shown in the surrounding countryside and correspond to known sites. The detail is precise and accurately surveyed.

James Archer's plan, *c.* 1682, shows the city defences including The Mount, a Civil War fortification on the Tadcaster Road, in great detail. Plot boundaries are shown but are rationalized into straight-sided units, and houses are shown conventionally. The suburbs were destroyed during the Civil War, so that Archer's plan shows their recovery, but their extent is not noticeably greater than in Speed's time. Though reasonably correct for the defences, the street plan suffers from serious errors.

Benedict Horsley's plan (Fig. 2), surveyed in 1694 and published in 1697, shows street frontages with considerable accuracy, but as the streets are shown dark against a light background outside the city walls, buildings in the suburbs are not differentiated from the fields. Individual churches and important structures are correctly shown in bird's-eye view. Considerable development is visible within the walls in the Bishophill, Hungate and Walmgate areas, and outside along Gillygate.

John Cossins's plan, *c.* 1727, follows Horsley's shading convention but extends it beyond the city walls. In addition to perspective drawings of churches on the plan itself, Cossins shows larger elevations of gentlemen's houses along the margins, an important innovation. The plan is in general cruder than Horsley's.

Francis Drake's plan, 1736, follows Horsley closely, especially in the use of conventions, the portrayal of The Mount earthwork, and the enriched cartouche containing the scale, which now contains an elevation

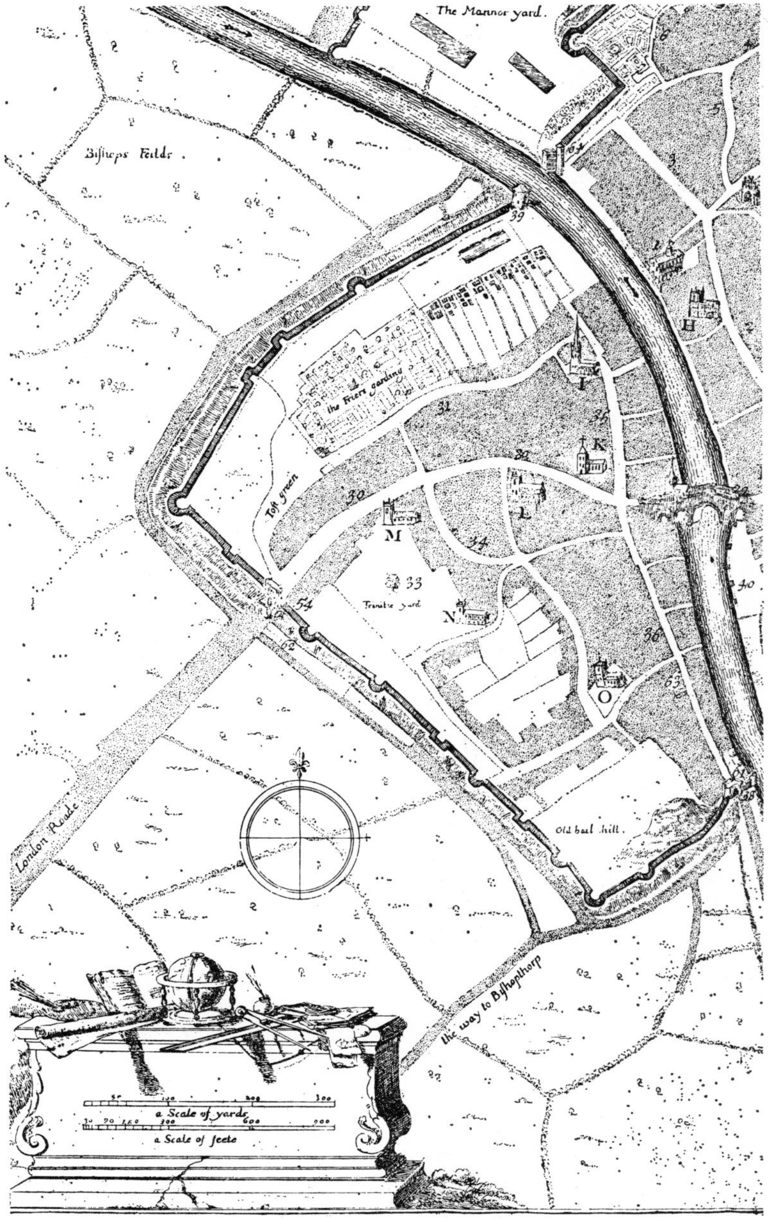

Fig. 2. From Horsley's plan of York, 1694.

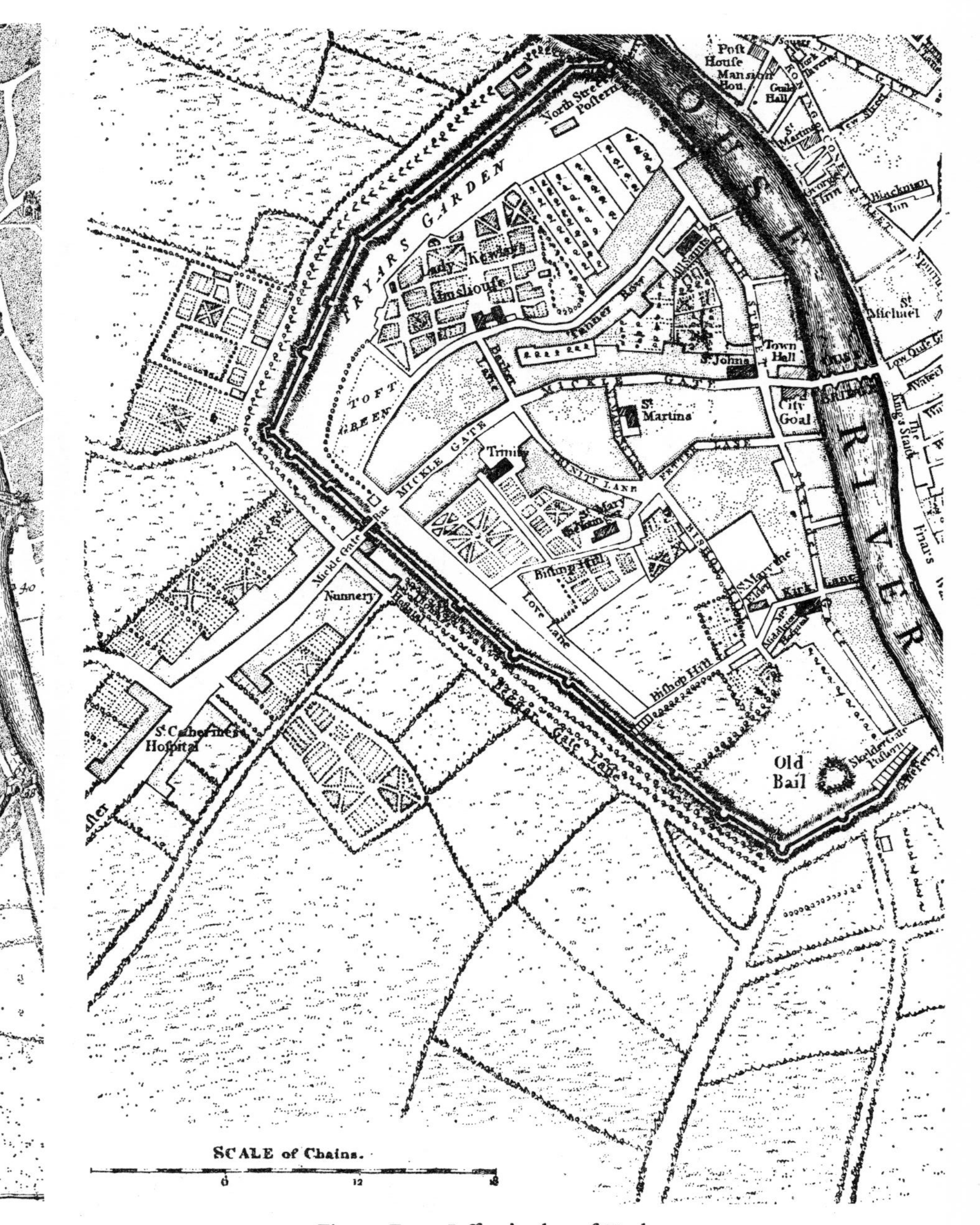

Fig. 3. From Jeffery's plan of York, 1772.

of the Mansion House. The new features shown include the Wigginton Road, the Long Walk and Blue Bridge, the new prison, and the Assembly Rooms; the sites of St. George's church and the chapel of St. Saviourgate are marked as vacant plots.

Peter Chassereau, 1750, shows little outside the city walls, but where fields appear they are shown in great detail. The Bishophill area, in opposition to Horsley's map and all its derivatives even after 1750, is now shown vacant. The development of the Monkgate area is clearly visible. The castle gatehouse has vanished. Chassereau shows all the churches in plan and, while following Horsley's stippling convention, shows the open areas within blocks.

Thomas Jefferys's plan, 1772 (Fig. 3), is similar to that of Chassereau but extends further to the south. Buildings not shown by Chassereau include St. Catherine's Hospital and the Nunnery (Bar Convent) in Blossom Street, the Town Hall and 'City Gaol'.

John Lund junior, 1772, produced four plans to show the pastures surrounding the city, providing information on the villages near York. The Micklegate Ward Stray plan (Plate 5) parallels that of the Manor of Dringhouses, produced by Samuel Parsons in 1624–9, and shows details of the new racecourse. Bootham Bar appears to have acquired a Classical pediment.

Ann Ward's plan in her edition of Drake's *Eboracum*, 1785, is Drake's plan almost unaltered, but it marks recent road widening, especially the formation of St. Helen's Square, King's Square, and the inclusion of St. Crux churchyard in Pavement. This is an exceedingly early example of the use of a plan for morphological purposes.

Francis White's plan, 1785, shows houses less accurately and churches more accurately than Chassereau. New development is indicated on the N.E. side of Thief Lane near the junction with Blossom Street. The Map of the Ainsty, to which it forms an inset, shows the layout of open fields in the villages, many not yet enclosed.

Alfred Smith's plan, 1822, shows the outline of individual buildings and plots. The castle moat has been filled in by an esplanade, but the gate to Castlegate is shown as still extant. This is a most important plan, showing York just before the great changes in the net of streets.

Robert Cooper's plan, 1832, although on a small scale, shows individual buildings as well as the street plan. Only the old part of the city to N.E. of the river is blocked-in solidly.

Bellerby's plan, 1847, is based on the 1822 Smith plan and is probably the same plate, altered to show new buildings. The railway has penetrated the city walls, with the station replacing the House of Correction at Toft Green. The wall has been pierced by the sweep of St. Leonard's Place. In the grounds of St. Mary's Abbey appears the new Yorkshire Museum.

Nathaniel Whittock's plan, 1858, is of considerable interest as a panoramic plan of the whole city Plate 8). It includes precise detail of the street elevations of the time and provides a unique record.

(J.H.H.)

SECTIONAL PREFACE

Pre-Roman Settlement

Most pre-Roman finds within the city have been made S.W. of the Ouse and for this reason it is convenient to consider the prehistoric evidence as a whole in this volume.

The importance of the geographical background as it affected Roman settlement has been summarised in *York* I (p. xxix) and similar factors controlled earlier settlement. The morainic ridge provided a natural causeway across the wide and often marshy Vale of York between, on the E., the chalk wolds which supported one of Britain's most significant prehistoric centres and, on the W., the Pennine foothills and the Dales which penetrated the higher land beyond.

This ridge provides an agriculturally attractive environment of well drained sandy and loamy soils, which in York reaches its greatest width about a mile S.W. of the river. The continuation E. of the river Ouse is narrower and lies S. of the junction with the river Foss. The Roman fortress was sited for tactical reasons on a small isolated block of moraine between the Ouse and Foss. As a result the river crossing has been moved N. from its natural site, and the centre of subsequent settlement lies N.E. rather than S.W. of the river. Significantly the Roman civil settlement remained S.W. of the Ouse and continued to form an important part of the mediaeval and later city. The bulk of the pre-Roman evidence comes from S.W. of the river (*see* Fig. 4).

There is no Palaeolithic or Mesolithic evidence from the Vale of York, although the area should have been attractive to the hunters and fishers of the latter period. The *Neolithic* period is mainly represented by axes and concentrations of struck flints with a distribution concentrated on the chalk and sandstone fringes of the Vale and to a lesser extent on the moraines, and absent in the low-lying areas. Within York at least twenty-three axes have been found of which all those with a sufficiently specific provenance come from S.W. of the river (*see* Table I, p. xxxix). Flint sites occur at Overton and Fulford just outside the city, and a unique hoard from the city was discovered during the erection of the N.E. Railway gasworks (582527) in 1868. The latter comprised at least forty-three implements including seven axes which were found in a compact group deep in the gravel terrace near the junction of Holgate Beck and the Ouse. The regular, sharp flakes and blades, and unused appearance of the finished blades suggest a merchant's hoard, whilst the inclusion of a barb and tang arrowhead could imply a late Neolithic–early Bronze Age context (*YAJ*, XLII, 131–2). Three finds of Beakers—a fine 'C' Beaker (*York Survey 1959*, fig. 11a, p. 87; YM, 1000, 1947) found near Bootham before 1842 (Wellbeloved, pl. XV, 15 and *YMH* 2nd ed. (1854)), and two sherds of 'B' Beaker from West Lodge Gate, probably Acomb Road, Holgate (583514; BM, 1853, 11–15, 18)—suggest late Neolithic burials within the city.

Bronze Age occupation of the Vale is limited to the dry ridges and the river banks. Round barrows occur E. and W. of York but none are proved within the city. The contracted burial in a cist under Clifford's Tower (*York* I, 69 *n.*1) need not be later than the early Bronze Age. No Food Vessels or later Bronze Age cinerary urns are known from the city. There is a marked concentration of bronze implements from the city (*see* Table II, below) although few have a known find spot. Bronzes also come from just outside the city, a looped spearhead from Heslington and a palstave from Bishopthorpe. The almost complete absence of bronze weapons so frequently found on the Trent together with the absence of burials

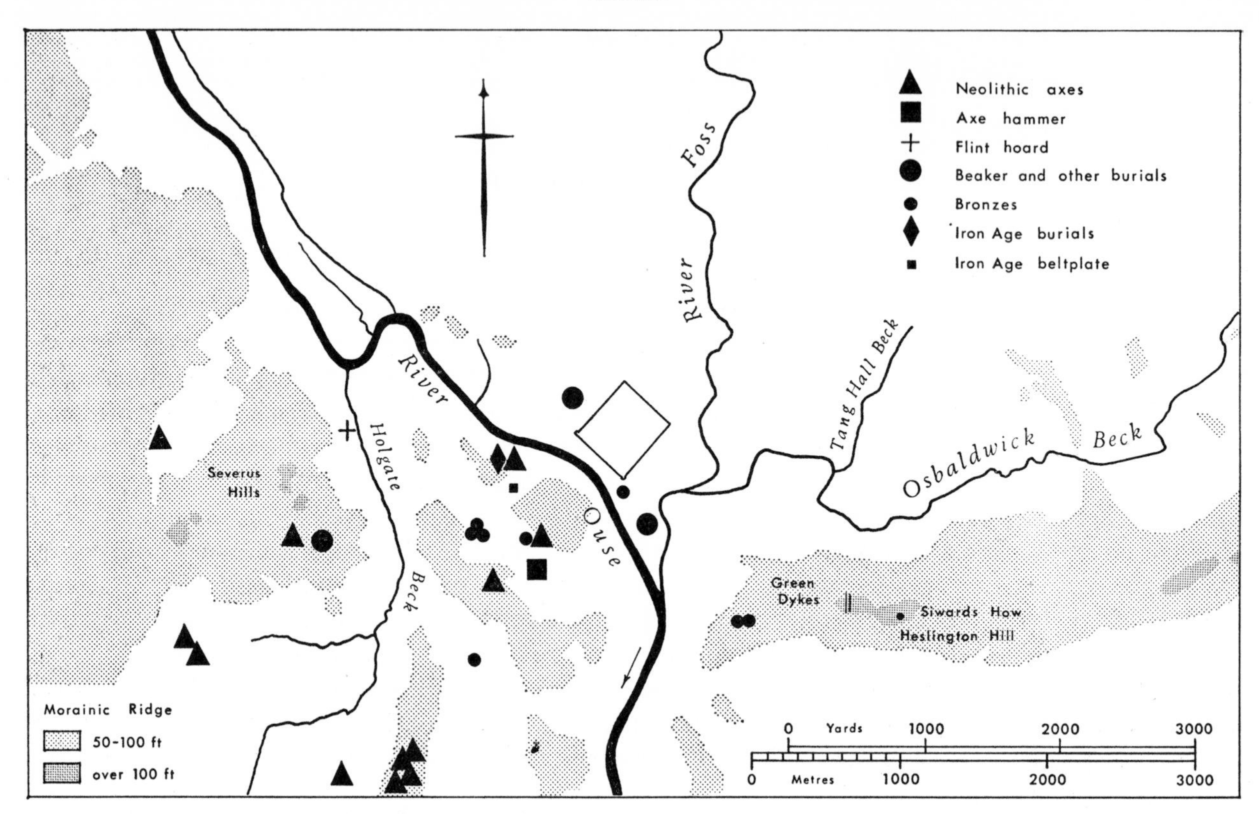

Fig. 4. Distribution of Pre-Roman Finds in York.

suggests that York had less importance as a centre in the later Bronze Age. The moraine was, however, part of a well defined E.–W. trade route from Irish metal sources to East Yorkshire and the Continent, as demonstrated by the presence of Irish gold ornaments in East Yorkshire and the numerous bronze hoards in the Vale. Stone axe-hammers, usually attributed to the Bronze Age, are less frequent than polished axes but there are four from the city (*see* Table III, below) and one from Poppleton (YM, 1052, 1948).

Iron Age acquaintance with the site of York is implied by the Celtic origin of the name *Eburacum* (Vol. I, xxx) and possible settlement on the present Railway Station site is implied by the find of contracted inhumations underlying Roman burials (Vol. I, 85, area f, (vi)) and of the well known enamelled bronze belt plate in the Yorkshire Museum (YM, 845, 1948; E. T. Leeds, *Celtic Ornament* (1933), 129; C. Fox, *Pattern and Purpose* (1958), 119, pl. 52). The latter is dated by Fox to the decade centring on A.D. 70 and the evidence, slight as it is, is consistent with a small agricultural settlement S.W. of the Ouse on the eve of the Roman conquest, its leaders having some share in the wealth that accrued to the Brigantes as a result of their Roman policy. A cross-ridge dyke formerly existed at Green Dykes on the E. side of the river, controlling the E.–W. ridge at its narrowest point and the approach to a river crossing well S. of the Roman and later bridges (*YAJ*, XLI (1966), 587 ff.).

Early History

The part of the city covered by this Inventory, regarded in mediaeval times as *ultra usam* and sometimes described as a suburb, was always subordinate to the area on the N.E. bank of the river. It had developed from the Roman *colonia*, of which structural remains survived above ground in many places until after

TABLE I

Neolithic Axes

1. Polished greenstone axe from Viking Road. Private collection.
2. Polished blade and butt end of opaque grey flint, probably parts of one axe, and possibly dumped. Found in a garden inside Micklegate Bar. YM, 1952. 19. 1, 2.
3. Polished greenstone axe, broken. YM, Cook MS., pl. 1, no. 1; 'from the railway diggings before crossing the Ouse–Scarborough Line'. 1847.
4. Polished greenstone axe, broken. YM, Cook MS., pl. 1, no. 7; 'nigh the Railway Bridge, Dringhouses, 1851'.
5. A stone axe from Holgate. Benson, *York*, I (1911), 5.
6. A stone axe from The Mount. Benson, *York*, I (1911), 5.
7. A possible sandstone axe or hammerstone from Dringhouses. YM, 349, 1948.
8. Three polished greenstone axes, found together at Dringhouses 1884. YM, 443–5, 1948. YPS *Report* (1905), 50, pl. 4, fig. 2.
9. Two polished stone axes from Gale Lane. One now lost. YM, 1948, 10–11.
10. Polished cherty-flint axe, damaged. From York. Hunterian Mus. B.1951. 2594.
11. A polished axe from York. YM, 477, 1948.
12. A polished stone axe from York. YM, 1022, 1948.
13. A polished stone axe from York. YM, 1565, 1948.
14. Hoard of axes, one of which is a polished greenstone and at least six more are polished flint. With these axes were found three arrowheads, nine ovoid spearheads, three scrapers, eleven blades and flakes, and two worked points, all of flint. N.E. Railway Gasworks, 1868. YM, 446–7, 1948; YM, FW 100. 1–18.

TABLE II

Bronzes

1. Found within York
   - A. *Flat axes*
     1. YM, 1033, 1948. From York.
     2. YM, 1183, 1948. From York.
     3. ? 'brass celt'. Knavesmire (Stukeley Letters, III, SS, LXXX (1885), 348).
   - B. *Spear*
     1. YM, 1171, 1948. Looped and damaged; High Ousegate.
   - C. *Socketed axes*
     1. YM, 1146, 1948. The Mount.
     2. BM, WG.2010. York Cemetery.
     3. BM, WG.2011. York Cemetery.
     4. Sheffield Museum. J.93.507. From York.
     5. ? 'bronze celt', L. 3⅜ in. *YMH* (1891), 205. From York.
     6. ? Sheffield Museum. J.93.505. 'Bought at York, from W. Cook's Yorkshire Collection.'
   - D. *Palstaves*
     1. YM, 1132, 1948. Looped. From York.
     2. *YMH* (1891), 204. Not looped. From York.
2. Found 'at or near York'
   - A. *Flat axe*
     1. YM, 1242, 1948. 'Vale of York.'
     2. BM, lost. Near York, decorated with chevrons. *Arch. J.*, XIX, 363.
   - B. *Winged axes*
     1. BM, 53, 11–15, 10. Near York.
     2. BM, 53, 11–15, 11. Near York.
   - C. *Socketed axes*
     1. BM, 63, 12–24, 1. With waste metal rammed into socket. At or near York.
     2. BM, Henderson Gift. At or near York.
3. *Bronze Hoards* (It seems probable that two groups of finds can be justifiably recorded as hoards.)
   1. BM, WG.2010–11 (*see* above), York Cemetery. It is reasonable to suppose that these were found together.
   2. A hoard of many socketed axes, found by George Milford, 1847, during the making of a railway cutting, was formerly in the Mayer Collection, Liverpool Museum; records of one socketed axe survive (M.6996).

TABLE III

Axe-hammers

1. YM, 1020, 1948. Found 1886. L. 4¼ in. Scarcroft Road (? same as Benson, I (1911), 5).
2. YM, 1022, 1948. L. 6¼ in. Label reads, 'found with a celt in York', possibly a battle-axe.
3. YM, 1032, 1948. L. 8 in.
4. YM, 1067, 1948. Fragment. L. 1⅝ in.

the Norman Conquest. This had been a walled town with a built-up area not completely identical with that of the mediaeval period, rising in a series of terraces up slopes steeper than at present; the streets formed a grid parallel to a main road N.W. of Micklegate leading to a bridge opposite the Guildhall; important groups of buildings stood on the sites of Railway Street and of the Old Station (Vol. I, xxxv–ix, 49–58).

Roman occupation continued into the 5th century, as shown by excavations near the church of St. Mary Bishophill Senior, and early Germanic cremation burials on The Mount are thought to be a cemetery of mercenaries rather than of invaders (J. N. L. Myres, *Anglo-Saxon Pottery and the Settlement of England* (1969), 73 ff.). Severe flooding during the 5th and 6th centuries rendered the riverside areas of the town uninhabitable and probably destroyed the Roman bridge (R. Cramp, *Anglian and Viking York* (1967), 3; more fully considered by H. G. Ramm in R. M. Butler (ed.), *Soldier and Civilian in Roman Yorkshire* (1971)). These areas were reoccupied by the 9th century, when metalwork was lost in Tanner Row (*Med. Arch.*, VIII (1964), 214–16), and the main road was deflected from the line Toft Green–Tanner Row to Micklegate, curving E. to the new river crossing on the site of Ouse Bridge.

Archaeological evidence for the Northumbrian period consists of meagre finds scattered widely over the *colonia* area, sufficient to indicate its continued occupation. Sculptured stones indicate at least two churches, at Bishophill Junior and on the Old Station site. The numerous 10th and 11th-century finds imply that the large and bustling city indicated by documentary sources extended onto the S.W. side of the Ouse over the whole of the later walled area. During the 10th and 11th centuries were built the churches of St. Mary the Old and St. Mary Bishop, both in Bishophill, of which the former may well have replaced a still older church. St. Martin, Micklegate was founded in the 10th or 11th century. This church-building activity indicates sufficient population to justify parochial organisation. The pre-Conquest churches, other than the two on Bishophill, and including Holy Trinity, St. Martin's and probably St. Gregory's, are sited along Micklegate, and the importance of this road linking the Roman road from Tadcaster with the new river crossing was both military and commercial. Its commercial use is indicated by the satellite settlement of Copmanthorpe, a name referring to merchants (EPNS, XXXIII, 227), and by the mention of road transport in 1070–80 (*YAJ*, XVIII (1905), 413).

North Street, first recorded in *c.* 1090, and Skeldergate, attested in the 12th century but of Scandinavian origin, follow the Ouse bank and indicate the importance of river-borne commerce to the life of York. Extension of occupation downstream by satellite settlements, as the -thorpe names imply (Clementhorpe, Bustardthorpe, Middlethorpe and Bishopthorpe), was probably also commercial. The archbishop's right to custom from ships berthed at Clementhorpe is specifically mentioned in a document of 1106 (A. F. Leach, *Visitations and Memorials of Southwell Minster* (1891), 195–6). York on the S.W. side of the Ouse was probably already fortified before 1066 and, although the extent of these defences is unknown, the line of the subsequent Norman ramparts was followed by the later mediaeval defences. The S. extremity of the town was occupied in 1068 or 1069 by a motte and bailey castle, the Old Baile. Most of the mediaeval street names are recorded in the 12th century: this fact and their Scandinavian origin suggests that by 1100 the city's later mediaeval plan was already established.

Domesday Book (VCH, *York*, 19–24) provides a detailed picture of York both before and after the Norman Conquest, but the statistics are hard to elucidate and the record excludes the property of St. Mary's Abbey. Administratively the city was divided into seven 'shires', of which one was cleared of houses when the two castles were built, and there was a complex division of jurisdiction between the archbishop and the king. The importance of the city as a centre of communications and commerce is implied. The rebellions of 1068 and 1069 followed by the Harrowing of the North resulted in devastation on both sides of the Ouse and one estimate based on Domesday suggests that the population of York was reduced by half as a result. However, the population must eventually have increased as survivors from the ruined villages for miles around made their way to York. It is therefore not surprising to find two more parish churches—All Saints', North Street, and St. John, Ousebridge End—being built in *c.* 1100, although these too may have had Saxon predecessors, indicating a substantial influx into the sparsely-populated areas N. of Micklegate. A set-back to growth must have been caused by the great fire of 1137 which spread to this side of the river. The fact that the fire could reach Holy Trinity Priory near the top of Micklegate suggests that the street was pretty well built up by that date.

Carved Saxon Stones

About thirty carved stones of Saxon date survive from York S.W. of the Ouse; with one exception they come from the restricted area of the two churches of St. Mary Bishophill and the neighbouring St. Martin's Church in Micklegate. The exception, of unknown original derivation, was found with a mixed cache of stones, including Roman material and a Norman cross base, in the rampart of the city wall by the Railway War Memorial. Part of an elegant and well-cut cross head, it also happens to be the only stone to which a pre-Danish date can be given with any assurance.

Some of the seven stones from Bishophill Junior were also dated by W. G. Collingwood before the Danish invasions. But although none show signs of Danish influence in their decoration, only three have significant ornament. Of these one is a hog-back (3) which must be post-Danish, and another was dated by Kendrick to the 11th century. The third, a fine figured cross (1), could indeed be of 9th-century date but in view of the persistence of Anglian ornament in the other stones and the fact that excavation (by L. P. Wenham, 1961–2) failed to reveal evidence of burials earlier than the 10th century, it is perhaps best to regard all three as 10th century or later.

None of the twenty-one stones from Bishophill Senior can be put before the 10th century. Some are clearly contemporary with the 11th-century church, in particular (3), (18), and (20). Only (19) can be proved demonstrably to precede it, although others certainly did. With the exception of the large cross-shaft, the stones were all probably sepulchral; six are grave covers, two certainly hog-back and two flat slabs, whilst a seventh (21) could also be part of a grave cover. Two stones could be either headstones or the bases of cross-shafts and the remainder are all from cross-shafts. A cemetery existed on the site in the 10th century and produced, beside the crosses, a strap end of probably 10th-century date and Scandinavian origin. No evidence was found for burials before the Danish invasions. The finest stone was the cross-shaft represented by (1) and (13), a fine figured shaft originally standing nearly 6 ft. high. It displays a flat linear style of which other examples can be seen at Shelford, Notts., and Leeds, and which represents a barbarisation of the pre-Danish style exemplified at St. Helen's, Auckland. The figures are reminiscent of those at Nunburnholme, but the carver of that cross has begun to move away from the linear treatment that is so marked a feature of this cross. The back of the stone shows a 'dragon' of distinctively Scandinavian type.

Another cross-shaft (2) has the fragment of what may be the nimbus of a figure in the same linear style. The only other stone with animal decoration (10) is part of a grave slab bearing a closely packed design of animals reminiscent of the more famous coped grave slab from St. Denys, Walmgate, and like it showing Scandinavian influence. Except for (21) with its crucifix and (12) which has a debased key pattern, all the other stones have an interlace pattern, usually simple but including in some cases later features such as the ring plait (15), (20) and chain interlace (16). Only (19) has any depth and roundness of relief. This stone also is the only magnesian limestone; (21) is an oolitic limestone; all the other stones from this side of the river are gritstone. With the exception of the hog-backs, the stones are similar in style and quality to those of the same date found recently in York Minster, where, in the S. transept, several graves still retained cover slabs, headstones and footstones in position.

Two stones from St. Martin are both probably from cross-shafts, one with a heavy late scroll showing on the only visible side and the other showing a late and barbarous human figure. Neither need be earlier than the 11th century. (H.G.R.)

Agrarian History

In 1297, arable strips were recorded outside Walmgate Bar in North Field, and no doubt some modified form of open field agriculture existed in other wards. Arable strips survived until 1444 in Bootham and the physical remains of early agriculture, represented by broad plough ridges, are preserved in several parts of the city. By 1546 the city appears to have rationalised its lands into forty-one closes which were leased to tenants, but strips may have survived longer in closes or groups in the hands of private individuals. Within the walls some closes were used for pasture and arable into the 19th century, for example near Walmgate, Bishophill and Peaseholme Green, but most had been built on or converted into gardens and orchards. Outside the walls the city was ringed with fields which often survived into the 20th century, contiguous with the former open fields in Acomb, Heworth and other townships, parts of which are now incorporated into the city.

By 1272 freemen and religious institutions had rights to keep a restricted number of animals, later limited to horses and cattle, and excluding pigs, sheep, and geese. Pastoral activities were always more important to the city than arable, and were carefully organised. Meadowland, or 'ings', usually on low ground, provided winter fodder, often supplemented with hay from lands held in outlying townships. The common pastures were divided into whole-year pasture on the four commons or strays, and half-year pasture, or average, on the closes which were thrown open each year from October to the end of March, representing a survival of similar rights in the former open fields.

The four strays, one for each of the present wards, formerly extended up to 6 miles from the city between the nearby villages with which they shared common grazing, and into the Forest of Galtres. By 1530 it was customary for a freeman to graze his quota of stock on the stray attached to the ward in which he lived, in the care of pasture-masters. Expansion in the city and adjacent townships created friction over grazing rights on the strays, often leading to agreements restricting numbers of animals. Bootham Ward shared its stray with Clifton, Huntington, Rawcliffe and Wigginton; Micklegate Ward with Middlethorpe and Dringhouses; Monk Ward with Heworth and Stockton-on-the-Forest; and Walmgate Ward with Heslington and Fulford. Not until the Parliamentary enclosure acts were passed did the city have exclusive rights to specific areas of stray, which are maintained by the city on behalf of the freemen down to the present time. (J.R.)

ECCLESIASTICAL BUILDINGS

York S.W. of the Ouse formerly comprised seven intramural parishes, of which three extended outside the walls and between them covered most of the suburban area. The extramural parish of St. Clement, which had given its name to Clementhorpe by the time of the Conquest, had become united to that of St. Mary Bishophill Senior for taxation purposes by the early 14th century; the benefices were formally united in 1586, when the small parish of St. Gregory was also merged with that of St. Martin in Micklegate, after a similar long period of effective union for taxation. There are now no monumental remains of the churches of St. Clement and St. Gregory, and St. Mary Bishophill Senior was demolished in 1963 (*see* Monument (9)). Within the suburban area were two chapels-of-ease, both to the parish of Holy Trinity (formerly St. Nicholas), Micklegate, namely St. James on The Mount, all trace of which has gone, and Dringhouses (*see* Monument (12)). The rural parish of Acomb was brought within the city boundaries in 1937 (*see* Monument (11)).

Mediaeval Churches

The churches in S.W. York are not spectacular but are of great interest. Three have towers which are specially noteable: St. Mary Bishophill Junior has a Saxon tower of exceptional size, comparable to that at Barton-on-Humber; St. John the Evangelist has the only 12th-century tower surviving in the City of York; the 15th-century tower and spire at All Saints', North Street, form a remarkable composition, comparable with that at St. Mary's, Castlegate (Plate 11), across the river. Both these last probably derive from the typical Friars' steeple and may be compared with examples like the Franciscan Friary of Christ Church, Coventry. The development of the plans of the parish churches is discussed below; Holy Trinity, having been a priory church, follows a different pattern. The development of masonry in the walling of the churches and the changes in design of window tracery are also outlined in separate sections. Among the fittings, the glass is exceptional both in quality and quantity and is discussed at some length (pp. lii–liv).

Development of the Parish Church Plan

Rectangular Cell without Structural Chancel. The simplest form of church plan recorded in this volume is the plain rectangular cell. Excavation has shown that the earliest building at St. Mary Bishophill Senior, erected early in the 11th century, was of this type. A similar cell was added soon after the Conquest to the 10th-century tower of St. Mary Bishophill Junior, the interior of which is exceptionally large and probably had had a small eastern arm attached to it, similar to that at Barton-on-Humber. The single cell plan became common after the Conquest and in the early 12th century York churches of this type included

Rectangular Romanesque Churches

Name	Length	Breadth	Wall thickness
In York S.W. of the Ouse			
All Saints', North St.		12 ft.	2 ft.
St. John the Evangelist	30⅔ ft.	17½ ft.	2⅓ ft.
St. Martin-cum-Gregory	32 ft.	18 ft.	2⅓ ft.
St. Mary Bishophill Junior	38½ ft.	20⅔ ft.	2½ ft.
St. Mary Bishophill Senior	36 ft.	19⅓ ft.	2⅓ ft.
In York E. of the Ouse			
Holy Trinity, Goodramgate	22¾ ft.	13¾ ft.	2½ ft.
St. Helen's	38 ft.	15⅔ ft.	2⅓ ft.
St. Martin's, Coney St.	41⅔ ft.	20½ ft.	2⅓ ft.
Outside the City of York			
Askham Bryan	58½ ft.	23 ft.	2¾–3 ft.
Askham Richard	44 ft.	25 ft.	
Copmanthorpe	48 ft.	21¼ ft.	2½ ft.
Fulford, St. Oswald	33¼ ft.	17¼ ft.	2¾ ft.
Murton	41 ft.	16⅓ ft.	2⅓ ft.
Osbaldwick	53 ft.	23¼ ft.	2⅓ ft.

Barwick-in-Elmet (W.R.) and Farlington (N.R.) also had churches of this type.

St. Martin-cum-Gregory, St. John the Evangelist, and All Saints', North Street. Other churches that were probably of the same type occur in York E. of the Ouse and within a few miles outside the City; they are listed above with their approximate internal dimensions and wall thicknesses.

Addition of Aisles. The next stage in development is usually the provision of an aisle or aisles to the simple cell. In order to enclose the sanctuary, the 12th-century arcade at St. Mary Bishophill Junior starts a few feet from the E. end leaving a length of wall flanking the altar; this was only pierced in modern times. Except in St. Mary Bishophill Senior, where the larger Saxon cell had a 12th-century three-bay N. aisle added, the aisles are of two bays. St. Mary Bishophill Junior had a N. aisle added in the 12th century; the S. aisle is of the 14th century. All Saints', North Street had a S. aisle in the late 12th century; that on the N. followed in the early 13th century. At St. Martin-cum-Gregory N. and S. aisles were added about the same time in the 13th century and at St. John the Evangelist the earlier S. aisle was of the 13th century and the N. aisle of the 14th century. Although later expansion may have enveloped the original cell, it is usually recognised by two-bay arcades inserted in rubble walls rather less than 2 ft. 6 in. thick.

Addition of Chancels. A general development in the 13th century was the extension of the chancel. Where this already existed, an extra bay was often added to the E. Where there was no structural chancel, as in so many early churches in the Vale of York, one was added *de novo*. At St. Oswald's, Fulford, a simple 12th-century rectangle of good magnesian limestone ashlar was given a late 12th-century chancel of rubble. In S.W. York a fine 13th-century chancel was built at St. Mary Bishophill Senior. Smaller ones were added at St. Mary Bishophill Junior, and probably at St. Martin-cum-Gregory and St. John's.

The addition of a new chancel to a simple rectangle can be identified by the fact that it is almost always of the same width as the original cell; there is rarely a chancel arch.

The Cruciform Plan. Apart from the Priory Church of Holy Trinity, Micklegate, the only example of a cruciform plan in the area surveyed is at All Saints, North Street, where this shape was reached early in the 13th century. The evidence for transepts of this date depends on inference and is set out in the account of the church (p. 3).

Chantry Chapels. From the 14th century onwards town church plans often became complicated through the foundation of chantry chapels in or adjoining both chancel and nave. They often formed added aisles, sometimes on a larger scale than the earlier building; their greater breadth might lead to the widening of adjacent aisles. At All Saints', North Street, Benge's chantry (1324/5), at the N.E. corner, was erected to align with an earlier transept, and in *c.* 1410 the N. aisle of the nave was widened to the same line; at the same time Adam del Bank's chantry (1410) in the S. aisle led to the same development on that side. At St. Martin-cum-Gregory the rebuilding of the N. aisle of the nave was connected with a chantry founded by Richard Toller in *c.* 1332. The width of the 14th-century N. aisle at St. John's may also have been due to chantries; that the present N. wall, remodelled *c.* 1500, represents an earlier one is shown by the survival of a 14th-century Founder's Canopy, now cut into by a later window.

A 14th-century chapel to the N. of the chancel at St. Mary Bishophill Junior looks like a chantry. It is true that none is mentioned in the Survey of 1546 (SS, XCI (1892–4), 80–2) but it is well known that these returns were not complete; moreover some chantries had become extinct before 1546. That there was a chantry of St. Katherine in the church of St. Mary Bishophill Junior seems to be clearly established by the detailed list of York clergy and their stipends compiled between 1522 and 1526 (BM, Lansdowne MS. 452, f. 1 *et seq.*). This names 'dominus Willelmus Richardson cantarista sancte Katerine in ecclesia beate Marie super Bishophill', whose chantry had a clear value of 20*s*. That this is not a chantry in the church of Bishophill Senior is proved by that church being always called 'ecclesia beate Marie veteris'; in it 'dominus

Willelmus Hoperton cantarista sancte Katerine' is named, with a valuation of £3. Both of the Bishophill churches, therefore, had chantries dedicated to St. Katherine, that of Bishophill Junior being the less valuable. At St. Mary Bishophill Senior the Basy chantry of *c.* 1320, built to the N.E., was the same width as the remodelled N. aisle of late 13th-century date.

Final Form and Position of the Entrance to the Chancel. The process of assimilation of earlier parts of the church into a logical plan was completed by *c.* 1500. The typical York parish church of that date is an aisled parallelogram.

The nave aisles at All Saints', North Street, were broadened *c.* 1410 in line with earlier parts of greater projection, and in the middle of the 15th century an extra western bay and a tower and spire, with aisles on either side, were added. The chancel and aisles were remodelled at St. Martin-cum-Gregory after 1425 and the S. aisle of the nave was rebuilt in the mid 15th century. The arcade was rebuilt and the S. aisle widened at St. John's in the late 15th century and the N. aisle was remodelled not long afterwards.

It cannot be assumed that the division between chancel and nave has always remained constant. When the 13th-century chancel was added at St. Mary Bishophill Senior it was larger in area than the original church and later the nave was extended into the W. bay.

The reverse happened at All Saints', North Street. The chancel through much of the Middle Ages was confined to the two E. bays of the church. It is likely that the chancel was extended one bay to the W. after the construction of the W. bay of the nave and the tower *c.* 1440–50. The chancel had been reduced to the original size before 1670, when Henry Johnston refers to the window of the Nine Orders of Angels as being 'the first south window of the body of the church' (Bodleian, MS. Top. Yorks. C14, f. 96); thus it remained until after 1860.

Walling

Pre-Conquest. When the tower of St. Mary Bishophill Junior was restored in 1908 the foundations were discovered to be of good rubble composed of Roman tile and bricks and broken stones. Otherwise the only footings examined were those of the 11th-century part of St. Mary Bishophill Senior. The footings, in a U-shaped trench, consisted of rubble and soil, with stones pitched in a rough herringbone fashion, without mortar but set in soil and occasionally clay. They narrowed to the width of the wall built on them, from a greater width below, and were wider on the side walls than the end walls. The average depth below ground level was just under 3 ft. (For a more detailed account *see* H. G. Ramm, 'Excavations at St. Mary Bishophill Senior' (MS. 1966), RCHM archive.) In general the Saxon walls are built of reused Roman material, large pieces of gritstone, *saxa quadrata* of magnesian limestone, and pieces of brick, very roughly coursed. The inner order of the tower arch at St. Mary Bishophill Junior does not have the characteristic through-stones. The middle section of the tower of St. Mary Bishophill Junior has occasional patches of reused Roman stone placed obliquely.

In general quoins are megalithic; one at the N.E. angle of the nave of St. Mary Bishophill Senior was formed of smaller pieces. The stone of the tower arch at Bishophill Junior may have been quarried by the Saxons; it is tooled in two directions (Plate 20) giving a rough herringbone finish. The ordinary wall thickness is about 2 ft. 4 in.

Norman. The walls of the rectangular churches are all thinner than the usual 3 ft. and are of haphazard material containing reused Roman gritstone and brick and some pieces of magnesian limestone. Only at St. John's was a plinth found, of square section and Saxon in character. The N. aisle wall at St. Mary Bishophill Junior is of 12th-century date and though of rubble is better coursed.

The only ashlar in a parish church is in the tower of St. John's and this is not normal in that some of the stones in the W. wall are markedly oblong and could be reused Roman material. The other walls use the normal square blocks of magnesian limestone and the tower had a plinth of two offsets. None of these walls have buttresses. The Priory Church of Holy Trinity is built of excellent late 12th-century ashlar with fine diagonal axing (Plate 20) with a plain chamfered plinth and a simple pilaster buttress.

13th century. The chancel walls at St. Mary Bishophill Junior are of early 13th-century rubble. The large chancel at St. Mary Bishophill Senior was of similar date and had an external facing of oblong blocks of gritstone, chamfered water-table and plinth, a keeled-roll string course and pilaster buttresses. The E. wall of the chancel at All Saints', North Street, of similar date, is also of squarish blocks with a chamfered plinth and pilaster buttresses with wall arcading internally. The remaining fragment of the N. aisle at Holy Trinity Priory (under the tower) has walls with skins of excellent oblong ashlar with rubble core, and a chamfered plinth; the buttresses were semi-octagonal in plan. Particularly valuable is the close proximity of 12th-century axing and early 13th-century chiselling on the inside of the W. end, where the masonry of the N. aisle wall abuts the pilaster buttress of the aisleless Norman nave; the N. aisle walls exhibit the bold claw tooling found after *c.* 1200 (Plate 20). The only example of late 13th-century walling was that of the N. nave aisle at St. Mary Bishophill Senior, built of large markedly long pieces of freestone with a chamfered plinth and deep buttresses of two orders with gabled weatherings.

14th century. The practice of building with skins of ashlar and a rubble core now ceases as walls become thinner and ashlar reaches its maximum size in the wall of the N. nave aisle at St. Martin-cum-Gregory, and the E. wall of the S. aisle at All Saints', North Street, contrasting with the E. wall of the chancel, with its earlier small blocks. Claw tooling is finer than that of the 13th century but still of the same type as on the N. arcade of St. John's (Plate 20). Not enough 14th-century buttresses survive to generalise about them; plinths and water-tables are chamfered as before.

15th and early 16th century. Walls are in general of good quality magnesian limestone but the stones are not so large as in the previous century and the tooling, where visible, has a pattern of very small squares; masons' marks are more in evidence. The chancel and aisles of St. Martin-cum-Gregory (after 1425), although very weathered externally, represent work of high quality; the buttresses are narrow and deep, and the plain parapet has a moulded string at the bottom and bold gargoyles; the S. nave aisle (*c.* 1450) is also of good quality but more robust. Parapets and plinths are moulded and there have been good carved gargoyles. At St. John's the S. aisle wall had deep three-stage buttresses surmounted by pinnacles and a battlemented parapet. Perhaps the finest ashlar is that of the late 15th-century clearstorey of St. Martin-cum-Gregory (*c.* 1470–80).

Fig. 5. (opp.) Stone Mouldings.

(5) Holy Trinity
- a. Crossing. N.W. pier, minor shaft, *c.* 1120.
- b. Nave. S. arcade, *c.* 1180.

(9) St. Mary Bishophill Senior
- c. S. doorway, E. jamb, *c.* 1180.
- d. N. arcade, 5th pier from E., *c.* 1180.

(4) All Saints', North Street
- e. Nave. S. arcade, 4th pier from E., *c.* 1190.
- f. Nave. N. arcade, 4th pier from E., early 13th century.
- g. Chancel. Arcading at S.E. corner, *c.* 1210.

(5) Holy Trinity
- h. Tower. Arcade at triforium level, *c.* 1230–40.
- i. Nave. N. doorway, *c.* 1220.

(7) St. Martin-cum-Gregory
- j. Nave. N. arcade, pier, early 13th century.
- k. Nave. N. arcade, E. corbel, early 13th century.
- l. Nave. S. arcade, pier, early 13th century.
- m. Nave. S. arcade, E. corbel, early 13th century.
- n. Chancel. N. arcade, pier, *c.* 1410.
- o. Chancel. S. arcade, pier, *c.* 1410.
- p. Nave. N. aisle. Tomb canopy, *c.* 1330.

(9) St. Mary Bishophill Senior
- q. S. doorway, arch, *c.* 1180.

(8) St. Mary Bishophill Junior
- r. Nave. S. arcade, western archway, early 14th century.

(6) St. John the Evangelist
- s. N. arcade, first pier, E. side, late 13th/early 14th-century.
- t. N. arcade, second pier from E., corbel of outer order, 14th century.

(7) St. Martin-cum-Gregory
- u, v, w. N. and S. chancel aisles. Brackets, 15th century.
- x. S. chancel aisle. Corbel under wall post.

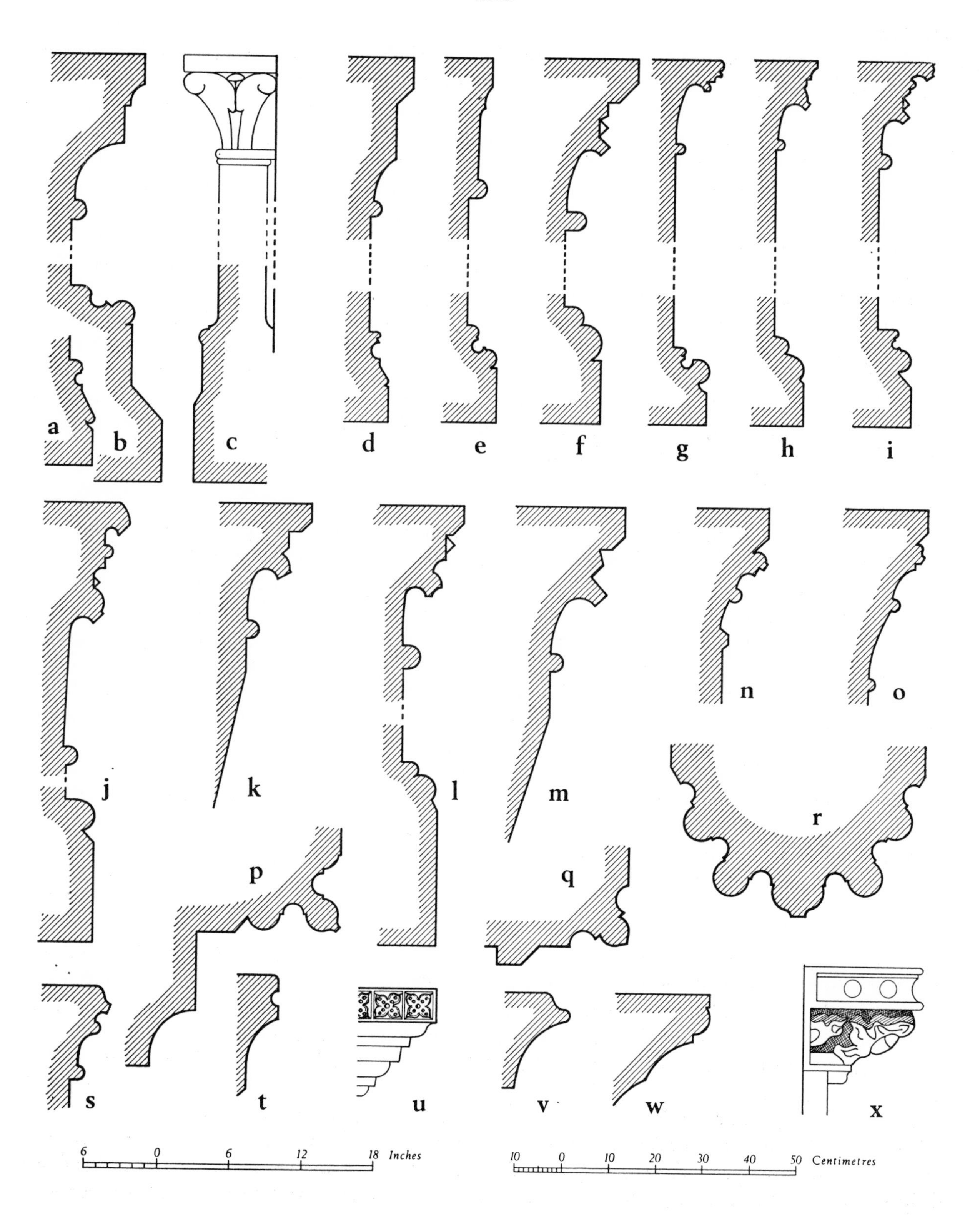
a
b
c
d
e
f
g
h
i
j
k
l
m
n
o
p
q
r
s
t
u
v
w
x
6 0 6 12 18 Inches
10 0 10 20 30 40 50 Centimetres

Windows

The churches show a sequence of windows covering the whole mediaeval period. The belfry windows at St. Mary Bishophill Junior with turned balusters between the openings illustrate a common pre-Conquest type. A Norman window in the tower of St. John the Evangelist is of interest as it has no provision for glass. Lancets of the 13th century are exemplified at St. Mary Bishophill Junior and (before its demolition) at St. Mary Bishophill Senior. Others occur at Holy Trinity, some elaborated by shafted jambs and dog-tooth enrichment. Early tracery, of the late 13th century or *c.* 1300, appeared in the nave of St. Mary Bishophill Senior. The E. window at St. John's has modern intersecting tracery reproducing the form of the previous window, of the beginning of the 14th century but with the addition of cusps. At All Saints' there are windows with reticulated tracery of the second quarter of the 14th century, as well as others of the same period; the E. window has rather later flowing tracery of the middle of the century. Straight-sided reticulation was used at the beginning of the 15th century in the S. wall of the nave of St. Mary Bishophill Senior. Typical of the early 15th century are the windows in the chancel aisles of St. Martin-cum-Gregory with gridiron rectilinear tracery in sharply pointed two-centred heads. Broader in proportion and in the best Perpendicular tradition are the windows of *c.* 1450 in the W. part of All Saints' and in the S. nave aisle of St. Martin-cum-Gregory. Windows at St. John's show a late 15th-century form with rectilinear tracery extending down below the springing of the low four-centred heads.

Church Building after the Reformation

There was little work of note carried out in the 16th century. The 17th century, however, saw extensive rebuilding and additions in three churches. In 1646 a timber-framed structure was placed on top of the remnant of the Norman tower of St. John's, and in 1659 at St. Mary Bishophill Senior a tower of brick was erected over the W. bay of the N. aisle and a roof fitted to the chancel on walls heightened in brick. In 1677 the tower at St. Martin-cum-Gregory was rebuilt in brick with stone dressings and a balustraded parapet (the latter removed 1844–5). In the 18th century St. Mary Bishophill Junior was given new aisle roofs.

A large number of restorations were carried out in the 19th century, the majority by Messrs. J. B. and W. Atkinson. To this period also belong the erection of two new churches, St. Edward's, Dringhouses (1847–9), and St. Paul's, Holgate Road (1850–1), and the rebuilding of a third, St. Stephen's, Acomb (1830–1). St. Edward's, Dringhouses, designed by Messrs. Vickers & Hugall of Pontefract, is a pleasing scholarly exposition of Decorated Gothic. The other two are less successful. St. Paul's, Holgate Road, built by Messrs. J. B. & W. Atkinson in the Early English style, is interesting for its use of cast iron for the piers; St. Stephen's, Acomb, rebuilt by G. T. Andrews ostensibly in the Early English style, shows no real appreciation of mediaeval architecture. (E.A.G.)

Non-Anglican Religious Buildings

By far the most interesting Roman Catholic building in York is the *Bar Convent* (13) in Blossom Street. The buildings are remarkably well documented and provide a gazetteer of well-known Roman Catholic architects and craftsmen. The various 18th-century buildings designed by Thomas Atkinson are all of distinction, and among them the chapel, based on a Roman model, is exceptional. The ranges dating from the first half of the 19th century by J. B. & W. Atkinson and by G. T. Andrews, the railway architect, are also noteworthy. The Convent, which has been on this site since 1686 and suffered remarkably little persecution throughout, comprises one of the oldest girls' boarding schools in England.

There were few Nonconformist chapels in the area covered by this volume and little of note survives of the only Primitive Methodist chapel (15), built in *c.* 1846 on Acomb Green. Albion Chapel (14) was designed for the Wesleyan Methodists by the Rev. John Nelson junior and opened in 1816 to serve the

(5) CHURCH OF THE HOLY TRINITY. The Trinity, 15th century.

(8) CHURCH OF ST. MARY BISHOPHILL JUNIOR, 10th and 11th century.

(7) CHURCH OF ST. MARTIN-CUM-GREGORY, 15th century.

(4) CHURCH OF ALL SAINTS, NORTH STREET, mid 15th century.

CHURCH OF ST. MARY, CASTLEGATE (E. of the Ouse), 15th century.

CHURCH INTERIORS

(5) CHURCH OF THE HOLY TRINITY, from W., 12th century and later.

(7) CHURCH OF ST. MARTIN-CUM-GREGORY, from E., 11th century and later.

(7) CHURCH OF ST. MARTIN-CUM-GREGORY. N. chancel aisle, early 15th century.

(9) CHURCH OF ST. MARY BISHOPHILL SENIOR. N. aisle, *c.* 1180 and later.

ARCHWAYS

(9) CHURCH OF ST. MARY BISHOPHILL SENIOR. S. doorway, *c.* 1180.

(8) CHURCH OF ST. MARY BISHOPHILL JUNIOR. Tower arch, 10th century.

(6) CHURCH OF ST. JOHN. Tower window 12th century.

(8) CHURCH OF ST. MARY BISHOPHILL JUNIOR. Coffin-lid, 14th century.

(5) CHURCH OF THE HOLY TRINITY. N. door, 15th century. Doorway 13th century and modern.

(8) CHURCH OF ST. MARY BISHOPHILL JUNIOR. E. window of S. aisle, 14th century.

(7) CHURCH OF ST. MARTIN-CUM-GREGORY. S. aisle, mid 15th century.

E. window, N. chancel aisle, early 15th century.
(7) CHURCH OF ST. MARTIN-CUM-GREGORY.

S. chancel aisle, 15th century.

(6) CHURCH OF ST. JOHN THE EVANGELIST. N. aisle, *c.* 1500.

(8) CHURCH OF ST. MARY BISHOPHILL JUNIOR. Nave, 15th century.

(7) CHURCH OF ST. MARTIN-CUM-GREGORY. S.E. corner of nave, 11th century and later.

(4) CHURCH OF ALL SAINTS. Corbel on N. arcade, 14th century.

(5) CHURCH OF THE HOLY TRINITY. Carvings in N. porch, 12th century, reset.

(6) CHURCH OF ST. JOHN THE EVANGELIST. Monument (1) of Sir Richard Yorke, *c.* 1498.

(4) CHURCH OF ALL SAINTS, NORTH STREET. Misericorde, 15th century.

MASONRY TOOLING

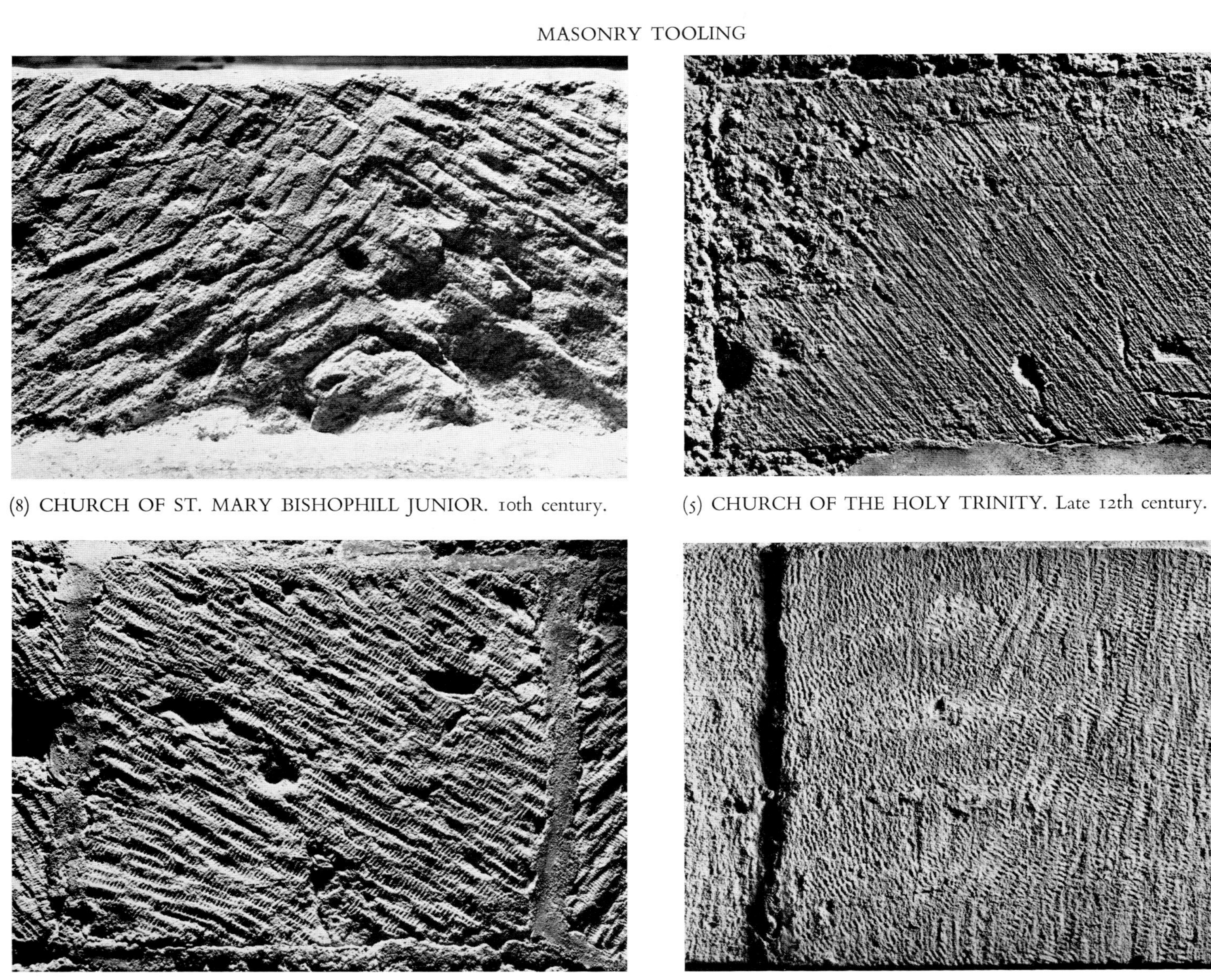

(8) CHURCH OF ST. MARY BISHOPHILL JUNIOR. 10th century.

(5) CHURCH OF THE HOLY TRINITY. Late 12th century.

(5) CHURCH OF THE HOLY TRINITY. Early 13th century.

6) CHURCH OF ST. JOHN THE EVANGELIST. 14th century.

needs of the area, and also to accommodate overflow from New Street Chapel (1805). It cost £6,000 and was to provide 1,600 sittings. It survives though sold in 1856, when it was replaced by the Wesley Chapel, Priory Street, a grandiose red brick building with stone dressings designed by James Simpson of Leeds, who was responsible also for Centenary Chapel, St. Saviourgate (1840). A Wesleyan Chapel in Acomb, now represented by 91–93 Front Street (16), was built in 1821; little of it but the shell remains.

Two noteworthy burial grounds are not attached to ecclesiastical buildings. The Friends' Burial Ground (18) in Bishophill, bought in 1667 and closed in 1885, contains headstones which, after 1818, record members of most of the eminent local Quaker families who contributed so much to social life in York. A burial ground near the railway station (17) was provided during the cholera epidemic of 1832.

Fittings

Bells. Three of the bells in the area covered by this volume are of the 14th century. One at St. John's, Ousebridge, has on it a shield charged with three helmets, which could be the arms of the benefactor or a device used by the maker. The treble at St. Mary Bishophill Junior has a Lombardic inscription also found at Church Fenton, and technical details identical with the later tenor in the same church. A bell at Holy Trinity Priory was made by John Potter (Free 1359) and has the same inscription as that on a bell at West Halton in Lincolnshire (G. Benson, 'York Bellfounders', YPS (1898), 7). The two largest bells at St. John's, Ousebridge, which came from St. Nicholas Hospital, Hull Road, and one of which is dated 1408, have inscriptions in Lombardic capitals and were said by the late H. B. Walters to be the work of John Potter also (MS. list prepared for the Council for the Care of Churches). A third bell from St. Nicholas Hospital and also now at St. John's has a black-letter inscription and is a little later.

The tenor at St. Mary Bishophill Junior is probably by John Hoton of York, who was paid for a bell at the Minster in 1473/4 (Browne, 254); there are similar bells at St. Nicholas, Newcastle (two), and at Heighington and Sedgefield in County Durham (Benson, 'York Bellfounders', *AASRP*, XXVII, pt. ii (1904), 641). John Hoton was Free as a potter in 1455, and the Rev. J. T. Fowler noted that inscriptions commonly put on bells by John Hoton were used by Nottinghamshire founders between 1450 and 1774 (Benson, YPS, 8; *AASRP*, 628).

A single bell of the 16th century remains at St. Martin's, the survivor of three given by John Beane and made in 1579 by Robert Mot, famous for his work at the celebrated Whitechapel Foundry, which still exists (Benson, YPS, 11) (Plate 21). The shield charged with a crown between three bells is his mark. He was a very successful founder and over fifty of his bells remain (H. B. Walters, *Church Bells of England* (1912), 216).

William Oldfield's bells are the most frequent in the 17th century: one of 1627 at All Saints', North Street, one at St. John's of 1633 and one at Acomb of the same date, a bell of 1640 at All Saints', North Street, and an undated one at St. John's are either definitely or probably by him. His favourite inscription is 'Jesus be our Speed' (St. John's and Acomb), but he also uses 'Soli Deo Gloria'. A bell, now recast, at Acomb and of 1660 was the only one by James Smith (Benson, YPS, 12), but the treble at St. Martin's was made by Samuel Smith senior (d. 1709) in 1697 and one dated 1731 at Holy Trinity by Samuel Smith junior (d. 1731), all members of the same firm whose shop was in Toft Green. Their favourite inscription is 'Gloria in altissimis Deo' (also found on the first bell formerly at St. Maurice's).

A fine peal of six made by Pack & Chapman of London in 1770 for St. Mary Bishophill Senior is now at St. Stephen's Church, Acomb. The partnership was formed in 1769 and was very important until the death of Thomas Pack in 1781 (Walters *op. cit.*, 217; J. E. Poppleton in *YAJ*, XVIII, 92).

Bell Frames. Two bell frames of the 17th century remain, one at St. John's of 1646 and another at St. Martin's of 1681 (Plate 21); they were almost certainly made by the same craftsman, Robert Rason

('Rinson') who was paid £17 for work on the St. Martin's bell frame and steeple (Benson in *AASRP*, XXXI, pt. ii (1912), 614).

Benefactors' Tables. There is a good series of Benefactors' tables and all entries have been fully recorded for the Commission archive, a practice which has proved its value as some have recently been destroyed. A number date from the 18th century and others from the early 19th century. Renewal of the painted entries is often evident, for example one table in All Saints', North Street, made in 1764, was relettered in 1804. The entries on a 19th-century table at Holy Trinity are in imitation black letter with red capitals. In general the tables are of planks with a moulded frame. The one at St. Martin's is surmounted by ball finials (Plate 22), and the framing of one which formerly belonged to St. Mary Bishophill Senior has some architectural pretension (Plate 22).

Brackets. Six large moulded stone brackets, oblong in plan, at St. Martin's are of a kind that occurs more often in secular than in ecclesiastical architecture. Their use is in some doubt. A similar bracket at Norrington Manor, Wilts., still retains spikes on which candles were fixed (*see* M. Wood, pl. LVII, G. H. for other examples), but at Bolton Castle, Wensleydale, lamp brackets exist alongside such brackets.

Brasses and Indents. A great number of indents remain in the parish churches, particularly in All Saints', North Street, and there were many in St. John's, Ousebridge, before the recent restoration.

Six brasses dating from before 1800 give a representative series. In All Saints', North Street, are two mediaeval oblong plates of practically the same size but not by the same craftsman. One, in memory of Robert Colynson, 1458, and William Stokton, 1471, both husbands of Isabel Stokton (herself commemorated, d. 1503), is in excellent black letter (Plate 36). The other, to Thomas Clerk, 1482, though not quite so well lettered, is perhaps the more interesting because of the signs of the Evangelists on the same stone, three of which remain (Plate 27). A brass to Thomas Askwith, 1609, in the same church has an inscription in capitals on an exceptionally heavy oblong plate and a coat of arms on another shield-shaped plate. It is effective, but the execution is not so professional as that of all the others (Plate 35).

In Holy Trinity, Micklegate, an oblong plate in memory of Alderman [Elias] Micklethwait, 1632, has merely his name in a good script. The only brass with a person represented on it is that of Thomas Atkinson, 1642, in All Saints', North Street; the demi-figure is well drawn but the lettering in capitals is not so competent. The whole is engraved on a large plate with the upper corners cut off which was originally set in the same stone as that of Thomas Clerk, 1482 (Plate 35). An oblong plate in All Saints', North Street, to Charles Towneley, 1712, is in good cursive script (Plate 36).

The few other plates are all of the early 19th century. Two, of 1810 and 1839, are on an interesting tombstone at Acomb to the Hubback family. The brass to Thomas Mosley, 1624, recorded in St. John's in 1904 had vanished by 1950, and another formerly in St. Mary Bishophill Senior, with initials of the Dawson family, *c.* 1813, has vanished recently.

There are 18th-century *Breadshelves* at St. Martin-cum-Gregory (Plate 23) and, from St. Mary Bishophill Senior, at St. Clement's, Scarcroft Road (Plate 23); both are designed as entities and have contemporary enrichments. A handsome brass *Candelabrum* of about 1715 is in St. Martin's (Plate 24), and in the Yorkshire Museum are two enamelled prick *Candlesticks* discovered under the floor of St. Mary Bishophill Senior (Plate 24). Two pewter candlesticks, given in 1754 to St. Martin-cum-Gregory, are now at Holy Trinity. *Chairs*, of 17th or early 18th-century date, include one of 'Yorkshire' type, from St. Mary Bishophill Senior (Plate 44), which is noteworthy and probably belonged to the church from the time it was made. Two late 17th-century sanctuary chairs with high backs with cane insets and elaborate enrichment are shown as belonging to All Saints', North Street, in J. B. Morrell, *Woodwork in York* (1949), Fig. 162, but have not been seen recently. Only one *Chest* is worth mentioning; this is in Holy Trinity, of wood and leather, and of *c.* 1600.

Coffin Lids. There is a good series of coffin lids in the area covered by this volume and in particular in All Saints', North Street. In general it would appear that the rougher coffin lid directly placed on a coffin just below floor level was supplanted by deeper burial and a floor slab in the mid 14th century. The floor slab to John Bawtrie (1411) in All Saints', North Street, is important in this respect as it has the type of cross to be expected on a coffin lid. Practically all the lids have a cross on them and in general the crosses are (A) incised, or (B) formed in relief but in a recessed circle. If these two groups can be regarded as different techniques, then the shapes of the crosses provide further classification, which however may have no true chronological significance.

(A) Incised Crosses. Two lids have crosses based on a circle; in All Saints', North Street, (1) has straight arms to the cross and trefoiled ends to the arms all conjoined with a circle, and (7) has petal-shaped cross arms within a circle, and both are probably of the 14th century. In the same church (4) has a cross with fleur-de-lis ends set in an incised circle (Plate 27). All the other lids have crosses with some variant of trefoil at the end of each arm. The round-lobed trefoil is perhaps the earlier, succeeded in the early 14th century by the fleur-de-lis. All Saints', North Street, (9) with straight arms and round-lobed trefoil is 13th century, and St. Mary Bishophill Senior (6) has straight arms and trefoils which have a round lobe but the outer ones are on drooping stalks and are tending towards the fleur-de-lis type.

All Saints', North Street, (5) has straight arms and fleur-de-lis ends but under each fleur-de-lis is a round knop. St. Mary Bishophill Senior (1) has a cross with straight arms and fleur-de-lis ends and the tooling on the lid is of early 14th-century type (Plate 27).

Next come two lids with crosses with curved sides to the arms, fleur-de-lis ends, and voided centres to the crosses. In All Saints', North Street, (2) and the two crosses on (10) are of this form.

The crosses on all these coffin lids have been similar, but two crosses on a lid at Holy Trinity (3) are quite complicated and may be compared with some in Lindsey, Lincolnshire, described as round leaf with cross band between the leaf and the butt, and of 1200–50 (L. Butler, *Arch. J.*, CXXI (1964), fig. 4D). Another similar cross is at South Leverton, Nottinghamshire (L. Butler, 'Mediaeval Cross Slabs in Nottinghamshire', Thoroton Soc. *Trans.*, LVI (1952), plate 1.j.).

(B) Crosses in Relief in a Recessed Circle. In All Saints', North Street, (2), a complicated cross with straight arms and the angles filled in with circles and with all ends trefoiled, is like one from St. Mary Bishophill Senior (3), and as the lobes are of fleur-de-lis type they may be slightly later than All Saints', North Street, (6) which has straight arms each with three ends all with round-lobed leaves; all may be of *c.* 1300.

South Derbyshire has a semi-relief group of this type as early as *c.* 1210–40 (Butler, *Arch. J.*, as above, fig. 6A) and a clustered trefoiled head like All Saints', North Street, (3) is found at Belvoir and Eastwell in Leicestershire in *c.* 1250/80 (*ibid.*, fig. 6D).

Knops on the stems can vary from round (All Saints', North Street, (11)) and oblong (St. Mary Bishophill Senior (1) and All Saints', North Street, (6)) to leaves (All Saints', North Street, (12)), but they have no dating value. All bases are of Calvary type except All Saints', North Street, (3) which only has a fleur-de-lis at the foot of the stem.

Various symbols are found, including two chalices (All Saints', North Street, (2), St. Mary Bishophill Senior (6)), a sword probably for a knight (Holy Trinity (3)), a mace and a sword (All Saints', North Street, (4)), a bow and arrow and sword perhaps for a man-at-arms (All Saints', North Street, (10)), and a cleaver (All Saints', North Street, (7)).

Only two lids have fragments of inscriptions, both in Lombardic capitals. The one at Holy Trinity has H(I)C JACE[T] and one from St. Mary Bishophill Senior (4) has PRIEZ PVR LEALME. . . .

There are two early 19th-century wooden *Collecting Shovels* from St. Mary Bishophill Senior at St.

Clement's, Scarcroft Road. The only early *Communion Rails* are the oak ones at St. Martin-cum-Gregory (Plate 127) for which Mr. Matthew Butler was paid £8 in 1753. They have the semicircular front feature which is almost peculiar to York. *Communion Tables* are generally of oak, have turned legs and strong rails, and are of 17th-century date. The one from St. Sampson's, now at St. Mary Bishophill Junior, is perhaps the best as it has carved arabesques on three of the upper rails. The *Door* (Plate 15) now placed in the north doorway at Holy Trinity is of the 15th century and was found in rubbish at the base of the tower in 1902–5; with its central wicket, it is like one at the Merchant Taylors' Hall.

Fonts and *Font Covers*. The fonts are hardly worthy of mention, but the four early 18th-century font covers in this area form an interesting group. In general they consist of enriched scrolls conjoined at the top and are of oak, and although the general effect is Jacobean, the carving is of Baroque character. The finest is the one now in Holy Trinity, from St. Saviour's, dated 1717; it differs from the others in having two tiers instead of one, but there is little doubt that the original font cover of 1717 is the top part and that the rest was added in 1794 (Plate 28); the upper part is very much like that at St. Martin's, Coney Street, in the degree of enrichment.

The three other covers are at St. Hilda's, Tang Hall (from St. John's), St. Martin's and St. Mary Bishophill Junior (Plate 28). They all have a dove set above an acorn finial, but whereas those from St. Martin's and St. Mary Bishophill Junior are virtually identical, with urn-forms set on pedestals on the backs of the scrolls, and may be by the same hand, the one from St. John's has richer scrolls like those on the Holy Trinity cover, is more competently carved and has flaming hearts as on the one at St. Martin's, Coney Street; it is identical with one at Bolton Percy (F. Bond, *Fonts and Font Covers* (Oxford, 1908), 312).

Glass. The glass in the area covered by this volume is exceptional not only in its quantity but also in its dating range: there are examples of all periods from the early 14th century to 1850 and later (Plates 29–31, 98–116, 122–3). The finest body of glass in a York parish church is in All Saints', North Street, where the early 15th-century sequence is probably the best of that period in any parish church in England,[1] for all the famous groups in other English churches are later. St. Martin-cum-Gregory also contains very good glass and that from St John's, Micklegate, almost equals it. Much of the glass had remained intact in its mediaeval form until *c*. 1730, but between 1730 and 1846 the glass in All Saints', North Street, was so moved about or even damaged that thereafter only one window remained intact in its mediaeval position. There is a similar story of movement and destruction in St. Martin-cum-Gregory, and the greater part of the glass formerly in St. John's, Micklegate, is now in the N. transept of York Minster. It is noteworthy that, whereas York churches had retained most of their glass unscathed through the religious and political troubles of the 16th and 17th centuries, they suffered very serious losses during a period of calm and of renewed antiquarian interest. Perhaps the brighter side of this picture is that there is also a record of restoration; many of the fragments too have been saved. The chief 18th-century glass-painter who not only produced new glass but restored the old was William Peckitt.

In the early 19th century William Wailes of Newcastle and Barnett & Son restored glass. J. W. Knowles of York repaired windows in All Saints', North Street, between 1861 and 1877 and in St. Martin-cum-Gregory in 1899; in 1965–7 the windows of All Saints', North Street, were re-arranged and again repaired by the York Minster glaziers O. Lazenby and P. Gibson.

The following account assesses the glass chronologically and gives a list of subjects, but details will be found under the respective church entries. The 14th-century painted glass is of good quality; it includes a quantity of pot metal. The canopies are of Geometrical type, and graphically there is a tendency toward minute detail.

---

[1] See E. A. Gee 'The Painted Glass in All Saints' Church, North Street, York' in *Archaeologia* CII (1969).

At All Saints', North Street, the glass in the E. window of the N. aisle was originally in the E. window of the chancel and, like the masonry there, may be of *c.* 1330; the very much restored glass in the E. window in the S. aisle of the same church has similar details. Two lights in the E. window in the nave N. aisle wall at St. Martin-cum-Gregory are from the same shop and as they were almost certainly made for Richard le Toller's chantry of 1332, the date becomes more definite. The centre figure-subject of the E. window of the S. aisle of the chancel in the same church is of similar character and date.

Donors in the E. window of the N. aisle of St. John's, Micklegate, came from windows of *c.* 1340; apart from the fact that the activities of these donors, Richard and Katherine Briggenhall, John and Joan Randeman and William and Agnes Grafton, suggest this date, Richard le Toller and his wife Isabella founded a chantry here in 1320 as well as in St. Martin's and may have employed the same glazier. Diaper backgrounds and other details of glass from the E. window of the S. aisle, belonging to the Shupton and Briggenhall chantry of 1319–38, appear to be identical with, or very closely related to, parts of the great W. window of York Minster by Robert (? Ketelbarn) and precisely dated to 1339. The St. John's window showed the story of St. John the Baptist.

The E. window of the S. chancel aisle of St. Martin-cum-Gregory has tracery lights, canopies and main panels in the outer lights probably given by Nicholas Fouke whose executors founded a chantry in 1367. If, as seems almost certain, he was the Nicholas Fouke Free in 1309/10 and Lord Mayor in 1340/1,[1] the window may be of the same date as or slightly later than the previous ones.

The whole character of the glass both in texture and drawing changed in *c.* 1400. The early 15th-century glass has much less pot metal, much more silver stain and the figures become naturalistic. The canopies are bigger and more architectural, and detail is not only larger in scale but is also given new forms.

Virtually the whole scheme of glazing at All Saints', North Street, as completed by *c.* 1450 survives: it consists of the 'Prick of Conscience' window of *c.* 1410 given by Roger and Cecilia Henrison, Abel and Agnes Hesyl and one Wiloby; the 'Corporal Acts' window also probably of *c.* 1410, the donor being Nicholas Blackburn; and the 'Nine Orders of Angels' window of *c.* 1410–20. The date of the present glazing in the E. window of the chancel may be narrowed down to 1412–27; it has Nicholas Blackburn senior and his wife Margaret and his son Nicholas Blackburn junior and his wife Margaret as donors. The third N. window was given by Reginald Bawtre after 1429; the second S. window is dated by the donors, James Baguley and Robert Chapman and his wife, to *c.* 1430–40, and the latest window of the series is the fourth on the S., which may have been made between 1436 and 1451 to commemorate Richard Killingholme and his wives Joan and Margaret.

There is relatively little painted glass of the later 15th century and by far the best is the former E. window of the N. aisle of St. John's given in memory of Sir Richard Yorke after 1498 and now in the Minster, in the W. aisle of the N. transept. Otherwise there are four panels in the second window on the S. side of the chancel at St. Mary Bishophill Junior and two remarkable small panels portraying the betrayal of Christ and David and Goliath, in the third window in the S. aisle of the chancel at St. Martin-cum-Gregory.

A shield of Bishop John Alcock in All Saints', North Street, must be of *c.* 1500. In the same church some strap work and the initials 'BB', as well as the vanished arms of Archbishop Harsnet, were probably painted in the 17th century by one of the Gyles family, and at St. Stephen's Acomb is a very good achievement-of-arms dated 1663 and probably by Edmund Gyles.

The important 18th-century glass painter, William Peckitt, is well represented and various fragments of this date from All Saints', North Street, include one with his signature. In St. Martin-cum-Gregory the first window in the N. nave aisle was painted by him in memory of two daughters and is dated 1792, and

---

[1] This Nicholas Fouke(s) was prominent in the parish and lived until shortly after 1362; there is no evidence of a namesake.

the fourth window in the N. chancel aisle consists of three main lights all from the Peckitt workshop; the outer lights have geometrical patterns and intricate medallions; the centre light is by Mrs. Peckitt in memory of her husband and dated 1796. Two pictures of dogs in Micklegate House dated 1756, are also by Peckitt.

The most important glass-painters at work in the early 19th century were Messrs. Wailes of Newcastle. At All Saints', North Street, they made new glass for the tracery and lower part of the E. window of the chancel and new tracery lights for the E. window of the N. aisle in 1844. Eight windows at St. Edward the Confessor's, at Dringhouses, one signed 'M.W. 1849', were made by the firm, and the E. window there won a first prize in the Great Exhibition of 1851. A window of 1850 by John Joseph Barnet, late of York, in Holy Trinity is very well drawn with crisp patterns in good glass.

The scenes depicted in the glass in York are many and varied. The most remarkable subject is the 'Prick of Conscience' in All Saints', North Street, and the 'Corporal Acts' and the 'Nine Orders of Angels', in the same church are scarcely less noteworthy; all three windows are of the early 15th century; their donors are named above. The only Old Testament subject, the slaying of Goliath by David, is in St. Martin-cum-Gregory (late 15th-century), though other scenes were before 1939 in the same church.

Christ is shown in Gethsemane (All Saints', 14th-century), is betrayed (St. Martin-cum-Gregory, 15th-century), shows himself to St. Thomas (All Saints', after 1429) and is shown crowned and enthroned in the Coronation of the Virgin (St. John's, 15th-century). The story of His life is the subject of the 14th-century glass in the E. window of the N. aisle in All Saints', North Street. The Virgin occurs twice in this same church (14th-century) and is enthroned in glory in St. Mary Bishophill Junior (15th-century). The Trinity is shown also in All Saints', North Street (1412–27), and in St. John's (15th-century). The various saints depicted are listed in the Index, under *Glass*.

The Mass of St. Gregory in the fourth window of the S. aisle in All Saints', North Street, is of iconographical interest. The concept is probably based on the original Image of Pity in the church of Santa Croce in Gerusalemme at Rome (Exeter Diocn. Arch. Soc., xv (1927), 22). The York panel lacks the *Arma Christi* and other elaborations found in later versions.

There is good heraldic glass at All Saints', North Street (15th to 16th-century), from St. John's, now in the Minster (after 1498) and at Acomb (royal arms dated 1663).

Hatchments never seem to have been very common in York churches and the only one remaining here is to Joshua Crompton (1832) in Holy Trinity.

Images. There is practically no good mediaeval sculpture, and the only interesting feature is a 14th-century corbel in All Saints', North Street, on the W. face of the third N. pier (Plate 18).

All the 15th-century examples are moveable and there is no proof that they belonged to the church in mediaeval times. All Saints', North Street, contains an attractive head and shoulders of a woman in limestone (Plate 39), a Nottingham-type alabaster panel of the Resurrection (Plate 39), and the wooden figure of a priest holding a chalice, all of which may have been bought by the late Rev. Patrick Shaw. A Trinity at Holy Trinity church, Micklegate (Plate 9), when bought in Holland was said to have come from the York priory. An 18th-century wooden figure of King David in All Saints', North Street (Plate 42), possibly belonged to the destroyed reredos and, if so, may have been made by William Etty in 1710.

The Bar Convent in Blossom Street has some exceptional later figures. Three Baroque alabaster statues, of St. Sebastian, St. Michael and St. Margaret (Plate 140), are probably Spanish and came to the Convent in 1805 through the Rev. Anthony Plunkett, who was the last Prior of the Dominican Priory of Bornheim in the Netherlands. They are probably of the early 18th century. Two white marble statues of the Virgin and of St. Joseph were bought in Florence in 1823 specifically for the Convent. Four figures of the Latin Doctors, St. Jerome, St. Augustine, St. Ambrose and St. Gregory (Plate 40), carved in solid oak in a

MASONS' MARKS

(See under Inscriptions etc. p. lvi.)

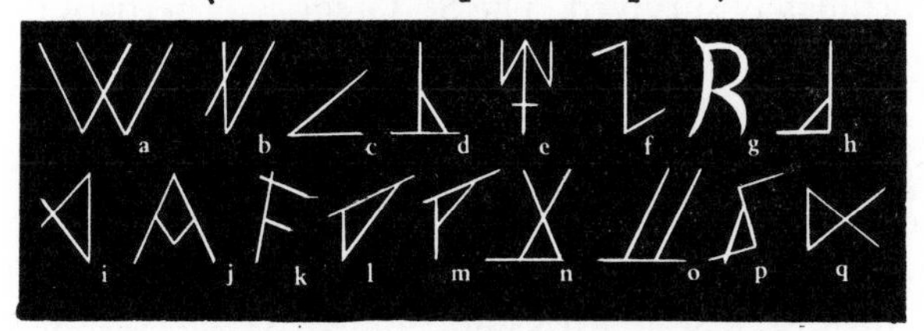

Fig. 6. (4) Church of All Saints, North Street.

a. N. arcade, second arch from E.
   S. arcade: first pier from E., N. side;
   second pier, N. side.
b. N. arcade, second arch from E.
c. N. arcade, second arch from E.
d. N. arcade, fifth arch from E., N. chamfer.
e. N. arcade, fifth pier from E., W. side.
   S. arcade, fifth pier, N. face.
f. S. arcade: first pier from E., N.W. side;
   second pier, E. side.
g. S. arcade, second pier, S.W. side (probably reused stone).
h. S. arcade, fifth pier, S. face.
i. Tower, N. pier, N. face.
j. Tower, N. pier, N.W. face.
k. Tower: N. pier, E. face, high up;
   S. pier, N.E. face;
   S. pier, S.E. face;
   Tower, S. arch.
l. Tower, N. pier, S. face.
m. Tower, S. pier, N. face.
n. Tower, S. pier, W. face.
o. Tower, S. pier, S.E. face.
p. Tower stair, S. side of doorway.
q. Tower stair, N. side of doorway.

Fig. 7. (5) Church of the Holy Trinity.
In porch, in arch in E. wall of tower.

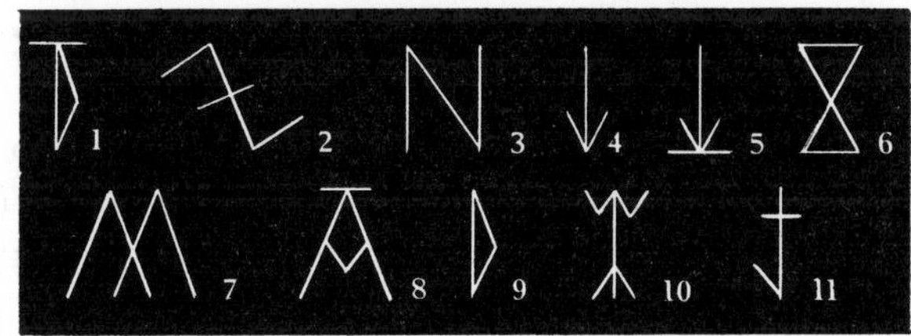

Fig. 8. (7) Church of St. Martin-cum-Gregory.

1. Chancel, N. arcade, W. arch, third stone on E. side.
   Chancel, S. arcade, W. respond pier.
   S. chancel aisle, S. wall, second window from E., W. jamb.
2. Chancel, N. arcade, central pier.
   N. chancel aisle, N. wall, W. window, W. jamb (twice).
   S. chancel aisle, S. wall, second window from E., W. jamb.
   S. chancel aisle, S. wall, third window from E., E. jamb.
   S. chancel aisle, S. wall, W. window, E. jamb, fifth stone up.
3. Chancel, S. arcade, central pier.
   Chancel, S. arcade, second stone on W. side.
   N. chancel aisle, N. wall, E. window, E. jamb.
   S. side, archway between chancel aisle and nave aisle, N. respond.
4. Chancel, S. arcade, central pier.
   N. chancel aisle, N. wall, W. window, W. jamb.
   S. side, archway between chancel aisle and nave aisle, N. respond.
5. Chancel, N. arcade, E. respond.
6. S. chancel aisle, S. wall, second window from E., E. jamb.
   S. side, archway between chancel aisle and nave aisle, N. respond.
7. Chancel, S. arcade, W. respond (twice).
   Chancel, S. arcade, W. arch, third stone on E. side.
8. Chancel, N. arcade, central pier.
   N. chancel aisle, N. wall, third window from E., W. jamb.
9. S. side, archway between chancel aisle and nave aisle, N. respond.
   (?) same archway, S. respond.
10. Tower arch, stone adjacent to, but not certainly part of, N. respond.
11. N. aisle, N. wall externally, on a stone next to the W. buttress of the N. chancel aisle.

distinctly Italian Baroque style, were once on top of an 18th-century reredos long since removed, and they and a pelican in piety fortunately survived. The St. Gregory is perhaps the best.

Inscriptions etc. Various signatures and scratchings appear on quarries in the windows of St. Martin-cum-Gregory.

There are some interesting masons' marks (Figs. 6–8), of various dates in the 14th and 15th centuries, in All Saints', North Street, and of the early 15th century in the chancel and chancel aisles of St. Martin. In general they are of simple forms and of little use for identification and dating, but one like a two-pronged fork, found on the fifth N. pier at All Saints', North Street, is found on the corresponding pier of the S. arcade, and both are associated with the tower. This latter point is significant, for a similar one is found on the S. pier of the tower of St. Mary's, Castlegate, and on the jamb of the arch to the Howme Chantry at Holy Trinity, Goodramgate (founded before 1396 and certainly built by 1433), and in St. Michael's, Spurriergate, on Perpendicular remodelling.

Two marks on the chancel of St. Martin-cum-Gregory are like ones on the S. pier of St. Mary's, Castlegate, and are also of the first half of the 15th century. St. Martin's, Coney Street, also has marks common to one or other of the above churches, and all can be assigned to the first half of the 15th century.

The only *Lectern* of note, of desk type, 15th-century, from St. John's (Plate 44) is now at Upper Poppleton.

Lord Mayors' Tables. The finest table of Lord Mayors is that from St. John's, Micklegate, now in York Minster (first entry 1708) (Plate 22); it is the only one to retain sword and mace rests. Tables also survive in All Saints', North Street (first entry 1723), St. Clement's (1704 and 1848) and Holy Trinity (1772); the two in St. Clement's are from St. Mary Bishophill Senior.

Monuments and Floor slabs. The only table tomb is to Sir Richard Yorke (1498) in St. John's, Micklegate. The mediaeval coffin lids provide a good series (*see* p. li). There are no effigies, mediaeval or later, and the wall monuments, although not of great quality, are nevertheless of interest. A list of signed monuments and tombstones follows this account.

The monuments designed as cartouches form a distinctive, and distinguished, group ranging in date from 1615 to 1766. The earliest to Ann Danby (1615) in Holy Trinity, Micklegate, of freestone with finely carved scrolls, is the only 17th-century one. The large cartouche of 1701 to Andrew Perrott in St. Martin-cum-Gregory has fine script and is of high quality; the small one below it to his widow, Martha (1721), is relatively crude in comparison. A cartouche to Thomas and Sarah Carter (1708) in the same church, and those to John Etty (1709) in All Saints', North Street, and Elias Pawson (1715) now in St. Clement's, are, like that to Perrott, excellently carved. Of these, the one to John Etty is particularly interesting not only in being the memorial to a very important local carpenter, to whom Grinling Gibbons may have been apprenticed, but because it could have been made by his son, William Etty, who was an equally important craftsman. The freestone cartouche to John Greene (1729) in Holy Trinity is severe compared with the previous ones, which are of white marble, and those to Anne Haynes (1747) and Elizabeth Potter (1766) in St. John's are small and show that the fashion is in decline. A lozenge in a framework (1760) in Holy Trinity is unusual.

Among other monuments noted, those of Henry Pawson (1730) and of Althea Fairfax (1744) from St. Mary Bishophill Senior, now in St. Clement's, have particularly good lettering. Two in Holy Trinity, of 1797 and 1807, are the only ones in the area embodying coloured marble; most are of black and white or veined marble. An extraordinary, and probably unique, design is that with a great charter and seal held up by two books and an urn against a Gothic frame to John and Mary Burton (1771) in Holy Trinity. The memorial to Thomas and Elizabeth Bennett (1825) in St. John's is exceptional in a different way: all the decoration, consisting of a weeping tree framing a sarcophagus fronted by a scroll, is in plaster.

Signed Monuments and Tombstones

Date	Name	Church	No.	Remarks
		Thomas Atkinson (1729–98)		
1789	Thomas Suttell	St. Mary Bishophill Senior	(7)	Destroyed
1793	Elizabeth Francis Dealtry	St. Edward's, Dringhouses	(2)	
1794	Joseph Harrison (1773) Helen Harrison (1794)	St. Mary Bishophill Senior		Destroyed
1797	Elizabeth Scarisbrick	Holy Trinity	(19)	(Plate 34)
		(Thomas) Bennett (*fl.* 1802–31)		
1825	Thomas Bennett (1773) Elizabeth Bennett (1825)	St. John's, Micklegate	(7)	(Plate 35)
1828	William Garforth	St. Martin-cum-Gregory	(11)	
		E.B.		
1798	Clarke Family (1788–1803)	St. Stephen's, Acomb		Tombstone. 'E.B. 1798'
		R. G. Davies (*fl.* 1820–57)		
1837	John Weatherill (1820) Jane Weatherill (1837)	St. Stephen's, Acomb	(9)	'Davies, Newcastle upon Tyne'
		Fisher Family		
1787	Thomas Rodwell	St. Clements, formerly in St. Mary Bishophill Senior	(14)	'Fisher' (John Fisher, 1736–1804
1808	John Atkinson	St. Martin-cum-Gregory	(13)	'Fishers York' (sons of John Fisher)
1813 (date of fitting)	Elizabeth Wood (1799) Catherine Dawson (1807) George Dawson (1812) Philadelphia Gore (1808)	St. Mary Bishophill Senior	(2)	
		Charles Fisher (1789–1861)		
1819	William Gage	St. Martin-cum-Gregory	(3)	
1821	Jeremiah Barstow Dorothy Barstow (1803)	St. Stephen's, Acomb	(7)	(Plate 33)
1830	Susanna Wray	St. Martin-cum-Gregory	(9)	
1830	Ellen Jones	St. Stephen's, Acomb	(12)	Tombstone
1832	Elizabeth Bridgewater	Cholera Burial Ground		
1838	Ann Collett (1829) Elizabeth Fletcher (1833) Sarah Collett (1838) Humphrey Fletcher (1838)	St. Martin-cum-Gregory	(14)	
1844	Rev. John Graham	St. Clements, formerly in St. Mary Bishophill Senior	(16)	
		Samuel Appleyard Lucas (*fl.* 1848–55)		
1848	Hannah, Mary and Henry Lucas	St. Mary Bishophill Senior	(18)	Tombstone, now destroyed
		William Palmer (1673–1739)		
1730	Henry Pawson	St. Clements, formerly in St. Mary Bishophill Senior	(17)	
		William Abbey Plows (1798–1865)		
1828	Ann Tomlinson (1825) Jonathan Tomlinson (1828)	St. Mary Bishophill Senior	(22)	Tombstone, destroyed
1836	James Bromley (1827) Henry Bromley (1828) Martha Bromley (1836)	St. Mary Bishophill Senior	(23)	Tombstone, destroyed

		William Abbey Plows (continued)		
1846	Mary Peckitt (1820) Mary Rowntree (1846)	St. Martin-cum-Gregory	(10)	(Plate 36)
1855	Frances Etridge (1825) Thomas Etridge (1855)	St. Stephen's, Acomb	(1)	
		Shaftoe		
1839	Thomas Hurworth (1830) Elizabeth Hurworth (1839)	St. Martin-cum-Gregory	(19)	
		Matthew Skelton (*fl.* 1825–*c.* 1877)		
1827	Henry Prest	St. Mary Bishophill Senior	(5)	Destroyed
1835	Christopher Brearey (1826) Jane Brearey (1835)	St. Mary Bishophill Senior	(10)	Destroyed
1835	Francis Smales (1834) Ann Smales (1835)	St. Mary Bishophill Junior	(6)	
1837	Mary Burgess (1829) John Burgess (1837)	St. Mary Bishophill Junior	(4)	
1839	William Duffin	Holy Trinity	(10)	
1840	Jane Yorke	Holy Trinity	(7)	
1843	Frances Beckwith (1818) Stephen Beckwith (1843)	St. Mary Bishophill Senior	(6)	Destroyed
1847	Margaret Anne Yorke	Holy Trinity	(9)	
1847	William Crummack	Holy Trinity	(13)	
1848	Joshua Ingram (1836) Elizabeth Ingram (1848)	Holy Trinity	(15)	
1849	Maria Dorothy Smales	St. Mary Bishophill Junior	(5)	
1855	Elizabeth Steward (1847) John Steward (1855)	Holy Trinity	(12)	
		William Stead senior (*c.* 1752–1834) William Stead junior (1781–1823)		
1780	Robert Stockdale	St. Mary Bishophill Junior	(1)	
1793	Henry Jubb (1792) Elizabeth Jubb (1793)	Holy Trinity	(11)	
1800	Samuel Francis Barlow	St. Edward's, Dringhouses	(3)	
1801	Margaret Benson (1795) Christopher Benson (1796) Christopher Benson (1801)	St. John's, Micklegate	(4)	
1822	Rev. Robert Benson	St. Martin-cum-Gregory	(2)	
1830	Lucinda Benson	St. Martin-cum-Gregory	(17)	
		Michael Taylor (1760–1846)		
1798	William Kay	St. Stephen's, Acomb	(6)	
1819	Jane Boulby (1803) Adam Boulby (1819)	St. Martin-cum-Gregory	(1)	
1821	Mary Strickland (1744) Jarrard Strickland (1791) Cecilia Strickland (1821)	St. Martin-cum-Gregory	(6)	
1827	William Philips	St. Stephen's, Acomb		
1832	Anna Maria Crompton (1819) Joshua Crompton (1832)	Holy Trinity	(16)	
1810–45	Smith Family	St. Stephen's, Acomb	(3)	
		(John) Waudby of Hull and York (*fl.* 1828–46)		
1828	John Stevenson	St. Mary Bishophill Senior	(21)	'Waudby Coney Street'. Tombstone, destroyed
		W. Williams of Huddersfield (*fl.* 1824–39)		
1834	Joseph Pickford (1804) Mary Pickford (1834)	St. Stephen's, Acomb	(2)	

Floor slabs. Numerous floor slabs have been noted, including many since destroyed at St. John's and St. Mary Bishophill Senior. All Saints', North Street, contains mediaeval ones to John Rothum (1390), John Wardalle (1395), John Bawtrie (1411) and Thomas de Kyllyngwyke (15th century). This last has an intricate cross and a good black-letter inscription. The slab in St. Martin-cum-Gregory to Henry Cattall (1460) is interesting because it has a black-letter marginal inscription within which are later inscriptions added to the Peckitt family (famous for their painted glass) and the Rowntree family who married the Peckitt heiress (Plate 36). A slab of 1599 to Joan Stoddart in All Saints', North Street, has rarity value as the only 16th-century one.

Armorial slabs of note (Plate 36) include those to Susanna Beilby (1664) in St. Martin's, Joshua Witton (1674) in All Saints', North Street, Andrew Perrot (1701) in St. Martin's and Frances Bathurst (1724) in the same church. In this context the tombstones with coats of arms of Richard Bealby (1805) and Robert Driffield (1816) at Acomb may be mentioned.

Panels. At the Bar Convent are two sets of 16th-century panels, each consisting of four panels, carved with Biblical subjects and allegorical themes respectively. The first is well carved, the second less competently done. They are both Flemish and in general consist of figures set within classical, round-headed archways. Two early 17th-century panels, dissimilar, are set in the restored, or even made-up, chairs at St. Stephen's church, Acomb; one is carved with Adam and Eve, the other with Hope with dove and anchor (Plate 42).

Plate. (Ref. T. M. Fallow and R. C. Hope, *YAJ*, VIII (1884), 300 *et seq.*) Nothing of mediaeval date remains, and only two or three pieces dating from before the Civil War survive. *Cups.* Perhaps the earliest cup (with cover) is one at St. Mary Bishophill Junior of 1570/1 by Robert Beckwith. Another at Acomb is said to be of *c.* 1570; the shape and decoration are similar to those of the foregoing. A cup at Holy Trinity of 1611/12 (Plate 37), made in London, has a simple bowl and attractive stem with pear-shaped knop. The cup (with cover) at All Saints', North Street, given by Archbishop Samuel Harsnett in 1630, is also like the St. Mary Bishophill Junior one, with a similar deep straight-sided bowl with angular base. Two cups from St. Martin-cum-Gregory (now at Holy Trinity), one of 1636 and the other a copy of 1818 again have bowls like that of the early St. Mary Bishophill Junior one and stems with thin round knops. There are some good *patens*, other than the covers specifically associated with cups. The best is the one from St. John's (1697) (now at Holy Trinity) which has a round plate with gadrooned lip and a good stem beneath. One from St. Martin-cum-Gregory of 1737, exceptionally elegant with an elaborate deckle-edged rim, and one of 1843 at Holy Trinity are domestic salvers turned to ecclesiastical use. The *flagons* present an interesting series. Two at St. Martin-cum-Gregory date from 1720–9 and 1740 respectively; each has a rounded body, with curved tapering neck and lid but no thumb-piece. The one of 1739–40 at Holy Trinity is much simpler, with plain straight sides tapering upwards, a shaped lid with thumb-piece and a graceful handle. Most of these were London made. The most impressive one, however, from St. John's and now at Holy Trinity, was made at York in 1790/1; it is very simple and heavy and has a plain straight-sided tapering body, flat lid and distinctive grid-iron thumb-piece (Plate 37). Two flagons of 1781/2 at All Saints', North Street, are elegant, with urn-shaped bodies and well executed inscriptions (Plate 37).

Base Metal. Some of the pewter is noteworthy (Plate 37). Perhaps most remarkable is the flagon from St. John's at Holy Trinity which has incised figures and decoration of *c.* 1620. Another flagon from St. John's of *c.* 1725 is of a relatively rare type with bulbous body, tapering neck, lid with acorn knob and thumb-piece.

Pulpits. A pulpit at St. Martin-cum-Gregory, of Jacobean type, was made in 1636 by John Harland; it is like one in All Saints', Pavement. One of 1675 in All Saints', North Street, is enriched but in more orthodox classical style than the foregoing and had figures painted in the panels. It may have been made by John Etty; his friend Henry Gyles, the glass painter, used similar emblematic figures (Plate 38).

Reredos. The only reredos noted is in St. Martin-cum-Gregory and is of 1749–51; the churchwardens were permitted to negotiate for a reredos in 1749 and sold the old one for £2 2*s.* in 1751. The new one (Plate 127), made by the local joiner Bernard Dickinson, is of oak, of a robust classical design, and still retains the Commandments, Paternoster and Creed; it forms an effective composition with the communion rails (made by Matthew Butler after Dickinson's death) and the steps on which they are set. One in All Saints', North Street, made by John Etty in 1710, remained until 1807.

Royal Arms. The arms of William and Mary are in St. Martin-cum-Gregory; some mid 18th-century ones are in Holy Trinity; those in St. Mary Bishophill Junior are dated 1793 (Plate 41), and the arms of William IV (Plate 41) remain at Acomb; those dated 1764 formerly in All Saints', North Street, have vanished in relatively modern times. In general the arms are painted on boards framed within a moulding rounded at the top. Both those in St. Mary Bishophill Junior and St. Stephen's, Acomb, have been cleaned and are in good order, and the latter is vigorously drawn, but the ones at St. Martin-cum-Gregory and Holy Trinity are very faded.

Seating. Only one mediaeval stall remains, in All Saints', North Street (Plate 19); it is closely dated to 1467–72 by the carvings on the misericorde, which would appear to represent the arms of the Rev. John Gillyot, then Rector, Master of the Corpus Christi Guild in 1472. It resembles one which survived the fire at York Minster and two from St. Saviour's.

Wands. Wands of office still remain at All Saints', North Street, and St. Martin-cum-Gregory (1678).

Weathervane. There is a good brass weathervane of 1759 on the spire at All Saints', North Street (Plate 92).

Miscellanea. Four leather fire buckets of 1794 at St. Martin-cum-Gregory are somewhat unusual survivals (Plate 24).

(E.A.G.)

SECULAR

Houses to 1650

The earliest domestic buildings surviving on the south-west side of the river date from the second half of the 14th century, and evidence for earlier occupation must be sought from archaeological, documentary and historical sources. Apart from some fragments of stone walling at the back of the Queen's Hotel, Micklegate (55), tentatively assigned to the 12th century but making no coherent building pattern, nothing survives from before *c.* 1350. Documentary evidence, however, shows that the district was growing in popularity as leading citizens moved in and better houses were built. In the second half of the 13th century Robert de Clairevaux settled in Micklegate and his son Sir Thomas acquired from Nicholas de Selby a house (the site of Nos. 35 and 37) near St. Martin's church (*YASRS*, LXXXIII, 182); this was called a 'stone house' in 1281 and a 'stone hall and cellars' in 1290. Hugh de Selby, Nicholas's grandfather, who had been Mayor in 1230, was living *c.* 1240 in the house called Mountsorrell on the corner of St. Martin's Lane (VCH, *York*, 45). Mountsorrell also is described as a stone house, and survived wholly or in part until 1866.[1] Old drawings show it as a stuccoed building with a jettied, and therefore timber-framed, upper storey. It is not clear in the references to stone houses whether they were fully stone-built (as the surviving ruined 12th-century house in Stonegate was) or had timber-framed upper storeys (as Mountsorrell appears to have had). This latter type was certainly still being built in York in the 14th century. The number of documentary references to stone houses, stone chambers and stone halls suggests that they were normal amongst the more wealthy merchants of the 12th and 13th centuries. Benedictus and Moseus the Jews were living in Micklegate in 1290, and probably occupied this same type of building, as Moseus's house had an annual value of 40*s.*, and that of Benedictus, owned by his father Bonamicus, 33*s.* 4*d.* (*YAJ*, III (1875), 193), the same value as the Clairevaux stone house.

The surviving mediaeval houses are not numerous, and are mostly so fragmentary as to provide no evidence except that some sort of mediaeval structure existed on the site. Those relatively intact are too few to provide a chronological sequence of timber-framed development, especially as none of them is precisely dated. Their dating in this Inventory is therefore based on the much greater number of houses across the river, and may, in the light of further research, need to be modified. To the 14th century can be ascribed the row of seven houses, Nos. 99–111 Micklegate (87) (Plates 49, 50 and Fig. 60), of which only the four numbered 99–103 now survive, the other three, heightened and refronted in brick in 1774, having been demolished in 1961. Like all the surviving mediaeval houses they are timber-framed, and exhibit features suggesting an early date. These include the isolated curved struts supporting the truss-rafters (Fig. 61), the tenoning of main struts into the stem of the crown-post below its enlarged head, and possibly the absence of any other strutted framing,[2] though the application of this last criterion should perhaps be regarded with caution in regard to town houses. The prodigal use of a long street frontage for a row of small tenements also hints at an early date, whilst their position contiguous with the gateway to Holy Trinity Priory (Cave, pl. XVII), and standing on the edge of the Priory precinct, leaves little doubt but that they were built to provide rental income for the Priory. In their disposition of seven tenements with a total length of 100 ft. they may well be a copy of the row in St. Martin's Lane, Coney Street, built in 1335 (Salzman, 430–2). Furthermore if they were Priory property, they probably date from before 1369 when the Priory was taken into the king's hands (*CFR 1369–77*, 14). The evidence suggests that these

---

[1] A statement that other stone houses stood near the church of St. Gregory in 1230 (G. Benson, *Notes on the Church & Parish of St. Martin-cum-Gregory*, pt. I (1901), 17) rests on a misreading by Dodsworth of an entry in the Healaugh Park Chartulary (*YASRS*, XCII, 155–6).

[2] Discussed as 'tension-braced' framing by Wood, 222–4.

rows of small '*domus rentales*' were a comparatively common feature of the street scene in the early 14th century. Possibly to the 14th century should also be ascribed the single-storeyed building of at least two bays, forming the back part of the Nag's Head, No. 100 Micklegate (88). Its framing has only in part survived and is possibly of more than one period. However, the present roof-truss at the south end has a steeply-cambered tie-beam carrying a crown-post and raking struts, all suggestive of an early date. If indeed 14th-century, it is the earliest single-storeyed building, whether a ground-floor domestic hall or an outbuilding for mercantile use.

No buildings dating from the first half of the 15th century have been found, and their absence seems to be echoed across the river, probably indicating slack building activity during that period of decline. Several houses, however, survive from the late 15th century, no less than three of them containing a ground-floor hall open to the roof. Possibly other ground-floor halls, set well back from the street, have been lost in alterations and redevelopment.

The most important of these late 15th-century houses is Jacob's Well (125) in Trinity Lane (Plate 191). This contains an open hall sited a few feet from the choir of Holy Trinity Priory church. At right-angles to the hall and lying along the street is a two-storeyed range of stone and timber framing jettied on two adjacent sides. Its importance lies in the fact that its history can be traced back in detail to 1549 when it was described as 'lately belonging and appertaining to the recently dissolved Priory of Holy Trinity, York' (*AASRP* (1902), 531). Its situation in or by the Priory precinct and its proximity to the Priory choir strongly suggest that it was the house of a chantry priest or of a person of similar standing.[1] As such its plan of an open hall and two-storeyed range is of considerable interest as showing the type of accommodation provided in York for parish or chantry priests. The original way in to the upper floor from the hall has disappeared entirely. There is a modern doorway in the upper partition wall against the hall, which must reproduce an opening of at latest 17th-century date, and may represent the original means of access approached by a stair against the end wall of the Hall. The insistence, in the early deeds, that the structure formed 'two tenements' probably indicates that the south half of the present long two-storeyed block, which had an internal staircase at the south end by the side of the stone chimney, was originally a separate dwelling. The general plan suggests a similar function for two other buildings of the same unusual type—No. 31 North Street (104) (*see* below) and, across the river, Bowes Morrell House, No. 111 Walmgate, perhaps a chantry priest's house associated with one of the nearby churches, St. Peter the Willows or St. Margaret.

The other two open halls are in the complex of buildings at the end of North Street near All Saints' church. The range nearest the church consists of three tenements, No. 31 North Street and Nos. 1 and 2 All Saints' Lane (104) (Plate 185 and plan Fig. 66, p. 98). No. 31 North Street consists of a single-bayed open hall and a two-storeyed corner wing jettied on adjacent sides. The other two tenements, Nos. 1 and 2 All Saints' Lane, form a continuation of the corner wing and are integrally built with it. Each of them has one room down and one room up, with access to the upper room by an internal staircase. Neither staircase is original, but the disposition of the ceiling joists in No. 1 shows that the present staircase replaces an original one, and there was no doubt an identical layout in No. 2. The simple framing to the upper storey facing the lane survives nearly intact and indicates that each room, of two small bays, was lit by an oriel window awkwardly cut in half by a vertical stud in the wall-framing. This, however, may indicate that the room could be divided into two by a hanging, with each half lit by half the oriel. The layout of the whole range is thus the same as at Jacob's Well, with the addition of an extra tenement. Here again proxi-

---

[1] In the above-mentioned article, Solloway identified the house as belonging to the Nelson Chantry in the Priory Church. On this basis he dated the building to 1474–80.

mity to the church suggests ecclesiastical occupation, and the number of wealthy chantries and benefactors in the first half of the 15th century in this church makes its function as a chantry priest's house very probable, especially as the old parsonage house, drawn by Cave (pl. VI) was elsewhere.

A few yards along North Street, at its junction with Tanner Row, stands another house of the same period also containing an open ground-floor hall. This house, No. 1 Tanner Row (120) (Plate 48), is in fact the survivor of at least a pair of Wealden houses (Wood, 218). Another example of this type can be seen across the river at No. 51 Goodramgate, and at least one other local example is known from old photographs. These appear to be the most northerly recorded occurrences of this type of house, common in Kent and Sussex. The design was probably brought here by a southern carpenter employed by one of the York merchants, whose connections with the Staple of Calais imply contacts with the south-eastern counties. Advances in standards of comfort, however, led to the hall of No. 1 Tanner Row being divided into two storeys in the early 17th century, and the upper storey was given a jetty to match the lateral blocks, thus disguising the original form, which only came to light recently when the framework was exposed after years of neglect. The simplicity of conversion into two storeys makes it possible that other houses were similarly treated. The building has now been thoroughly renovated and the disguise is effective once more.

The influence of Holy Trinity Priory can again be seen in the range of three houses Nos. 85–89 Micklegate (79) (Plate 174), the date of which has been disputed. Stylistically it belongs to *c.* 1500, although it is not shown on Speed's map of York dated 1610. Like Nos. 99–109 Micklegate it was built along the edge of the precinct where building in depth was not possible, the only rearward projections having been small staircase annexes, and was probably designed to provide rental income for the Priory after the poverty-stricken years of the previous century (*CPR 1446–52*, 69, and *YCR*, I, 26–7). In this connection it is significant that there were no windows in the rear wall where they would have overlooked the Priory grounds. Furthermore in the early 17th century, by which time the Priory had fallen into secular hands, a timber-framed wing was built at the back of No. 89, projecting well into the former Priory precinct.[1] It is interesting to note that the top-floor rooms of the main range were still open to the roof. Owing to the paucity of houses dating from the first half of the 16th century, it is not possible to say when this 'open-roof' feature went out of fashion. The disappearance, as here, of the central purlin (Fig. 13e), followed elsewhere by the crown-post, probably marks the beginning of this development, and the next century may have provided ceiled roofs of a simpler construction, reflecting the demand for economy in materials, and less draughty top-floor accommodation. Probably Nos. 85–89 Micklegate is one of the last examples of this mediaeval type.

The considerable wealth still available in the city in the late 15th and early 16th centuries is indicated by large houses such as Nos. 17–21 Micklegate (59), of the late 15th century. This is of three storeys, originally jettied to the street, and with an L-shaped extension at the back probably marking the site of a staircase annexe. The plan has been much obscured by later alterations, but there is no evidence that the house comprised more than one tenement, whilst the original roof-structure, although mutilated, shows that the top-floor rooms were all open to the roof. The provision of a staircase situated in an annexe at the back of the house seems to be a new feature. It marked an improvement on the inconvenient internal staircase which took up living space, and can be seen as part of the trend towards greater comfort, though in this case it complicated the constructional problems and was ultimately replaced by the internal stairwell.

That the more normal two or three-storeyed house on a deep and narrow messuage continued to be built is clear from the numerous fragmentary examples, No. 112 Micklegate (92) being one of the best

---

[1] The same pattern of development can be seen in No. 111(C) Micklegate (87).

preserved. This, on a site which belonged to the Vicars Choral from the 13th century, shows signs of influence from an exotic timber-framed tradition in its use of principal rafters. This feature is exceedingly rare in York, though other examples can be seen in No. 32 Micklegate (61) of the mid 16th century and in the Manor Farm at Acomb (137), a late 15th or early 16th-century building, now greatly altered, which belonged to the Treasurers of York Minster. Lying on the western periphery of York it might have been open to the infiltration of a more westerly carpentry tradition. No. 112 Micklegate had a markedly cambered tie-beam in its central truss indicating that the top floor was open to the roof. Were it not for the incorporated alien elements, it might be regarded as one of the transitional types prevalent in the early Tudor period before the demand for attics spurred the carpenters on to further constructional experiments.

Another large house dating from the second half of the 16th century is the recently demolished block, Nos. 2, 4 and 6 Micklegate (54), of three storeys jettied along the street front. This property appears from grants of the late 12th and early 13th centuries to have belonged to St. Peter's Hospital (*EYC*, I, 176–8), and therefore presumably came on to the market in the 1540s, after which it was rebuilt by one of its new owners. Here again details of the layout have been lost in alterations, but the height of the top-floor rooms suggests that they were ceiled off from the roof. The wing at the back by St. John's church is an addition or replacement of the early 17th century. These large blocks of houses, clearly combining several smaller messuages, possibly reflected and accentuated the growing disparity between the few very wealthy merchants and the increasing number of impoverished traders (VCH, *York*, 122 *et seq.*).

The late 16th century saw the introduction of attics as usable rooms, partly for storage but also for habitation, since they were provided with both headroom and fenestration. The earliest examples of this development made use of the sole-piece (*see* under *Roof Structure*), the side-walls being carried up some 3 to 4 ft. above the level of the top floor. This construction allowed adequate headroom and light from the gable-ends, and the building required only a small increase in height. Of the four recorded examples in this area, three have recently been demolished: Nos. 16–18 (58) (Plate 52) and 111 Micklegate, House (C) (87) and a warehouse (?) behind No. 10 North Street (101). Evidence for gable windows survived in the two Micklegate examples. The one surviving example is a timber-framed wing added at the back of Nos. 17–21 Micklegate, of two unjettied storeys and attic. Originally at least five bays long, the attic must have been a gloomy space as, with its north end abutting the earlier house, it was lit only by a window in the southern gable. There were no structural partitions, and it may have been used mainly for storage, though, as the well-lit floor below was also unpartitioned, it may possibly have been a hostelry with extensive sleeping accommodation (*cf.* the illustration of the sleeping quarters in a French inn of *c.* 1462 in Glasgow University Library MS. Hunter 252[1]). Nos. 16–18 Micklegate (Plate 52) was remarkable not only for the attic construction but also as a large building probably designed from the beginning as an inn. Double-gabled and jettied both at the front and the back, it provided three full storeys as well as attics. It has further importance in that there is evidence for each half of the building having had an original chimney between the front and back rooms. By leaving a space between the front and back rooms, the builder made the construction of the central chimneys very easy and economical, and it seems likely that the staircases also were incorporated into this central area. It is surprising that this type of plan, so well-suited to the long narrow messuages, was not adopted more readily. It is possible that the erection of such large houses as Nos. 2–6 and 16–18 Micklegate, taking in several mediaeval messuages, was facilitated by the devastating plagues of 1550 and 1551, which decimated the lower Micklegate area (VCH, *York*, 120), and must have pushed a lot of tenements onto the market all at the same time, with a depressing effect on prices. These had already been artificially depressed, first by the statute of Henry VIII in 1540 ordering

---

[1] Reproduced in CIBA Review 62 (1947), 2288.

(9) CHURCH OF ST. MARY BISHOPHILL SENIOR. 1770.

(7) CHURCH OF ST. MARTIN-CUM-GREGORY. Tenor, 1579.

(7) CHURCH OF ST. MARTIN-CUM-GREGORY. 1681.

(6) CHURCH OF ST. JOHN. 18th century.

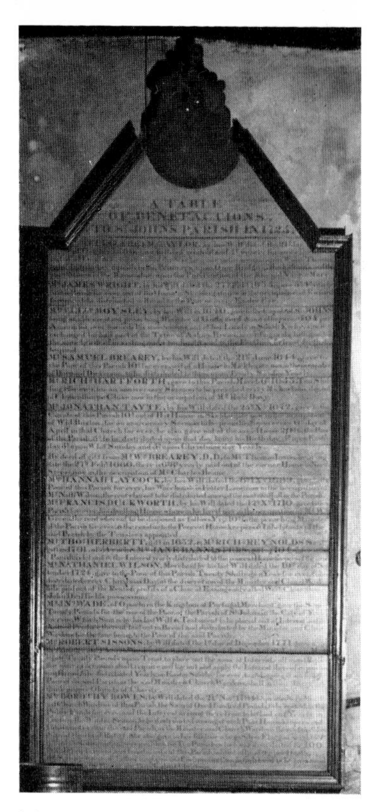

(6) CHURCH OF ST. JOHN. 1725 and later.

(9) ST. MARY BISHOPHILL SENIOR. 18th century.

(9) ST. MARY BISHOPHILL SENIOR. 18th century.

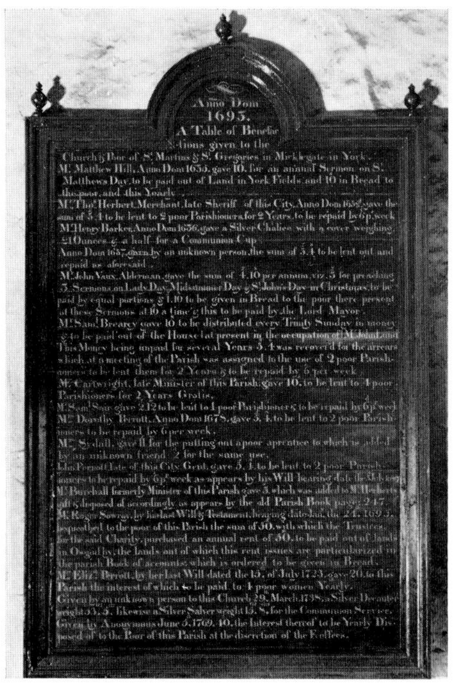

(7) ST. MARTIN-CUM-GREGORY. Late 18th century.

(9) CHURCH OF ST. MARY BISHOPHILL SENIOR. Collecting shovels, 18th century.

(7) ST. MARTIN-CUM-GREGORY. 18th century.

(9) CHURCH OF ST. MARY BISHOPHILL SENIOR. Mid 18th century.

(7) ST. MARTIN-CUM-GREGORY. Poor box, 18th century and later.

CANDLESTICKS AND FIRE BUCKETS

(9) ST. MARY BISHOPHILL SENIOR. 14th century.

(7) CHURCH OF ST. MARTIN-CUM-GREGORY. *c.* 1715.

(9) ST. MARY BISHOPHILL SENIOR. 14th century.

(7) CHURCH OF ST. MARTIN-CUM-GREGORY. 1794.

Cross-shaft (1).

Cross-shaft (3).

Tomb slab (10).

Tomb slab (20).

(9) CHURCH OF ST. MARY BISHOPHILL SENIOR. 10th–11th century.

PRE-CONQUEST AND LATER CARVED STONES

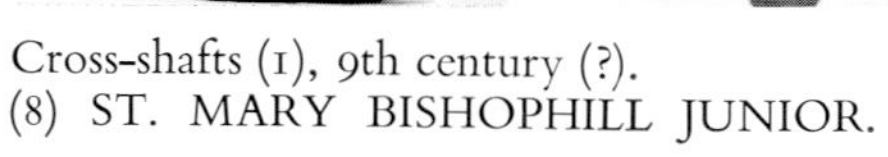

Cross-shafts (1), 9th century (?).
(8) ST. MARY BISHOPHILL JUNIOR.

(2), 10th century (?).

(7) ST. MARTIN-CUM-GREGORY. Cross-shaft (3), 11th century.

(5) HOLY TRINITY, St. Nicholas Chapel. 11th century.

(9) ST. MARY BISHOPHILL SENIOR. Hogback (17), 10th–11th century.

(8) ST. MARY BISHOPHILL JUNIOR. Hogback (3), 10th century.

(28) OLD RAILWAY STATION. Cross-head, 8th century.

COFFIN LIDS AND FLOOR SLABS

(4) ALL SAINTS, NORTH STREET. Coffin lid (4), 13th century

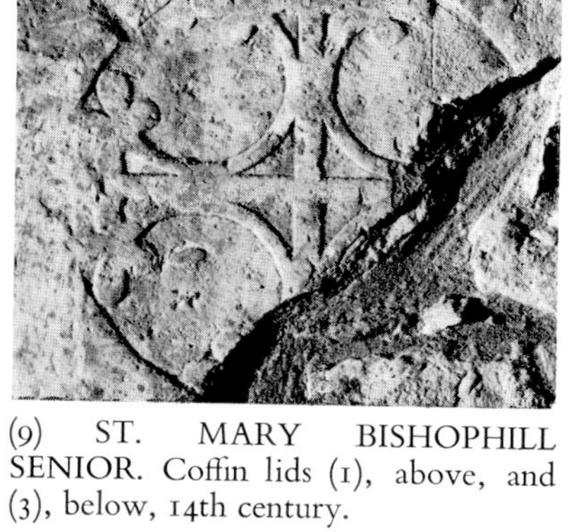

(9) ST. MARY BISHOPHILL SENIOR. Coffin lids (1), above, and (3), below, 14th century.

(4) ALL SAINTS, NORTH STREET. Brass (2) of Thomas Clerk, 1482.

(8) CHURCH OF ST. MARY BISHOPHILL JUNIOR. *c.* 1700.

(5) CHURCH OF THE HOLY TRINITY. 1717 and 1794.

(7) CHURCH OF ST. MARTIN-CUM-GREGORY. *c.* 1700.

(6) CHURCH OF ST. JOHN THE EVANGELIST (now at St. Hilda's). *c.* 1638.

(6) CHURCH OF ST. JOHN THE EVANGELIST. Window II. 14th century. © *Dean and Chapter*.

Window VIII.

Window VI.

(7) CHURCH OF ST. MARTIN-CUM-GREGORY. 14th century.

STAINED GLASS DETAILS

The Betrayal and David and Goliath.
(7) CHURCH OF ST. MARTIN-CUM-GREGORY. Window in S. aisle, late 15th century.

(4) CHURCH OF ALL SAINTS, NORTH STREET. Window XI, before restoration, 15th century.

St. John the Baptist.

St. Catherine.

(7) CHURCH OF ST. MARTIN-CUM-GREGORY. Window VI, *c.* 1335.

Beheading of St. John the Baptist.

(6) CHURCH OF ST. JOHN THE EVANGELIST. Window II, 14th century. © *Dean and Chapter.*

(4) ALL SAINTS, NORTH STREET. Monument (1) of John Etty, 1708.

(7) ST. MARTIN-CUM-GREGORY Monument (5) of Thomas Carter, 1708.

(7) ST. MARTIN-CUM-GREGORY. Monuments (15, 16) of Andrew Perrott, 1701, and Dame Martha Perrott, 1721.

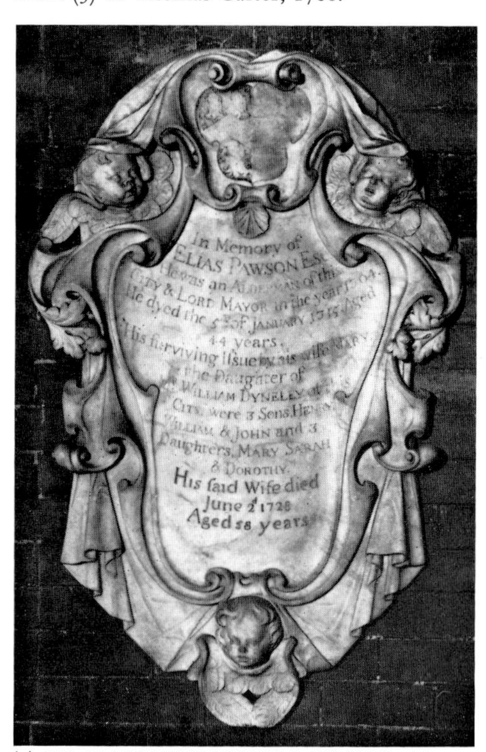

(9) ST. MARY BISHOPHILL SENIOR. Monument (15) of Elias Pawson, 1715.

(11) ST. STEPHEN, ACOMB. Monument (7) of Jeremiah Barstow, 1803.

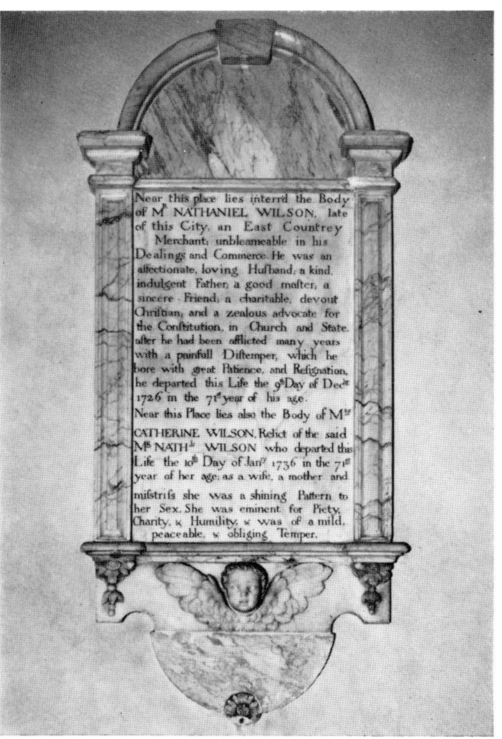

(6) ST. JOHN THE EVANGELIST. Monument (2) of Nathaniel Wilson, 1726.

(7) ST. MARTIN-CUM-GREGORY. Monument (18) of Samuel Dawson, 1731.

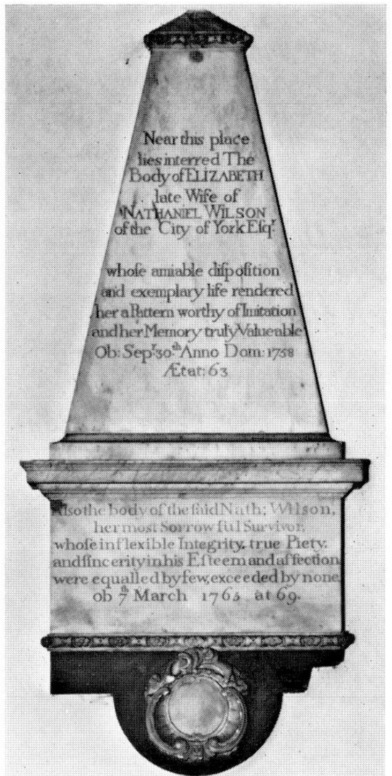

(11) ST. STEPHEN, ACOMB. Monument (8) of Elizabeth Wilson, 1758.

Monument (17) of Elizabeth Ann, 1760.

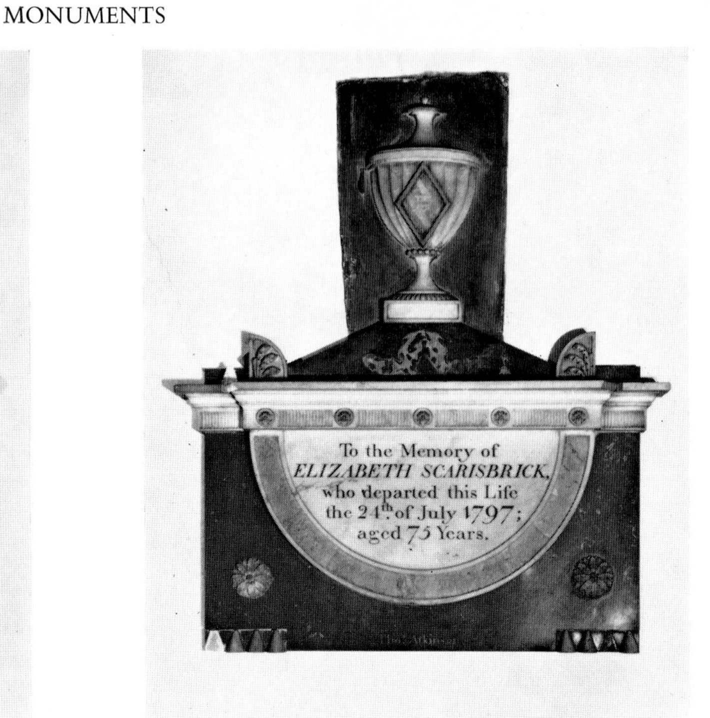

Monument (19) of Elizabeth Scarisbrick, 1797.

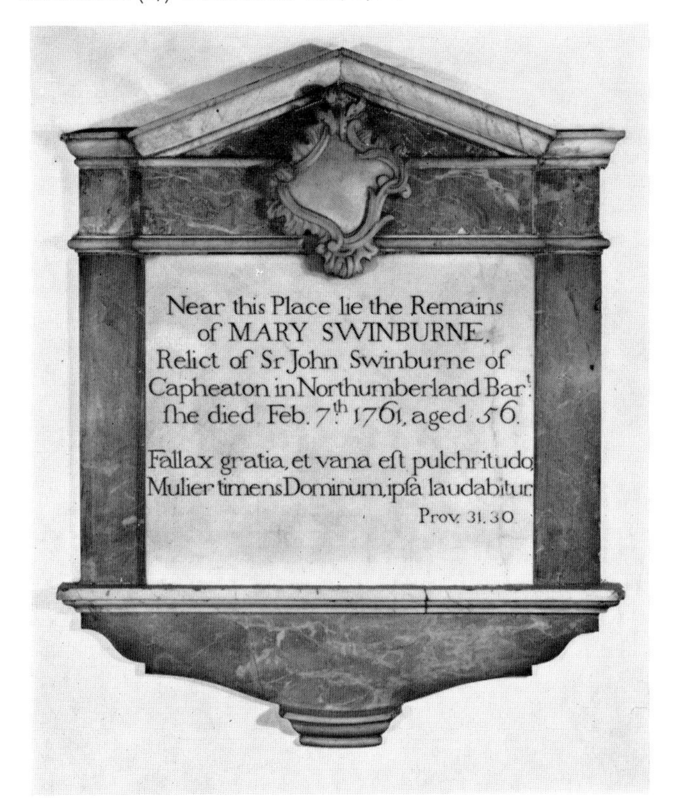

Monument (14) of Lady Swinburne, 1761.

Monument (2) of Dr. John Burton, 1771.

(5) CHURCH OF THE HOLY TRINITY.

(4) ALL SAINTS, NORTH STREET. Brass (4) of Thomas Askwith, 1609.

(4) ALL SAINTS, NORTH STREET. Brass (3) of Thomas Atkinson, 1642.

(6) ST. JOHN THE EVANGELIST. Monument (7), 1825.

(11) ST. STEPHEN, ACOMB. Monument (3) of Smith family, 1805.

Floor slabs (28) of Mrs. Frances Bathurst, 1724, (23) of Robert Benson, 1765.
(7) CHURCH OF ST. MARTIN-CUM-GREGORY.

(4) ALL SAINTS, NORTH STREET. Floor slab (2) of Joshua Witton, 1674.

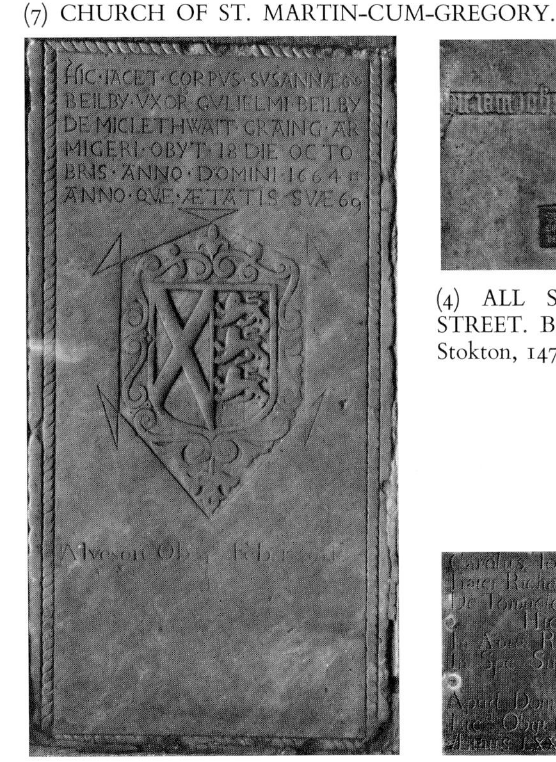

(7) ST. MARTIN-CUM-GREGORY. Floor slab (10) of Susanna Beilby, 1664.

(4) ALL SAINTS, NORTH STREET. Brass (1) of William Stokton, 1471.

(4) ALL SAINTS, NORTH STREET. Brass (5) of Charles Towneley, 1712.

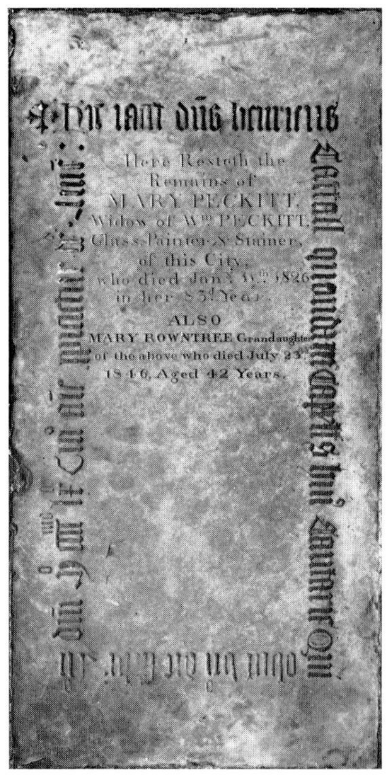

(7) ST. MARTIN-CUM-GREGORY. Floor slab (20) of Henry Cattall, 1460 (reused 1826).

(5) THE HOLY TRINITY. Cup, 1611/2.

(4) ALL SAINTS, NORTH STREET. Paten, 1630.

(6) ST. JOHN THE EVANGELIST. Flagon, 1791.

(4) ALL SAINTS, NORTH STREET. Cup, 1630.

(4) ALL SAINTS, NORTH STREET. Flagon, 1781/2.

(6) ST. JOHN THE EVANGELIST. Pewter flagon, *c.*1725.

(6) ST. JOHN THE EVANGELIST. Pewter flagon, *c.* 1620.

PULPITS

(4) CHURCH OF ALL SAINTS, NORTH STREET. 1675.

(7) CHURCH OF ST. MARTIN-CUM-GREGORY. 1636.

Female figure, 15th century (?).
(4) CHURCH OF ALL SAINTS, NORTH STREET.

The Resurrection; alabaster, 15th century.

St. Jerome.

St. Ambrose.

St. Augustine.

St. Gregory.

(13) THE BAR CONVENT. Reredos. The four Latin Doctors, 18th century.

(8) ST. MARY BISHOPHILL JUNIOR. 1793.

(11) CHURCH OF ST. STEPHEN, ACOMB. *c.* 1832.

(11) CHURCH OF ST. STEPHEN, ACOMB. Glass in N. transept, by Edmund Gyles (?), 1663.

WOODWORK. CARVINGS

(4) ALL SAINTS, NORTH STREET. King David, 18th century.

(11) CHURCH OF ST. STEPHEN, ACOMB. Panels on chairs. Adam and Eve, and Hope, early 17th century.

(13) THE BAR CONVENT. Charity, 16th century.

Annunciation.

Adoration of Shepherds.

Crucifixion.

Resurrection.

(13) THE BAR CONVENT. Staircase panelling, 16th century.

N. side (1).

S. side (6).

S. side (2).

S. side (5).

N. side (6).

N. side (5).

N. side (3).

N. side (4).

S. side (3).

(4) CHURCH OF ALL SAINTS, NORTH STREET. Angels on Chancel roof, *c.* 1470.

WOODWORK. FURNITURE

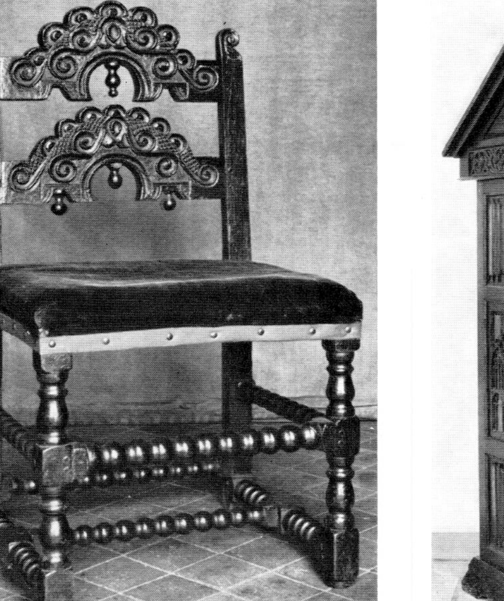

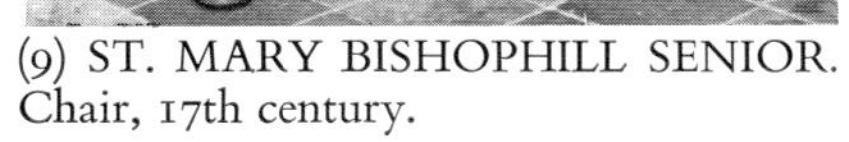
(9) ST. MARY BISHOPHILL SENIOR. Chair, 17th century.

(6) ST. JOHN. Lectern, late 15th century.

(4) ALL SAINTS. NORTH STREET. Chair, early 18th century.

(8) ST. MARY BISHOPHILL JUNIOR. Chair, 17th century (?).

(8) ST. MARY BISHOPHILL JUNIOR. Communion table, 17th century; top modern.

(9) ST. MARY BISHOPHILL SENIOR. Communion table, 17th century.

the rebuilding of decayed houses and secondly by the enormous grants of properties, previously owned by the various religious houses, to Leonard Beckwith in 1543, most of which he seems to have sold very quickly.

Another important building was the Plumbers' Arms (118) in Skeldergate of *c.* 1575 (Plate 46), demolished in 1964. This made a faint concession to the Renaissance by sporting a moulded and carved bressumer to the jetty with egg-and-dart motifs. It also had two large original chimneys with weathered offsets, each containing two flues. These were engulfed in modern buildings, but can be seen in an old photograph of *c.* 1855 (NMR, BB/57/1517). Although on the ground floor each fireplace served a separate room, on the first floor there was no structural partition, and the significance of the two fireplaces is not clear. Possibly the importance of this large first-floor area as a living or reception-room was enhanced by the dual heating arrangements. The original fenestration had all been replaced, but a small annexe, added on the north-west side some thirty years after the building of the main house, retained, blocked up, all its old windows on first and second floors. These window frames were of timber, constructed for glazing, and were also interesting as shop-made units ready to be jammed into position between the pegged framing. As they were moveable they could legally be treated as tenant's fixtures (Salzman, 185). That the style of decoration on the main house was still in fashion in the early 17th century is shown by the egg-and-dart motif on the bressumer being copied on the annexe jetty. This annexe with its two extremely well-lit rooms must have provided a high degree of extra comfort and privacy, far beyond that available in most houses of the city. On the ground and first floors the space between the two chimneys was filled in, probably when the annexe was built, to form a closet on each floor, 5 ft. by 3 ft. under a pent roof. Each closet was lit by a two-light wood-mullioned window. The first floor was reached by an 18th-century staircase in a block to the east, almost certainly occupying the site of an earlier block containing the original staircase and some additional accommodation. The attic was contained in the triangle of the roof, and was probably lit from both gable-ends.

The problem of the lack of partitioning is not confined to the Plumbers' Arms, as it occurs in several other houses of this period, and has already been noticed in the late 16th-century wing behind Nos. 17–21 Micklegate. It may sometimes have been to provide larger rooms for social occasions, but it often occurs on the ground floor where the layout must have included shops or workrooms. Equally it is clear that not all the examples can be dismissed as warehouses or other non-domestic forms of building. It may be an indication of the same casual attitude to privacy found in the 15th century when the ground-floor rooms were not separated from the entrance and through-passage. The same problem appears in The Old Rectory in Tanner Row (122) (Plate 190), a house of the early 17th century. This is of two storeys, jettied to the street, and with a usable attic completely within the triangle of the roof. The gypsum plaster floor in the attic is possibly an original feature. The building is three bays long, and has no structural partitions on either ground or first floor. The first bay was lit from the front, whilst the second and third bays each had a glazed three-light window to each floor in the side wall. A large chimney-breast was inserted into the middle of the building probably in the late 17th century. The glazed windows suggest that it was primarily a domestic building, but this is strongly counterbalanced by the lack of an original chimney-breast, which was probably almost universal by this date in new houses.

The increased demand for comfort is well illustrated by alterations carried out *c.* 1600 in Nos. 17–21 Micklegate in conjunction with the long rear wing (*see* p. 72), added at the same time. The old staircase annexe was filled up with a large chimney to heat the eastern end of the house, and the staircase moved, probably to the more convenient position in a new annexe to west of the present staircase, where it may also have provided access to the new wing. At the same time the large room on the first floor was given a handsome plaster ceiling and had its walls lined with panelling. Similar improvements were made in the early 17th century at No. 32 Micklegate where a large chimney and staircase were inserted in the middle of the house, and the roof altered to allow the roof-space to be used as rooms.

These early 17th-century buildings are the latest surviving examples of timber framing on this side of the river apart from fragmentary survivals almost lost in later remodellings. Timber-framed building continued spasmodically throughout the city, but the rising demands for comfort, warmth and privacy, and the slow change of taste under the late-coming Renaissance influence led to impatience with the old layouts and appearance of timber houses. The Civil War provided a pause in building activities, and the Corporation, by its resolution of 27 January 1644/5, 'It is moved to Common Council on Monday next for making an order for building houses upright from the ground in brick' (YCA, B.36, f. 122v.), put an end to the traditional methods of building just as thoroughly as the Great Fire did in London twenty years later. That so many timber-framed buildings have survived to our own day is due mainly to the fact that it was so much cheaper in the not very prosperous 18th and 19th centuries to cut off the front, if one wished to modernise, and replace it with a brick façade than to rebuild entirely, so that each generation had merely to make what *ad hoc* internal alterations suited the tastes and pockets of its own time.

It is unfortunate that 'le read brick house', built in Micklegate by Thomas Waller probably just before 1600 (Davies, 137) no longer exists, as it must have been almost the first brick house built. (It is possible that some part of it is incorporated at the back of Nos. 142–146 Micklegate (100), which seems to have been the Waller house.) By contrast, the first surviving brick building in Shrewsbury is said to be of *c.* 1670 (*Arch. J.*, CXIII (1956), 187), and in Exeter 1680 (*ibid.*, CXIV (1957), 170). The trade-name 'tiler', standard in York throughout the later Middle Ages, was rapidly displaced in 1590–1620 by 'bricklayer'. References in documents, however, suggest that brick was occasionally being used structurally as early as 1578 (*YCR*, VII, 177; *YCR*, VIII, 35). Brick was apparently becoming fashionable in the late 1630s, when two major buildings were erected in the Bishophill area: Fairfax (later Buckingham) House (begun 1638), and Towers' Folly (*c.* 1640) (129), but until this date it seems to have been used in positions of minor visual importance (*see* p. xxviii). Brick building, however, did not become general until the Corporation in 1645 forced the issue, probably as a measure of protection against fire, after the burning of the suburbs in the previous year's siege. Even then the numbers of bricklayers in the Freemen's Rolls did not increase as much as might be expected, a fact significant in the light of the scarcity of late 17th-century buildings. From a total of 86 tilers and bricklayers admitted to the Freedom in the half-century 1597–1646, the number admitted (of bricklayers only) increased to no more than 109 in the years 1647–96.

Types of Sites

The commonest type, as elsewhere in York, was the long narrow messuage with a street frontage of from 12 to 20 ft., stretching back some considerable distance—in the case of the riverine areas of North Street and Skeldergate, from the street back to the river. These latter properties had wharves and warehouses at the river end of the plot, but most of them have disappeared in recent redevelopments. No. 10 North Street was a good example. The narrow messuages survive mostly in Micklegate now, where the high value of sites tended to fossilise this arrangement. These plots usually possessed, as well as the various buildings, a garden or orchard, and in the case of the more important properties, stabling as well; all these usually survived up to the 19th century. An alternative type was the long shallow site along the street front. In the 14th and 15th centuries in this area this type seems to be confined to sites developed by the religious houses. Many of these, whether local, as St. Peter's and St. Leonard's Hospitals, or remote, owned much property in the district. The principal local house, Holy Trinity Priory, clearly saw the pecuniary disadvantages of having a plain precinct wall to the most important street in York, and adopted the idea of shallow development to use the street frontage without encroaching more than was necessary on its grounds. Its privacy it guarded by not providing any fenestration on the Priory side. The 14th-century development along Micklegate here probably extended nearly to the present churchyard, as an early 19th-century engraving shows a small tenement on the eastern side of the Priory Gateway similar to those still existing on the western (Cave, pl. XVII). Another example of the long street frontage is Nos. 17–21 Micklegate, a late 15th-century range in an area in which the religious houses owned much property. Other examples belong to the late 16th or early 17th century—e.g. Nos. 2, 4 and 6 Micklegate—and probably bear witness to the emergence of a particularly wealthy elite of merchants, who, especially as the Reformation had thrown so much religious property on to the market, were able to buy up two or three adjacent tenements for redevelopment.

Plans

Domestic layouts have nearly all been modified so that their original forms are no longer ascertainable, except perhaps for those of the simplest type. Such are the two ranges, Nos. 99–109 Micklegate (partly demolished; Fig. 60) and Nos. 1 and 2 All Saints Lane (Plate 185, Fig. 66). In these ranges, each tenement contains two rooms: a ground-floor room, which, in the Micklegate range at any rate, was no doubt a shop, and an upper room, jettied to the street and open to the roof. In Nos. 99–109 Micklegate each pair of houses has a central passageway on the ground floor, and the end house, No. 109, had a passage along its far end. In their present form these passages are of 18th-century date, but the arrangement of brackets under the jetty of Nos. 99–101 suggests that the position of the openings may have survived from the original layout. Presumably all these houses had communal wells and soil-houses in the narrow yards behind. Nos. 85–89 Micklegate also contain three separate tenements with one room to each floor, the top floor again open to the roof. The ranges fronting the street are either of two storeys, as Nos. 70–72 Micklegate (74), or of three, as No. 112 Micklegate, and are usually two bays deep, though the Nag's Head, No. 100 Micklegate, is of three bays. Both of these arrangements provide only one room in depth. Most of the back ranges are of later date than those at the front, and probably represent rebuilding of earlier structures. The Nag's Head, No. 100 Micklegate, is alone in preserving an earlier structure at the back, and the fact that this rear building is of only one storey suggests that it may have been the original hall. If this was a common arrangement in the earlier mediaeval period (*cf.* W. A. Pantin, 'Domestic Architecture in Oxford', *Ant. J.*, XXVII (1947), 133), the rebuilding of out-of-date hall-blocks would explain the existence of so many late 16th and early 17th-century ranges behind the front ranges. Plans in the late 16th and early 17th centuries were tailored to the re-emergence of the long street frontage incorporating several mediaeval messuages, and this resulted in unusually large buildings or pairs of houses, as in Nos. 16–18 Micklegate (Plate 52) and probably the houses now numbered 67, 69 and 71 Micklegate (71, 73) (Plate 158). In a category by itself must be placed the late 15th-century block Nos. 17–21 Micklegate, lying along the street front and containing lengthwise at least four rooms. Three are of one bay each, 9 ft. 6 in. to 10 ft long, and the fourth has two bays, one 10 ft. 6 in. and the other only 6 ft. This last may have been the principal room on each floor, the small bay housing whatever heating arrangements were in use.

Modifications to layouts, apart from the rebuilding of front or rear blocks, were mainly directed to making the houses more comfortable for living. Cutting down the vertical size of rooms was a great step to this end, as well as, in some cases, providing extra accommodation. The ground-floor halls of Jacob's Well, No. 1 Tanner Row, No. 100 Micklegate, and probably No. 31 North Street were all converted into two storeys in the late 16th or early 17th century, and many of the open-roof structures must have been ceiled off in the 17th century, as Nos. 99–101 Micklegate. Heating improvements are discussed under *Fireplaces*.

Insufficient evidence is available to equate the layouts and rooms with inventory descriptions. Some inventories have indeed been published,[1] but most of these deal with high dignitaries of church and state, and where they do concern York merchants they are highly selective, dealing with stocks-in-trade more than domestic arrangements. In so far as any conclusions can be drawn from a random number of published inventories of between 1390 and 1450, there seems to be a basic minimum of four rooms, of which the kitchen (*coquina*), which is always mentioned, was no doubt a separate building at the back. The hall (*aula*) always appears, and frequently has a spere in it, probably a moveable screen to protect against draughts. One inventory describes the spere as 'wooden', which suggests perhaps that others, less elaborate, were merely a framework for a cloth hanging. Other rooms mentioned regularly include the buttery (*pinserna*), the bakehouse (*pistrinum*) and the store-room (*salarium*).

Materials

Oak was the most common constructional material, and none of the early stone houses recorded in documents have survived. Chamfered stone base courses to carry the timber ground-sill survive at No. 1 Tanner Row, and a single course of chamfered rubble still existed at No. 31 North Street as recently as 1953. The only stone building in the area is the south wing of Jacob's Well, but this was so heavily restored in 1905 that it is now impossible to decide its original date.

Brick, like stone, is rarely found. The 'wall-tile' infilling to timber framing was superseded in the late 16th century by brick walls on to which panelling was being fastened by the early 17th century. As a constructional material, however, brick only became fully established after the Civil War, though the moulded brick windows inserted into the stone wing of Jacob's Well must belong to the first half of the 17th century. Drake's remarks about York in 1736 are an interesting comment on the rarity of brick buildings even at that date (Drake, 279–80).

Timber Framing

The timber framing is plain and simple, and the houses are all jettied on at least one side. The unjettied walls are framed on full-height posts with enlarged heads, and occasionally, as in No. 112 Micklegate, with a small shoulder to carry the first-floor transverse rail or beam. This same house is unusual in that there is no enlargement at the post heads. The height from the ground to the wall-plate varies from 13 ft. 6 in. in No. 111 Micklegate (House 'B') to 22 ft. in No. 95 Micklegate (85), but most of the 15th and 16th-century houses tend to lie in the bracket of 15–18 ft. The main posts to each floor on the jettied fronts also have enlarged heads—enlarged on the external face under the jetties (Fig. 55), and on the inner face at tie-beam level. The framing within the bays and along partition walls consists of plain studs with curved braces or struts at one or both ends. In the earliest surviving houses, Nos. 99–109 Micklegate, the central stud in each partition wall is wider than its fellows, though in later buildings all the studs tend to be of the same width. In this same range of houses all the oblique members are braces, no struts being used (Plate 49). By the second half of the 15th century, however, nearly all oblique members are struts and only rarely, as in parts of Nos. 31 North Street (Plate 185, Fig. 66) and 1 Tanner Row, are braces used in the earlier fashion.[2] Braced framing reappears occasionally late in the period, and can be seen in the rear wing of Nos. 17–21 Micklegate, and in the open bay on the first floor of The Plumbers' Arms, where it was presumably considered less of an obstruction than a strut down to the floor. An unusual brace with inverted curvature appears at the end of the 16th century in No. 111 Micklegate (House 'C'). Struts return towards the end of the period and ogee struts appear in The Plumbers' Arms. They can have had little structural

[1] Mostly in volumes of the Surtees Society; *cf. Med. Arch.*, VI–VII (1962–3), 207, n. 10.

[2] These two methods of strengthening appear to correspond with Rigold's 'arch-braced' and 'tension-braced' framing (Wood, 222–4).

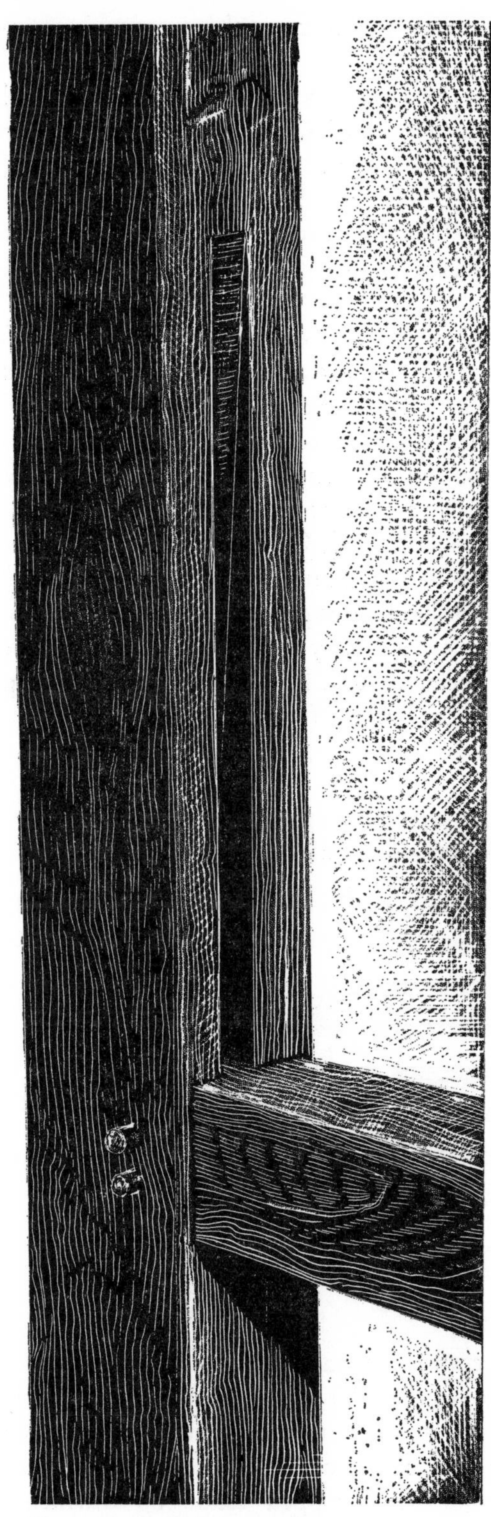

Fig. 9. Vertical Chase-mortice.

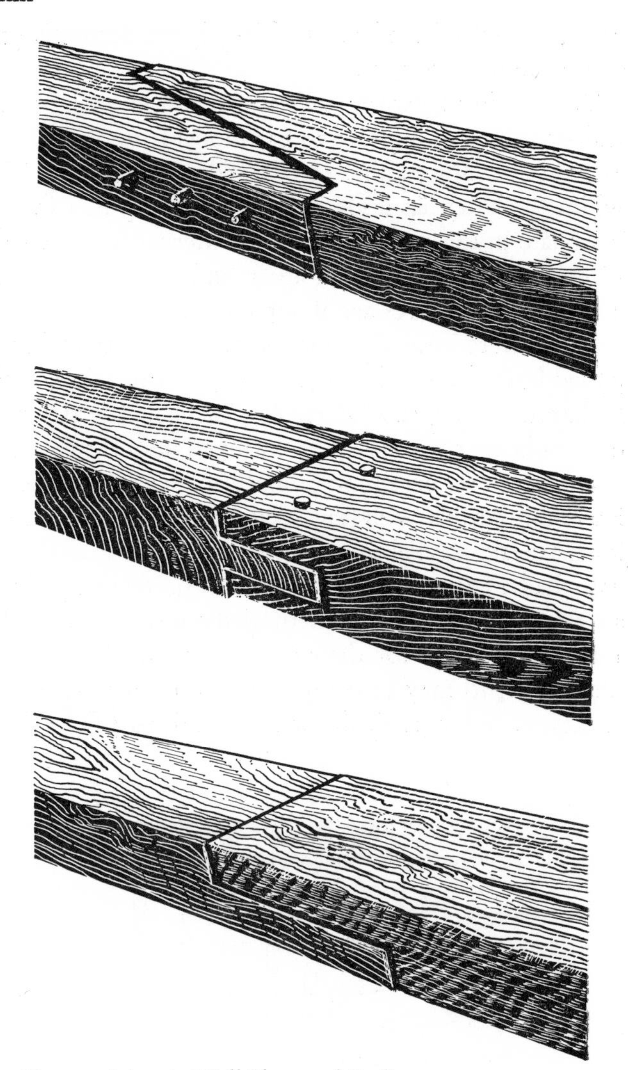

Fig. 10. Joints in Wall Plates and Purlins.

Fig. 11. Carpenters' Marks.
a. (120) No. 1 Tanner Row; b. (54) Nos. 2, 4, 6 Micklegate and (126) Nos. 2, 4 Trinity Lane; c. (54) Nos. 2, 4, 6 Micklegate; d. (58) Nos. 16, 18 Micklegate; e. (54) No. 2 Micklegate.

value, and must have been used largely for decorative purposes. In The Old Rectory the struts are practically straight (Plate 190).

The spacing between studs varies considerably, but a change can clearly be distinguished in external wall-studding (as opposed to partitions) from the late 14th-century houses Nos. 99–109 Micklegate to the early 17th-century annexe to The Plumbers' Arms. The former in a 12 ft. bay has one central stud only, whilst the latter has studs varying from 9 in. to 1 ft. 5 in. apart. At an intermediate period, No. 1 Tanner Row late in the 15th century had external studs at 1 ft. 3 in. to 1 ft. 7 in. apart, whilst those in No. 111 Micklegate were 1 ft. 2 in. apart, and those in the rear wing of Nos. 17–21 Micklegate 1 ft. 4 in. apart. The studs themselves are about 5–6 in. in width.

Jetties were invariably constructed to the street front, but only one house, Nos. 16–18 Micklegate, was jettied both front and rear (Plate 52). Corner sites, such as Nos. 31 North Street and 1 Tanner Row, and L-shaped houses, such as Jacob's Well, had jetties on two adjacent faces, and made use of the normal dragon-beam construction at the junction of the two walls (Plate 185). The joists running from the dragon-beam are all set on an increasingly acute angle to the beam in the late 15th-century houses No. 1 Tanner Row and Jacob's Well (Fig. 12a), whereas in the early 17th-century annexe to The Plumbers' Arms, the joists are laid parallel to each other on each side of the dragon-beam (Fig. 12b). One house, No. 95 Micklegate, had a jetty to the first floor, but no further jetty above, although the house was of three storeys and an attic. The projection of the jetty seems to have been standardised throughout the period at about 1 ft. 6 in., except for the early 17th-century Old Rectory, which has a jetty of about 1 ft. 2 in. These measurements may be compared with projections generally of 1 ft. 6 in. to 2 ft. 6 in. in various building contracts, and in particular with the 2 ft. to 4 ft. of the Coney Street contract (Salzman, *passim*).

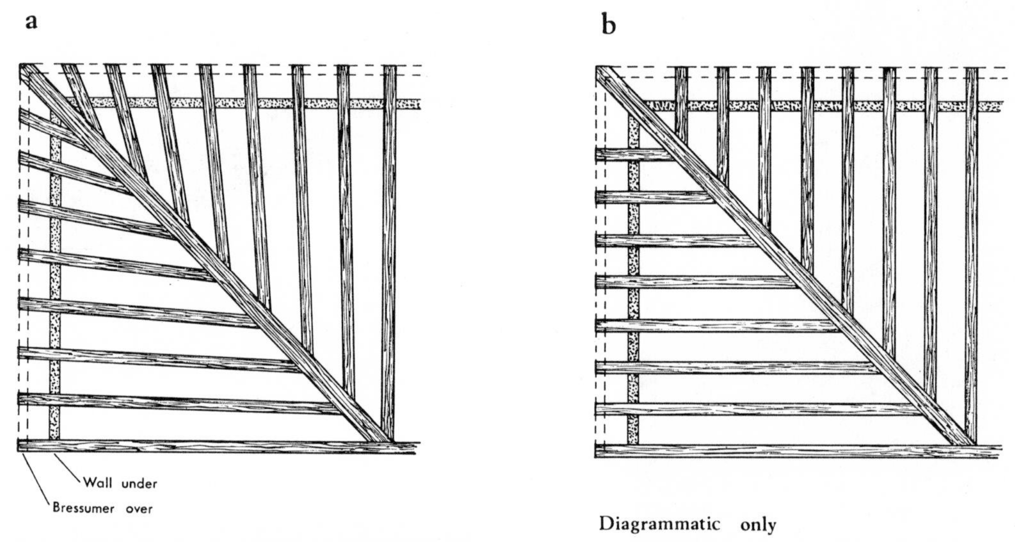

Fig. 12. Dragon Beams and Joists forming jetties.

The construction of the jetty can usually only be seen in the course of repairs or demolition. In No. 17 Micklegate the truss-joist carried at its end the horizontal sill-plate of the floor above. This was presumably pegged vertically on to the joist. The post of the upper floor was notched to rest mostly on this plate, but also on the joist, though the methods of fastening could not be seen. The construction is slightly different again in Nos. 85–89 Micklegate (Fig. 55). Here the joists end in horizontal stopped[1] tenons on to which is

[1] I.e. the tenon does not run through to the face; the stopped tenon is the only form found in this area.

morticed and pegged the moulded sill-plate which projects above them by 2 in. On this plate stands the post, notched, as in No. 17, to rest on the truss-joist also, but thickened out at this point and cut with a tenon which is pegged into the joist. This method cloaks the joist-ends and provides a place for moulded decoration.

The infilling of the framing is almost always of thin bricks, set on edge longways in the wall and referred to in documents as 'wall-tiles'. The only exception to this is in Nos. 99–109 Micklegate where Nos. 105–107 had, as infilling in the spandrel between a post and a curved brace, a series of riven baulks of wood with the bark left on—mostly logs of 2 in. diameter chopped in half and flattened at each end. These were set horizontally 2–3 in. apart, and covered up, and the spaces between filled in with daub. It is not, however, certain that this was the original infilling, as the baulks were nailed at each end. However, the partition between Nos. 99 and 101 has a brown infilling of daub plaster containing hair, though the framework on which it is spread could not be ascertained. A similar daub infilling was recorded at one area in No. 1 Tanner Row, but this may have been a late 16th or early 17th-century repair, as the same wall elsewhere had wall-tile infilling. From the 15th century onwards wall-tiles were used, standardised within very narrow limits to 10½ by 5½ by 1½ in. In Nos. 17–21 Micklegate they were jammed between framing pegs—a series of pegs let into the sides of the timbers and projecting out a few inches. Most of the wall-tiles, however, were anchored by means of shallow grooves cut in all the members, the purpose of the groove being to provide a key for a layer of plaster or mortar against which the wall-tiles were bedded. This system lasted to the end of the period, being found in the late 16th century in Nos. 2 and 4 Trinity Lane and The Plumbers' Arms (Fig. 71), and in the early 17th century in No. 6 Trinity Lane.

To join together pieces of timber for long lengths, as in purlins and wall-plates, two methods were in use. The earlier method was the splayed lap, found in No. 103 and Nos. 85–89 Micklegate (Fig. 10, top). In No. 103 three pegs were driven up vertically through the two pieces, and the same probably occurred in Nos. 85–89, though this was not seen. The later method, found in Nos. 16–18 and No. 100 Micklegate (Fig. 10, middle) was a forked tenon joint, pegged horizontally, with two pegs in No. 100 and four in Nos. 16–18. These joints were often sited over or near to the supporting wall-posts or crown-posts to provide extra stability. There is only one example of a halved joint, in a reset wall-plate in No. 1 Tanner Row (Fig. 10, bottom) where it may be a later repair; in the same house is an example *in situ* of the splayed lap joint.

A feature recorded more frequently across the river is the method of inserting internal beams after the outer timber-walls had already been erected, but the only example recorded in this area was in Nos. 2 and 4 Trinity Lane. A long vertical chase-mortice was cut in the inner face of a main post (Fig. 9), increasing in depth to 4½ in. at the bottom. There it corresponded with the base of an ordinary 4½ in.-deep mortice in the opposite main post. The horizontal beam was lowered at an angle into position, one tenon being placed against the ordinary mortice. As the other tenon was lowered down the chase, the first tenon sank home until the beam was horizontal and both tenons were in position. They were then pegged to the post. The rest of the chase was usually filled up with fragments of wall-tile and plastered, the plaster being sometimes painted to match the woodwork. Joists were laid on top of the side-rails without any fastening, as in The Plumbers' Arms, and No. 111 Micklegate (House 'C').

Floors have almost always been renewed, though a few survive with elm planks 1 in. thick and about 10 in. across, which may be original. In The Plumbers' Arms, the front room on the attic floor had flooring made up by a layer of close-set laths nailed on to the joists. On these and at right-angles to them was placed a thick layer of reeds, and then gypsum was spread over the top (Fig. 71). The similar floor in the attic of The Old Rectory has already been noticed (*see* p. lxv). This type of flooring was used only in attic rooms of late 16th and early 17th-century date, and was probably an attempt at insulating the room below from the cold and draughty attic. The same principles seem to have been applied as were adopted contemporaneously for sound insulation (*see* W. Allen and R. Pocock, 'Sound Insulation: Some Historical Notes', *JRIBA* (March 1946), 183–8).

Roof Structures

Roof structures provide a more reliable guide to dating than does wall-framing, as after *c.* 1500 roofs of crown-post construction disappear. The crown-post roofs have coupled rafters, but no ridge-beam or principals. The crown-post, except in the latest example, Nos. 85–89 Micklegate, carries a central purlin in its enlarged head. Across the purlin lies the collar to each pair of rafters; supporting members include longitudinal braces from the crown-post up to the purlin, and a curved strut on each side of the crown-post down to the tie-beam (Fig. 13b). The latter is cambered, and all joints are of mortice-and-tenon type, pegged together. The height of the crown-post varies from 4 ft. 9 in. in No. 111 Micklegate (House B) (Fig. 62) to 7 ft. 8 in. in Nos. 17–21 Micklegate. Nos. 99–109 Micklegate (Fig. 13a) shows some archaic features such as the steady taper of the crown-post in No. 103 (Plate 50 and Fig. 61, bottom), the fastening of the crown-post struts below the enlarged head (Fig. 13a), and the separate curved raking struts from the tie-beam to the rafters (Plate 50 and Fig. 61 top). In the 15th century these raking struts were incorporated into the framing by moving their feet nearer to the crown-post and halving them across the crown-post struts (Fig. 13c), and this development is associated with the appearance of side purlins. These were carried in an enlargement of the strut where it was tenoned to the rafters. This arrangement is more common on the other side of the river, but obtained in the now altered roof of Nos. 17–21 Micklegate (Fig. 13d). The central purlin continued in use at the same time. By the turn of the 15th century, Nos. 85–89 Micklegate had dispensed with the central purlin whilst illogically retaining the crown-post for which the central purlin was the *raison d'être* (Fig. 13e). It seems that the introduction of the side purlins was a safety device for those houses with a wide span of 20 ft. or more and a high roof pitch, for they do not occur in the smaller houses such as Jacob's Well, or No. 31 North Street (Fig. 66). The latter, however, has the curved purlin struts in the eastern gable (Fig. 66) but not elsewhere, so they may also have been regarded as a decorative feature towards the street. Another feature which appears late in the 15th century is the oblique purlin-brace—a longitudinal curved brace from the purlin strut to the side purlin[1] (Fig. 13d, e). These purlin-braces can be seen reused in Nos. 17–21 Micklegate (Plate 50) and *in situ* in Nos. 85–89 Micklegate. This latter house is the last case of crown-post construction, and the disappearance of the crown-post together with the two large struts which supported it clearly lightened the roof structure without weakening it, and effected great economy in the use of timber.

The intermediate stage before the utilisation of the roof space is not well represented, but appears to have been an experimental one directed to finding better methods of supporting the side purlins. The solution seems to have been either in the continued use of curved struts from the tie-beam directly to the purlin and rafter, and without any enlargement of the strut head, or else in curved or straight kerb-principals rising from the tie-beam up to a collar, and supporting the purlin in a notch half-way up. The purlin-strut was used in The Nag's Head, No. 100 Micklegate, and both types in No. 112 Micklegate (Fig. 13f, g), but this house is atypical in many ways, as in its use of principal rafters, and the survival of the earlier type of brace from horizontal members down to posts. Both types were also used in the wing behind No. 89 Micklegate (Fig. 13e), in alternate bays and in conjunction with two purlins each side. The use of the kerb-principal to support the purlin seems to have filtered down from use in buildings of large span. It occurs across the river in St. Anthony's Hall as early as 1453, and is found on this side in the roof of Holy Trinity Church which cannot be earlier than 1551. It continued in use until late in the century,

[1] This is a form of wind-bracing in which the braces run in an oblique plane opposed to the slope of the roof whose purlin they abut.

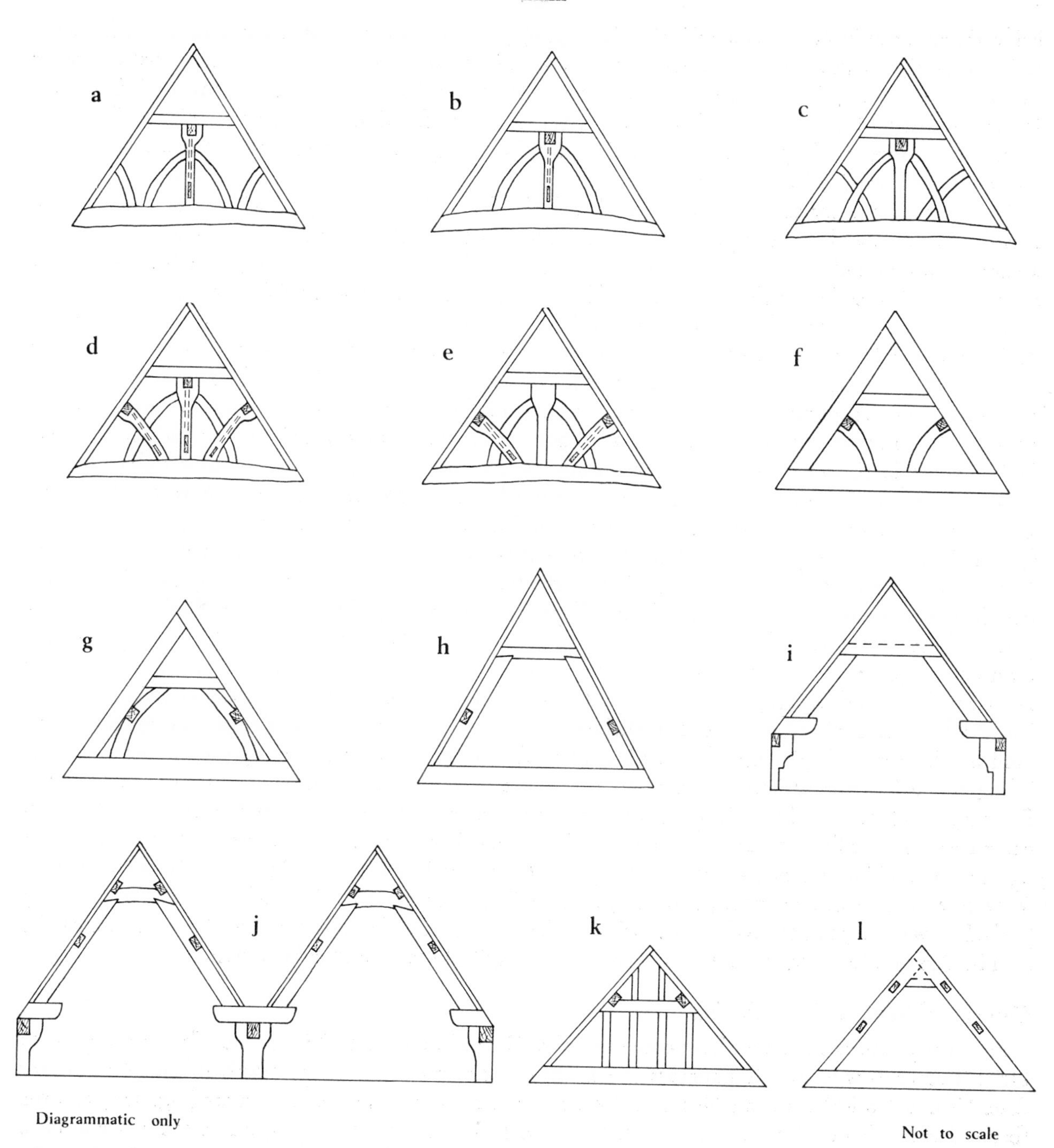

Fig. 13. Roof Trusses.

a. (87) Nos. 99–109 Micklegate, early–mid 14th century; b. (125) Jacob's Well, Trinity Lane, late 15th century; c. (104) Church Cottages, No. 31 North Street, late 15th century; d. (59) Nos. 17, 21 Micklegate, late 15th century; e. (79) Nos. 85–89 Micklegate, *c.* 1500; f, g. (92) No. 112 Micklegate, 16th century.

h. (126) Nos. 2, 4 Trinity Lane, late 16th century; i. (87) No. 111C Micklegate, late 16th–early 17th century; j. (58) Nos. 16, 18 Micklegate, late 16th–early 17th century; k. (127) No. 6 Trinity Lane, early 17th century; l. (122) The Old Rectory, No. 7A Tanner Row, early 17th century.

being found again in Nos. 2 and 4 Trinity Lane (126) (Fig. 13h) where it probably reflects the influence of Holy Trinity Church. In the mid-16th century also, internal tie-beams tend to become flat instead of cambered, suggesting that the top-floor rooms were being ceiled at this period. This feature is illustrated in the unusual roof construction of No. 32 Micklegate where flat tie-beams carry principal rafters. The principals are supported by struts from the tie-beam, fossilising the system of purlin-struts, although the purlins are here carried by the principals.

The late 16th-century carpenters were faced with the demand for more space within the same ground-area. Rather than add an extra jettied storey, they turned the roof-space into attics, in accordance with a country-wide trend. Their first solution was to heighten the building by some 3 to 4 ft., flooring the attic at that distance below the wall-plate, whilst subsidiary transverse and axial beams and longitudinal side rails were incorporated to carry the attic joists. Unimpeded circulation between the inner trusses of the attic was then achieved by means of sole-pieces[1] which were in effect junction-pieces between posts and wall-plates below and common rafters and kerb-principals above (Fig. 63). The kerb-principals were carried up to support a collar high enough to give full headroom. A variant of this in the outbuilding behind No. 10 North Street combines the common rafter and kerb-principal into a principal rafter cut back at collar-level, and then continued to the apex in the same scantling as the common rafters (for this 'diminishing principal' *see* J. T. Smith, 'Medieval Roofs', *Arch. J.*, CXV (1958), 124–5). Nos. 16–18 Micklegate (Fig. 13j) and No. 10 North Street have two purlins each side, but No. 111 Micklegate (House 'C') (Fig. 13i) can only have had one—at collar level—as the gable has common rafters with no principals. Nos. 16–18 Micklegate is also interesting in that the twin-gabled construction required a double-headed post in the centre of the building with the valley-plate housed centrally in it, and a double-ended sole-piece to serve both roofs (Fig. 13j). The double-headed post also occurs in No. 67 Micklegate for a twin-gabled house, and in Nos. 17–21 Micklegate for a rear extension. The sole-piece system was rather clumsy in design and extravagant of timber, and was possibly not structurally secure, as No. 10 North Street has had the collar replaced, and No. 111 Micklegate (House 'C') has had an extra collar inserted just above head-height. By the beginning of the 17th century the sole-piece was replaced by the simpler system of lifting the attics entirely into the roof, flooring them at wall-plate level. The gables had simple stud framing, as in No. 6 Trinity Lane (127), with a collar which carried the side purlin (Fig. 13k). On the internal trusses, support for the purlins came either from the collar without the stud framing of the gable, as again in No. 6 Trinity Lane, or from housings in the principal rafters, as in The Old Rectory (Fig. 13l). Very few internal arrangements, however, survive from this period. The gable-end was occasionally pushed forward beyond the wall below, with brackets supporting the ends of the wall-plates. This can be seen in The Old Rectory, and seems to have existed originally in The Plumbers' Arms.

Carpenters' Marks and Numbering

Not many carpenters' marks have come to light. The earliest (Fig. 11a) is on one of the front posts to No. 1 Tanner Row on which it appears twice. The other recorded marks are mostly of late 16th-century date. Nos. 2–6 Micklegate and Nos. 2 and 4 Trinity Lane both have the same mark (Fig. 11b) and the former has an additional mark or doodle as well (Fig. 11c). Similar to the mark in Fig. 11b is one in Nos. 16–18 Micklegate, where it is repeated three times (Fig. 11d). To the early 17th century must belong the mark on the added back range to No. 2 Micklegate (Fig. 11e). Numbering of timbers was done in Roman numerals and was confined to roof members and tie-beams until the late 16th century when studs began to be numbered as well. Examples of the latter are in Nos. 4 and 6 Micklegate and Nos. 2 and 4 Trinity

[1] *See* R. A. Cordingley, 'British Historical Roof-types and their members: a Classification' in *Transactions of the Ancient Monuments Society*, N.S., IX (1961), 113.

Lane. Roof trusses are usually numbered as self-contained units, but in the 14th-century range, Nos. 99–109 Micklegate, the roof timbers are numbered continuously along the north side and back along the south side from I–XIIII. There was apparently no numbering on the truss at the west end.

Carvings

Decorative detail in wood is extremely scarce, in stone and plaster entirely absent (except for plaster ceilings, *q.v.*). The earliest surviving examples are the corner posts carrying the dragon-beams of Nos. 31 North Street and 1 Tanner Row. Each of these has a carved and moulded cap below the springing of the post-head, and both have the same basic design, although differing in detail. Each cap contains a top and bottom set of mouldings sandwiching a central decorative panel sunk back to the surface of the main post. In No. 31 this panel has, on the southern face, a row of three rosettes carved in relief, and on the eastern face a very worn row of four sunk quatrefoils, whilst the upper set of mouldings is crowned by battlementing. In No. 1 Tanner Row (Plate 48) the central panel has a row of square compartments each containing a pointed-lobed quatrefoil with sunk spandrels. There were three quatrefoils on each face originally, but now there are only two on the northern side, and two and a fragment of a third on the eastern face, the post having been considerably mutilated and cut back. On both caps the mouldings are of late 15th-century style, though those of No. 31 are very badly worn. Other carvings of this period are to be found at Jacob's Well, where a beam across the western end of the hall has its chamfered lower edge divided into five sections by four carved petal-form quatrefoils, and where the reset brackets of the canopy to the doorway display motifs of an eagle, a tudor rose and conventional foliage (Plate 191). The only moulded bressumers to survive are to be found on Nos. 85–89 Micklegate (Plate 174) on both the first and second-floor jetties. These are very badly worn and mutilated, but can be recognised as a hollow moulding surmounted by battlementing, probably not much later than *c.* 1500. The Plumbers' Arms of *c.* 1575 had a moulded fascia board fastened to the jetty and carved with egg-and-dart, copied in the early 17th century for the two fascia boards of the annexe. Also to the early 17th century belongs the large oak staircase inserted into No. 32 Micklegate (Plate 83). Its string is carved with stylised roses and thistles, and its newel-posts with crude jewel ornaments.

See also under *Doorways*, *Plasterwork* and *Staircases*.

Doorways

Alterations and rebuildings have played havoc with mediaeval doorways, and none now survives intact. In two cases, No. 1 Tanner Row and Jacob's Well (Plate 191), the jamb posts of an original entrance doorway remain, and in both houses pegholes in these posts suggest that a shaped wooden doorhead was pegged in between them. No other original doorways survive, but Jacob's Well preserves the only surviving evidence for the canopied doorways which must once have been a normal feature of the street scene. Reset in 1905 under a modern canopy are two elaborately carved and moulded brackets (Plate 191), from the demolished Old Wheatsheaf inn in Davygate. The moulding of the bracket suggests a date at the end of the 15th century.

Fireplaces and Overmantels

Contemporary arrangements for heating are extremely rare, and most houses must have had either portable systems such as braziers, or very simple non-structural fireplaces with plaster cowls, all of which could be removed without leaving any trace of their presence. The earliest surviving heating arrangement is the stone-built chimney in the south wing of Jacob's Well, but any original fireplace was lost in 1905. It is not clear now whether this is contemporary with the late 15th-century timber-framed house adjoining the wing, or whether it belongs to an earlier building.

From the late 16th century the large double house, Nos. 16–18 Micklegate, has lost both its original chimneys, removed *c.* 1760 to make space for new staircases, but the evidence survives to show their original position centrally between the front and rear rooms. Of the same date were the two massive chimneys in The Plumbers' Arms. These projected externally on the side of the building (Fig. 70), and were built of brick with stone dressings and weathered offsets. Internally each of them contained a fireplace to each floor. On the first floor these had been completely disguised by late 18th-century grates and surrounds, but during demolition the original late 16th-century brick fireplaces were uncovered. These had shallow three-centred heads with a continuous chamfer to the head and jambs. The large brick chimney in No. 1 Tanner Row is associated with the insertion of an intermediate floor into the hall and the construction of two newel stairs, all of which must date to the early 17th century. A large chimney inserted in No. 32 Micklegate is also of this date. This exiguous list is completed by an early 17th-century carved overmantel in Nos. 2–6 Micklegate. Of a type fairly common throughout the city, it is the only one recorded on this side of the river, and has now been destroyed.

Plasterwork and Ceilings

Decorative plaster ceilings do not seem to have achieved the popularity they enjoyed in some parts of the country, and only three, all of late 16th or early 17th-century date, have survived. This is possibly a reflection both of a local indifference to embellishments, and of the general poverty of the period, whilst the very localised situation of the three in this area—all within about 100 yds. of each other—hints at a single source of influence and, perhaps, execution.[1] In this context it may be of significance that no plasterers are recorded in the York Freemen's Rolls in the relevant century 1550–1650.

The plainest of these ceilings is in No. 17 Micklegate. It has a running pattern applied to the pre-existing ceiling beam, and a large fleur-de-lis in each corner and flanking the ends of the beam. Each fleur-de-lis is enclosed by a moulded rib in the form of a square. There is a much richer design in No. 18 Micklegate (Plate 51), but the date is the same as No. 17, for the soffit of the ceiling beams in both houses has an almost identical arabesque pattern. The arabesques together with the fauna in the frieze provide a design half-way between the simplicity of No. 17 Micklegate and the florid exuberance of the remaining ceiling at No. 6 North Street. This has recently been removed, on the demolition of the building, and has been re-erected in The King's Manor. It has four large identical panels (Plate 186) formed by intersecting ceiling beams, and has a remarkable collection of decorative motifs. Beasts' heads, arabesques, foliate bosses, stylised and naturalistic vine leaves with bunches of grapes are woven into an intricate pattern with a deeply-moulded border. Centrally on the wall-plate cornice of each compartment there is a projecting section supported by a man's head—a feature which suggests a date well into the 17th century.

Ornamented timber ceilings have rarely survived. The earliest survival is in the suburbs at No. 4 Front Street, Acomb (132), where there is a ceiling-beam with paired ogee mouldings of the early 16th century. A very plain example existed in No. 111 Micklegate (House 'C'), whilst intersecting ceiling-beams in The Plumbers' Arms had a simple quarter-round and hollow moulding with run-out stops. Both these date from the late 16th century. To the early 17th century belong ceilings in No. 32 Micklegate, with ovolo-moulded beams and joists, and in The Old Rectory where the existence of beams having only plain chamfers with run-out stops is perhaps another argument for its non-domestic function.

Staircases

Of all the mediaeval fitments, staircases have perhaps suffered most from modernisation, both for ease

[1] A fourth example in the same area existed at Mountsorrel on the site of No. 1 St. Martin's Lane; fragments are preserved in the Yorkshire Museum, Basement, Case 4, nos. 1 and 9 (from 'house of the Metham family').

of usage and under the pressure of fashion. Consequently no complete examples have survived, though a fair amount of evidence for their positions can be found. In No. 1 All Saints' Lane, the arrangement of the joists carrying the first floor shows that there was a short straight staircase going up against the east wall almost exactly where the present staircase is. It seems a fair inference that in the adjoining cottage, No. 2, there was a similar arrangement where the present straight staircase is situated against the west wall. Similarly in Jacob's Well, Trinity Lane, there is evidence that there was a staircase in the south-east corner between the front wall and the side of the stone chimney.

Nos. 17–21 Micklegate had a timber-framed annexe at the rear in the north-east corner, the most likely explanation of which is that it was to house the staircase. The same type of external staircase annexe must have been provided for each of Nos. 85, 87 and 89 Micklegate for ascending to the first floor, as gaps are left in the first-floor stud-walls at the same place in each house, and the rooms are still entered at these points. The continuations up to the second floor, however, were inside the rooms, and in No. 87 there still remains a short straight staircase against the east wall for reaching the top floor. This staircase is opposite the entrance doorway to the room, and has timber-framed construction on its open side.

In the late 16th century the new wing behind Nos. 17–21 Micklegate was provided with an internal staircase, though this has now been replaced by a later one in the same position. Similarly, though lacking the structural evidence which survives in Nos. 17–21 Micklegate, the original staircase at The Plumbers' Arms was no doubt in the eastern annexe where the 18th-century staircase eventually replaced it. Two staircases in No. 1 Tanner Row, both with octagonal newel posts, are possibly associated with the extensive remodelling of the early 17th century, when the hall was divided into two storeys and a large chimney inserted. The provision of two separate staircases suggests that the eastern wing was converted to a self-contained tenement at that date. Also of the early 17th century is the moulded and carved oak staircase with symmetrically turned balusters in No. 32 Micklegate (Plate 83). It was altered in the 19th century, but is the only surviving example of the period (*see* also under *Carvings*). In a class by themselves, and unrelated to the local development of staircases, are the richly-carved newel posts and handrails of a grandiose staircase of *c.* 1640, reset in The Old Rectory. They are said to have come from Alne Hall in the North Riding (Plate 82).

Wall-coverings

Panelling came into fashion here in the early 17th century, and all the surviving examples are of this date. It probably became fashionable once the brick-built fireplace was established as a normal part of design, replacing the woven hangings which must have preceded it in unheated rooms. Esdras Browyas and Anthony Rayskaert, described as 'Dochemen, arresworkers' were admitted to the freedom of the city in 1570, and were lucky in that a whole generation elapsed before their trade was put out of business by the new fashion of wood panelling. There is little variety in the local panelling, and most of it has vanished in recent demolitions. Square or rectangular oak panels were formed by 'run-through' rails and short styles. The rails were moulded on the lower edge—that is, at the top of the panel—and chamfered on the upper edge, forming the bottom of the panel. The styles were moulded on both edges, and the ogee was the moulding most in favour in the early period.

Windows and Glazing

The earliest surviving windows are the two three-light mullioned windows of oak in Jacob's Well, of late 15th-century date (Plate 191). The windows were originally unglazed, with mullions of a fairly broad diamond section, set 6–7 in. apart. They were probably closed by internal shutters and now have 20th-century glazing. Evidence for windows of the same date exists at Nos. 1 and 2 All Saints' Lane, and at

No. 1 Tanner Row. At Nos. 1 and 2 All Saints' Lane and No. 1 Tanner Row the upper storey seems to have had a series of paired oriel windows separated internally by a wall-stud. No. 1 Tanner Row had a large opening on the ground floor, but it seems more likely that this was a shop opening with perhaps a ledge for serving, rather than a window (*cf.* MS. of *c.* 1450, Bibl. Nat., Paris, illustrated in *CIBA Review*, 47 (1943), 1694; and 'The Wool Hall', Lavenham of *c.* 1500, Wood, pl. XXXIV). In the top floor of Nos. 85–89 Micklegate the evidence shows that there were two windows in each bay, probably paired oriels, as in Nos. 1 and 2 All Saints' Lane and No. 1 Tanner Row, with a central dividing wall-stud, and presumably the same arrangement obtained on the first floor. The lack of infill-grooving in the upper half of some of the studs in the top floor of Nos. 2 and 4 Trinity Lane suggests that there was a row of small windows just under the eaves along the south wall. The early 17th century provides examples of windows in The Old Rectory and The Plumbers' Arms. The Old Rectory has at least four three-light windows on the west front, of which the two ground-floor ones are completely blocked and the upper southern one has one light blocked. The remaining lights have ovolo-moulded mullions, and the shape of these together with their wide spacing show that they were designed for glazing.

The Plumbers' Arms in its early 17th-century annexe exhibits a most remarkable series of original windows. On both first and second floors the annexe had a two-light window in the south wall, set at the western end to clear the chimney, and a three-light window in the north wall. On the first floor, almost all the west wall was taken up by a 'picture-window' containing two rows of five lights each, the rows separated by a large chamfered transom. Above this window on the second floor was a three-light window. All the windows were shop-made, ready to jam into position in the existing timber-framework. The picture window, at any rate, was designed to be glazed, as the glazing groove was visible, and presumably the smaller windows were as well. All the mullions were lozenge-shaped, but very much slimmer than in the 15th-century houses. There was also a two-light window on each floor to the closet formed between the chimneys in the early 17th century. These had wooden mullions and jambs, but as they were blocked, the type of moulding could not be seen. At that date they would probably have been made for glazing. A similar type of shop-made window was probably used in No. 111 Micklegate (House 'C'), but was taken out in the 19th century to make way for a staircase. Mention should also be made of two brick-built windows, each of two lights, in the back wall of Jacob's Well. The chamfered mullions and jambs have been heavily and inaccurately restored, but a photograph taken during the restoration of 1905 shows that they were originally shaped with rebates for glazing. They probably date from the first half of the 17th century and illustrate the pattern wherein brick was at first used structurally in the less important elevations away from the street.

The scarcity of original windows makes an assessment of early glazing very difficult. The oriel windows in Nos. 1 and 2 All Saints' Lane were probably unglazed with shutters fastening to the central stud, and unglazed windows were certainly being put up in the late 15th century at Jacob's Well. There are documentary records of glazing: in a house belonging to Symond Stele, in 1422 (SS, LXXXV, 16); and in a 16th-century house in Micklegate when payment to John Calbeck, glazier, who rented the house from 1517 till 1533, included 'five shillings and fourpence paid in glase yt he leyfft in ye howse in Mykylgatt when he whent frome it' (YAYAS, *Report* (1950–1), 19). Other entries show that domestic glazing was being carried out extensively at this latter period.

(T.W.F., J.H.H.)

HOUSES 1650–1800

Plans and Planning

No houses of the latter half of the 17th century remain intact in this part of the city, and therefore no analysis of development of the buildings of the period is possible. The plans and dimensions of the York town houses of the 18th century were governed by the mediaeval (and earlier) development of the city. The houses within the boundary of the city walls were generally built on sites recently occupied by earlier structures with the building set forward to the road frontages of long narrow curtilages. Such a confined site for the average-sized house conditioned the ground plan to one room to the front and one to the rear, with a side entrance passage and staircase; some minor variations, however, are possible, such as the placing of the staircase transversely between the back and front rooms (*see* Fig. 64, (95)). Some of the properties had access to the service quarters at the rear from a subsidiary road, but others had to incorporate a through passage for the access of servants and goods to the back of the house from the main thoroughfare, and this arrangement is well illustrated by monument (68).

Business and trade expansion in the main thoroughfare of Micklegate, particularly in the 19th century, has led to the conversion of private residences into shops, with considerable mutilation of the original ground-floor arrangements and the despoiling of pleasant, though modest, Georgian street elevations. Many a Georgian front masks a mediaeval structure, the jettied timber-framed building having been trimmed back and a brick façade erected. The recently demolished range of buildings to the west of St. John's Church, Nos. 2–18 (54, 56), and the range opposite to them, Nos. 17–19 (59), are examples of this Georgian 'improvement'. It is fortunate that a group of town mansions survive, and they have passed into the ownership of enlightened institutions: Micklegate House (81) and the adjacent Bathurst House (80) (University of York) and Garforth House (66) (Chartered Accountants' Offices), while the Bar Convent (13) still occupies its original premises in Blossom Street. The house numbered 53, 55 Micklegate (65) has been sub-divided but little internal alteration has been made; 56 Skeldergate (117), however, has suffered considerably with a carriageway driven through half the ground floor to give access to a builder's yard at the rear.

On the rising ground away from the River Ouse, some of the larger houses have cellars and basements provided for storage and larders. Small closet wings are noted on houses as early as *c.* 1700; they do not appear in houses of the second half of the 18th century. The service quarters, including kitchens, are generally on the ground floor to the rear within the main house block, and only in one or two houses on the higher ground away from the river are they in semi-basements. Domestic quarters in a rear wing appear to be limited to some of those houses which incorporate older structures at the rear, though one or two of the larger houses had service wings provided from the start. By contrast, some of the most important houses, Micklegate House (81), Garforth House (66) and 53/55 Micklegate (65) (*see* plans, Figs. 58, 53, 51), have no such domestic annexes. The placing of the service range at the rear became more common practice in the 19th century. Throughout the 18th century most houses had a reception room on the first floor overlooking the street—the saloon. In the grander houses the saloon is reached by a staircase placed in a spacious stairhall at the back with a decorated window on the half-landing and an ornamental ceiling. A fine example is at Garforth House where a Venetian window occupying the full width of the half-landing bears testimony to the high quality of York craftsmanship. These grand staircases were not continued above the first floor, access to the full height of the house being provided by a secondary staircase rising from the ground floor or the basement, to the attics.

External Features

Brickwork. Up to the 17th century York buildings, apart from churches, were constructed in timber framing, but this form of building within the city boundaries was terminated by order of the Council on 27 January 1644/5: 'It is moved to Common Council on Monday next for making an order for building upright from the ground in brick' (YCA, B.36, f. 122v.).

From the middle of the 17th century onwards brick is the accepted building material for houses of all grades, though from before *c.* 1700 none remain complete. Bricks were made locally from the clay that abounds in and around York. Projecting at the rear of 104 Micklegate (89) is a late 17th-century brick wing (Plate 54), and moulded brick bands and various blocked windows of the period survive at the back of 35/37 Micklegate (63). All brickwork of the late 17th century is irregular in the rise of courses and random bonded. A warehouse in Skeldergate (29), now reduced from three parallel ranges to one,[1] incorporates bricks measuring 11 in. by 2½ in. by 5 in., four courses rising 11¼ in. (Fig. 47).

Most accomplished brickwork occurs in the important houses of the 18th century and none better than in Middlethorpe Hall of 1699 (163). This is in Flemish bonding, as are most Georgian fronts in York; it is further refined at Middlethorpe by tuck pointing. Of the early 18th-century town houses, fine examples of brickwork are seen at the Queen's Hotel (55) and the Bathurst House, 86 Micklegate (80). Usually the street elevations are in a rich warm brown brick of fine quality and uniformity with deep red gauged rubbed brick window arches. A paler stock brick occurs in a few main elevations with red brick dressings as at 118 Micklegate (95) (Plate 184). This treatment is also adopted for the rear elevations of some major houses, such as the Garforth House (66), 53/55 Micklegate (65), and Micklegate House (81). These same houses have best quality regular brickwork of uniform colour in their principal elevations; it is a standard of craftsmanship which occurs also in a later building, the Bar Convent (13) of 1786. These Micklegate houses have bricks measuring 9 in. by 4½ in. by 2¼ in., five courses rising 1 ft. 1 in. As the 18th century advanced the bricks became larger but maintained a general red-brown hue where used in principal elevations; the paler, variable coloured brick is more usual in houses of the following century.

Fig. 14. Stone Quoins. Mid 18th century.

The use of stone dressings is limited to a few important houses, presumably because of its relatively high cost; Middlethorpe Hall (163) has all openings, bands and rusticated quoins in dressed ashlar. Quoins appear again on a small group of mid 18th-century houses, and they are of a distinctive form: the quoin blocks are five brick courses in depth, with no bevel to the edges, the longer blocks project further from the face of the wall than the shorter ones and, moreover, the longer blocks on the face are also the longer on the returns and the shorter blocks the shorter on the returns, contrary to usual practice (*see* Fig. 14). This idiosyncrasy appears on 53/55 Micklegate (65) and Garforth House (66), on Micklegate House (81) and two other buildings in the city, Peaseholme House of 1752 and 39–45 Bootham of 1747–8 (*see* p. lxxxi).

The earlier elevations are commonly divided by projecting horizontal bands. The best examples of the

[1] Finally demolished 1970.

(36) ROWNTREE'S PARK. Memorial Gates, *c.* 1715.

HOUSES

(118) SKELDERGATE. No. 61, 16th century and later.

(78) MICKLEGATE. No. 83. 18th century.

(123) TANNER ROW No. 16. *c.* 1820.

(42) BISHOPHILL. The Old Rectory, Victor Street. Late 17th century.

(50) THE MOUNT. Nunroyd, No. 109. Early 18th century and later.

(145) ACOMB. Tuscan House, No. 31 Front Street. 18th century.

MOUNT PARADE. No. 19. *c.* 1830.

Exterior.

Interior at first floor.

(120) TANNER ROW. House, No. 1. N.W. angle, late 15th century.

(120) TANNER ROW. House, No. 1. Corner post, late 15th century.

(118) SKELDERGATE. The Plumbers' Arms, No. 61. Addition, *c.* 1600.

(87) House, No. 99. Late 14th century.

(87) House, No. 99. Late 14th century.

(87) House, No. 111(B). 15th century.

(87) The Coach and Horses, No. 103. Crown-post, 14th century.

(59) House, No. 17. Late 15th century.

(87) The Coach and Horses, No. 103. 14th century.
MICKLEGATE

(58) House, No. 18. First-floor ceiling, early 17th century.

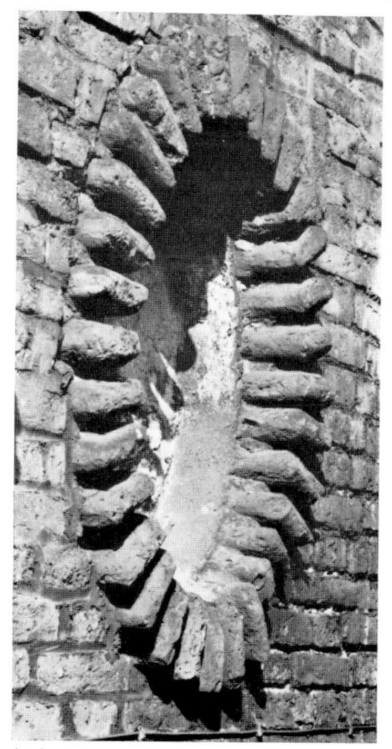

(63) House, No. 37. Rear wing. 17th century.

(89) House, Nos. 102, 104. Rear gable, late 17th century.

MICKLEGATE

Street front.

Back.

(58) Nos. 16, 18. *c.* 1600 and later.

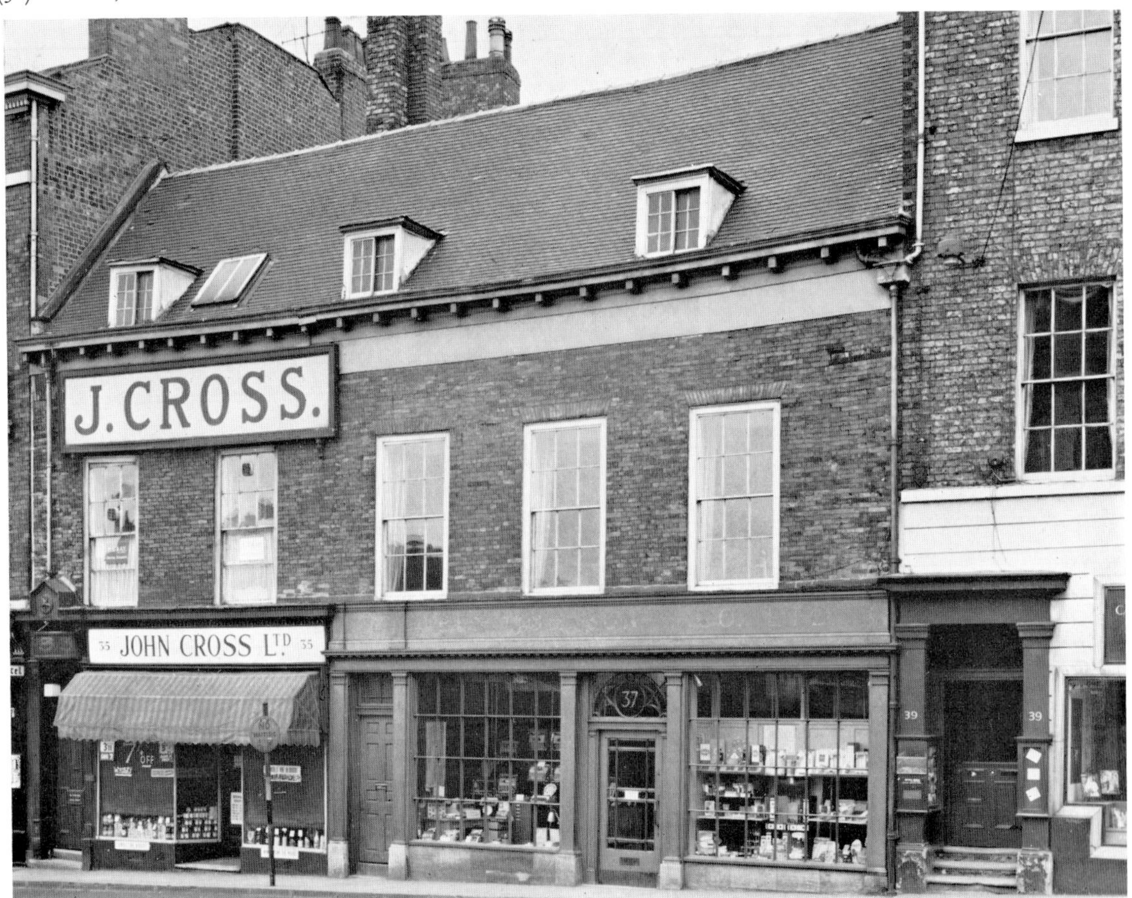

(63) Nos. 35, 37. Early 18th century.

MICKLEGATE

(57) Nos. 11, 13. *c.* 1740.

(64) Nos. 42, 44, 46, 48. 1747.

(83) No. 92. *c.* 1798.

(94) No. 114. Late 17th century and later.

MICKLEGATE

(100) Nos. 142, 144, 146. 17th century and later.

(89) Nos. 102, 104. 17th century.
MICKLEGATE

(63) Nos. 35, 37. 17th century.

(98) Nos. 134, 136. 1740.

(66) Garforth House, No. 54. *c.* 1757.

(81) Micklegate House, Nos. 88, 90. *c.* 1750.

(97) Nos. 128, 130, 132. *c.* 1755.

MICKLEGATE

(52) HOLGATE ROAD. Holgate House, No. 163. Summerhouse removed to The Mount School, 1774.

SKELDERGATE. John Carr's house, 1765–6. *(Photo. Northern Echo)*

(38) BISHOPHILL. Bishophill House, Nos. 11, 13. 18th century.

(14) SKELDERGATE. Albion Chapel, 1816, and John Carr's house, 1765–6. (*Photo. Northern Echo*)

HOUSES. 19th CENTURY

THE MOUNT. Nos. 144–36 (even). 1824.

DOVE STREET. S.E. side. 1827–30.

TADCASTER ROAD. No. 306. 1833.

THE MOUNT. No. 119. 1833.

THE MOUNT. Nos. 77, 79. 1831.

THE MOUNT. Nos. 104, 102, 100. 1807.

MOUNT PARADE. 1823–30.

MOUNT VALE. Nos. 147, 149, 151. 1823–7.

(74) MICKLEGATE. Nos. 70, 72. *c.* 1823.

THE MOUNT. Nos. 94, 92. 1821.

THE MOUNT. Nos. 132, 130. *c.* 1830.

MICKLEGATE. Nos. 84, 82, 80. 1821.

late 17th century are at the rear of 35/37 Micklegate (63); they are of four oversailing brick courses incorporating moulded brick (Plate 54). The houses of pre-1727 date now numbered 3–9 Micklegate (55) have a double band above the first-floor windows: the upper is of five courses with the lowest course of hollow chamfered moulded or rubbed brick, and the lower of three courses. Of this early 18th-century period, 86 Micklegate (80) has a first-floor band of four courses. Throughout the first half of the 18th century bands of oversailing brick courses, varying from three to five courses in depth, are to be found, and from the mid century plain ashlar bands are used on some of the better houses. Several houses have the distinctive arrangement of a broad band at the first-floor level, with a subsidiary band above serving as a continuous sill band to the first-floor windows. This Palladian feature appears or appeared on houses in York designed by John Carr, such as Castlegate House and Fairfax House, Castlegate, and his own house, 58 Skeldergate, demolished some years ago.

Two important houses, Micklegate House (81) and Garforth House (66), have this feature, but with a further band at the second floor and, associated with it, the unusual rustication described above. Micklegate House, built for John Bourchier of Beningbrough by 1752, has long been ascribed to Carr (George Benson, in the *Architectural Review*, II (Jan./Nov. 1897), 111, states that John Carr of York was the architect, but gives no documentary evidence). Should the designer of Micklegate House be established, to him also must probably be ascribed several other major York town houses with the same features, namely: the range of four houses, 39–45 Bootham, 1748; 53/55 Micklegate (65), *c.* 1755; 54 Micklegate (Garforth House) (66), 1757, and Peaseholme House, 1752. The last named house was built as a speculation by Robert Heworth, a joiner and carpenter, and it may well be that he was its designer. No. 56 Skeldergate (117) has much about its elevation that is found on houses certainly designed by Carr.

Cornices. Most eaves cornices of the period are of wood; only one of moulded ashlar (Middlethorpe Hall (163)) and a very few of brick remain. In the second half of the century bricks set on edge, or set obliquely, are used as cornices to subsidiary elevations. Cornices in association with parapets are exceptional at any date in York. A late 17th-century example with a shallow parapet and oversailing brick courses with some moulded brick incorporated is the range at the rear of 37 Micklegate (63) (Plate 54). On the side elevation at the rear of 5 Micklegate (55) is a heavy moulded and modillioned cornice of timber of late 17th-century type, with considerable projection from the façade. This type of cornice, but more moderate in scale, is used on houses of the early 18th century, such as on the front elevations of 3–9 Micklegate (55) and Bathurst House (80) (Plate 175).

The modillioned cornice is also employed on most of the more important buildings in the third quarter of the century: Micklegate House (81) of *c.* 1752 (Plate 177), 53/55 Micklegate (65) of *c.* 1755, and John Carr's residence of 1765 which formerly stood in Skeldergate; the same feature is used on the Bar Convent front (13) of 1786 by Thomas Atkinson (Plate 138). Garforth House (66), 1757, has a distinctive cornice treatment with a Roman-Doric entablature.

During the second half of the century the moulded wooden cornices with modillions and dentils gradually decreased in scale and projection. Typical of the cornices applied to buildings according to the stipulation of the City Council Order of 1763 (*see below*), to carry rainwater off the roofs by gutters, was the example of 1764 on the recently demolished houses, 16 and 18 Micklegate (58) (Plate 52). Holgate House (52) of *c.* 1772 has the modillions reduced to simple shaped angle-pieces of little projection, and 57/59 Micklegate (68) of *c.* 1783 further illustrates this tendency to reduce scale and projection of cornices.

Doorcases (external). No examples of 17th-century main entrances remain *in situ* in this part of the city, though it is rich in Georgian ones of all categories, from the mansion entrance to the more modest front doorway of the smaller house, inspired by one of the many copy-books available, and this despite wholesale destruction due to conversion of ground floors to shop and other commercial premises in the 19th and 20th centuries.

The entrances at Middlethorpe Hall (163), *c.* 1700, with bolection-moulded eared architraves in ashlar are stylistically late 17th-century (Plate 62); the doors and fanlights are later insertions. The early 18th-century entrance to 19 Bishophill (41) has a simple moulded wood surround with ears, a fanlight and a flat hood supported on consoles. Micklegate House (81), 1752, has possibly the finest Georgian entrance in York, standing upon stone steps with sweeping iron railings on either side. The freestanding Corinthian columns flanking the rusticated surround to the round-headed doorway support an entablature and triangular pediment (Plate 178). Of the Doric doorcases so popular throughout the country in the second half of the 18th century, probably the earliest dated example in York is the one remaining of a balancing pair to the Garforth House (66), 1757. There are similar doorcases to 106 Micklegate (90) and 118 Micklegate (95).

As the 18th century advanced there was a general tendency towards alteration of the proportions of entrances and the use of Adam-style applied decoration. The matching entrances to 57/59 Micklegate (68) of 1783 illustrate the fashion of the period, and so too does the simple entrance to No. 92 Micklegate (83), which is emphasised by raising the threshold on a flight of steps. Central emphasis is given to the Blossom Street front of the Bar Convent (13) of 1786 by the entrance and the first-floor window above it, but the design of the porch, with round-arched opening set in a rusticated surround framed by paired Doric columns supporting an entablature and triangular pediment, is rather old-fashioned. The only concession to current trends is the employment of close-set fluting to the entablature and tympanum.

Fenestration. No original late 17th-century windows remain intact; the few examples have been either blocked or reformed in later times or had hung sashes inserted in place of the wooden transomed and mullioned or casement windows. Of this period, however, a bull's-eye window formed with rubbed headers remains *in situ* in the rear wing of 35/37 Micklegate (63) (Plate 51). Possibly the earliest example of a hung-sash window in Yorkshire occurs at Middlethorpe Hall (163), *c.* 1700, where no fewer than twenty such to each of the two main elevations exemplify the new form of fenestration. The earliest sliding-sash window, commonly called the Yorkshire sash, is at 1 Tanner Row (120) (early 18th-century). Throughout the 18th century most window openings to the street fronts have flat arches formed with gauged rubbed bricks; some of the more modest houses have segmental arches of brick headers or cut stock bricks. Throughout the period, the rubbed brick voussoirs are single narrow bricks with saw cuts on the face simulating joints. The depth of the arches of the late 17th and early 18th century is generally shallower than in the mid-century and later, a maximum depth of five-and-a-half courses being reached at the Garforth House, 1757. Ashlar key-blocks are rare; only a few larger houses, Micklegate House (81), 53/55 Micklegate (65), and Acomb House (142), have them.

The early 18th-century window openings had tall narrow proportions, well illustrated by those on the two main floors of the Bathurst House (80) and the first floor of 3–9 Micklegate (55); both have windows three panes wide by six tall. By the mid-century the proportions had changed, with a marked decrease in the height, to three panes wide by four tall. This proportion was maintained until the end of the century.

Hung-sashes of the early 18th century had heavy glazing bars and the sash boxes were exposed, flush with the face of the wall. Most of them have been replaced at later times, but here and there some remain *in situ*, as those on the second floor of 11/13 Micklegate (57), and some on the top floor to the front of 3–9 Micklegate (55). The sashes in a group of houses of the mid-century are set back to leave 4½ in. reveals, but as late as 1783 Nos. 57/59 Micklegate (68) has almost flush moulded frames.

Round-arched windows only occur on staircases or in Venetian windows, as in Garforth House, Micklegate (66), and No. 56 Skeldergate (117).

Leadwork. Many good lead rainwater heads still survive on York buildings though the lead fall-pipes have very often been replaced in cast iron. A City Corporation Order of 15 June 1763 (YCA, M.17) states, 'That the spouts of the City's houses and places to be put up and fixed for bringing down the rain and other water pursuant to the Act of Parliament [3 Geo. III, c. 48] in that case made be provided by the City Steward by the Consent and approbation of the Alderman Wardens of the Ward in which such houses and places respectively are situate'.

The finest and earliest head is that to the Bathurst House (80); its survival, together with the cornice and eaves, is remarkable, since the house was raised by a storey in the early 19th century. It has a projecting centre section enriched with a winged cherub's head and swags and the initials of Charles and Frances Bathurst (Frances died on 24 January 1724) (Plate 81). The earliest example incorporating a date, 1752, in the part of the city under review is on the rear elevation of Micklegate House (81); the tapered angular form incorporating classical mouldings is typical of the general shape of rainwater heads of the first half of the 18th century. Characteristic of the later designs, to the end of the century, is the inverted bell shape; it is well illustrated by the dated examples, of 1763 at 67 Micklegate (71) and of 1783 at the rear of 57 Micklegate (68). The shape is simplified to a fluted bowl form in the early 19th century. Surviving fall-pipes from the first half of the 18th century are of square section, with the bracket-junctions enriched with decorative motifs or the owners' crests (Plate 81).

Internal Features

Doors and Doorcases. Good examples of doorways in the late 17th-century idiom survive at Middlethorpe Hall (163) of *c.* 1700 and the Queen's Hotel, Micklegate (55).

At the beginning of the Georgian period bolection-moulded panels to doors gave way to recessed and fielded panels, six to a door, as illustrated by the fine example in Acomb House (142) (Plate 67). Of the mid 18th century, there are many good examples of the carpenter's craft in major houses, such as 53/55 Micklegate (65) and Garforth House (66).

The influence of the Adam fashion was well illustrated by the interior doorcase to the saloon of 113 Micklegate (93) (Plate 66), now demolished, and a first-floor room at 53/55 Micklegate was refurbished in the late 18th century, the entrance from the landing having sunk panels to the architrave, enriched with applied arabesques and lion-mask paterae from the moulds of a York craftsman, Thomas Wolstenholme, one of a family of carvers and gilders. (For architrave mouldings, *see* Fig. 15.)

Fireplaces. Though many fine fireplaces of all periods of the 18th century remain *in situ*, none of a late 17th-century date exists in this part of the city. In the style of the late 17th century, however, is the chimney-piece in the dining room of Middlethorpe Hall (163) (Plate 200).

Angle fireplaces occur in three buildings of the early 18th century: the Queen's Hotel, Micklegate (55), in the room adjacent to the saloon (Plate 166), and in the rear wing of the Bathurst House (80). Although the first is compact and small in scale, the carved enrichment to the overmantel itself shows that it heats an important room. The second, in a rear wing, is clearly in a room of secondary importance. A third early 18th-century example, in association with heavy bolection-moulded panelling, is in the 'Blue Room' on the first floor of Middlethorpe Hall (163) (Plate 201). Small angle fireplaces appear later in the century in upper rooms and attics of modest houses and are not overtly dateable.

In the idiom of the first half of the 18th century is the fireplace in the ground-floor front room of Acomb House (142), with a bolection-moulded marble surround to the fireplace, a shelf moulded and enriched, and a panel in the overmantel with an eared surround terminating in volutes at the top (Plate 194). On the first floor of the same building, in probably the original saloon, is a fine carved fireplace and overmantel in a more rococo style (Plate 72). Examples of a more modest type of fireplace of the second quarter of the

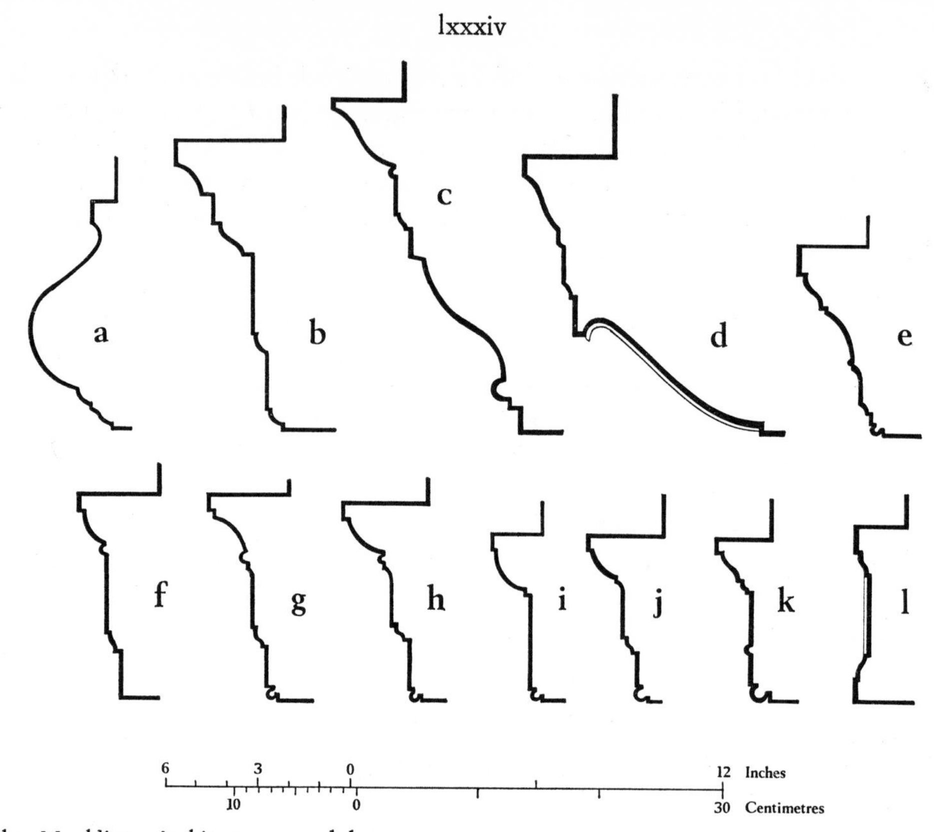

Fig. 15. Timber Mouldings. Architraves around doorways.

a, b. (163) Middlethorpe Hall, *c.* 1700; c, d. (55) Queen's Hotel, Nos. 7, 9 Micklegate, before 1727; e. (80) Bathurst House, No. 86 Micklegate, *c.* 1727; f. (81) Micklegate House, Nos. 88, 90 Micklegate, 1752; g, h. (66) Garforth House, No. 54 Micklegate, 1757; i. (65) Nos. 53, 55 Micklegate, *c.* 1755; j. (117) No. 56 Skeldergate, *c.* 1765; k. (68) No. 57 Micklegate, 1783; l. (65) No. 53 Micklegate, *c.* 1800.

18th century are found in smaller houses, such as 12 Front Street, Acomb (136), and 122 Micklegate (96) of *c.* 1740, and in the less important rooms of larger houses. They are of white marble, with simple jambs generally with fielded panels, a segmental arched opening with a key-block and a plain shelf. The spandrels are usually filled with fielded panels, and the key-blocks are decorated with a variety of motifs, such as a scallop shell or conventional foliage.

Many fine examples dating from the mid 18th century survive; they reflect the use of builders' pattern-books by the local craftsmen and are, therefore, following fashion rather than setting it. The chimney-piece in the ground-floor front room of 118 Micklegate (95) (having survived in shop premises recently reconverted to living accommodation) is of *c.* 1742 (Plate 73). One of the fireplaces from the saloon of Micklegate House was removed some years ago and reused in the Princess Victoria Room of the Treasurer's House, York; it is a good example of the style of the mid 18th century (Plate 71). All fireplaces of this category, with an overmantel, are in rooms that have some kind of panelling. In rooms that are without panelling where the walls were covered originally with paper or fabric, the fireplaces are much reduced in importance, though the decoration may be similar to that on the more important type of chimney-piece found in the same house.

Garforth House of 1757 (66) shows the variation in design of fireplaces in one house according to their situation. In the front room on the ground floor, which was probably the dining room, the design is bold and impressive (Plate 71); a first-floor room above has a fireplace of delicate design with rococo arabesques, suitable for a ladies' withdrawing room (Plate 73) and an adjacent room has a fireplace of similar character but less elaborate; on the second floor the fireplaces are very much simpler showing the economy of workmanship which sufficed for the furnishing of secondary rooms (Plate 71).

The general decoration of chimney-pieces with carved arabesques remained in fashion until the late 1760s, when the new, more delicate style introduced by Robert and James Adam had reached York. The Adam influence in the last quarter of the century is seen in applied composition decoration on fireplaces in houses of all categories, and many an earlier building has had an original fireplace replaced with a 'modern' one in the fashionable style. Local craftsmen using copy-books even produced their own moulds for casting composition ornaments.

The finest example of the conventional Adam-style fireplace complete with firebasket is in Middlethorpe Manor (164) (Plate 75). Closely dated examples are to be seen in Holgate House of *c.* 1773 (Plate 75) and 57/59 Micklegate (68), a house of 1783. Those in the latter have ostensibly Adam urns, swags and transverse flutings, but they are most probably by a local designer–craftsman and not direct copies of Adam motifs. This is suggested by, for example, the decoration in the central tablet on the freize of the saloon fireplace, an urn form with lush growth of foliage, which is too undisciplined to be an accurate copy of an Adam design. The making of applied composition decoration for the enrichment of fireplaces, doorcases, etc., by local craftsmen more or less in the Adam fashion is well illustrated by the work of the Wolstenholme family of carvers and gilders, notably Thomas Wolstenholme (1759–1812), which can be identified in 118 Micklegate (95), 53/55 Micklegate (65) and in the range added to Middlethorpe Hall (163).

Iron grates cast at the Carron works near Falkirk to Adam designs are numerous, though many that matched a fireplace surround have been removed in the cause of fuel economy, to be replaced by a slow-burning grate or other form of heating. One of the most complete surviving *ensembles* is in the flat at No. 5 Micklegate (55), which also has applied pewter and cast-lead decoration on the surround (Plate 61).

Panelling. Bolection-moulded panelling, so characteristic of the late years of the 17th century and the early 18th century, is to be found in a number of houses. The finest example, in oak, is in the dining room at Middlethorpe Hall (163) (Plates 200–2). This elegant room, which contains a fireplace and shallow round-headed niches all with flanking Ionic pilasters and bold enriched entablatures, has an unusual arrangement of the two adjacent panels by the corners at one end of the room. The surface plane is advanced slightly and emphasised by extending the main cornice down to an enriched entablature over the sunk fielded panels. This contrivance is introduced to balance the entrances at the opposite end of the room. Panelling of early 18th-century date of equal quality lines a suite of three rooms on the first floor of the Queen's Hotel, Micklegate (55) of *c.* 1720. Similar boldly projecting bolection-moulded panels occur in the first-floor room to the rear of 122–126 Micklegate (96) of *c.* 1740 (Fig. 16) and the front room on the first floor of 35 Micklegate (63). Acomb House (142) of *c.* 1745 has a fully panelled front room on the ground floor, with a plain dado and tall bolection-moulded panels above the rail and an entablature all in pine (Plate 194). This is probably the latest example of bolection-moulded panelling in the area under review.

By the fifth decade of the century simpler forms of panelling were being developed. In the saloon of 118 Micklegate (95), there is a plain dado with tall panels above; the moulded surrounds project only slightly from the wall surface and the fielded panels are recessed (Fig. 16; Plate 183). A second, more modest type of panelling is employed in a house of 1747, Nos. 42–48 Micklegate (64); the walls are lined with softwood boarding and a simple raised moulding is fixed over it to form rectangular-shaped panels (Fig. 16; Plate 69). In houses of the following decade panelling is becoming unfashionable, and in some of the major houses, such as the Garforth House, Micklegate (66) of 1757, only the front room on the ground

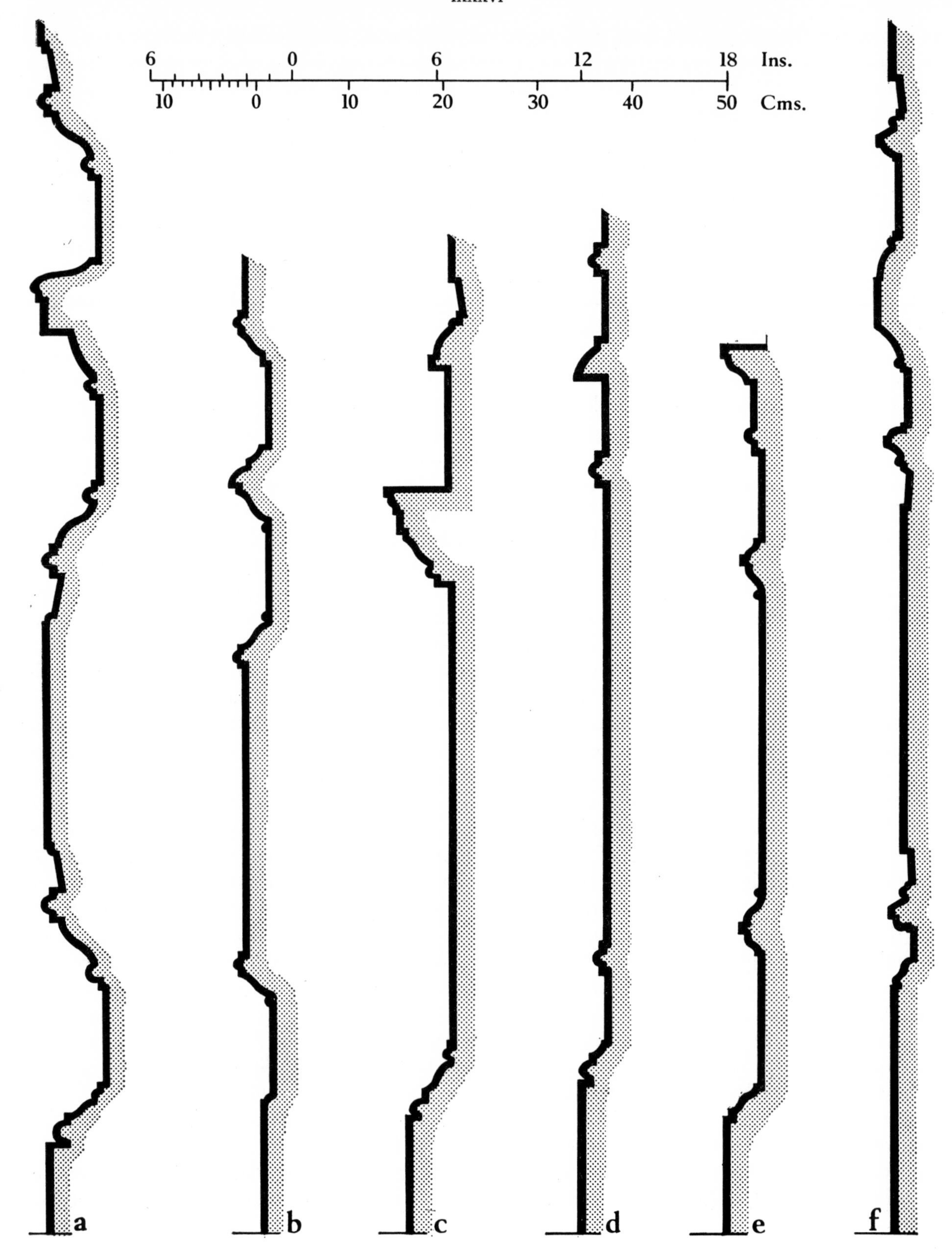
6
0
6
12
18
Ins.
10
0
10
20
30
40
50
Cms.
a
b
c
d
e
f

floor, probably the dining room, is fully panelled, with a plain dado and tall sunk fielded panels above with moulded surrounds (Plate 71). Elsewhere in this house panels with carved enrichments to the surrounds are used in the dado, and the walls above are plain, presumably intended to be papered or lined with fabric. Micklegate House (81) of 1752, the largest of the town houses in this part of the city, originally had the saloon panelled, but the panelling has since been removed and some of it reused in the Princess Victoria Room of the Treasurer's House; in this the moulded panel-surround has been abandoned and the panelled effect is achieved simply by raising the panels very slightly from the background (Plate 71). The decoration of 53/55 Micklegate (65) of *c.* 1755 is archaic; the front ground-floor room has sunk fielded panels, the larger panels having raised eared surrounds enriched with Greek fret ornament, and similar panels occur in the staircase hall. In a back room the walls are lined with pine planks, and panels are delineated by applied mouldings (Fig. 16).

Plasterwork. No elaborate late 17th-century or early 18th-century plasterwork is to be found in the houses in the area under review. The ceiling over the staircase of the Bathurst House (80) has a plain cove above the deep cornice (Plate 90) with a simple geometrical design formed with raised mouldings. In the eastern house of the Thompson pair of houses, 3–9 Micklegate (55), of *c.* 1720, the ceiling over the stairhall, from which the original staircase has been removed, remains *in situ*; it is elliptical on plan with a heavy modillion cornice with some enrichment, the spaces between the modillions being enriched with well modelled fruit, floral motifs, and a family crest; the cove above is quite plain (Plate 165). The finest examples of the plasterer's art in York date from the middle of the century; possibly the best is at Bishophill House (38). In the first-floor saloon, which has a large apsidal bay at the eastern end (Plate 61), the decoration of the ceiling is rococo and includes an abundance of finely modelled foliage, fruit and flowers (Plates 154, 155); it bears a close resemblance to a ceiling by Giuseppe Cortese at Newburgh Priory in the North Riding.

Micklegate House (81) of 1752 has lost most of the plaster enrichment of the ceilings in the main rooms; the fragments that remain in the saloon are rococo and incorporate animal figures and profile heads (Plate 77). The ceiling over the main staircase remains intact; it is in a severer geometrical style than that in the saloon, but the free-flowing arabesque enrichments in the compartments are again in the rococo fashion. Busts of Shakespeare and Newton (?) within roundels are incorporated, and stylised 'sea-god' masks appear in the spandrels (Plate 180). The Garforth House (66) of 1757 and 53/55 Micklegate (65) of *c.* 1755 also contain fine displays of enriched plaster ceilings (Plates 169, 180).

Examination of this group of mid 18th-century ceilings reveals no obvious common feature, such as the use of the same moulds. If it be assumed that this means that no single craftsman had a monopoly, then the same would demonstrate the ready availability of highly accomplished artist craftsmen at this period to accept commissions in York. The decline in the building of larger town houses in the city during the 1770s shows in the comparative dearth of plasterwork in the Adam style. One ceiling in Acomb House (142) is clearly influenced by Adam decoration in the general design, though none of the motifs, apart from the radial treatment of the centre circle, is found precisely in work known to have been designed by Robert Adam (Plate 77). (J.E.W.)

Staircases. Only two staircases dating from before 1650 have been found, and only one of these is *in situ*;

Fig. 16 (opp.). Timber Mouldings. Wall Panelling.

a. (55) Queen's Hotel, Nos. 7, 9 Micklegate, before 1727; b. (96) Nos. 122, 126 Micklegate, *c.* 1730; c. (95) No. 118 Micklegate, *c.* 1745; d. (64) No. 48 Micklegate, 1747; e. (65) Nos. 53, 55 Micklegate, *c.* 1755; f. (94) No. 114 Micklegate, early 18th century.

for convenience they are included in the following account. Staircases dating from after 1800 are described elsewhere (p. ciii).

In the 'Back Part' behind 26–28 Micklegate (61), now No. 32, part of an early 17th-century staircase survives; it has an elaborately carved newel-post and turned symmetrical balusters. The pre-1650 stair at the Old Rectory, Tanner Row (122), was brought from elsewhere, and only the carved newel-posts and handrails are old.

There are about a dozen staircases of the later 17th-century type, but some do not appear to be in their original positions. Three are in 18th-century houses, where they are reused as balustrades in attics or upper flights. It is unlikely that any is earlier than *c.* 1650. The latest dated example is at Middlethorpe Hall, a stylistically advanced house built shortly before 1703, but it would seem probable that in more modest buildings the same type of staircase continued to be used during the earlier part of the 18th century. Characteristic features of the type are the balusters of heavy bulbous shape, closed strings and square newel-posts, usually with attached half-balusters. The stairs are generally arranged in short flights with quarter-landings about an open well. The earliest example, stylistically, is at 68 Micklegate (72) (Plates 82, 176; Fig. 17a), which is constructed of plain but massive timbers, the heavy string being especially notable. All the other examples show a later development in which the balusters are slightly more slender, and more complex in profile, as in the secondary stair at Middlethorpe Hall (163) (Plate 83; Fig. 17b) and at 48 Skeldergate (112) (Fig. 17c), just after 1700. In the later staircases of this type the newel-post does not rise above the level of the handrail of the ascending flight but has a moulded cap in continuation of the handrail. The thick string at 68 Micklegate (72) is exceptional; in all the other, later, examples the string is much thinner, but since a broad housing for the balusters was still necessary the string capping had to be wide and so projected outwards from the string, sufficiently far to have a deep moulding on the under side.

Two examples of a different type of late 17th-century stair are in 95 Micklegate (85) (Fig. 17d) and Middlethorpe Grange (162) (Plate 83). These have balusters of thin planks cut to the silhouette of turned shafts.

The turned balusters in all but the very earliest 18th-century staircases are very different from those of the late 17th century. The main feature takes the form of a classical column standing on a round or square 'knop' above a 'pedestal' of various contrasting shapes, the shapes being more irregular in the earlier examples; by the second half of the century the form had become more standardised and simpler. Again, these balusters are not as heavy as those of the late 17th century and do not normally exceed a maximum breadth of 2 in. as against 3 in. in the earlier period. Two staircases with this type of baluster are probably earlier than 1700. That in 114 Micklegate (94) (Plate 84), of oak, has extremely thin square knops, very similar to those at Holy Trinity Church, Goodramgate, recorded to have been installed in 1675. The other is at 102 Micklegate (89). An early 18th-century example with a very thin knop was at 4–6 Micklegate (54) (Plate 161), now demolished.

The principal staircase at Middlethorpe Hall (163) (Plate 202) is very much grander than the others considered here. It dates from *c.* 1700 and is advanced in style, for the balusters have fluted shafts and carved enrichment to patterns which are not found in the city until a later period. It is the earliest stair without a closed string, though it still has, in the older manner, three flights about an open well with square newel-posts at the angles. The treads are designed in imitation of stone cantilevers: they have a considerable overlap to allow sloping bearers to be concealed within them. In York this form of 'false-cantilever' stair preceded the open-string stair.

There are very few staircases that can be placed with certainty within the period 1700–30. Nos. 3–9 Micklegate (55) and 86 Micklegate (80) (Plates 82, 84) were both built about 1720, and 10 North Street (101) (Fig. 17f) (recently demolished) was probably a little earlier. This last provided a good illustration of the early 18th-century treatment of the swept handrail. The staircase at 86 Micklegate (80) is the earliest in which the balustrade is without newel-posts and is curved at the corners. The balusters still have the square knop and complex shape below, but they show a tendency towards being a little less robust; there are three to each tread, two with fluted shafts and the third with two interlacing twisted stems, a form almost unique in York.

During the second quarter of the 18th century the open or cut string was introduced. The earliest surviving staircase with it is probably that in 122–126 Micklegate (96) (Plates 84, 87; Fig. 18m), probably of about 1740, though difficult to date accurately because it is an insertion into a 17th-century building. Rectangular cheek-pieces on the face of the string maintain an illusion of monolithic treads, but the overlap is now reduced to a minimum. The balusters of this staircase have a complex outline below the square knop, though the shafts are more slender and the details of caps and bases are more refined than those of the two stairs of *c.* 1720 discussed above. It also exemplifies the tendency in the middle of the century towards greater density in the spacing of the balusters; there are three to each tread, and the space between them is distinctly less than the width of the individual balusters. In this stair all the balusters have plinths of equal height, and the necessary adjustment in the total height of the balusters, due to the slope of the handrail, is made in the shaft. This treatment also occurs in the adjoining house, 118 Micklegate (95) (Plate 87), but in nearly all other staircases with open strings the plinths are graded in height to allow the turned parts of the balusters to be kept equal. This produces a more even line in a rising balustrade, which is particularly noticeable with closely set balusters.

From the period *c.* 1750–60 there are several staircases that can be considered to be among the finest in York. They represent the culmination of the early 18th-century development of the stair as one of the principal ornaments of the house; attention is focused mainly on balustrades of varied design with the use of twisted and fluted shafts, carved enrichment, etc. The opportunity for their creation was given by the erection at this time of several large town houses for county families. The principal ones are 88–90 Micklegate (81) (Plate 182), 54 Micklegate (66) (Plate 88), and 53–55 Micklegate (65) (Plate 171). To these may be added Bishophill House (38) (Plate 82), a smaller and slightly earlier building but with a good stair, and also 40 Blossom Street (47), demolished in 1965. Before this period perhaps only Middlethorpe Hall and 86 Micklegate have staircases which are comparable in their spaciousness and elaboration. These staircases rise only to the first floor, to which they provide an imposing approach. The secondary staircases in these houses are more modest but are nevertheless good and up-to-date examples of the kind usually found in middle-sized houses.

Nos. 88–90 Micklegate (Micklegate House) (81) (Fig. 18o) has probably the best of this group of ornate staircases. There is more carved enrichment than on any other, and the three balusters on each step have plain, fluted and twisted shafts respectively. No. 54 Micklegate (Fig. 18p) is only a few years later in date than the foregoing, but the staircase has rather more elegance. This is due partly to the lack of the broad stepped plinth-blocks used in the other staircases in this group, but also the balusters are thinner and their mouldings are extremely delicate for work in pinewood. The treads are true cantilevers in timber, connected by short risers, which greatly contribute to the lightness of the stair. A later example of the cantilevered timber tread occurs at Holgate House (52) (*c.* 1775).

Of the 18th-century stairs examined in this area of York more than half have balusters which, as a result of increasing standardisation from *c.* 1740, by *c.* 1760 may be grouped into two types. Of the two, the one which has a column above the knop and a bulb-shaped feature below is the earlier, though none can be firmly dated before the example of 1747 at 32–36 Blossom Street (46); previous examples appear transitional, as in the upper flights of 134–136 Micklegate (98) (Plate 85) where the balustrade has rather sturdy shafts and square newel-posts with attached half-balusters, a pattern more common earlier in the century.

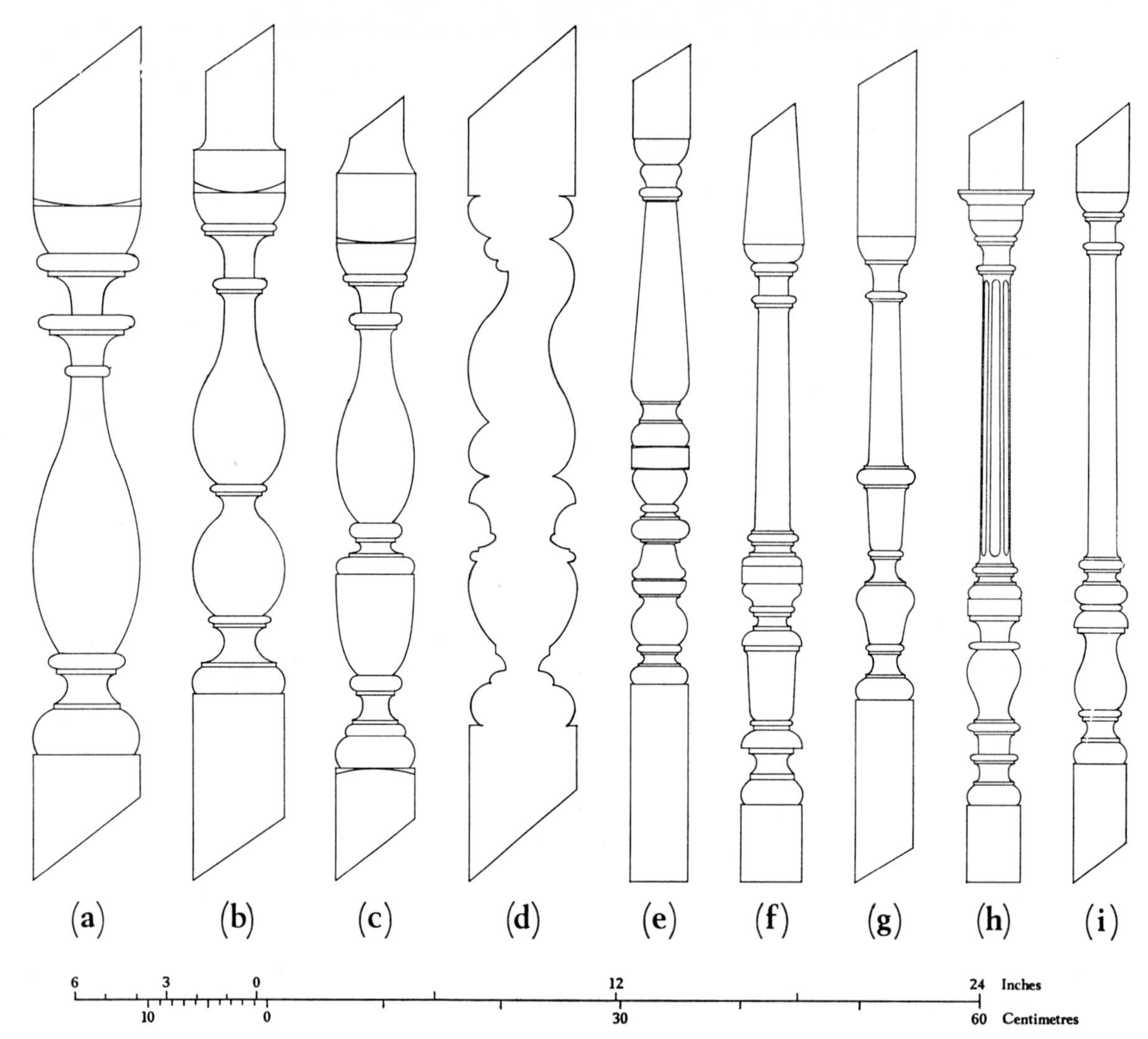

Fig. 17. Staircase Balusters.

a. (72) No. 68 Micklegate, mid 17th century; b. (163) Middlethorpe Hall (servants'), *c.* 1700; c. (112) No. 48 Skeldergate, *c.* 1700–10; d. (85) No. 95 Micklegate, late 17th century; e. (54) Nos. 4, 6 Micklegate, early 18th century; f. (101) No. 10 North Street, early 18th century; g. (118) Plumbers' Arms, No. 61 Skeldergate, early 18th century; h. (80) Bathurst House, No. 86 Micklegate, early 18th century; i. (75) Nos. 73, 75 Micklegate, early 18th century.

At 113 Micklegate (93) (Plate 86; Fig. 18j) the shafts have deep 'capitals' which are typical of 1720–30, though it is not possible to conclude that this stair is so early.

The second type utilises the 'urn' motif below the knop. The earliest dated example in fully developed form is of 1757, in the secondary stair at 54 Micklegate (66) (Plate 88; Fig. 18p). During the last quarter of the century all balusters are much alike, the only differences are in the proportions and the multiplicity of small mouldings. Dated examples, differing but little, are of *c.* 1777 at 26 North Street (103), 1783 at 57–59 Micklegate (68) (Fig. 18q), and 1789 at 22–26 Blossom Street (45). Most of the staircases of this period have an open string with two balusters to each tread, and the cheek-pieces on the face of the string, previously rectangular, after 1760 are usually shaped. The use of the closed string is mostly restricted to

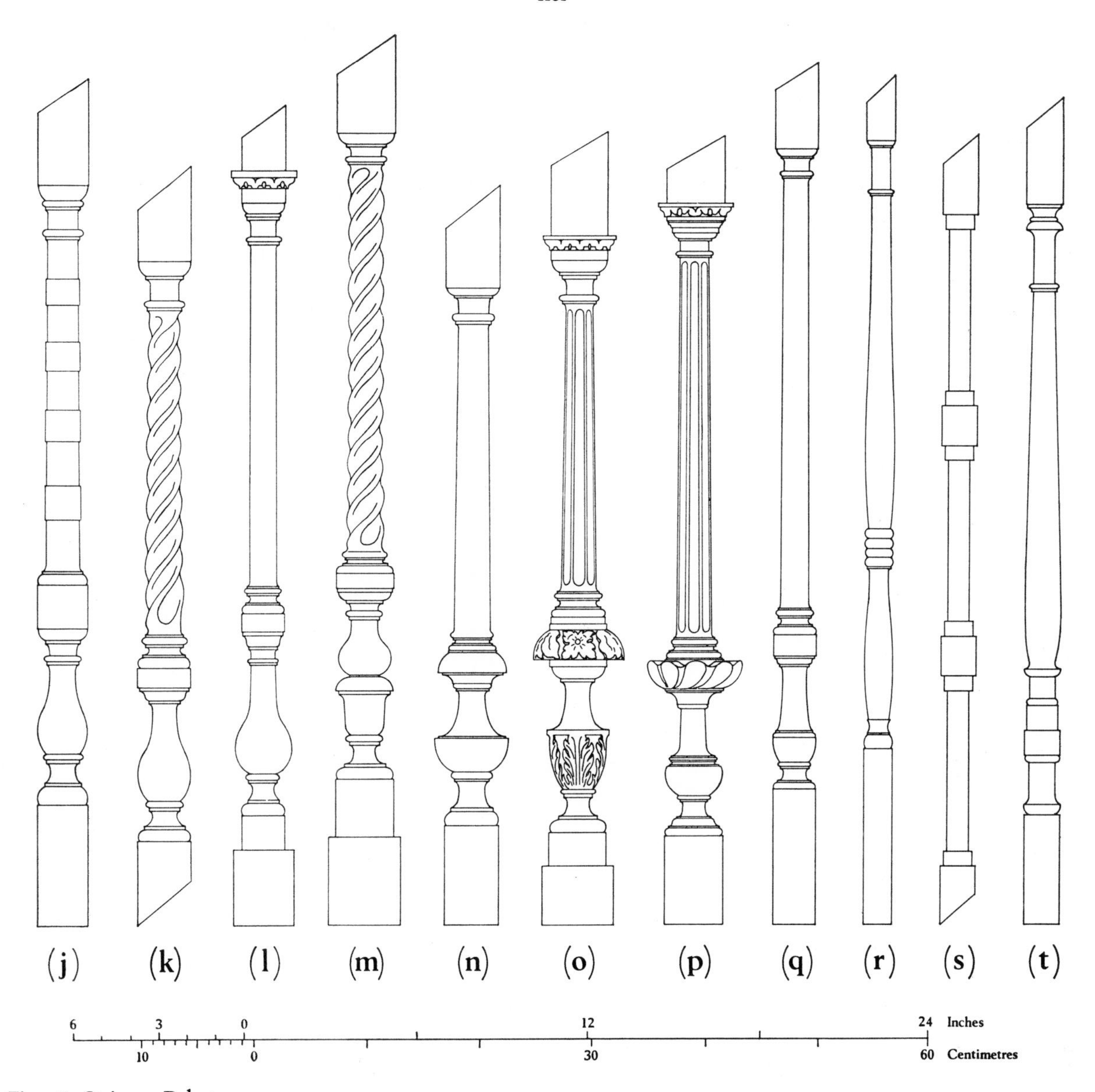

Fig. 18. Staircase Balusters.

j. (93) No. 113 Micklegate, *c.* 1740; k. (59) No. 17 Micklegate, 18th century, before 1750; l. (98) Nos. 134, 136 Micklegate, 1740; m. (96) Nos. 122, 126 Micklegate, *c.* 1738; n. (41) No. 19 Bishophill, early 18th century; o. (81) Micklegate House, Nos. 88, 90 Micklegate, 1752; p. (66) Garforth House, No. 54 Micklegate, *c.* 1750–7; q. (68) Nos. 57, 59 Micklegate, 1783; r. No. 6 South Parade, *c.* 1825–8; s. (112) No. 48 Skeldergate, *c.* 1820; t. (115) No. 53, Skeldergate, *c.* 1840.

staircases in small houses of cheaper quality or where the space is restricted and winders are necessary.

The urn-type baluster with square knop is found occasionally in early 19th-century staircases, though with the very slender proportions typical of that period; nevertheless the general change to the Regency types of slender turned and square balusters occurs quite suddenly about, or just before, 1800.

One staircase of interest which does not fit into the general 18th-century sequence is in 83 Micklegate (78), a small house of early or mid 18th-century date. To economise space, the stair follows the much earlier practice of having all the treads housed into a single central newel-post which rises through three floors.

(D.W.B.)

Three Micklegate Houses ascribed to John Carr

Three houses in Micklegate: Nos. 53/55 (65) (Plate 170); No. 54 (Garforth House) (66) (Plate 172), and Nos. 88, 90 (Micklegate House) (81) (Plate 177), are clearly by the same designer, who must also have been responsible for Peaseholme House and Nos. 39–45 Bootham across the river. The group can be closely dated to the years 1748–57 and probably to 1748–54. Among the distinctive features shared is a type of masonry quoin consisting of plain ashlar blocks in alternate sizes, the larger, square on plan, oversailing the smaller (Fig. 14). This treatment is very unusual, but significantly it occurs at Arncliffe Hall in the North Riding completed by 1754 to the designs of John Carr (APS, *Dictionary of Architecture*).

Both Micklegate House and Garforth House have long been credited to John Carr (G. Benson in *Architectural Review*, II (1897), 111; J. W. Knowles, 'York Artists', MS. in York City Library). Carr had settled in York before October 1751, when he bought a property in Skeldergate, describing himself as 'mason' (YCA, E.93, 268), and he was working for the Garforth family at Askham Richard Hall about 1750, on the evidence of his own reminiscences (R. Davies in *YAJ*, IV (1876), 212a). It would be a reasonable assumption therefore that the family would again turn to him for the building of their York town house within the next few years. Carr later remodelled Wiganthorpe Hall for the same family (Colvin, 125).

The group of five buildings, associated by common stylistic features, do not display the sophisticated Palladianism of Carr's certain works of later date in York: No. 47 Bootham (1753–4); Fairfax House, Castlegate (1755–62); and Castlegate House (1759–63). A certain solidity of massing does, however, connect them with the design of his own house in Skeldergate, built in 1765–6 (Plates 56, 57), and with some of his other works. It is probable that they constitute an early period in his stylistic development belonging to the time when he was a master mason rather than an architect, and that in later life he avoided claiming them as his own. The list of Carr's York town houses in APS, *Dictionary of Architecture*, provided about 1850 by his successor in practice, J. B. Atkinson, is improbably short, leaving an otherwise inexplicable void in his career at the very period when he had begun to win distinguished clients. It is possible that Atkinson omitted houses of that early phase, knowingly or in ignorance, because Carr himself had regarded them as unworthy of his later reputation when he was a renowned architect and an accredited representative of the Palladian school.

(J.E.W.)

Houses 1800–50

Until 1800 development outside the city walls was mainly confined to Bootham and Blossom Street. During the period of great expansion between 1800 and 1850, building necessarily beyond the restricted area enclosed by the walls resulted in the creation of comparatively extensive suburbs. At first, isolated houses for professional men built alongside the main roads created sparse ribbon development, which was soon followed by more intensive development comprising housing in specific areas for railway workers and speculative housing schemes and estates.

Little of the early 19th-century housing was of high quality but none was so poor as to deteriorate into slums. That in the Holgate Road area was of working-class type with rarely a servant in the house, so too was that in the Nunnery Lane area, but the houses in Blossom Street, Park Street, The Mount and Mount Vale were homes of the servant-keeping class (F. Seebohm Rowntree, *Poverty, A Study of Town Life* (1901), frontispiece map).

Sources for dating are plentiful. The newspapers, the *York Courier* and the *Yorkshire Gazette*, provide information chiefly in the form of advertisement of newly built houses for sale or to let. W. Hargrove's *History of York* mentions properties developed before 1818, the date of its publication, and Directories give a terminal date for the formation of streets and the occupation of houses in them. Interest in archaeology provides valuable dating evidence, for the digging of foundations preparatory to building often produced artefacts duly noted in the Proceedings of Societies, Museum catalogues or the newspapers. Property deeds and surviving parish rate books also provide information.

A good map of York published by E. Baines in 1822 is the earliest to show buildings in the modern manner, and the Ordnance Survey map published in 1852, but surveyed in 1850, is invaluable. It is to a scale of 1/1056 and covers the whole city in twenty-one sheets, being large enough to show seating inside churches and the shape of houses, and affords contemporary evidence of the houses existing at the terminal date of this Inventory.

The houses may be considered in the five groups described below.

(1) Large detached houses are few in number; two call for mention. An impressive town house Nos. 77, 79 The Mount (Plate 59) was built in 1831 on the corner of Park Street for Alderman Dunsley; the architect was Peter Atkinson (II). The building has been much altered in conversion to a hotel. Mill Mount House, now Mill Mount School (Plate 196) was erected in 1850 for Charles Heneage Elsey, Recorder of York, to the designs of J. B. & W. Atkinson.

(2) Villas, medium-sized detached houses for the upper middle class, usually with a symmetrical front, appear in some numbers after 1830. Mount Terrace House, similar to the villas but attached to the end of a terrace, was completed by 1827, and typical groups of villas were erected on The Mount and the Tadcaster Road between 1830 and 1835, and Holgate Villa, designed for a senior railway official, probably by G. T. Andrews, was up by 1846 (Directories).

(3) Groups of two or three houses designed to form one more or less symmetrical architectural composition occur all through the early 19th century. Nos. 100, 102, 104 The Mount, completed in 1808, make no attempt at symmetry in the placing of the doorways but the middle house is stuccoed and has rusticated quoins to mark it as the centre-piece of the group. Nos. 130, 132, 134 (*c.* 1830) presented a similar emphasis on the central house but the effect has been lost through drastic alterations to No. 134; the central house has first-floor windows recessed under segmental arches. Nos. 92, 94 The Mount (1821) form an almost symmetrical pair with the doorways placed together to form an impressive central focus. Nos. 39, 41 Micklegate built in 1835 by J. B. & W. Atkinson and Toft Green Chambers (*c.* 1845) are symmetrical pairs with doorways at the outer ends of the front, and Nos. 120, 122 Holgate Road (*c.* 1840) have the doorways in recessed side wings. One of the Micklegate pair became the Atkinsons' office.

(4) Rows of terrace houses on a common frontage and with a common roof line, were built from 1823 onwards, but each house is of individual design. Much of Holgate Road was developed in this way from 1823 and in the same year Mount Parade was projected; houses on Mount Parade were for sale in 1828 and Pigot's Directory for 1829 shows six of the houses occupied by gentry. The adjoining Mount Terrace was begun in 1827 and a house there was described as having 'two kitchens, a servants' hall, drawing room, two parlours, and four lodging rooms with attics'. The development was of superior middle-class type and the houses are of different designs with great variety of detail. Larger, rather featureless terrace houses followed in Park Place and Park Street, off Blossom Street, *c.* 1835, also separately designed. More successful were terrace houses built on The Mount after 1840.

(5) Terrace houses built to a uniform design form the largest part of the housing of the period. Such terraces had been built in York from mediaeval times and examples survive from the late 17th century at Precentor's Court and from *c.* 1745 in New Street and St. Saviourgate. These last were more imposing

buildings than most of the early 19th-century terraces, of which the first was probably in St. Mary's Row, Bishophill, later called Victor Street, and was put up by Thomas Rayson in 1811 (E.96, f. 169, and Deeds); it has now been demolished. Some of the small terrace houses in Albion Street, off Skeldergate, were mentioned as 'lately built' in 1818. These were small houses erected as investment property for letting. A terrace of three larger houses, Nos. 78–84 Micklegate, was erected by Peter Atkinson junior in 1821. Numerous terrace developments followed; among the earlier ones the largest and most imposing was South Parade which was being built with finance provided by forty subscribers in 1825. Most of the houses were occupied by 1829 (Pigot's Directory) and a number of well-known people lived there, including W. J. Boddy, the artist. From 1846 terraces were being erected between Holgate Road and the railway to house railway staff of all grades. G. T. Andrews was probably the architect chiefly responsible, as he controlled railway building at the time.

On 8 September 1850 the Town Clerk was directed to issue a notice that plans of new buildings must be submitted to the local Board of Health and on 27 January 1851 the City Surveyor was instructed to file such plans (York City Library, Council Minutes IV). Houses being built in 1851 in the area covered by this volume and the persons who submitted the plans were: *Cambridge Street*, Mr. Mosley; *Clementhorpe*, seven houses, Mr. Nicholson; *Holgate Lane*, Mr. Moseley, Henry Todd; *Holgate Road*, J. C. Cooke; *The Mount*, Mr. Coleman; *Mount Vale*, Richard Snowdon, James Guy; *Nunnery Lane*, eight cottages, R. Bainbridge; *Railway Street* (then Hudson Street), J. Brown, Noah Ackroyd, two houses and warehouse, Mr. Varvill, Mr. Young, six houses, Mrs. Eskelby, two houses, John Brown; *Rougier Street*, ten cottages, Mr. Jackson, and *Tanner Moat*, two cottages, Edward Calvert.

Architects and Contractors

The firm of architects founded by John Carr continued to flourish after 1800. *Peter Atkinson* II (1776–1842/3), son of John Carr's assistant Peter Atkinson I, designed Nos. 78, 82, 84 Micklegate in 1821 and Nos. 77, 79 The Mount in 1831/2, and his sons, *John Bownas Atkinson* (1807–74) and *William Atkinson* (1810/11–86), had a flourishing practice. In this area they designed Nos. 39, 41 Micklegate (1835), No. 17 Bridge Street (1837), a warehouse in North Street (1837/8), No. 2 Bridge Street (1842), four houses in Skeldergate (1843), Nos. 43, 45 and 47 Trinity Lane (1846), a warehouse on the Queen's Staith (1849), Mill Mount House (before 1850), and St. Paul's Vicarage (1850/1). (Mill Mount House is called Mill Field House in the architects' records.)

George Townsend Andrews (1804–55), as chief architect to the Railway, probably controlled the building of Cambridge Street, Mount Ephraim and Oxford Street, with the terraces in the vicinity, all *c.* 1846; and Holgate Villa, Holgate Road, had a staircase like one in the Railway Station by him. He definitely rebuilt St. Catherine's Hospital (1834/5) and 21 Blossom Street.

The most successful contractor was probably *Thomas Rayson* senior (1763/4–1836), who was witness to Peter (the elder) Atkinson's will (27 April 1805) and certainly worked with Peter's grandsons. He built some houses called St. Mary's Row, Bishophill (later Victor Street) in 1811, perhaps erected South Parade where he lived in No. 16 from 1828 to 1836, and certainly built No. 7 Park Street for himself in 1836.

Fig. 19 (opp.).
(1) Mill Mount, 1850.
(2) 100, 104 The Mount, 1807–8.
(3) 92, 94 The Mount, 1820–1.
(4) 84 Micklegate, 1821–2.
(5) 120 The Mount, 1842–3.
(6) 13 Clementhorpe, 1823.
(7) 136 The Mount, 1824.
(8) 116 The Mount, *c.* 1840.
(9) 147 Mount Vale, 1827.
(10) 9 Albion Street, 1815–20.
(11) 12 Cygnet Street, 1846.
(12) 6 Mount Terrace, *c.* 1827.
(13) 306 Tadcaster Road, 1833.
(14) 20 Mount Parade, 1834.
(15) 65 Acomb Road, *c.* 1828–9.
(16) 121 The Mount, 1833.
(17) 117 The Mount, 1833–4.

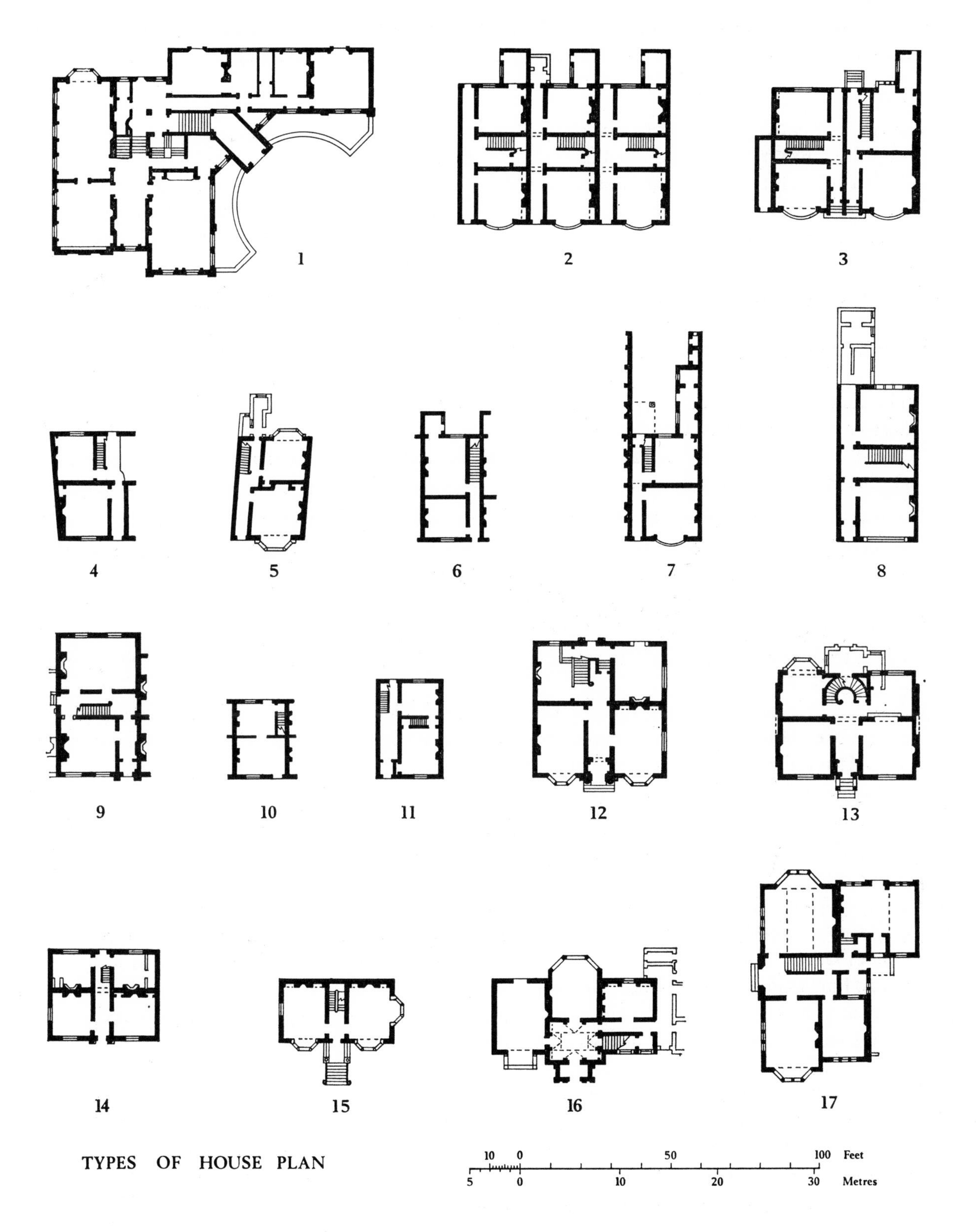

TYPES OF HOUSE PLAN

George Eshelby (*c.* 1788–1861), joiner and builder, put up 302 Tadcaster Road in 1833 and undoubtedly was the contractor for the other Mount Villas, 300, 304 and 306 Tadcaster Road.

The Plan

The types of house plans described below follow no historical sequence; different types were in use at the same time. Examples are illustrated in Fig. 19, p. xcv.

In a class by itself is Mill Mount House (now Mill Mount School) (Fig. 19: 1) with the entrance placed diagonally across the re-entrant angle between two wings. Most of the smaller detached houses, or villas, are symmetrical and have a central entrance passage and staircase with rooms to front and back on each side:

(i) The passage leads to a wider stairhall at the back, as at Mount Terrace House (Fig. 19: 12) and 127 The Mount. At 306 Tadcaster Road the staircase is elliptical (Fig. 19: 13).
(ii) The entrance passage continues to form a stairhall of the same width, as at 20 Mount Parade (Fig. 19: 14).
(iii) The central entrance and staircase have only one room each side, as 65 Acomb Road (Fig. 19: 15). Service rooms are in a basement below.
(iv) The plan is asymmetrical, as 121 The Mount (in classical style) (Fig. 19: 16) and 117 The Mount (in Gothic style) (Fig. 19: 17).

The plans of houses in groups and terraces conform to a few regular types:

(i) A side passage leads to stairs on the same alignment at the rear and flanks two rooms arranged one behind the other (Fig. 19: 6). The passage may widen out to accommodate the staircase (Fig. 19: 4, 5), and there may be a basement.
(ii) Almost as common as (i) is the type with a staircase placed between the front and back rooms at right angles to the entrance passage (Fig. 19: 2, 3, 8). In this the entrance passage does not always continue to the back (Fig. 19: 9).
(iii) A scullery projects at the rear, opening off the stairhall (Fig. 19: 3, 5) or off the back room (Fig. 19: 2, 6). Other houses have a long range of service rooms forming a single-storey wing, as at 136–144 The Mount (Fig. 19: 7) where the wing consists of scullery, wash-house, closet and coal house. These houses are as early as 1824.
(iv) The smallest houses have no passage (Fig. 19: 10).

Variations of these types occur, one of which is shown in Fig. 19: 11, where stairs between the rooms lead down to a cellar.

The Use of Rooms. A description dated 1815 of a large mid 18th-century house is given under Monument (81) with two elegant drawing rooms on the first floor, but after 1800 the upstairs drawing room or saloon went out of fashion. A description of one of the villas on The Mount, built in 1833, is given in an advertisement of 20 June 1846 (*Yorkshire Herald*) and lists a dining room, drawing room, breakfast room, library and butler's pantry on the ground floor; three spacious lodging rooms, a dressing room, and a water closet with large bathroom on the first floor; and servants' rooms and attics above. The kitchens were probably in the basement.

A description dated 1850 of accommodation at 70, 72 Holgate Road, a medium-sized house which had been built a few years before, lists the following rooms: in the basement is a kitchen, back kitchen, larder and coal store; the ground floor has a dining room with a children's room behind it to the one side, and a drawing room with parlour behind it on the other side of an entrance hall; on the first floor are three bedrooms, a night nursery and some closets, and at the top are two good attics (YCL, YL/Gray).

(117) SKELDERGATE. House, No. 56. Saloon ceiling, late 18th century.

(38) BISHOPHILL. Bishophill House. Ceiling of bay, mid 18th century.

(55) MICKLEGATE. House, Nos. 3, 5. Detail of fireplace with metal enrichment, *c.* 1780.

(38) BISHOPHILL. Bishophill House. Mid 18th century.

(163) MIDDLETHORPE HALL. *c.* 1700 and 19th century.

(65) MICKLEGATE. House, Nos. 53, 55. Mid 18th century and later.

(68) MICKLEGATE. House, No. 57. 1783.

(95) MICKLEGATE. House, No. 118. *c.* 1742.

(112) SKELDERGATE. House, No. 48. Late 18th century.

(117) SKELDERGATE. House, No. 56. *c.* 1773.

(66) MICKLEGATE. Garforth House, No. 54. *c.* 1757.

(47) BLOSSOM STREET. House, No. 40. *c.* 1750.

(46) BLOSSOM STREET. House, No. 32. *c.* 1748.

TADCASTER ROAD. House, No. 306. 1833.

THE MOUNT. House, No. 140. 1824.

THE MOUNT. House, No. 100. 1807.

THE MOUNT. Houses, Nos. 94, 92. 1821.

MICKLEGATE. House, No. 84. 1821.

THE MOUNT. House, No. 132. *c.* 1830.

(66) MICKLEGATE. Garforth House, No. 54. Dining Room. *c.* 1757.

(93) MICKLEGATE. House, No. 113. Second floor. *c.* 1740.

(163) MIDDLETHORPE HALL. Dining Room. *c.* 1700.

(93) MICKLEGATE. House, No. 113. Saloon. *c.* 1740 and later.

Saloon.

Drawing Room.

(66) MICKLEGATE. Garforth House, No. 54. *c.* 1757.

(52) HOLGATE ROAD. Holgate House. *c.* 1770.

142) ACOMB. Acomb House. Early 18th century.

(65) MICKLEGATE. House, No. 53.

TADCASTER ROAD. House, No. 302.

(44) BLOSSOM STREET. House, No. 19.

MICKLEGATE. House, No. 84.

Early 18th century.
(55) Queen's Hotel, Nos. 7, 9.

Corner cupboard, early 18th century, reset.

(64) House, No. 48. *c.* 1747.
MICKLEGATE

FIREPLACE

(128) TRINITY LANE. House, No. 27. *c.* 1735.

(81) Micklegate House, Nos. 88, 90. Best Bedroom (now in Treasurer's House). *c.* 1750.

(66) Garforth House, No. 54. Dining Room. 1757.

(67) House, No. 56. First floor. *c.* 1750.

(66) Garforth House, No. 54. Second floor. 1757.

MICKLEGATE

(95) MICKLEGATE. House, Nos. 118, 120. Ground floor. *c.* 1742.

(65) MICKLEGATE. House, No. 53. Ground floor. *c.* 1755.

(142) ACOMB. Acomb House. Saloon. Early 18th century.

(68) MICKLEGATE. House, Nos. 57, 59. Saloon. 1783.

(81) MICKLEGATE. Micklegate House, Nos. 88, 90. Entrance hall. *c.* 1750.

(95) MICKLEGATE. House, Nos. 118, 120. First floor. *c.* 1742.

(101) NORTH STREET. House, No. 10. Ground floor (after removal). Early 18th century.

(117) SKELDERGATE. House, No. 56 (now removed). Late 18th century.

(66) MICKLEGATE. Garforth House, No. 54. First floor. 1757.

(68) MICKLEGATE. House, Nos. 57, 59. Ground floor. 1783.

FIREPLACES

(81) Micklegate House, Nos. 88, 90. Dining room. *c.* 1750.

(96) House, Nos. 122, 126. First floor. Early 18th century.

(70) House, No. 61. Ground floor. Late 18th century.

(55) House, Nos. 3, 5. First floor. *c.* 1780.

MICKLEGATE

First floor, E. room.
(52) HOLGATE ROAD. Holgate House, No. 163. *c.* 1770.

First floor, W. room.

Entrance hall. *c.* 1700.
(164) MIDDLETHORPE. Middlethorpe Manor.

Ground floor, S.E. room. *c.* 1770.

FIREPLACES

(134) ACOMB. House, No. 8, Front Street. Ground floor. Late 18th century.

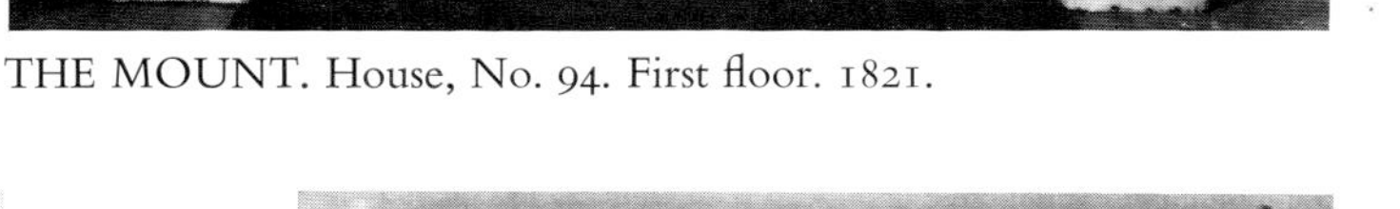

THE MOUNT. House, No. 94. First floor. 1821.

THE MOUNT. House, No. 127. Ground floor. 1833.

THE MOUNT. House, No. 138. Second floor. 1824.

Basements, which provided not only cellarage but also one or more rooms lit from outside, had been built in the mid 18th century (Monuments (81) and (66)) and basement kitchens became common in larger 19th-century houses, especially those built after 1830. In the smaller houses the kitchen remained on the ground floor with a larder, coal-house and W.C. in a single-storey wing at the back. Two-storey wings at the back were introduced after 1850.

Elevations

The early 19th-century houses in the part of York covered by this volume are not very impressive. There is little use of stucco and there are few iron balconies and no ornate iron porches as, for example, at Cheltenham. The homely adaptation of the Classical style shows signs of change towards neo-Classicism after 1830, at a time when one or two Gothic villas and a 'cottage orné' (the Herdsman's Cottage at the entrance to the Knavesmire) were also being built.

In the larger detached houses it is evident (allowing for irregularity resulting from alteration and addition) that symmetry was not strictly followed, as at Mill Mount House where the two wings provide balance, though one is of two storeys above the basement and the other of three. Similar balance without exact symmetry is seen in the group of three houses Nos. 100, 102, 104 The Mount.

The elevations of terrace houses are regular, but the openings on the ground floor do not always align with those above. Most of the houses in Mount Parade (Plate 59) have doorway and window with two windows aligned above but houses in South Parade and on The Mount (Plates 58, 157, 159) depart from this regularity in order to keep the front door nearer to the side of the house than would be convenient for the window above. In the smaller Dove Street houses (Plate 58) this difficulty is overcome by the omission of the second upper window in each house.

Detached villas and 'double-fronted' houses are generally symmetrical with a central porch between two bay windows and with three hung-sash windows above. Two variations on this theme are shown on Plate 58 where No. 306 Tadcaster Road has the design of the porch echoed in the window surrounds and No. 119 The Mount is designed in Gothic style. No. 117 The Mount, also in Gothic style, studiously avoids symmetry for a 'picturesque' design.

Eaves cornices (Plate 79). At the beginning of the 19th century cornices follow the late 18th-century pattern, and the moulded and dentilled cornice of 100–104 The Mount (1807/8) could stylistically be twenty years earlier. Some of the later cornices make use of earlier forms: at 92, 94 The Mount (1821) and at 128, 130 The Mount of *c.* 1840 are moulded cornices on square modillions reminiscent of those of the early 18th century.

The most popular early 19th-century cornice, however, and that most characteristic of the period, consists of a fairly ornate gutter carried on paired brackets, as at 136–144 The Mount (1824); this type can be found all through the first half of the 19th century. An Italianate version of it occurs at 306 Tadcaster Road (1833), where a boldly oversailing roof is carried on shaped brackets arranged in pairs, and a similar treatment is found at 60 York Road, Acomb, and 124 Holgate Road.

Many cornices of *c.* 1830–5 have simple classical mouldings as at 77 The Mount (Abbey Park Hotel) but as early as 1835 a new type appeared at 39–41 Micklegate, probably by J. B. & W. Atkinson and for many years their own office; it has a gutter supported on shaped brackets with vertical incised lines on the outer face; another version of it is seen at 94–96 Micklegate (1842/3) and this particular type of cornice became characteristic of the decade 1860–70. At the beginning of the century the gutters were often still of wood lined with lead, and round lead fallpipes with opposed fleurs-de-lis on the astragals were in general use in York, usually associated with a simple fluted rainwater head, as at 102 The Mount (1807/8), but cast-iron gutters, heads and fallpipes soon became common.

Materials

Brick is the principle walling material, and common stock bricks of rather pale mottled appearance were generally used for façades as well as for less conspicuous work. Through the first two decades of the 19th century there is little change in colour and size from the bricks in use at the end of the 18th century. 'White' bricks appear in Mount Terrace House in 1827 and in three villas on The Mount (Nos. 117, 119, 127) after 1835 but it is unlikely that they were produced locally, and they never became popular in York, but the taxation of bricks by number, introduced at the end of the 18th century, led in the end to the manufacture of bricks of larger size, and these larger bricks continued in use for some time after the repeal of the tax in 1850.

Stucco, which was so popular in the early 19th century elsewhere, is little used here, but it occurs in 1807/8 on No. 102 The Mount, in 1833 on No. 121 and *c.* 1843 on Nos. 124 and 128 in the same street. For the ground floor only it is used with horizontal rustication at 39, 41 Micklegate (1835), and the entrance to Mill Mount is similarly treated. At 89 The Mount and Acomb Park stucco is used to unify building of different periods.

Stone was used for the whole façade of 122 The Mount (1848/9) (Plate 159) and elsewhere sparingly for dressings, as for the pilasters which flank the stucco front of 21 Front Street, Acomb (Plate 193). The lower part of an ashlar front which survives at 16 Tanner Row may be of the late 18th or the early 19th century.

Welsh slate was the general roof covering, and roofs were mostly of low pitch. Westmorland slate had replaced the traditional clay tile on Fairfax House, Castlegate, in 1755, and by the early years of the 19th century tiles had almost completely given way to slate.

A good description of the materials to be used in a better class house in 1850 is given in a lease of 7 March 1850 of the site for Mill Mount House and others. It rules that '. . . the several houses shall be built and finished in a good substantial manner, the Walls being of good sound Brick, the Whole of the Timber Deals and Battens of Baltic Red Wood, except the Inner Panels and Mouldings which may be of American Pine and the Roofs covered with the best Bangor Welsh Slate nailed with Copper Nails; the principal outer Walls in which are placed the Doors and Windows, are not to be less than fourteen inches in thickness with proper footings, and all Windows and openings to be revealed or recessed four and a half Inches; all flues to be fourteen inches by ten inches inside Measure, well lined with lime and hair; All principal inner walls on basement and ground floors to be nine inches in thickness, the Lead for Gutters to be seven pounds weight to the square foot, and for flashings five pounds weight; Deals eleven inches by two inches to be used for floors not exceeding sixteen feet in bearing, and Battens seven inches by two and a half inches for floors under ten feet bearing, the Joists not being more than twelve inches apart, and on bridged or bolted Wall-plates under joist ends; All trimming Joists to be half an inch thicker than common joists; The Main Roofs to be framed with tie beams, principal Rafters and King Posts and the Spars not to be more than twelve inches apart . . .'.

Doorways

External doorways after 1800 have a horizontal moulded entablature at the top: the open pediment, so popular in the late 18th century, disappears. Up to 1820/30 doorways are tall and narrow and the detailing is Classical but used in a free manner. Pilasters are nearly always reeded, have no entasis, and are of the same width from top to bottom. A 'milled' enrichment of Adam type persists for a time and the Adamesque door with six flat panels and applied mouldings is common at the beginning of the century, but later the shapes vary and ultimately a four-panel door with raised mouldings supplants the long six-panel tradition.

A good example of a doorway of the beginning of the period is at 100 The Mount (1807/8) (Plate 65); it has a radial fanlight of 18th-century type set between slender foliated consoles with lions' masks at the

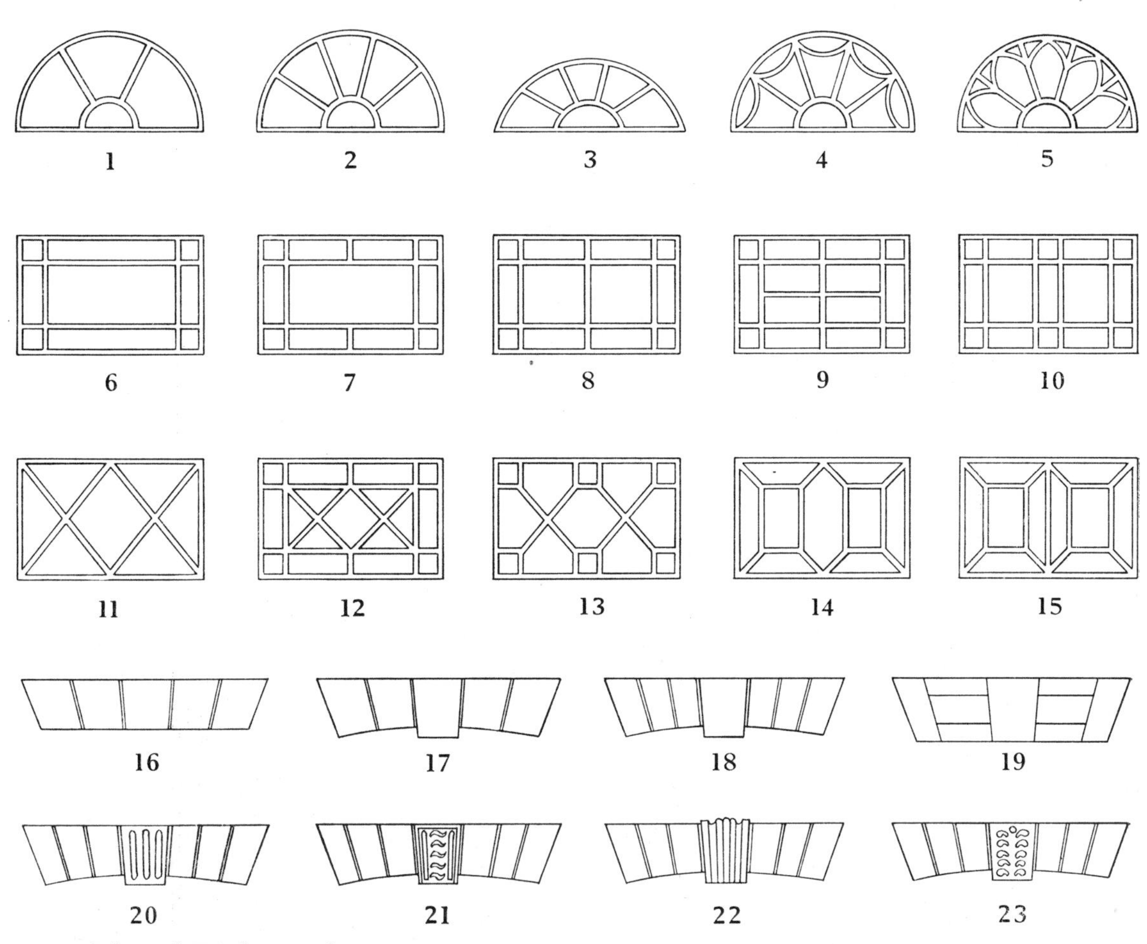

Fig. 20. Fanlights and Window Heads.

(1) 136–142 (even) The Mount, 1824.
(2) 132 The Mount, *c.* 1830.
(3) (123) 16 Tanner Row, *c.* 1830–40.
(4) 100 The Mount, 1807–8; 6 Mount Terrace, 1834.
(5) 147–151 Mount Vale, 1823–8.
(6) 5–8 Dove Street, 1827–30.
(7) 7 Bishopgate Street, *c.* 1830.
(8) 13 Clementhorpe, 1823; 226, 228 Mount Vale, *c.* 1830–40; 74, 76 Holgate Road, *c.* 1845–50; 43 Tanner Row, *c.* 1845–50.
(9) 25, 26 Dove Street, 1827–30.
(10) (157) 34 Tadcaster Road, Dringhouses, *c.* 1830; 26, 28–34 Holgate Road, *c.* 1845–50; 28, 30 Cambridge Street, 1846.
(11) 1–16 Rougier Street, 1842–3.
(12) 92, 94, 96 The Mount, 1821; 71 Acomb Road, 1828–9; 19 Mount Parade, 1830; 10 Mount Ephraim, 1846–51.
(13) South Parade, 1825–9; 14, 16, 31, 32 Dove Street, 1827–30; 17, 19, 21 Mount Ephraim, 1846–51
(14) 6 Bishopgate Street, *c.*1830; 126 Holgate Road, with lamp, 1835–40.
(15) 6 Mount Terrace, front door, 1834; 45 Holgate Road, *c.* 1845–50. 306 Tadcaster Road similar but larger, 1833.
(16) 147–151 Mount Vale, 1823–8; (157) 34 Tadcaster Road, Dringhouses (some of 9 and 10 voussoirs), *c.* 1830.
(17) 74–86 (even) Holgate Road, *c.* 1845–50. 147–151 Mount Vale similar but with segmental top and key-block projecting, 1823–8.
(18) 21–26 Dove Street, 1827–30; 70, 70a, 72 Holgate Road, 1846–7; 28–34 Holgate Road, 1845–50. 66 Holgate Road similar but with key-block projecting above top, 1845–50.
(19) 10, 12 Holgate Road, *c.* 1840
(20) 65 Acomb Road, 1828–9; 63, 65 Holgate Road, *c.* 1830. 147–151 Mount Vale similar but with upper side segmental and key-block projecting upwards, 1823–8.
(21) 43, 44 Dove Street, 1827–30; 19 Mount Parade, *c.* 1830.
(22) South Parade, 1824–8.
(23) 39 Blossom Street, 1828; 77, 79 The Mount, annexe, *c.* 1835–40.

top and slender reeded pilasters below. The six panels of the door have applied mouldings. A doorway at 84 Micklegate (1821) (Plate 65) by Peter Atkinson has no consoles and bold reeded pilasters support the entablature directly; the door has six fielded panels. The doorways of 136–144 The Mount (1824) (Plate

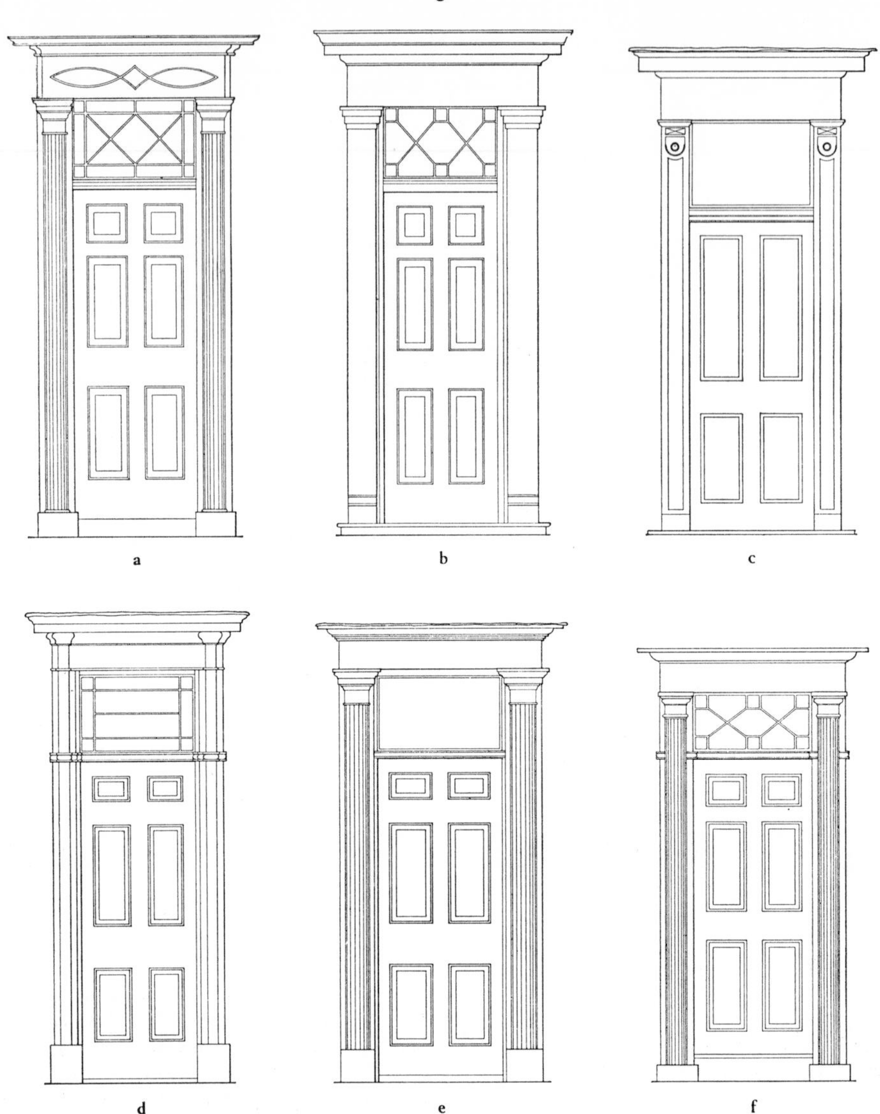

Fig. 21. Entrance Doorways.
a–e. Dove Street, 1827–30; f. Mount Ephraim, 1846–51.

64) are of the same type, but in each frieze is an oblong panel with the corners set in. There is a similar doorway at Clementhorpe (1823/40). Doorways at South Parade (1824/8) and at 34 Tadcaster Road are of the same sort, but have oblong fanlights with diagonal glazing bars producing a fret, which are like similar features in contemporary corner cupboards and library fittings and which became a standard design of the decade 1830–40. (For fanlights of this form *see* Fig. 20.)

The coupled doorways of 92, 94 The Mount show a tendency towards broader openings and a more conventional use of Classical detail. The columns are fluted, have correct entasis and support a moulded and modillioned entablature with triglyphs on the frieze, but the enrichment on the cornice is unorthodox. The doors have six fielded panels and there is an oblong fanlight with diagonal glazing bars. Doorways in Newington Place, 147–151 Mount Vale (1823–8), also have fluted columns supporting entablatures with a triglyph above each column, modillioned cornices and radial fanlights. Similar doorways are found at

Cumberland House, 20 Mount Parade (after 1834), where the frieze is fluted, and at the back of Mount Terrace House (by 1827), where the frieze has triglyphs and the fanlight is of the 18th-century form already noticed at 100 The Mount. Often the whole surround is recessed, as in a series by the Bar Convent.

Some doorways have an individuality which excludes them from the ordinary categories. Each of those of *c.* 1830 at 130–132 The Mount (Plate 65) has a moulded entablature with square modillions which is supported by attached columns with strong spiral reeding and with well carved Composite capitals and Attic bases. Above the door is a radial fanlight and the door has four panels with projecting moulding, a type which became common after 1850. The only doorway which shows strong Soane influence is at 16 Tanner Row (*c.* 1830); it has very slender reeded pilasters with lions' masks at the top and a door recessed between two glazed lights; it has the linear decoration and flush panels associated with Soane, and the fanlight is segmental with radial glazing bars.

From *c.* 1835 many doorways had plain panelled pilasters and a simple entablature, sometimes with modillions as at 39, 41 Micklegate (1835) by J. B. & W. Atkinson, in Cygnet Street (1846/51), 28 and 30 Cambridge Street (1846) and 89 The Mount. Towards 1850 doorways commonly retain the plain pilasters but the projecting entablature is supported by coarse brackets or consoles. Early examples are at 120 and 126 The Mount (1842/3 and *c.* 1840); there are others at 8–10 Cambridge Street (1846) and Bishopgate Street, where the houses were being erected just before and after 1850.

Porches with freestanding columns are rare. One at 306 Tadcaster Road (1833) (Plate 64) has good Roman-Doric columns and an elaborate entablature with small guttae at the bottom of the plain frieze. Over the doorway is a fanlight with marginal panes and the door is of false two-leaf type with six panels. A porch to Mount Terrace House (by 1827) has Ionic columns and a similar fanlight. The Elephant and Castle Inn in Skeldergate (*c.* 1840/50) had plain columns supporting an entablature with modillions. Other porches have solid side walls; a type embodying pairs of plain pilasters on either side and a plain entablature is seen at 121 The Mount (first rated 1834/5) and 127 The Mount (first rated 1839/40).

Two recessed porches with Doric pillars *in antis* are found at 123 The Mount (1833) and at 77, 79 The Mount (1831/2) (Plate 59).

Internal doorways are usually surrounded by simple mouldings butted against square blocks at the angles (Plate 68). An early doorway of this type at 53 Micklegate (Plate 68) has guilloche decoration on jambs and lintel and lions' masks on roundels at the angles; the decorations are by the composition manufacturer and carver Thomas Wolstenholme (1759–1812). In general, however, the jambs and lintel are simply moulded and the angle piece has a round moulded patera as at 84 Micklegate (1821) by Peter Atkinson (II). A more ornate version of the same theme in which the moulding is more complicated and the paterae foliated is at 302 Tadcaster Road (1833), and there is a similar doorway used externally at 98 The Mount.

An original drawing by J. B. & W. Atkinson of 77, 79 The Mount dated 1831 shows some doorways with mitred architraves and others flanked by plain pilasters with brackets under a plain entablature decorated with five paterae.

Windows

From 1800 to 1850 hung-sash windows were in general use. The openings had flat arches but there was a tendency towards the end of the period to make the under side of the arch slightly segmental. Flat arches in ordinary common brick were built from the late 18th century onwards and common bricks were used for window arches all through this period in cheaper building and in subsidiary situations. For better quality work arches of special rubbed brick were used till about 1830; from 1820 natural and artificial stone lintels were introduced, grooved to simulate the voussoirs of an arch. Where rubbed bricks were used they were the full depth of the arch but incised to give the impression of jointing. There is effective contrast between rich red rubbed brick arches and pale buff walling at 130, 132 The Mount (*c.* 1830) and elsewhere (Plate 60). Artificial stone lintels often have the centre emphasised by an enriched 'key-stone'. Window heads of natural stone occur at 147–151 Mount Vale (1823–8) (Plate 59, Fig. 20), and of artificial stone at South Parade (1824–8) (Plate 157, Fig. 20) and in many other later houses, particularly in the area of Holgate Road and Mount Parade. A few houses including 122, 124, 128 The Mount and two villas in Park Street, have moulded stone architraves in the style of the first half of the 18th century.

Round-headed windows were used for lighting staircases but only occasionally for rooms, as at 122, 124 and 128 The Mount, all of *c.* 1830–40.

Tripartite sashed windows with narrow lights flanking the main light, commonly used by James Wyatt in the late 18th century (*see* A. Dale, *James Wyatt* (1956), where they are dubbed 'the Wyatt window') appear in York only rarely and after 1800 as at 17 Micklegate, 19 Park Street, and 89 The Mount. A variation with a Gothic flavour, from 130 The Mount, is illustrated in Plate 91.

The horizontally-sliding sash window, which is used extensively in the country around York in the early 19th century, is uncommon in the City. There are some in Acomb, which was a separate village at this time; there is an example in the town in the back kitchen of 100 The Mount, added in 1808/13.

Bay windows are used from 1800 to 1850. Two polygonal bay windows with rubbed-brick heads at Nunroyd, 109 The Mount, are of late 18th-century date but were not copied. At the beginning of the 19th century shallow bays, segmental in plan and framed in wood, were fashionable, nearly always with a curved version of the tripartite sash window with narrow side lights. None of this type in the area covered by this volume is of proven late 18th-century date, although there are some of 1797 across the river at 3–5 Gillygate, built by Thomas Wolstenholme. Good examples at 100, 102 The Mount (1807/8) (Plate 59) have all the frame members reeded, a slender leaf enrichment above each jamb and mullion, picking up a similar detail in the doorways, and moulded cornices. Others at 92 and 94 The Mount (1821) are much plainer; further examples occur at 136–144 The Mount (1824) (Plate 58), at South Parade (1825/8) (Plate 157) and at 34 Tadcaster Road, Dringhouses.

About 1830 the segmental bay fell into disfavour and a three-sided bay with canted sides and angles built up in timber or artificial stone supplanted it. Perhaps the earliest examples are at Mount Terrace House (1827); those at 127 The Mount (1833) are certainly original.

The introduction to England in 1832 of a new process for sheet glass and improvements in the technique of making plate glass made possible the use of larger window panes, free of the distortions to which crown glass had been subject. The increase in size of pane is noticeable from 1833, and the large plate-glass panes at 306 Tadcaster Road may be original (also 1833). Generally large panes were not at first used to fill a whole sash but they were surrounded by a border of narrow marginal panes which combined the advantages of an uninterrupted outlook with the texture given by glazing bars. All the fenestration of 46 St. Paul's Square is treated in this way and in particular the three-sided bay (Plate 91) is a fine example of both bay and glazing; there is a similar one at 117 The Mount. 46 Holgate Road and 54–60 Holgate Road also have marginal panes. 7 Park Street, erected by the builder Thomas Rayson for himself in 1836, has such a good display of such windows to the garden front as to suggest that he built 46 St. Paul's Square also.

When used in stair lights the marginal panes were often made of white glass flashed with ruby, blue or green, and then cut with patterns through to the white. An exceptional French window at 302 Tadcaster Road (1833) (Plate 91) has marginal panes combined with a Gothic treatment of the upper glazing bars.

Most of the windows were fitted with shutters; where they were external they have mostly been removed but they still remain at 120, 125 and 127 The Mount (1833); internal shutters, such as some decorated with Soanesque grooves at 86 Micklegate, were also still in use.

Staircases

Early 19th-century staircases (Plate 89) were not so elaborate as those of the previous century but often still formed architectural features of some importance. They usually had cut strings with decoration applied at the end of each step. Handrails were slender and at first moulded like those of the late 18th century but later were reduced to a simple round section. Newels were commonly omitted. The commonest form of baluster had a simple square section, at first very slender but later more robust; often they were of wood with iron ones introduced at intervals for strength and rigidity. Examples are found at 74, 78–84 Micklegate and 94, 100–104, 136–144 The Mount. Turned balusters similar to those of the late 18th century continued till 1850 but are usually more spindly (Mount Terrace House, and Acomb Park). Others show large numbers of small roll mouldings, sometimes with a small leaf-like decoration. Staircases at 89 and 92 The Mount retain many 18th-century features; typical of the 19th century are those at 127 The Mount (after 1833), 302 The Mount and 304 Tadcaster Road.

Iron balusters, very rare in York before 1800, were sometimes used. A simple form with decoration consisting of a hollow-sided diamond at 125 The Mount and 306 Tadcaster Road, both of *c.* 1830–40, are very like balusters elsewhere by Peter Atkinson. Less elegant balusters, round in section with a roundel at the middle, were designed by G. T. Andrews; an example from the Old Railway Station is illustrated. On plan staircases were generally rectangular but those at 306 Tadcaster Road (1833) and at 126 The Mount were elliptical or semicircular (Fig. 19: 13).

Fireplaces

In the illustrations on Plate 76 fireplace surrounds are shown with contemporary grates. No. 8 Front Street, Acomb (*c.* 1800), shows a continuing use of late 18th-century Adamesque elements in the urns and the central panel with figures, but the reeding of the sides and head is characteristic of much work of the first half of the 19th century. The inner marble slip and the surround of the iron grate itself also show the use of roundels at the angles, similar to those used on door and window architraves, which are also characteristic of the period. A surround at 100 The Mount (1807/8) with fluted pilasters and festoons on the head is rather closer to the Adam style. Elsewhere Adamesque elements are used in an inconsequent manner before being abandoned altogether. The fireplace illustrated from 92, 94 The Mount shows a very simple surround with the basic elements of side and head cut to a symmetrical moulding and not mitred at the corners but butted against an angle piece. The motif is repeated in the ironwork of the grate. At 136–144 The Mount simple reeding is used and again it is stopped at the angles. At 127 The Mount (after 1833) a Victorian type is shown with very plain side pilasters and head and florid foliated ironwork to the grate. In the latest fireplace before 1850 the mantelshelf was carried on coarse console brackets, and the grate often had a semicircular-headed opening.

The iron grates were all fixtures unlike the free-standing fire-baskets of the 18th century. Large numbers of them were made by Messrs. Carron of Falkirk. Of those illustrated, that from 94 The Mount is signed Carron, that from 138 The Mount is signed Low Moor Company, Bradford (Plate 76).

Other Interior Details

Walls and *Ceilings* were generally quite plain. Of the panelled wall treatment of the 18th century only the chair rail remains in the entrance hall. Ceilings were only decorated with foliated centre-pieces from which light fittings were suspended. These centre-pieces, often made of cast-iron and very florid in design, were particularly associated with gas lighting, introduced to York in 1824, but were being put up before gas came into use. Ceilings in Gothic style with simple barrel and ribbed vaults appear in 117 and 119 The Mount and groined vaults in 121.

Cornices used in the first half of the 19th century to effect the transition from wall to ceiling mostly show a complete departure from the Classical models used in the 18th century and consist of a series of mouldings, often incorporating reeding, in the plane of the wall and in the plane of the ceiling only. Where Classical forms are followed the mouldings are softened and the clear-cut vertical face of the fascia in a Classical cornice never appears.

Recesses flanking chimney-breasts were commonly finished with segmental or semicircular arched heads.

Ironwork

Balconies are little used in comparison with many other towns and show the same designs that are to be found in other areas (Plate 80). Railings are simple and sometimes associated with heavy cast-iron gate piers enriched with honeysuckle ornament (Plate 80). Bootscrapers were commonly provided. At the Bar Convent, railings and bootscraper were made by William Haxby in 1815 and, together with an iron gate and stonework, cost £54 (Bar Convent Archives, 7 B 3(10)). Trellis-work for porches only appears at 151 Mount Vale.

Lighting and Plumbing

Progress in domestic services is exemplified by the fitting at the Bar Convent of gas pipes in 1834/5 and water closets in 1844; these last had zinc and copper tubes, ball cocks and waste pipes and were fitted by William and Thomas Hodgson (Bar Convent Archives, 7 B 12(15)).

(E.A.G.)

Postscript

Among the many houses built in the S.W. part of the City after 1850 one is of outstanding importance as being illustrative of contemporary culture. Elm Bank, a large house on the N.W. side of The Mount, occupied from 1870 by William Benson Richardson, a solicitor, was acquired by Sidney Leetham, miller, who in 1898 employed Messrs. W. G. and A. J. Penty as architects and George Walton (1867–1933) as interior decorator to remodel the interior. Walton was a pioneer of the *Art Nouveau* movement and Elm Bank was one of his earliest works outside his native Glasgow.

The house, built of white brick with stone dressings, with an Italianate tower at one corner, has been converted to a hotel and considerably enlarged. Inside, the staircase is contained in a large hall with a gallery giving access to the first-floor rooms. In this stairhall and in the principal rooms on the ground floor much of Walton's decoration still remains, including fireplaces and an overmantel inlaid with glass and ceramics, stained glass, and wall paintings (these last restored) (Plates 205, 206).

For early photographs of the Elm Bank interiors see *The Studio* XXII (1901), 36. For George Walton see N. Pevsner in the *Royal Institute of British Architects Journal* XLVI 3rd Ser. No. 11, 3 Apr. 1939, 537–48. See also John Betjeman in *Daily Telegraph Magazine*, 7 May 1965, and Patrick Nuttgens, *York* (1971), 74.

Walton was a contemporary of James Rennie Mackintosh, who also worked in Glasgow, and who was a pioneer of modern architecture. The influence of Mackintosh's work can be seen in two schools designed by W. H. Brierley, at Scarcroft Road (1896) and Poppleton Road (1904). The functionalism of these two red brick buildings contrasts with the florid eclecticism of the railway offices adjoining the old station, designed by H. Field and W. Bell in the 'Queen Anne' style of Norman Shaw, and completed in 1906.

(R.W.McD.)

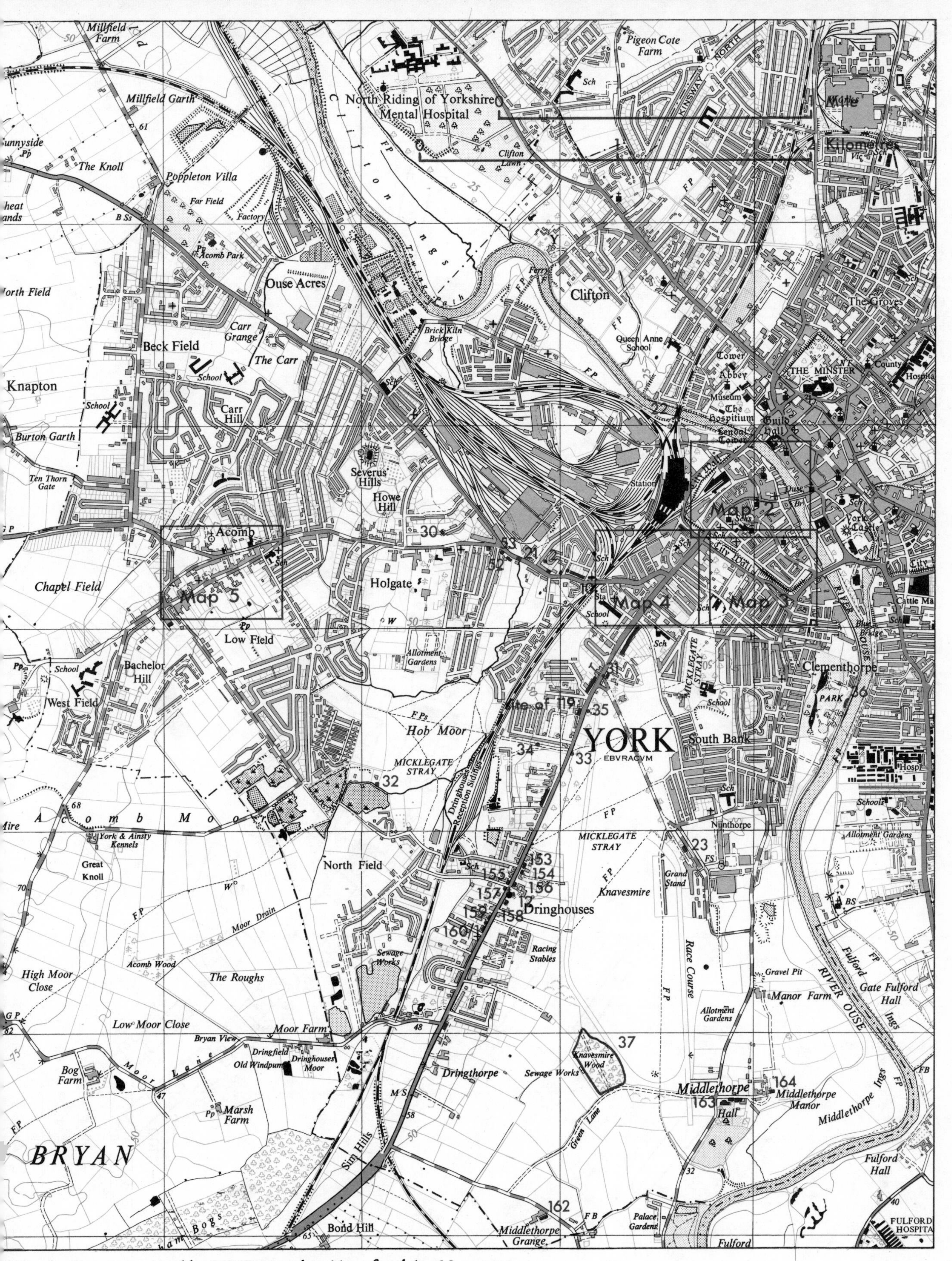

MAP 1 showing areas covered by MAPS 2–5 and position of outlying Monuments.

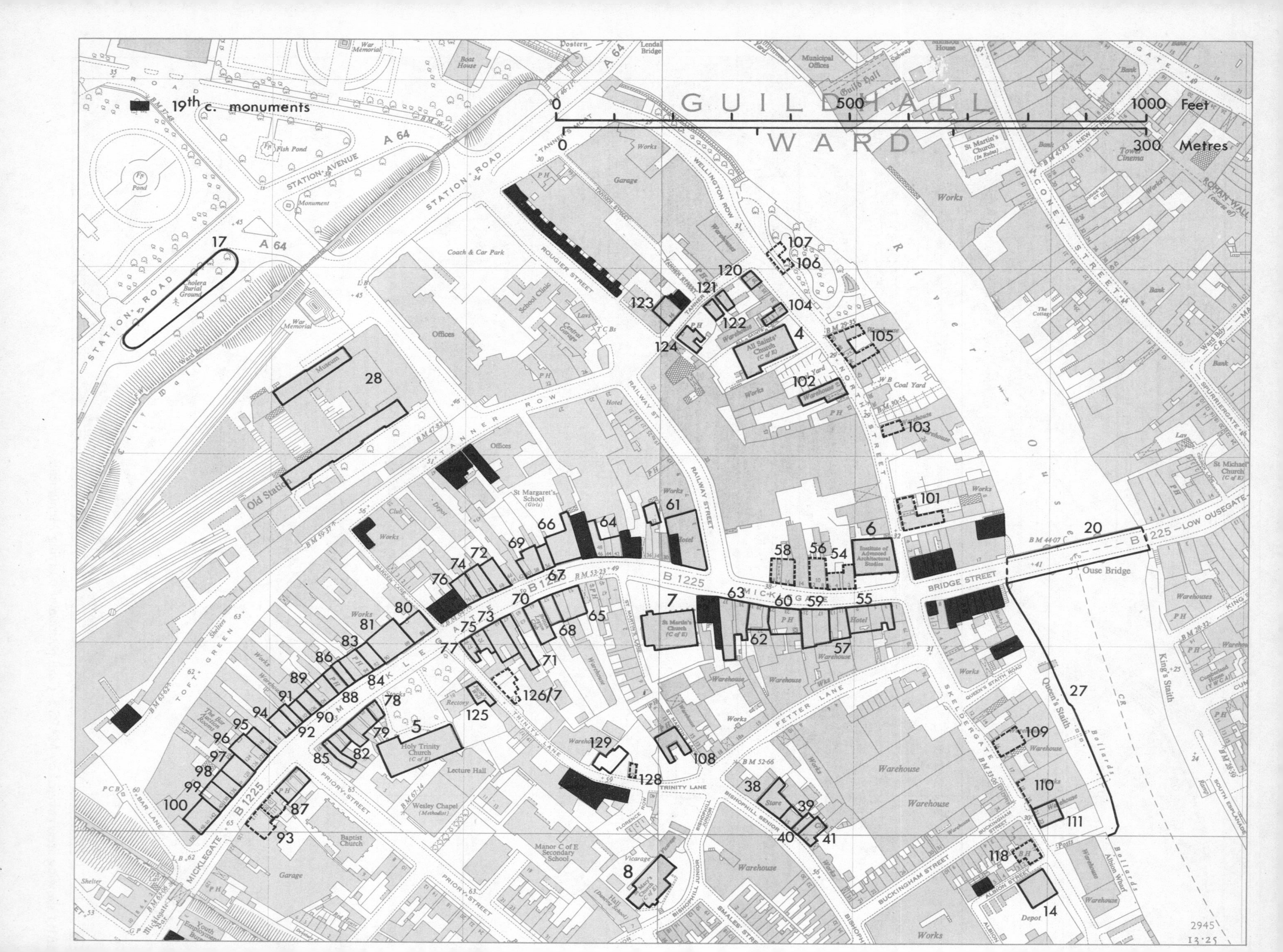

19th c. monuments
GUILDHALL WARD
0 500 1000 Feet
0 300 Metres
STATION ROAD
STATION AVENUE
A 64
Cholera Burial Ground
War Memorial
Fish Pond
Pond
Monument
Boat House
Postern
Lendal Bridge
Municipal Offices
Guild Hall
St Martin's Church (In Ruins)
Town Cinema
CONEY STREET
NEW STREET
SPURRIERGATE
ROMAN WALL (course of)
St Michael's Church (C of E)
LOW OUSEGATE
Ouse Bridge
BRIDGE STREET
King's Staith
Queen's Staith
Bollards
Albion Wharf
SKELDERGATE
QUEEN'S STAITH ROAD
BUCKINGHAM STREET
ALBION STREET
Depot
FETTER LANE
BISHOPHILL SENIOR
BISHOPHILL JUNIOR
SMALES' STREET
St Mary's Church (C of E)
Vicarage
Hall (Dancing School)
TRINITY LANE
FLORENCE ROW
Holy Trinity Church (C of E)
Lecture Hall
Rectory
Wesley Chapel (Methodist)
Manor C of E Secondary School
Baptist Church
PRIORY STREET
MICKLEGATE
B 1225
BAR LANE
Micklegate Bar
TOFT GREEN
The Bar Auction Rooms
Garage
Old Station
Museum
Offices
TANNER ROW
Coach & Car Park
ROUGIER STREET
TANNER STREET
TANNER'S MOAT
WELLINGTON ROW
School Clinic
Central Garage
RAILWAY STREET
St Margaret's School (Girls)
Club
Works
St Martin's Church (C of E)
ST MARTIN'S LANE
Institute of Advanced Architectural Studies
NORTH STREET
Coal Yard
All Saints' Church (C of E)
River Ouse
Warehouse
Hotel
17
28
123
121
120
107
106
104
124
122
4
105
102
103
101
6
20
61
64
66
69
72
74
76
67
58
56
54
7
63
60
59
55
57
62
65
68
70
73
75
77
71
80
81
83
86
84
89
88
91
90
94
92
95
96
97
98
99
100
78
79
82
85
5
125
126/7
129
128
108
38
39
40
41
8
87
93
109
110
111
118
14
27
2945
13.25

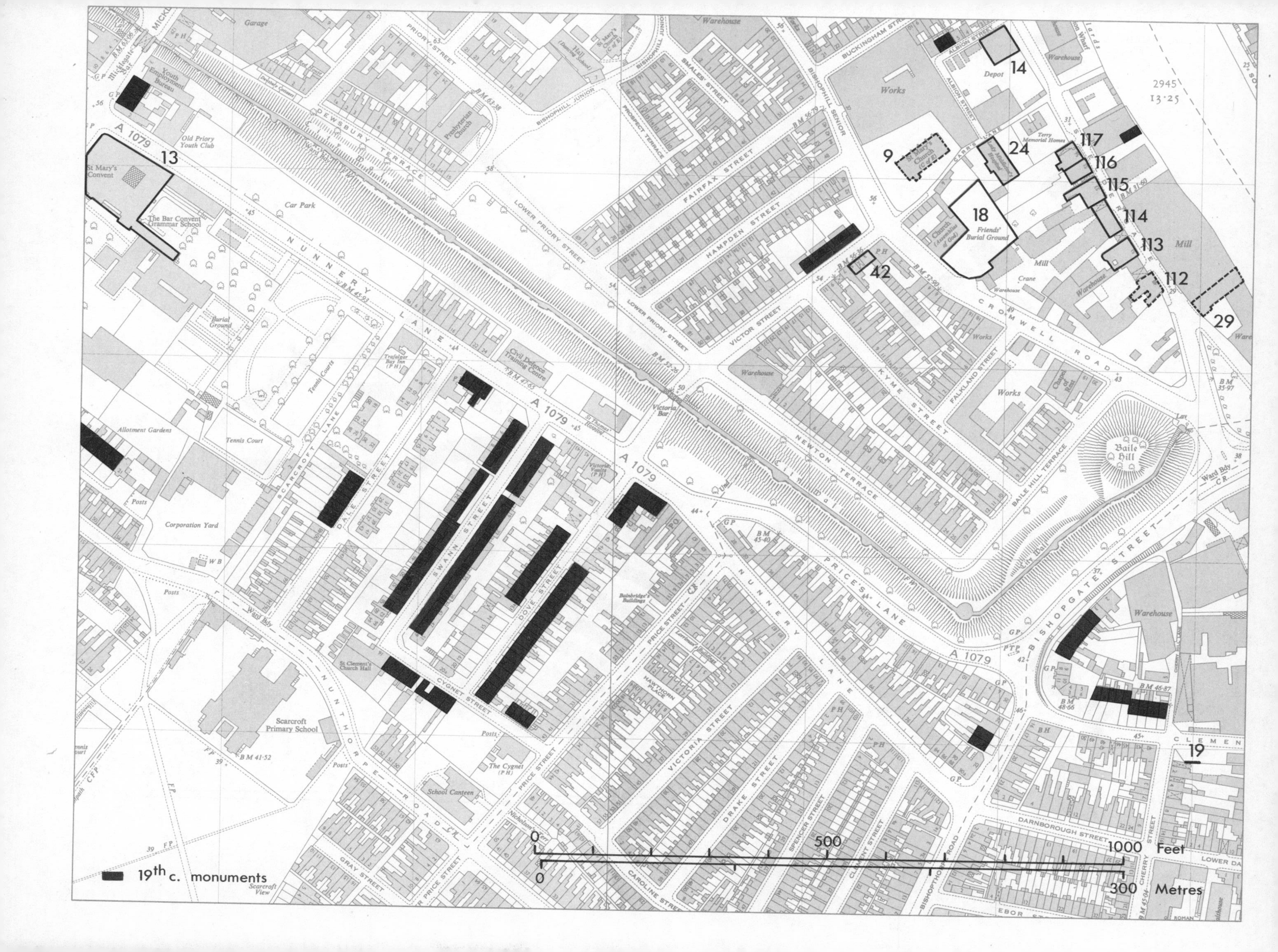
19th c. monuments
0
500
1000 Feet
0
300 Metres
13
9
14
18
24
29
42
112
113
114
115
116
117
19
St Mary's Convent
The Bar Convent Grammar School
Youth Employment Bureau
Old Priory Youth Club
Car Park
Garage
NUNNERY LANE
A 1079
DEWSBURY TERRACE
PRIORY STREET
Presbyterian Church
LOWER PRIORY STREET
BISHOPHILL JUNIOR
BISHOPHILL SENIOR
PROSPECT TERRACE
FAIRFAX STREET
SMALES' STREET
HAMPDEN STREET
VICTOR STREET
KYME STREET
NEWTON TERRACE
FALKLAND STREET
BAILE HILL TERRACE
CROMWELL ROAD
BISHOPGATE STREET
PRICE'S LANE
VICTORIA STREET
PRICE STREET
DRAKE STREET
SPENCER STREET
CLEMENT STREET
BISHOPTHORPE ROAD
CAROLINE STREET
DARNBOROUGH STREET
CHERRY STREET
HAWTHORN PLACE
Bainbridge's Buildings
DOVE STREET
SWANN STREET
CYGNET STREET
DALE STREET
SCARCROFT LANE
NUNTHORPE ROAD
GRAY STREET
The Cygnet (P.H.)
School Canteen
St Clement's Church Hall
Scarcroft Primary School
Scarcroft View
Tennis Courts
Tennis Court
Burial Ground
Corporation Yard
Allotment Gardens
Civil Defence Training Centre
Victoria Bar
Baile Hill
Friends' Burial Ground
Terry Memorial Homes
Works
Depot
Warehouse
Mill
Crane
ALBION STREET
BUCKINGHAM STREET
Hall (Dancing School)
Posts
2945
13·25

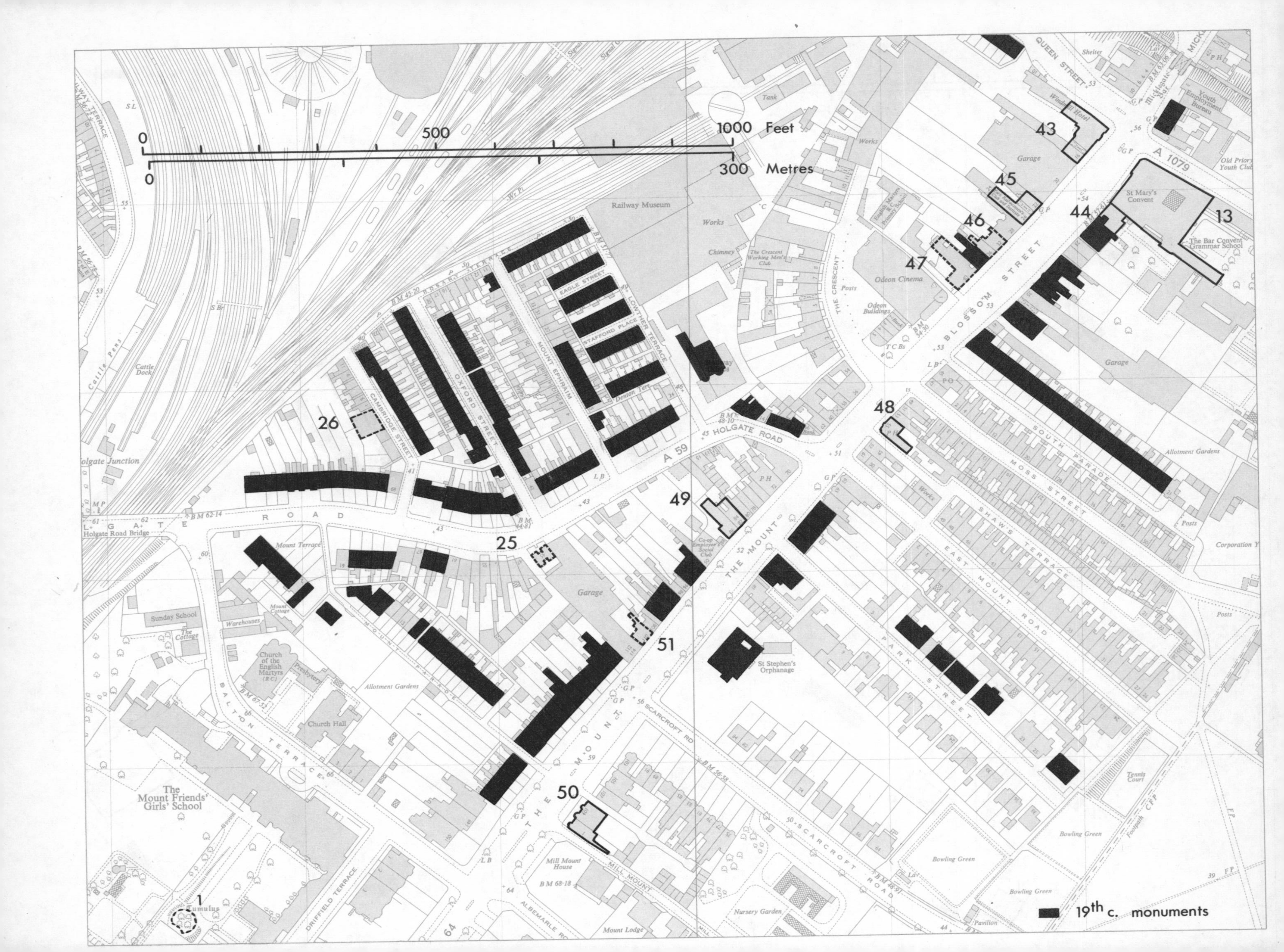
0
500
1000 Feet
0
300 Metres
19th c. monuments
Railway Museum
Works
Chimney
The Crescent Working Men's Club
THE CRESCENT
Odeon Cinema
Odeon Buildings
Garage
Windmill Hotel
QUEEN STREET
Youth Employment Bureau
Old Priory Youth Club
A 1079
St Mary's Convent
The Bar Convent Grammar School
BLOSSOM STREET
SOUTH PARADE
MOSS STREET
SHAW'S TERRACE
EAST MOUNT ROAD
PARK STREET
Allotment Gardens
Corporation Y
Tennis Court
Bowling Green
Pavilion
SCARCROFT ROAD
SCARCROFT RD
Nursery Garden
Mount Lodge
MILL MOUNT
Mill Mount House
ALBEMARLE RD
THE MOUNT
St Stephen's Orphanage
Co-op Employees' Social Club
HOLGATE ROAD
A 59
EAGLE STREET
STAFFORD PLACE
LOWTHER TERRACE
Denton Terr
MOUNT EPHRAIM
OXFORD STREET
CAMBRIDGE STREET
ROSARY TERRACE
Railway Institute
Cattle Dock
Cattle Pens
Holgate Junction
Holgate Road Bridge
HOLGATE ROAD
Mount Terrace
Mount Cottage
Warehouses
Sunday School
The Cottage
Church of the English Martyrs (R.C.)
Presbytery
Church Hall
MOUNT PARADE
BALTON TERRACE
The Mount Friends' Girls' School
DRIFFIELD TERRACE
Tumulus
1
13
25
26
43
44
45
46
47
48
49
50
51

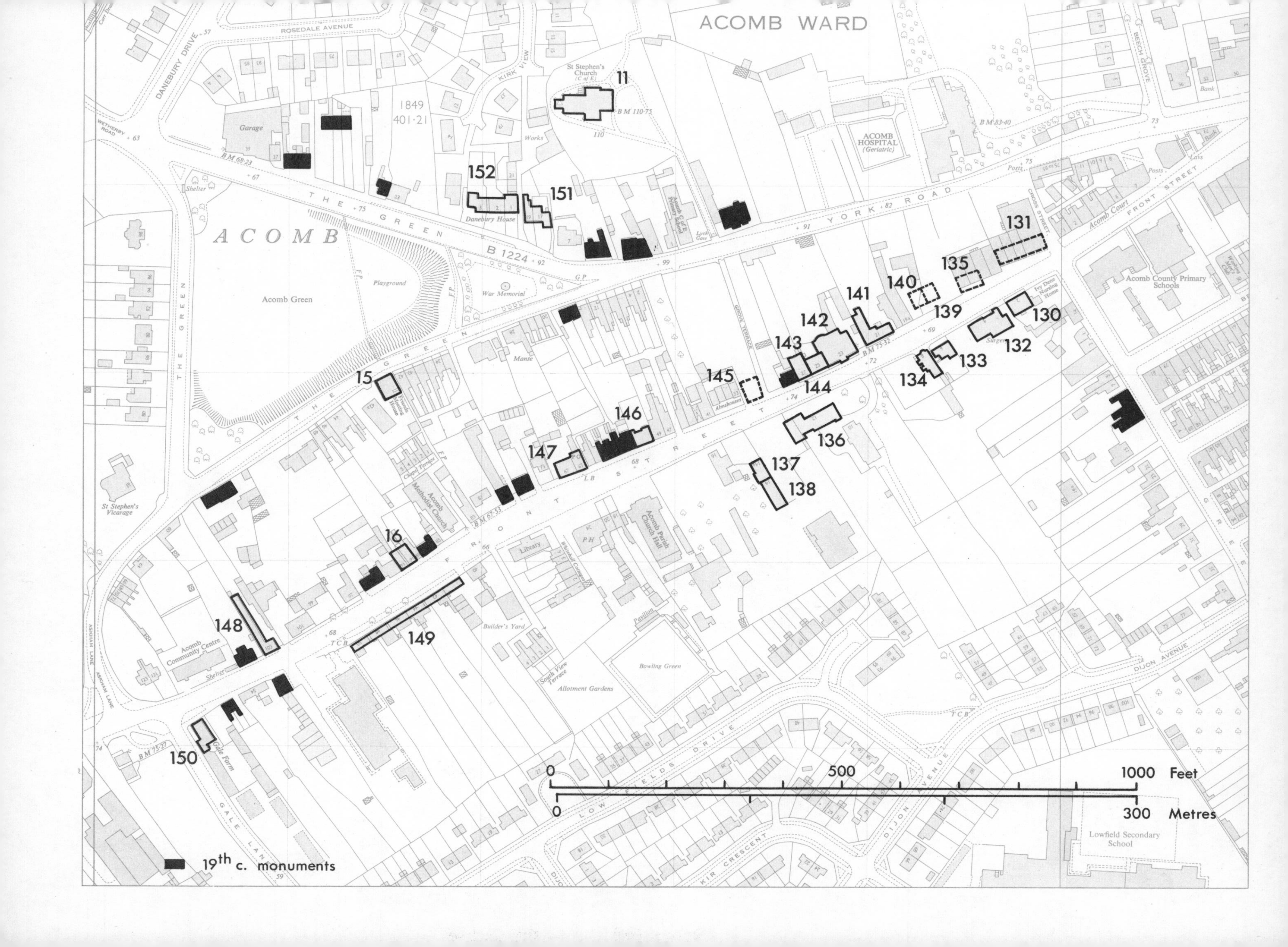

ACOMB WARD
ACOMB
St Stephen's Church (C of E)
ACOMB HOSPITAL (Geriatric)
Acomb County Primary Schools
Acomb C of E Primary School
Acomb Green
Playground
War Memorial
Manse
Friends Meeting House
Acomb Methodist Church
Acomb Parish Church Hall
Library
Bowling Green
Allotment Gardens
Builder's Yard
Acomb Community Centre
Gale Farm
St Stephen's Vicarage
Lowfield Secondary School
Garage
Danebury House
ROSEDALE AVENUE
DANEBURY DRIVE
WETHERBY ROAD
KIRK VIEW
BEECH GROVE
YORK ROAD
THE GREEN
B 1224
FRONT STREET
CROSS STREET
GROVE TERRACE
DIJON AVENUE
LOWFIELDS DRIVE
KIR CRESCENT
ASKHAM LANE
GALE LANE
1849
401·21
11
15
16
130
131
132
133
134
135
136
137
138
139
140
141
142
143
144
145
146
147
148
149
150
151
152
0
500
1000 Feet
0
300 Metres
19th c. monuments

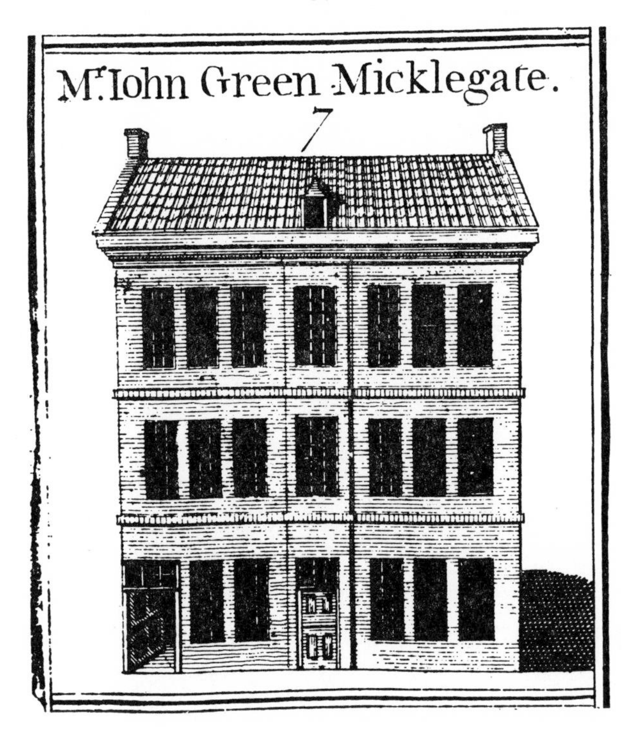

Fig. 22.

Fig. 23.

Drawings of houses from John Cossins's Map of York, *c.* 1727.

AN INVENTORY OF THE HISTORICAL MONUMENTS IN THE CITY OF YORK SOUTH-WEST OF THE RIVER OUSE

(The dimensions given in the Inventory are internal unless otherwise stated, and read first from E. to W. The National Grid References are in 100-kilometre square SE. The dates given in the descriptions of memorials are usually of the death of the persons commemorated. Numbers following unidentified shields of arms refer to their blazons, which are listed at the end of the Armorial Index.)

EARTHWORKS AND ALLIED STRUCTURES AND CULTIVATION SYSTEMS

(Defensive earthworks and fortifications are described in *York* II. *The Defences.*)

EARTHWORK:

(1) Mound (59245102), Mount School, now approximately circular, with a flat asphalted top and surrounded by an asphalt path, is 4¼ ft. high and 67 ft. in diameter. In 1852 the OS map shows the diameter as 80 ft.; it was then part of a landscape garden. Though then marked, as on all subsequent OS maps, as a tumulus, evidence of its original purpose is lacking.

It could well be the steading for a post mill, but its position cannot be identified with that of any mill shown on early maps of the area. The flimsy basis for the identification of this mound as a 'tumulus' is set forth by Hargrove (*History*, I, 245).

ENCLOSURE:

(2) Enclosure, Enfield Crescent, Holgate Hill (58955133), now built over, though fragments survive in house gardens. The plan was recorded on the large-scale OS maps 1852–1936, and the site was excavated in 1936 by P. Corder (YAYAS *Procs.* (1951–2), 31, with plan).

The enclosure of rectangular plan (160 ft. by 148 ft.) was defined by a rampart 3 ft. 2 in. high and 20–25 ft. wide with a hollow on the inner side from which the material for the rampart had been quarried; there was no evidence for any entrance. The work is on the steep slopes of a hill whose summit has been levelled. Here excavation revealed a small area of cobbling 11 ft. by 4½ ft. with 14th-century occupation debris and roof tiles, but no walling. Two sherds of 14th-century pottery were found in the make-up of the bank, but association between the bank and the occupation of the hill-top was not proved. Corder suggested a 14th-century military lookout post; but if the enclosure is to be dissociated from the hill-top feature, it could be of 17th-century date (compare RCHM, *Newark on Trent* (1964), monuments 12, 16 and 17) since it is known that the Scots captured a work at Holgate in 1644.

CULTIVATION SYSTEMS:

(3) Open Fields of mediaeval type, survive as traces or are visible on recent air photographs. They consist of scattered parcels of broad plough ridges (usually 28–30 ft. wide, but sometimes up to 40 ft.), separated by narrow furrows. Though now within the city boundaries they belong in part to older townships now incorporated, and are listed below accordingly.

The pattern of land use seems to have been 'ings' or water meadows by the river or its tributaries, moor or 'stray' on the lower-lying lacustrine clays, and arable on glacial deposits or lacustrine sands and gravels.

(a) *York.* Two city pastures S.W. of the river survive intact, Hob Moor and Knavesmire (Plate 4). On the former are narrow plough ridges about 6 ft. wide which represent temporary ploughing during the Napoleonic Wars. At enclosure the Knavesmire was enlarged to compensate for loss of pasture rights on the surrounding arable after harvest, and broad ridge-and-furrow is visible on its margins on both sides of the stray, E. of the Tadcaster Road in York (590500 and neighbourhood) and in Dringhouses and Middlethorpe (*see* below). Ridge-and-furrow is also visible in the former Campleshon Field, S. of Campleshon Road (600500 and neighbourhood); in the field E. of Bishopthorpe Road, between the road and Nun Ings (601497 etc.); and in an orchard adjacent to Albemarle Road (595508 etc.). York Fields, part of the city's former open fields, are now obliterated by Victorian housing S.W. of Scarcroft Road.

(b) *Acomb.* Field names surviving from before enclosure were recorded on the OS map of 1853 (6 in. scale, sheet 174). The moor lay at the S. end of the parish, the ings at the N., beside the river. A long stretch of marshy land known as the Carr ran N. from the village, separating the Beck Field (the N. part of which was known as Far Field) from a long field subdivided into Ouse Acres, Mill Field, Low Field and Hob Moor Field. West Field lay S.W. of the village and the land N. of Grange Lane was known as Chapel Field. Rapid expansion of housing since 1952 has obliterated traces of these fields, but air photographs taken then show ridge-and-furrow S. of York Sugar Factory (Ouse Acres, 577526), and towards Askham Lane (565510). Here the ridge-and-furrow turns into low terraces, whose lines are continued by garden boundaries on N. of Askham Lane (Chapel Field), S. of the houses in Front Street (near 575510) disappearing into the housing estate at Tudor Road (Low Field), and W. of Hob Moor (580305, in Hob Moor Field). Broad and narrow ridge-and-furrow to E. of Acomb Wood and N. of Moor Drain (near 573496) probably represent post-enclosure ploughing on Acomb Moor (Enclosure Award, 1776; reprinted in H. Richardson, *A History of Acomb* (1963), 49 ff.).

(c) *Dringhouses.* The village, or township, lies athwart the Tadcaster Road on a narrow glacial ridge. As shown on the 1853 OS map, the territory of the village, which lay in three parishes (Holy Trinity, Micklegate; St. Mary Bishophill Senior; and Acomb), was on both sides but mainly W. of the road. It also included a narrow strip crossing the Knavesmire and incorporating the ancient manor of Bustardthorpe on the W. bank of the Ouse. There were several detached portions on Middlethorpe Ings. Dringhouses has now become a separate ecclesiastical parish, and has lost its detached portions and the strip across the Knavesmire.

A manor map of 1629 (York City Library)[1] (Plate 4) shows three areas of fields: North Field (still marked on modern OS maps), West Field and 'Streate Lands' on the E. of the main Road. North and West Fields were still largely open and unconsolidated in 1629, whereas 'Streate Lands' seem to have been already consolidated and enclosed. Between 1772 and 1838, 'Streate Lands' was transferred to Middlethorpe township (map of Micklegate Ward Stray, by John Lund jun., 1772, YCA, D/Vv.; Tithe Award Map, 1838, 284S, Borthwick Inst.). Dringhouses Moor lay to the S.W. of the village in relatively wet, low-lying land. 'The Roughs', still marked on modern OS maps, are in the W. corner of Dringhouses Moor (in 1629 this area lay outside the manor); there are considerable traces here of narrow ridge-and-furrow, running mainly N.N.W.–S.S.E., representing a late plough-up of marginal land.

North Field is now completely built over, though considerable stretches of ridge-and-furrow show on recent air photographs (E. of railway 584496; W. of Eason View 579495; and S. of Hob Moor 583501). These ridges compare closely with the original disposition of strips in furlongs, as shown on the 1629 map. Even now, after the area has been developed, some hedges and roads preserve the lines of the original furlongs. The 1853 OS map shows the North Field enclosed, but still preserving in detail the furlong structure.

No cultivations survive on the former West Field and 'Streate Lands', but closely-spaced, parallel hedges represent the enclosure of consolidated holdings of strips. Both of these open fields, by the nature of the narrow, glacial ridge on which they lay, were very long and thin, with all the strips running one way (roughly W.N.W.–E.S.E.).

Ridge-and-furrow 30–35 ft. wide is visible E. of the Tadcaster Road in the pasture fields N. and S. of Cherry Lane, and next to the Knavesmire. These presumably represent original open-field strips behind the crofts and tofts. Although they were enclosed at an early date, they were still used as half-year commons in the 18th century (YCA, D/Vv., Micklegate Ward Stray, 1772).

In Bustardthorpe ridge-and-furrow can still be seen on the low, gravel ridge (between the Knavesmire and the Ings), on the E. margin of the Knavesmire (598493), and E. of Bishopthorpe Road, where ridge-and-furrow 36–45 ft. wide is particularly well-marked E. of the old gravel workings (N. and S. of 601491). All this ridge-and-furrow runs E.–W. On the 1629 map, the area is marked 'Yorke Feilde' subject to the Manor of Dringhouses. There is a close correspondence between the ridges and the consolidated strip-holdings of the open-field. A large enclosed area immediately to the S. of 'Yorke Feilde' is called Bustard Hall Garth, and is obviously the site of the former Domesday manor of 'Torp' (Bustardthorpe).

A long, narrow strip of ground, now part of the Knavesmire (centring on 590500) also bears broad ridge-and-furrow, originally part of the Manor of Dringhouses. By 1629 the area was already enclosed and known as 'The Flatts'. The 1772 map shows that it was still subject to common rights, as were most of the ancient enclosures in the manors of Dringhouses and Middlethorpe.

(d) *Holgate.* No field names survive, and the area is almost entirely built over. The ings were on the alluvium by the Ouse and the moor on the lacustrine deposits S. of the village, while the arable was on the glacial deposits around the village, here reaching their maximum height of 125 ft. above OD on Severus Hills. Ridge-and-furrow, forming terraces on the steeper slopes, survived until recently on the hills, particularly How Hill (580515), where fragments still remain, and W. of Holgate Beck, S. of Hamilton Drive (587508).

[1] The map, drawn in 1629, was surveyed in 1624 by Samuel Parsons; it is described as 'The Plott of the Mannor of Dringhouses lyinge within the Countie of the Cittie of Yorke'. The draughtsmanship is of fine quality and the plan extremely accurate, showing individual buildings and garths.

(e) *Middlethorpe*. This township extended from the R. Ouse to the boundary of Dringhouses on the W. The arable land stretched in a crescent from N. of Middlethorpe Manor and Hall in the N.E. to the edge of the Common Moor in the S.W. N.W. of the present village ridge-and-furrow apparent on air photographs taken in 1952 is now built over. On the N. ridge-and-furrow does survive N. and S. of Knavesmire Lane (597488), and N. of the Hall and Manor (600488). The widths of the ridges are, respectively, 33–40 ft. and approximately 30 ft. These traces on the ground compare quite closely with the disposition of the strips in the open-field, as shown on the 1629 map.

Middlethorpe Ings are outside the boundary of cultivation on the W. bank of the Ouse. Middlethorpe Common Moor lay to the W. of the arable land, and was almost completely enclosed between the surveying and drawing of the map (1624–9). Only a small piece of land at the S. end of the Knavesmire is now called Middlethorpe Common. Two small parcels of land, called 'The Moore Lands' and 'Honger Hills' on the 1629 map, on the N.W. side of the Common, probably represent arable intakes at some time before the 17th century.

'Streate Lands', the arable area S. of Dringhouses village and E. of the Tadcaster Road, is shown as part of the Manor of Dringhouses in 1629, yet in 1838, when the Middlethorpe Tithe Award Map was surveyed, it was part of Middlethorpe township. A dyke which forms the boundary between the Knavesmire and Middlethorpe Common is referred to in a lease of 1567 (YCA, YC/DA, G.5), when it was accepted by the City and the Lord of the Manor of Middlethorpe as their common boundary.

ECCLESIASTICAL

York S.W. of the Ouse formerly comprised seven intramural parishes, of which three extended outside the walls and between them covered most of the suburban area. The extramural parish of St. Clement, which had given its name to Clementhorpe by the time of the Conquest, had become united to that of St. Mary Bishophill Senior for taxation purposes by the early 14th century; the benefices were formally united in 1586 when the small parish of St. Gregory was also merged with that of St. Martin in Micklegate, after a similar long period of effective union for taxation. There are now no monumental remains of the churches of St. Clement and St. Gregory, and St. Mary Bishophill Senior was demolished in 1963 (*see* Monument (9)). Within the suburban area were two chapels-of-ease, both to the parish of Holy Trinity (or St. Nicholas), Micklegate, namely St. James on The Mount, all trace of which has gone, and Dringhouses (*see* Monument (12)). The rural parish of Acomb was brought within the city boundaries in 1937 (*see* Monument (11)).

(4) PARISH CHURCH OF ALL SAINTS (Plate 93), stands in a churchyard of some extent on the W. side of North Street; it is built partly of rubble, partly of magnesian limestone ashlar, and has roofs of modern tile.

The church, which later belonged to the Priory of Holy Trinity (Monument (5)) in Micklegate, is not mentioned in the foundation charter of *c.* 1090–1100. It is listed among the possessions of the Priory in a letter of Pope Alexander III (1166–79) and in the charter of Henry II (1175–88) (*EYC*, VI, 76–7, 84–5). The original church was a simple rectangular cell, a local type found in the late 11th century, to which a S. aisle was added in the later 12th century. In the 13th century it was enlarged as a cruciform building with aisleless *Chancel* and aisled *Nave* of three bays.[1] The E. end was partly rebuilt in the first half of the 14th century when chapels flanking the chancel were added. The N. Chapel was built in *c.* 1324/5 when John Benge, chaplain, founded at the altar of St. Mary a chantry for the souls of John and Hugh Benge and their ancestors (*CPR, 1324–7*, 31). The E. window of the chancel, of which the contemporary glass is now in the E. window of the N. aisle, dates from *c.* 1320–40. The E. window of the S. Chapel, probably the choir of St. Katherine mentioned in a will of 1406 (Raine, 253), has reticulated tracery of *c.* 1340. Much work, including the widening of the *Aisles*, was being carried out between *c.* 1390 and *c.* 1410. In 1394 Richard Byrd of North Street, tanner, left 6*s.* 8*d.* to the new fabric of the church (Shaw, 83). In 1407 William Vescy, mercer, left 100*s.* for improving and ornamenting the choir and founded, at the altar of St. Thomas the Martyr, a chantry which was licensed in 1410 (*CPR, 1408–13*, 162). This altar was probably in the N. aisle by the third window from the E., which had a figure of St. Thomas. An inventory made in 1409/10,

[1] A cruciform plan is indicated by the relatively narrow central span on the N. side, by the pairs of buttresses flanking the penultimate bay of each choir aisle before 1860 and by the scale and arrangement of the N.E. chapel, which must originally have had a roof running from N. to S. The view of the church from N.E. by W. Monkhouse and F. Bedford (*The Churches of York*, 1843) shows that the aisle roof changed in height above the second buttress from the E., indicating that a former transept had been absorbed in the chancel aisle.

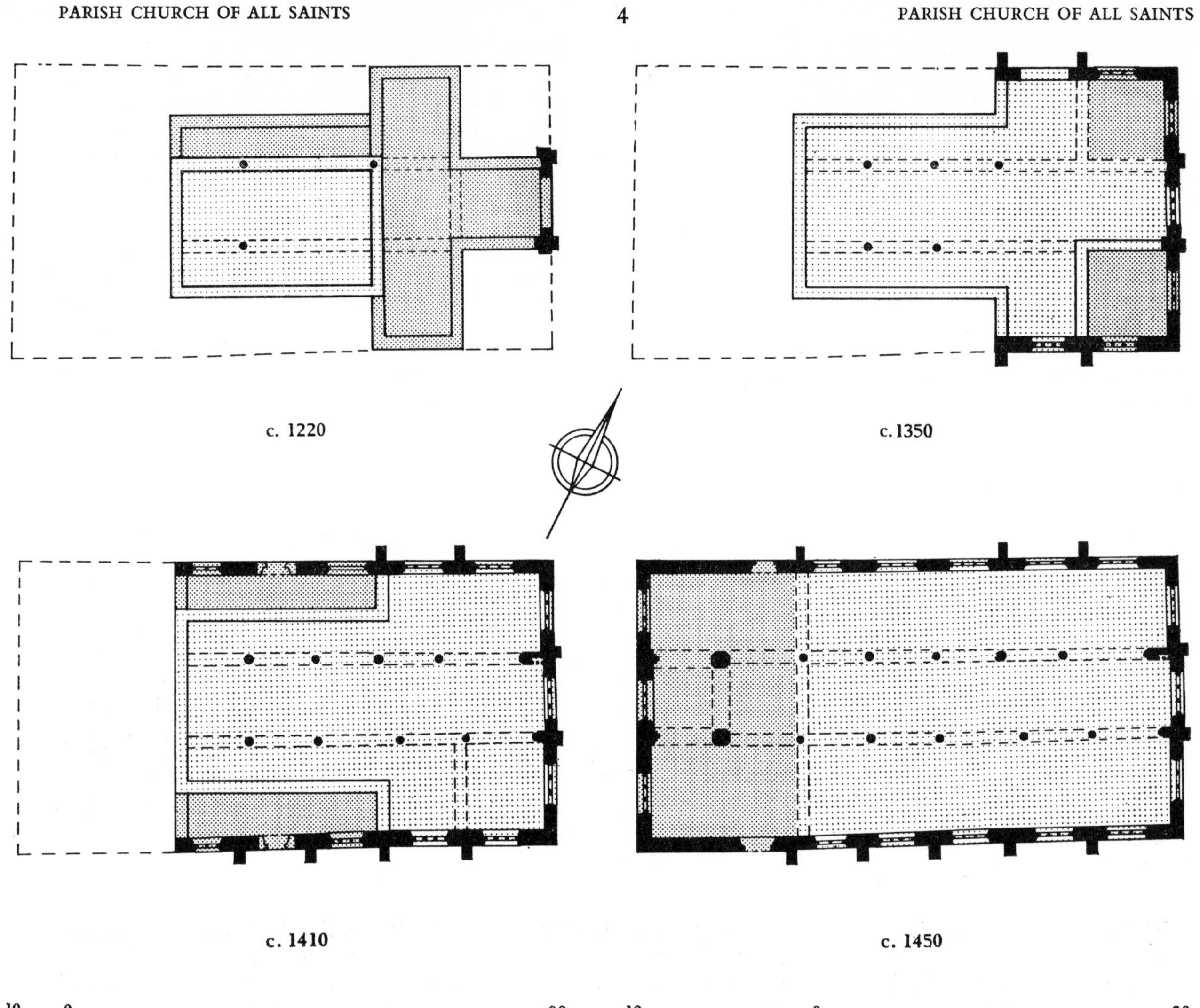

Fig. 24. (4) Church of All Saints, North Street. Architectural development. (New work in darker stipple.)

after the death of Hugh Grantham, mason, records that he owed John Ebirston 6*s.* 8*d.* for stone for a window in All Saints, North Street, and that he was owed 40*s.* for the window. Grantham was also owed £4 by John Thornton and William Pontefract; this may connect John Thornton with glass that would on stylistic grounds be associated with him. In 1410 Adam del Bank, dyer, founded a chantry at the altar of St. Nicholas[1] in the S. aisle and left 10 marks to repair the altar of St. James in the same aisle and the stonework of a window there (Raine, 254). In 1429 Reginald Bawtre, merchant, left 100*s.* to the fabric of a new glass window in the N. aisle, and glazing of the windows in both aisles continued until *c.* 1440 (*see* Fittings: *Glass*).

The church was extended W. by two bays in the second quarter or middle of the 15th century with a *Tower* in the W. bay of the nave, and had been entirely reroofed by *c.* 1475. Mention of an anchoress in 1430 (Raine, 254) suggests that the S. extension of the S. aisle (*see* Architectural Description) was already complete. The W. piers of the nave arcades both have a rare mason's mark found on other early 15th-century buildings in York. In 1467 Thomas Howson, Vicar, desired to be buried opposite the image of Our Lord crucified, which stood above the central window of the

[1] In 1458 there is reference to 'cancello cantariae S. Nicolai et B. Kat. virg'. (*TE*, II, clxviii.)

W. front. In 1482 Margaret Clerk gave 6*s.* 8*d.* towards the consecration of a bell in the tower (Raine, 254). In 1444 John Sharpe, tile maker, left 3*s.* 4*d.* to the fabric and 500 'thack tile'. In 1448 Richard Toone, tanner, left one fother of lead for the roof, if begun within a few years. The arms of Gilliot on a boss on the chancel roof refer to John Gilyot, Rector 1467–1472/3 (Borthwick Inst., Register of George Neville, ff. 15, 155); they occur again on a misericorde (*see* Fittings: *Seating*). A major restoration, including the rebuilding of the S. wall, was carried out by J. B. and W. Atkinson in 1866–7 (APS *Dictionary of Architecture*, VIII (1892), York, 5; Borthwick Inst., Faculty Papers, 1866/1) at a cost of £1,600. Further work was carried out in 1884 and in 1907–8. The plan prepared by J. W. Tate in 1866 and redrawn by E. R. Tate in 1908, shows the chancel one bay further E. than at present (C. Kerry, 'History and Antiquities of All Saints' Church, North Street, York', in *AASRP*, IX, pt. 1 (1867), 57–69; P. J. Shaw, *All Hallows, North Street* (1908)).

Architectural Description—The church is an aisled quadrilateral without structural division between chancel and nave; the tower at the W. end of the nave is set within the quadrilateral. The present division between chancel and nave is modern. The plan prepared in 1866 and redrawn in 1908 shows the nave extending one bay further E. as in the mediaeval arrangement.

The *Chancel* and *Nave* (79½ ft. by 12 ft.) (Plates 94, 95, 96) have the base of a 13th-century E. wall of small blocks of ashlar flanked by angle buttresses of which those to the E. were refaced in 1867 and those to the N. and S. now show only as slight projections from the added E. walls of the aisles. Internally the E. wall is plastered but the base of one bay of a 13th-century wall arcade is visible on the S.; it consists of a two-centred arch with moulded label containing a heavy trefoiled head on attached shafts with bell caps. On the S. it abuts on a similar bay in the S. wall; between the bays is a band of dog-tooth ornament. The E. window, an insertion of *c.* 1320–40, has original jambs, but the mullions are recut or modern. Externally the head has a wavemould, which merges into the chamfered jambs; the label has been trimmed back. Internally all members are chamfered; at springing level on either side is a small demi-figure with long hair and bonnet and holding a spray.

The lower part of the N. wall has exposed rubble incorporating an aumbry at the E. end. The first arch of the N. arcade is two-centred, of one chamfered order and built of freestone. The E. respond and the first pier each have a chamfered cap with octagonal shaft; the respond retains part of a chamfered base and the pier has a moulded octagonal base on a square plinth. Original tooling and masons' marks like those on the piers opposite indicate an early 15th-century date. The second pier has a hollow-chamfered abacus cut off to E. and W. and with claw tooling probably of *c.* 1200; it has a crudely moulded cap and bold necking. The column is monolithic Roman shaft reused with a top course of fine gritstone. The

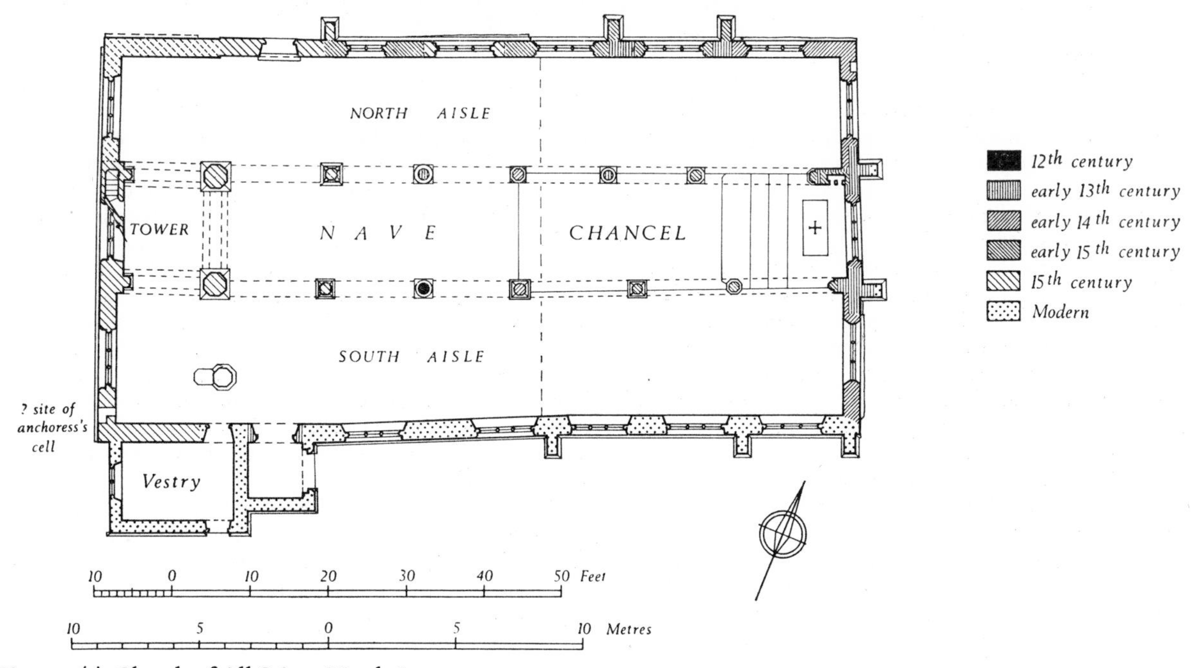

Fig. 25. (4) Church of All Saints, North Street.

second and third arches are identical, with small voussoirs and of one large chamfered order. Masons' marks identical with those in the arcade opposite suggest a date in the early 15th century, perhaps incorporating reused material. The third pier has a square abacus with octagonal capital and similar base with bold angle stops. A corbel on the W. face has a demi-figure like those in the E. window (Plate 18). The pier, like that opposite, is probably of the 14th century.

The fourth pier has a circular chamfered abacus and bell cap with nail-head ornament and a bold necking. The base has two flattened rolls on a square plinth. Cap and base are in magnesian limestone and have 13th-century tooling. The fourth and fifth arches are two-centred, of two chamfered orders, the fifth arch being higher than the fourth. Both have claw tooling of Early English character and of the early 13th century contemporary with the pier. Large springing courses at the E. end of the fourth arch indicate a 14th-century modification; the fifth arch was modified at the W. end when a pier replaced the earlier W. wall. The fifth pier resembles the third. A mason's mark associates it with the pier opposite and other York work of the first half of the 15th century. The contemporary sixth arch is higher than those to the E., of two chamfered orders and with very large voussoirs which bond into the pier of the tower.

The E. respond and the first two piers of the S. arcade have square abaci with plain octagonal caps and octagonal shafts. The second pier has a moulded base set on an older plinth; it has finer detail than the first. The first and second arches have large voussoirs. The piers have masons' marks that associate them with the first pier on the N. and indicate an early 15th-century date. The third pier resembles that opposite and is probably of the 14th century; the plinth is set awkwardly on a larger and older one. The third arch is higher than those to the E.; it has very slight chamfers, stopped to the piers, of one square order and is probably of the 14th century. The fourth pier, of the late 12th century, has a square abacus, hollow-chamfered on the lower edge and swept inwards to a round necking; the whole is one piece of gritstone. The top course and the three lowest courses are of brown gritstone with haphazard tooling; the other courses are of magnesian limestone finely axed. There is a round water-holding base and a bold oblong plinth projecting to N. and S. The fourth and fifth arches, like the third, are each of one square order with a slight chamfer. The fifth pier and fifth arch resemble those opposite but a straight joint on the N. side suggests that the arch is earlier than the pier of the tower.

The *North Aisle* (13½ ft. wide) (Plate 96) has an E. wall with large blocks of ashlar below and smaller, less regular pieces above. The first bay of the N. wall is similar with a chamfered plinth. The E. window of *c.* 1340 has three ogee-headed trefoiled lights and reticulated tracery. A similar window in the first bay of the N. wall had the head cut off, presumably when the present 15th-century roof replaced an earlier one running N. and S. To N. of the E. window externally is a small niche with four-centred head, blocked with brick before 1908. The second, third and fourth bays each have a window of three cinquefoiled lights under a square head; the window in the fifth bay is of two lights. Much of the walling is restored but the masonry with large blocks at the base and a chamfered plinth is original, of the 15th century. W. of the fifth bay is a 15th-century three-stage buttress with oversailing plinth. The sixth bay is of ashlar in small blocks with a two-centred doorway, probably inserted as there is brick on either side. The seventh bay sets back at a straight joint; the base is of large ashlar with very fine joints, the upper part of brick. The W. wall is of good magnesian limestone and has a boldly moulded plinth. The W. window has three cinquefoiled lights with vertical tracery. The *South Aisle* (15 ft. to 16 ft. wide) has an E. wall of the 14th century incorporating some very large blocks of magnesian limestone at the base. The heavily restored E. window has three trefoiled lights and reticulated tracery. The S. wall was entirely rebuilt in 1867. The present S. doorway in the sixth bay was formerly between the second and third windows; it was widened in 1908. It is of the 13th century, and has a two-centred head and continuous reveals with a band of nail-head ornament between two rolls. Much mediaeval stone, including many coffin lids, was used in the rebuild. A modern porch and vestry mask the sixth and seventh bays. The W. wall is of one build with the rest of the W. end; the three-light window has vertical tracery. At the S. end internally is a small oblong opening with chamfered reveals; beyond is an archway with a four-centred head. High up in the wall is another small opening like the first. All three were probably associated with the cell of the anchoress mentioned in 1430.

The *Tower* (10 ft. square) (Plate 11) is of three stages surmounted by a spire, in all 120 ft. high. The two-centred tower arch of two chamfered orders springs without capitals from two octagonal piers (Plate 95). The lower N. and S. arches of two hollow-chamfered orders merge into the piers; on the W. sides the inner orders form responds with square bases and bold stops. A doorway with chamfered reveals at the N. end of the W. wall leads to a newel stair, which is corbelled out above. Outside, above the W. window, is a small niche, now with ogee trefoiled head, but originally a two-light window. The second stage is octagonal with weathered angles and a two-light window of 15th-century type in each of the cardinal faces. The third stage, also octagonal, has a tall two-light transomed window in each cardinal face. Above the third stage is an openwork parapet with three lights to each side. The octagonal spire is of ashlar with the window heads and mullions largely renewed.

The timber *Roofs* of the chancel and chancel aisles (Plate 97) are each of five bays. The trusses have moulded principals with collars and hammer beams, with arched braces between. The hammer beams are carved as figures of angels (Plate 43). There is a moulded collar purlin and a single moulded purlin on each side. There are carved bosses under the collar and side purlins; the arched braces of the chancel have foliated spandrels. The chancel roof timbers are heavier than those of the aisles and in general the carvings in the N. aisle are better than those in the chancel; those in the S. aisle are rather crude. The angels are generally represented winged and with flowing hair. They are shown playing musical instruments or holding emblems or, in three instances, a church or shrine. Many of the wings and other details are missing or damaged. The bosses

(142) ACOMB. Acomb House, No. 23, Front Street. First floor. Late 18th century.

(142) Acomb House. First floor. Detail. Late 18th century.

(81) MICKLEGATE. Micklegate House, Nos. 88, 90. Drawing Room. Detail. *c.* 1750.

18th-CENTURY CORNICES

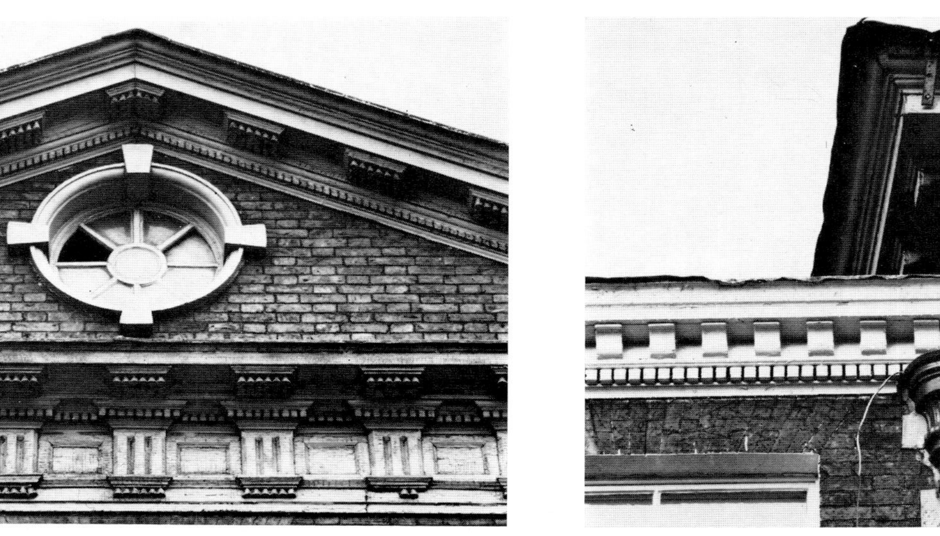

(66) Garforth House, No. 54. *c.* 1757.

(83) House, No. 92. Cornice and rainwater head. *c.* 1798.

(65) House, Nos. 53, 55. *c.* 1755.

(68) House, Nos. 57, 59. 1783.

MICKLEGATE

TADCASTER ROAD. House, No. 306. 1833.

THE MOUNT. House, No. 77. 1831/2.

THE MOUNT, House, No. 140. 1824.

MICKLEGATE. House, Nos. 39, 41. 1835.

THE MOUNT. House, Nos. 92, 94. 1821.

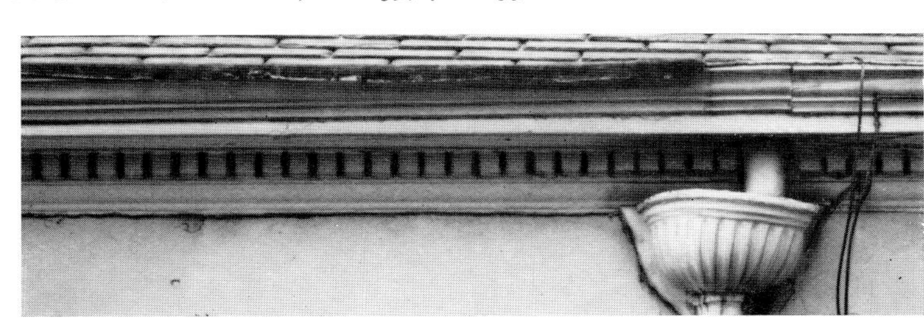

THE MOUNT. House, No. 102. 1807/8.

IRONWORK

TADCASTER ROAD. House, No. 302. Balcony, 1833.

THE MOUNT. House, No. 132. Balcony, *c.* 1830.

THE MOUNT. House, No. 122. Balcony, 1848.

ACOMB ROAD. House, No. 11. Railings, *c.* 1840.

LEADWORK. RAINWATER HEADS

(80) Bathurst House, No. 86. Early 18th century.

(81) Micklegate House, Nos. 88, 90. 1752.

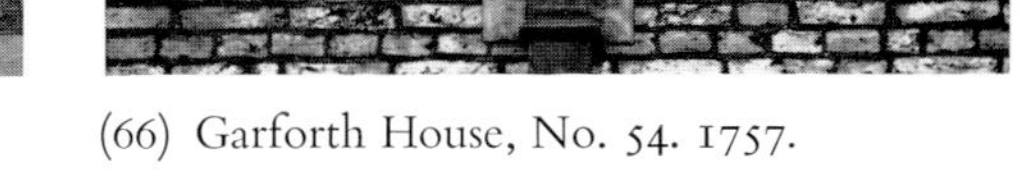

(66) Garforth House, No. 54. 1757.

(73) House, No. 71. *c.* 1750.

(71) House, No. 67. 1763.

(68) House, No. 57. 1783.

MICKLEGATE

(122) TANNER ROW. The Old Rectory, No. 7A. *c.* 1640.

(72) MICKLEGATE. House, No. 68. *c.* 1650.

(80) MICKLEGATE. Bathurst House, No. 86. Early 18th century.

(38) BISHOPHILL. Bishophill House, Nos. 11, 13. Mid 18th century.

(61) MICKLEGATE. House, No. 32. Early 17th century.

(162) MIDDLETHORPE. Middlethorpe Grange. Late 17th century.

Main staircase.

Servants' staircase.

(163) MIDDLETHORPE. Middlethorpe Hall. *c.* 1700.

STAIRCASES

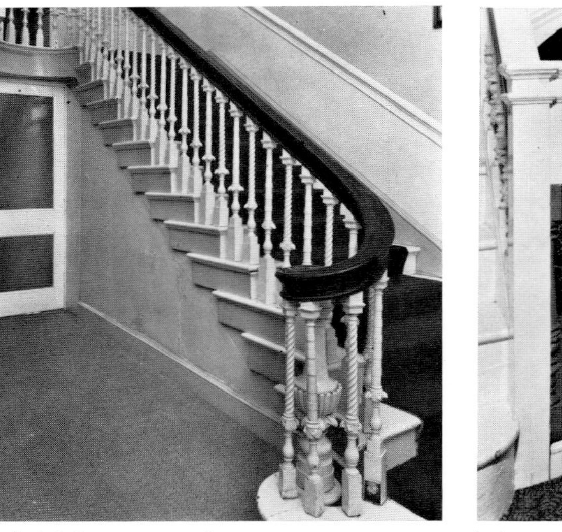

(142) ACOMB. Acomb House, No. 23 Front Street. Early 18th century.

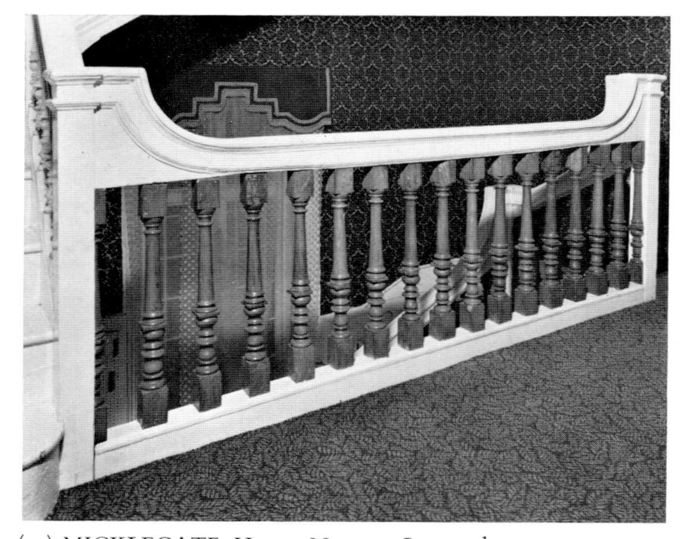

(94) MICKLEGATE. House, No. 114. Late 17th century.

(96) MICKLEGATE. House, Nos. 122, 126. Mid 18th century.

(98) MICKLEGATE. House, Nos. 134, 136. 1740.

(80) MICKLEGATE. Bathurst House, No. 86. Early 18th century.

(81) MICKLEGATE. Micklegate House, Nos. 88, 90. *c.* 1750.

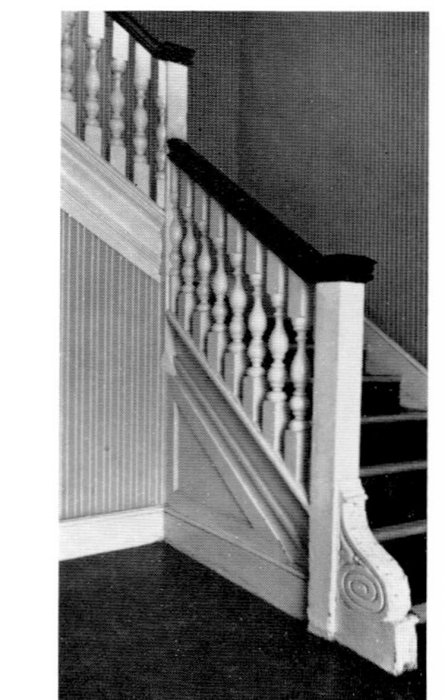

(141) ACOMB. The Lodge, No. 21 Front Street. *c.* 1700.

(98) House, Nos. 134, 136. First floor to attics. 1740.
MICKLEGATE

(59) House, No. 17. Early 18th century.

(56) MICKLEGATE. House, No. 8. Second floor. *c.* 1700.

(119) TADCASTER ROAD. The White House, Nos. 238, 240. Early 18th century.

(58) MICKLEGATE. House, No. 16. Second floor. *c.* 1764.

(93) MICKLEGATE. House, No. 113. *c.* 1740.

(98) MICKLEGATE. House, Nos. 134, 136. 1740.

(73) MICKLEGATE. House, No. 69. Mid 18th century.

(95) MICKLEGATE. House, Nos. 118, 120. *c.* 1742.

(96) MICKLEGATE. House, Nos. 122, 126. 17th and mid 18th century.

(96) MICKLEGATE. House, Nos. 122, 126. Mid 18th century.

(141) ACOMB. The Lodge, No. 21 Front Street. *c.* 1700.

(41) BISHOPHILL. House, No. 19. Early 18th century.

(64) MICKLEGATE. House, No. 42. Mid 18th century.

Main staircase.

Servants' staircase.

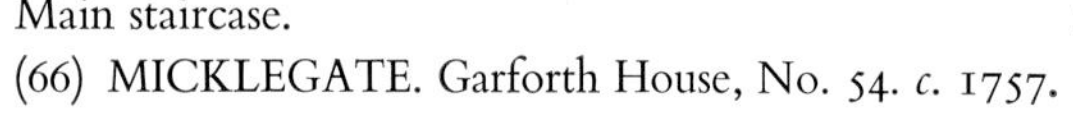

(66) MICKLEGATE. Garforth House, No. 54. *c.* 1757.

THE MOUNT. House, No. 94. 1821.

THE MOUNT. House, No. 127. 1833.

TADCASTER ROAD. House, No. 306. 1833.

(28) OLD RAILWAY STATION. 1840.

(65) House, No. 53. *c.* 1755.

(66) Garforth House, No. 54. *c.* 1757.

(80) Bathurst House, No. 86. Early 18th century.

(81) Micklegate House, Nos. 88, 90. *c.* 1750.

MICKLEGATE

THE MOUNT. House, No. 140. 1824.

ST. PAUL'S SQUARE. House, No. 46. *c.* 1835.

THE MOUNT. House, No. 130. *c.* 1830.

TADCASTER ROAD. House, No. 302. 1833.

(31) BOUND STONE. 18th century.

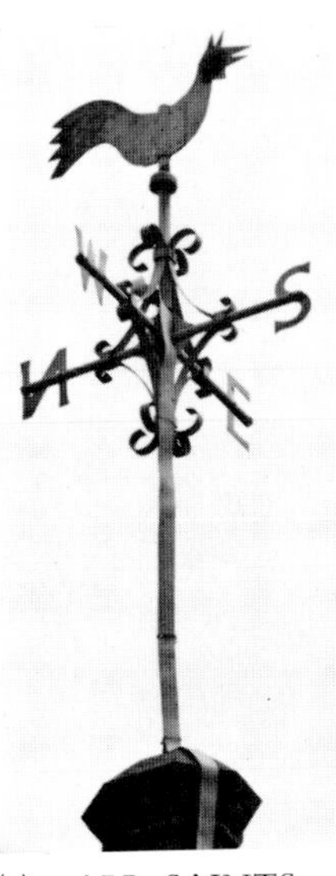

(4) ALL SAINTS, NORTH STREET. Weathervane. 1759.

(84) MICKLEGATE. Falcon Inn. Sign. *c.* 1800.

(81) MICKLEGATE. Micklegate House, Nos. 88, 90. Range and spit. *c.* 1750.

include human and animal heads, grotesques, and foliage. The two E. bosses in the centre of the chancel portray Christ, bearded, with head dress and a cord round the brow, and an angel holding a shield with the arms of Gilliot.

The N. aisle of the nave is covered with a barrel vault of plaster with 19th-century cased ties forming four bays. Above the N. nave arcade three stone corbels, wall posts and a chamfered wall plate of the mediaeval roof remain; the wall plate stops short over the fifth post from the E., where the original nave ended.

Fittings—*Aumbry*: in chancel in N. wall, with two trefoiled openings under square head, mediaeval. The church plan of 1866 (Shaw, opp. 14) shows two other aumbries, both in N. aisle: at E. end of the N. wall, and under third N. window where an altar of St. Thomas the Martyr stood.

Bells: three; (1) 'Soli Deo Gloria 1640', (2) 'God send vs all the blisse of heaven Anno D(omi)ni 1627', both probably by William Oldfield; (3) small sanctus bell with 'ihc' (G. Benson, *The Bells of the Ancient Churches of York* (1885), 10). *Benefactors' Tables*: two, both in N. aisle, on N. wall, large panels with round pediments and bold moulded frames; entries now indecipherable but *see* Shaw, 70–1, (1) to E., probably made 1764 ('Painting and Lettering new Benefaction Table...' with a reference to Widow Wade's gift—Church Wardens' Accounts in Shaw, 69), (2) to W., perhaps made in 1764, relettered in 1804, among gifts recorded is Samuel Harsnett, Archbishop of York, 1630, silver chalice and cover with his Arms (see *Plate*). *Bracket*: on N. arcade, on W. of third pier, supported on small figure with close-fitting tunic and tonsure, 14th-century.

Brasses and *Indents*. *Brasses:* in S. aisle of chancel, (1) rectangular plate (Plate 36) with inscription in black letter 'Orate sp(ec)ialiter pro a(n)i(m)abus Will(el)mi Stokton et Rob(er)ti Colynson quondam maior(um) ciuitatis Ebor. & Isabelle vxoris eoru(n)dem quor(um) a(n)i(m)ab(us) propicietur deus Amen' (William Stockton, Lord Mayor 1446, d. 1471; married secondly Isabella, widow of Robert Colynson, Lord Mayor 1457, d. 1458); set in large freestone slab commemorating John Wardall; W. of (1) at entrance to aisle, (2) (Plate 27) oblong plate inscribed 'Oret q(ui)sq(ue) speciali(ter) p(er)t(ra)nsie(n)s p(ro) a(n)i(m)ab(us) Tho(m)e Clerk quo(n)da(m) cl(er)ici ciuitatis Ebor. & toci(us) com(munitatis) & Margar(e)te vx(oris) q(u)i obieru(n)t xvj dieb(us) ffebr. & Marcij A(nn)o do(min)i M°cccc°lxxxij° q(u)or(um) a(n)i(m)ab(us) p(ro)piciet(ur) d(eu)s Amen' (Thomas Clerk, clericus, free 1449; *Freemen*, I, 170), and Evangelists' symbols (N.E. for St. Mark missing), set in small marble slab. On same slab, indent for brass (3) Thomas Atkinson, 1642, now fixed on wall of N. aisle (Plate 35), lettered in capitals and also inscribed 'Vixi dum volui, volui dum Christe volebas. Mortuus et vivus sum moriorq(ue) tuus' (Thomas, tanner, son of Henry Atkinson, tanner, 1589; *Freemen*, II, 32). In S. aisle, on S. wall, (4) large oblong plate (Plate 35) with shield-of-arms of Askwith differenced with a crescent and inscription to Thomas Askwith (Sheriff 1592), 1609, and wife Anne; below (4), (5) oblong plate, Charles Towneley, 1712 (Plate 36). *Indents*: in N. aisle, (1) oblong, in grey marble; (2) small, in tapered coffin lid of freestone; W. of (2), (3) similar, in coffin lid; at entrance to chancel, (4) in marble slab, and (5) oblong, in worn stone; in S. aisle (6) for figure and inscription plate together, in marble slab.

Chairs: three; (1) with straight back with two shaped and carved horizontal members, turned front legs and rail, 17th-century; (2) similar but broader and heavier, probably reproduction; (3) (Plate 44), early 18th-century. *Chest* (of drawers): in S. aisle, of oak, with two drawers above two recesses with fielded-panelled doors and all with contemporary brass handles, late 18th-century. *Coffins* and *Coffin Lids* (see also *Brasses* and *Indents*, Indents (2) and (3)): In N. chancel aisle, (1) freestone, tapered, with incised cross, 14th-century (perhaps of Margaret Etton (1391) or William Meburn (1394), both of whom wished to be buried before altar of Blessed Virgin Mary. In S. aisle, at E. end, (2) freestone, with incised foliated cross and a chalice below. In tower, against W. wall, (3) thick lid with foliated cross and multi-stepped base; (4) with foliated cross (Plate 27), 13th-century. In porch, (5) top of small freestone lid with foliated cross; (6) part of lid with large foliated cross in relief in round recess; (7) complete lid with simple cross, hatchet by cross shaft; (8) part of decayed lid with foliated cross; (9) upper part of lid with simple foliated cross; (10) lid with two crosses, bow and arrow under one and sword under other. Lids (5), (10) found in S. wall in 1867. Outside, (11)–(15) built into S. wall; W. of fourth window, (11) lower part of cross with stepped foot; under fourth window, (12) part of cross; under same window and to W., (13) part of lid with stepped bottom foot to cross; near E. side of fifth window, (14) piece of coffin lid with black letter inscription; under same window, (15) incised shaft of cross; built into S. wall of porch, (16) stone with part of one incised cross shaft; to W. of church, (17) stone coffin, tapered and with shaping for head, lidless; all 13th or 14th-century. *Font*: plain octagonal bowl curving inwards to octagonal shaft resting on moulded base, bowl with mediaeval tooling, shaft and plinth with modern tooling, base perhaps reused 15th-century capital, steps and cover modern.

Glass[1] (described from sill of window upwards and from spectator's left to right in each range of panels): In chancel, E. window, I (Plate 98), Lower range, (a) kneeling figures of Nicholas Blackburn junior (Lord Mayor 1429, d. 1448) and his wife Margaret (d. 1454) beneath shields bearing letter 'B' and arms of Blackburn differenced with mullet; (b) seated Trinity; (c) kneeling figures of Nicholas Blackburn senior (Lord Mayor 1412, d. 1432) and his wife Margaret (d. 1435), beneath shields bearing 'B' and arms of Blackburn. Inscriptions in part original but greatly falsified in 1844. Main range, glass mostly original, (a) St. John Baptist (Plate 104) bearing Agnus Dei on book; (b) St. Anne (Frontispiece), teaching the Virgin to read in a book inscribed 'D(omi)ne exaudi or(ati)onem mea(m) aurib(us) p(er)cipe ob(secre)ti(onem meam)' (*Psalm cxliii*, 1); (c) St. Christopher (Plate 104), bearing the Infant Christ, round his head a scroll inscribed 'Cristofori d(omi)n(u)s sedeo qui crimina tollo'. Glass originally in second window in N. wall; date probably between 1412 and 1428; moved by 1846, restored in 1844 by Wailes of Newcastle,

[1] See E. A. Gee 'The Painted Glass in All Saints' Church, North Street, York' in *Archaeologia* CII (1969).

who supplied in new glass most of lower part of window and all tracery lights; cleaned and releaded 1966.

In N. Aisle, E. window, II (Plate 99). Lower range: (a) Annunciation (Plate 102), much restored; (b) Nativity, much restored; (c) Christ rising from the tomb, angel and soldiers original. Upper range: (a) Adoration of the Magi, largely original; (b) Crucifixion, restored except for figure of St. John; (c) Coronation of the Virgin, largely original (Plate 102); tall canopies at heads of main lights, including geometrical traceries, and borders largely of old glass (Plate 107); tracery lights almost entirely 19th-century but some of background of oak leaves and part of figure of St. Michael in topmost light original. Glass originally in E. window of chancel, dating from *c.* 1320–40, the earliest in the church, moved to present position by 1846, restored 1844 by Wailes, releaded 1877, cleaned 1967.

N. wall, first window from E., III (Plate 98). At base of each light a panel containing three kneeling figures of donors, all looking to E.; inscriptions now lost formerly recorded the names of Roger Henrison of Ulleskelf, freeman, 1401, and Abel de Hesyl, living in parish in 1327 (*YASRS*, LXXIV (1929), 168), chamberlain 1329–30, and bailiff 1336–7 (*YASRS*, LXXXIII (1932), 192). Above, portrayal in individual panels (from *l* to *r*, upwards) of the signs of the end of the world as narrated in the 14th-century poem, 'The Pricke of Conscience' (ed. R. Morris (Berlin, 1863), 129–31), though the inscriptions beneath the panels differ from the text (restorations below from Henry Johnston's[1] record of the window in 1670): (1) Rising of the Sea, '[Ye first day fourty] cubetes [certain Ye see sall] ryse vp [abowen ilka mountayne]'; (2) Subsiding of the Sea, 'Ye seconde day ye see sall be so lawe [uneth] men sall it cee'; (3) The Waters return to their former level, 'Ye iij day yt sall be playne And stand as yt was agayne'; (4) Fishes rise out of the Sea (Plate 108), '[Ye fforth day] fisches sal mak[e a roring Hideus & hevy] to mannes [heryng]'; (5) The Sea on fire, 'Ye fift day ye sea sall bryn And all ye waters that my ryn'; (6) Bloody dew on Trees, 'Ye sext day sall [herbes &] trees Wyth blody dropes . . . grysely bees'; (7) Earthquakes (Plate 108), 'Ye seventh day howses mon fall Castels and towres and ilk a wall'; (8) Rocks and stones consumed, '[Ye viij day] ye roches & stanes [Sall bryn] togeder all at anes'; (9) Earth noises everywhere, '[Ye ix day] erth dyn [sall be Seve]rally in ilk [contry]'; (10) Earth level again, '[Ye tende day for [to] neven Erthe sall be playne & even'; (11) Men come out of holes, '[Ye xj day] sall men come owte [Of their] holes & wende a bowte'; (12) Dead men's bones arise, 'Ye xij day sal dede mens banes Be sumen sett & ryse all at anes'; (13) Stars fall from Heaven (Plate 103) 'Ye thirtend day suthe sall Stevyns fra the heuen fall'; (14) Death of all living (Plate 103) 'Ye xiiij day all yat liues yan Sall dy bathe chylde man & woman;' (15) Universal Fire, 'Ye xv day yus sall betyde Ye werlde sall bryn on ilk a syde'. In the Tracery lights: (W.) reception of the Blessed by St. Peter; (E.) damned dragged by demons into hell; in topmost light, fragments of seated Majesty recorded in 1670. Glass in original position, almost certainly by John Thornton of Coventry, and donors' names suggest early 15th-century; restored 1861 by J. W. Knowles of York, releaded 1877, cleaned 1966.

N. wall, second Window, IV (Plate 109). At base, in E. and W. lights, panels showing donors, the former in 1670 in a S. window (? second), the latter of Reginald Bawtre (d. 1429) formerly in next window W.; in middle light, two shallow arches, presumably from original canopies (*see below*). In main panels, six of the Corporal Acts of Mercy (Plates 111–13). In lower range: (1) Clothing the Naked; (2) Visiting the Sick; (3) Relieving those in Prison. In upper range: (4) Feeding the Hungry; (5) Giving drink to the Thirsty; (6) Entertaining the Stranger. Main Canopies, not belonging, perhaps slightly later than main panels (Plate 107). Glass (excepting main canopies) originally in fifth N. window, of two lights, implying three panels in each light; probably dating from between 1410 and 1435; former arrangement included figure of Nicholas Blackburn and undifferenced arms of Blackburn, probably indicative of memorial to the father of Nicholas Blackburn senior (d. 1432), namely Nicholas Blackburn of Richmond, senior, freeman, 1396. Present arrangement of 1846 or earlier, when presumably canopies were changed for height; restored by J. W. Knowles 1861, releaded 1877 and 1966.

N. wall, third Window, V (Plate 100). Three large figures: St. Thomas the Apostle, with scroll inscribed 'D(omi)n(u)s meus et deus meus'; Christ bearing cross-staff with pennon, with scroll inscribed 'Thoma [ten]dite manu(m) manu(m) i(n) latus meu(m) qui no(n) viderunt'; archbishop, probably St. Thomas of Canterbury, whose altar was in this bay in 1407. Borders include niches containing small figures of prophets (Plate 110); canopies with pinnacled turrets with pairs of figures. Reginald Bawtre (*see* foregoing) bequeathed £5 in 1429 (Shaw, 90). Glass releaded 1877; after exchanges, figures restored to original positions during cleaning and releading 1966.

N. wall, fourth Window, VI. In lowest range: arms of (1) John Alcock (1430–1500), bishop of Ely; (2) France Modern and England quarterly, mutilated and remodelled; (3) Beauchamp. In middle range: (4), (6) roundels; (5) arms perhaps of Percy, inserted 1966 (in E. window of chancel in 1659: BM, Lansdowne MS. 919, f. 14v.). In top range: arms of (7) Luttrell (?); (8) (unidentified 1); (9) (unidentified 2)—(8), (9) said to have come from Winchester. Some patterned background quarries and canopies, 15th-century. All of plain glass in 1730 (Gent, 163); by 1877, when releaded, containing heraldry from E. window of N. aisle, transfer having probably been made *c.* 1845 at restoration by Wailes; again releaded and extensively rearranged 1966.

N. wall, fifth window, VII. In main lights, tops only of elaborate canopies, 15th-century (for rest of original glass, *see* N. wall, second window, IV).

In S. Aisle, E. Window, VIII (Plate 99). Glass including remains of original glazing, *c.* 1340, drastically restored by Wailes 1844. Lower range: (1) female donor; (2) the Agony, Christ kneeling before the Cup; (3) female donor. Upper range: (4) St. Mary; (5) Crucifixion; (6) St. John. The main panels set in background of quarries within wide borders, an unusually early example of such treatment; middle light with border of castles and cups, and outer lights with vine

[1] *See* Sectional Preface, p. xxxii.

scrolls; quarries, some original, with oak sprays. Medallions containing angels and grotesque figures playing musical instruments in upper parts of main lights and in tracery.

S. wall, first window, IX (Plate 100). At foot, in side lights, groups of kneeling donors: E., priest, civilian and woman, scroll inscribed 'libera nos' and, over the second, a shield with 'R' impaling a *bend*, recorded in 17th century as James Baguley, Rector (1413–40), and Robert Chapman (Free, 1423) and wife (the arms of Baguley of Baguley survived in 1659); W., woman between two men. In main lights: (1) St. Michael in plate armour, with scroll inscribed 'laudantes a(n)i(m)as suscipe [san]cta Trinitas'; face stolen 1842 and replaced in plain glass; (2) arms of Whytehead, perhaps old, reset 1861; (3) St. John the Evangelist (Plate 105) in richly embroidered garment powdered with letters 'J' and 'M', with scroll inscribed 'benedictus sit sermo oris tui'. Borders and canopies of side lights original. Probably the original glazing of 1425–40 (between 1846 and 1966 the figure of St. Michael was in E. light of the third N. window); restored by J. W. Knowles 1861, at expense of rector Robert Whytehead, cleaned, releaded 1965–6. (At base of window inscription recording the work of 1861.)

S. wall, second window, X. The original glazing was probably that now in the fourth S. window and two groups of donors below; of the latter that from the E. light is now in the E. light of the second N. window; in the W. light the group included a man and two women with an inscription to Richard Killingholme (Free 1397, d. 1451) and wives Joan (d. 1436) and Margaret (see *Floor Slabs* (15)).

S. wall, third window, XI. The Nine Orders of Angels (Plates 101, 106). (The following description starts from the top and reads from left to right, range by range, downwards.) In cusped head of each light, three square quarries set diagonally and each within a sun. Top range: (i) Seraph leading archbishop, cardinal and bishop, inscribed above '[Sera]phyn amore [arden]tes [et d(eum circumamb)ulantes];' (ii) Cherub holding book leading doctors and clerks, inscribed above '[Cherubyn (jus)] scient[es] et recte dispone[n]tes'; (iii) Throne leading group of civilians in enriched and furred robes, formerly inscribed '[Throni ...... sub (iugan)tes]'. Middle range: (iv) Domination bearing sword leading emperor, king and pope, inscribed above 'D(omi)nac(i)ones humilit(er) d(omi)nant[es et b]enigne castiga(ntes)'; (v) Principality bearing cross and sceptre, leading noblemen and bishop, inscribed above '[Principatus] bonis succure(n)tes p[ro in]ferio[ribus o]r[dinantes]'; (vi) Power, in plate armour and holding staff bearing banner of the sun, leading group of clergy(?) and a woman, inscribed above '[Pote]stat[es (e celo) egre]dientes [malignos succumbentes]'. Lowest range: (vii) Virtue bearing spear, leading group of well-dressed men and a woman, inscribed above '[Virt]utes [(mira)cula fa(cientes deum) ita] reuelantes'; (viii) Archangel in cap and holding trumpet, leading group of men, one of whom bears a metal-shod spade, inscribed above '[Archangeli mortales (om)nes] deo [(conducen)tes]'; (ix) Angel in deacon's robe and holding staff, leading three men, one wearing glasses, two women and child, inscribed above '[Angeli] mestos consolantes [diu]ina [annunciantes]'. Such of the glass as is ancient is part of the original glazing of this window; it was complete, apart from slight damage to inscriptions, when drawn by Henry Johnston (p. xxxii) in 1670 (Bodleian, MS. Top. Yorks. c14, f. 96) (Plate 101) and still recognizable in 1730 (Gent, 163). Serious damage occurred later (Plate 30). In 1965 the glass was releaded and restored according to Johnston's drawing.

S. wall, fourth window, XII. In main range, under original canopies (Plate 107): (i) St. James (?), wearing skin robe, head by Wailes; (ii) Crowned Virgin, standing with Child (for border figure *see* Plate 110); (iii) Mass of St. Gregory (Plate 105); the saint, shown as archbishop with nimbus, holds the Host and adores the half-length figure of Christ emerging from the tomb, on missal quotation from Canon 'Simili modo p(os)tquam cenatum est accipiens et hunc p(re)clarum' and on scroll proceeding from head of Christ 'Accipe hoc care me(us) p(ro) qui(bus)cu(mqu)e pecieris impetrabis'. Glass of *c.* 1440 originally in second S. window but by 1730 moved to fifth S. window. Cleaned, rearranged and releaded, and many fragments from other windows incorporated to fill lower parts, 1966.

Images: In N. aisle: (1) in fine white limestone, head and shoulders of woman (Plate 39) with traces of blue paint on undergarment and red on outer garment, perhaps 15th-century; (2) Nottingham alabaster carving of Resurrection (Plate 39) with traces of gilding and red paint, 15th-century, set in 19th-century wooden frame; (3) priest pouring wine into chalice, carved oak figure with traces of colour, probably mediaeval, face decayed; (4) King David (Plate 42), carved in soft wood, heavily stained, possibly from 18th-century reredos; (5) St. Lucy (?), figure of woman with sword through throat, much restored.

Inscriptions and *Scratchings*: for masons' marks, *see* Fig. 6, p. v. *Lord Mayors' Table*, in N. aisle long panel with square panel above, both with moulded frames and, at top, third square panel with round pediment, with two maces in saltire and 'G.R.', in top panel and below in oval frames arms of York and following names: Chas. Parrot, Lord Mayor 1723; Thos. Kilby, Lord Mayor 1784; Jno. Kilby 1804; Sir Jno. Simpson, Lord Mayor 1836.

Monuments and *Floor slabs*. *Monuments*: on S. wall, between third and fourth windows: (1) John Etty, 1707/8, carved cartouche (Plate 32), inscription, scarcely legible, recorded as 'Nigh to this lyeth John Etty, Carpenter, who By the strength of his own genius and application had acquired great knowledge of Mathematicks especially Geometry and Architecture in all its parts, far beyond any of his Contemporaries in this City, who died the 28th of Jan. 1709 Aged 75. His Art was great, his Industry no less What one projected, th'other brought to pass'. (Etty was buried 30 January 1708); between fourth and fifth windows: (2) Margaret Pennington, 1753, freestone tablet. *Floor slabs*: all of freestone except where otherwise stated. In N. aisle: (1) Joan Stoddart, 1599, inscribed in ligatured capitals; James, son of Thomas Pennyman, D.D., 1699, Esther, wife, 1745, on same slab in script; and 'Iohn Stoddart· clerke / Parson · of · this Rectory · induct here of Marche 1593' in small incised compartment in lower corner; (2) Joshua Witton, 1674, black marble, with arms of Witton impaling Thornton (Plate 36); (3) Richard Wilson, 1742, Elizabeth, widow, 1766; (4) John Rothum, 1390 (will proved

1 May 1390; Shaw, 45); (5) '[Orate] p(ro) a(n)i(m)a Will(el)mi L[on]disdall de [Ebor tanner et pro animabus El]ene et Alicie uxor(um) ei(us) a(nn)o d(omi)ni M°CCCC° [lxxx] septemo', slab in four pieces (William Lonesdale, barker, Free 1454 (*Freemen*, I, 175); will proved 4 March 1487/8 (Wills, vol. v, f. 325)); (6) 'Orate pro a(n)i(m)a Ioh(ann)is de Coupland civis et tannator' (John Coupeland, barker, Free 1425 (*Freemen*, I, 138); will proved 8 June 1469 (Wills, vol. IV, f. 135)); (7) Ann Dawson, 1730, Ann Pick and Susannah Cass, 1780, granddaughters; (8) 'D.D.', small stone with large initials. In nave: (9) Mary Mason, 1718/19; (10) Mary Milner, 1783, George, husband, 1789. In S. aisle: (11) Thomas de Kyllyngwyke (Free 1360; living 1381) and wife Juliana, upper half of fine slab with intricate cross-head and part of shaft, black-letter inscription 'hic · iacent · thomas · de [K]yllyngwke · quondam [ci]uis ebor · et Juliana uxor eiusde(m) q(uo)r(um) a(n)imab(us) p(ro)picie(tur) d(eu)s am(en)'; slab palimpsest, on under side a fulling-bat and some shears (Shaw, 45); (12) John de Wardalle (John de Weredale, barker, Free 1355 (*Freemen*, I, 50); will proved 30 Dec. 1395 (Wills, vol. I, f. 90); (13) Anna Clarke, 1795, John, husband, 1800; (14) John Bawtrie [1411] (succentor to the Vicars Choral 1388 (Shaw, 44); will proved 24 April 1411 (Wills, Dean and Chapter, vol. I, f. 157. Minster Library)); (15) Richard de Killingholme and Joan, Margaret his wives (his will proved 11 June 1451 (Wills, vol. II, f. 223), black-letter inscription, now mostly indecipherable, 'Orate pro animabus Ricardi Killingholme et Johanne et Margarite uxorum eius' (Shaw, 47); (16) Sarah Grainger, 1825, William, husband, 1830; (17) Elizabeth Harrison, 1772, Alexander and Richard, sons; (18) Susannah Clarke, 1788, Sarah Clarke, 1792; (19) Ann Harrison, relict of James, 1792; (20) Elizabeth Harrison, 1762, James Harrison, 1771.

Plate includes: cup given by Samuel Harsnett, Archbishop of York, 1630, and stand paten, with arms of See of York impaling Harsnett (Plate 37), both with York mark with date letter 'Z' in a pointed shield (1630/1) and maker's stamp 'T.W.' for Thomas Waite, goldsmith of York, Free 1613 (*Freemen*, I, 62); paten with a Glory, given by William Orfeur, with York mark for 1782/3 and makers' stamp for John Hampston and John Prince, and two flagons (Plate 37), also given by William Orfeur, each with arms of Orfeur of Cumberland, York mark of 1781/2 and makers' stamp as before; alms dish, given in 1698 by Thomas Simpson (*see* T. M. Fallow and H. B. McCall, *Yorkshire Church Plate*, I (1912), 6–8; and Shaw, 55 *et seq.*).

Pulpit (Plate 38): of oak, hexagonal, panels with painted figures on pedestals in moulded panels on five sides, of Hope with anchor, Our Lady with Infant Christ for Charity, Faith holding cup, a woman, and Peace holding doves, and, on frieze, 'and how shall they preach except they be sent', on lower base member, 'Anno Dom. 1675', set on modern stone base. *Reredos*: between first and second N. windows, war memorial made of carved enrichment probably from early 18th-century reredos. Mr. Etty (probably William, son of the John Etty buried in 1708/9) was paid £8 in 1710 for making and setting up a reredos and Mr. Graime received 10*s*. 6*d*. for painting a dove on it (Shaw, 67); noted in 1857 as 'altar piece of oak with pilasters of the Ionic order and gilt capitals, (Sheahan and Whellan, I, 504). *Royal Arms* (?): formerly at W. end, large panel with moulded frame and shaped top, destroyed *c*. 1962; in 1764, the king's arms were renewed and varnished (Churchwardens' Accounts, in Shaw, 69). *Stall*: one only remaining, misericorde (Plate 19) carved with pelican in piety flanked, on left, by letters 'GIM', for John Gilyot, Master (of Arts), on right, by arms of Gilliot, probably presented by John Gilyot, rector 1467–72/3. *Stoup*: by N. tower pier small round stoup on square block. *Tables* of the Creed etc.: In N. aisle E. wall, (1) Commandments, in existence *c*. 1730, but battlementing probably *c*. 1860. In S. aisle, W. wall, (2) Creed and Lord's Prayer, similar to (1).

Tiles: with impressed patterns and dark brown, yellow and green glazes, mediaeval, found 1867, said to be in Yorkshire Museum (Shaw, 16). *Wand*: of mahogany with silver head and silver band. *Weathervane* (Plate 92): on spire, elongated cock cut from flat brass sheet, with open beak, flared tail and applied eye; payment to churchwardens, 1759, for 'brass weathercock, Flemish brass for the same, gilding same' (Shaw, 69). *Miscellanea*: Stones—by N. tower pier, (i) voussoir with chevron and paterae on under-surface, perhaps from chancel arch or arcade *c*. 1150; (ii) large moulded voussoir, late 13th-century; (iii) large nook-shaft capital, 13th-century; (iv) piece of window tracery with cusp. Table, small, frame and top of soft wood, mainly 17th-century, with two consoles, early 18th-century, and some tracery, possibly mediaeval.

CHURCH OF THE HOLY REDEEMER, Boroughbridge Road. This church of 1962–4 includes re-erected parts of the structure of the demolished church of St. Mary Bishophill Senior, *q.v.*

(5) PARISH (former PRIORY) CHURCH OF HOLY TRINITY (Plates 12, 117), stands on the S. side of a large churchyard, the mediaeval layfolks' cemetery, fronting on Micklegate. The mediaeval walls are of brownish limestone, white magnesian limestone, and a little gritstone; the post-mediaeval walls are of dressed ashlar, with some brick; the roofs are of modern tiles and Welsh slates.

In *c*. 1090–1100 Ralph Paynel gave to the Benedictine Abbey of Marmoutier, near Tours, the church of Holy Trinity in York together with other properties which had belonged to a wealthy pre-Conquest minster of canons (*EYC*, VI, 66–9). The Domesday entries show that this minster had been known as Christ Church (VCH, *Yorkshire*, II, 192, 274). Of this minster there are no remains *in situ*, but the 11th-century architectural fragment (*see* Fittings, *Architectural Fragments* (1)) interpreted as part of a tympanum indicates a building of distinction. A church with short choir, partly recovered by excavation, aisleless transepts and nave, was built soon after the founding of the priory; only the two western piers of the *Crossing* and the N.W. angle of the *Nave* remain. A fire on 4 June 1137 destroyed the

PARISH CHURCH OF HOLY TRINITY

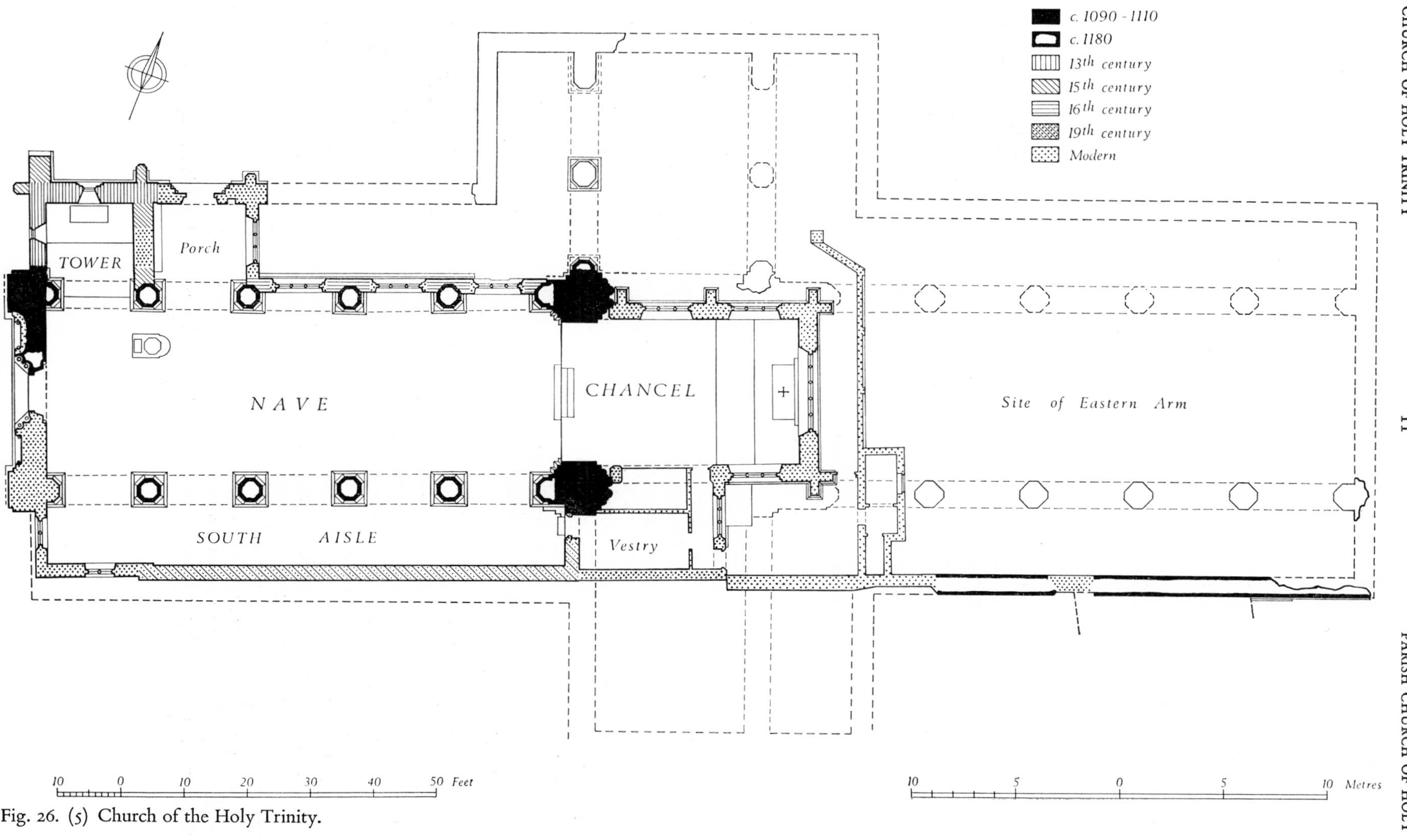

Fig. 26. (5) Church of the Holy Trinity.

Minster, St. Mary's Abbey, 39 parish churches and 'Holy Trinity in the Suburbs' (*YAJ*, XLI (1965), 367). The rebuilt church had an aisled eastern arm of five bays, a crossing of *c.* 1180, a N. transept with aisles on both sides and an aisled nave of five bays. By *c.* 1210 the W. end was remodelled, at least in the lower part, in the Gothic style, but completion of the upper parts and of the nave aisles was delayed, and gifts of oaks between 1235 and 1255 indicate the period of construction of the roofs (*CCR*, *1234–7*, 315, 432; *1237–42*, 264; *1251–3*, 270; *1254–6*, 140). The 'new chancel' mentioned in wills of 1459 and 1466 (Raine, 227–8) may have had relevance to the serious decay noted in 1446 when the priory was exempted from taxation on the grounds of poverty (*CPR*, *1446–52*, 69).

Pastoral duties presumably formed part of the pre-Conquest church of Christ Church and these would normally have passed to the priory, the laity enjoying certain rights in the nave of the new church. A survey of *c.* 1225 speaks of the 'parochia sancte Trinitatis' (PRO, E.135/25/1), the designation used in the taxations of 1327 and 1381. In 1304 Gilbert de Gaudibus, priest, was inducted to the vicarage of the altar of St. Nicholas in the church of Holy Trinity, and in 1402 William Byrsgrefe and his wife, Alice, asked to be buried in St. Nicholas, before the altar of St. Thomas. A chantry founded by Thomas Nelson at the same altar in 1474 is stated to be in Holy Trinity Priory (Drake, 264). Documents of 1452 and 1455 describe the parish church of St. Nicholas as adjoining ('iuxta') or annexed to the priory. In 1453 the parishioners of St. Nicholas had permission to set up their steeple upon the gable on the N. side of the priory church (SS, LVII (1871–2), 273).

In 1537 when Miles Walshforth was presented to the Nelson chantry it is described as being 'in the late conventual church of Holy Trinity' (J. Solloway, *The Alien Benedictines of York* (1910), 315), but in the Chantry Survey of 1548 he is shown holding the preferment in the church of St. Nicholas.

In 1543 Leonard Beckwith was confirmed in his possession of the priory. On 15 February 1551/2 a gale brought down the central tower and the fall probably reduced the choir to ruins and damaged the clerestory and triforium of the nave. In 1564–6 stones were taken from the 'defaced walls' for the repair of Ouse Bridge (YCR, VI, 73, 116) and in 1603 for repair of the city walls (YCA, C. vol. 11, ff. 69–72). By that date the nave aisles had become ruinous and blocking walls had been built in the arcades. In 1722 a vestry was built in the W. bay of the nave. The roof was ceiled in 1732 and a gallery built in 1755–6. Further alterations were made in 1829 (Lawton, 18; Faculty 1829/2 Borthwick Inst.).

Major restorations began in 1850 (*Yorkshire Gazette*, 10 Aug. 1850), when a new *South Aisle* was built and the church as a whole was repaired and refurnished by J. B. and W. Atkinson. A new *Chancel* and *Vestry* were built in 1886–7 by Charles Fisher and William Hepper, in 1894 a pinewood reredos designed by C. Hodgson Fowler and carved by G. W. Milburn was set up, and in 1898 an original lancet window in the W. wall of the N. aisle was reopened. In 1902–5, the nave ceiling was removed, the W. gallery taken down, and the W. bay of the nave rebuilt, the S. aisle being extended to correspond. On the N. a *Porch* was erected, partly on the old foundations of the N. aisle. The architect for these works was C. Hodgson Fowler. (T. Stapleton, 'Holy Trinity Priory, York', Archaeological Institute at York, 1846, *Procs.* (1848), 1–231; J. Solloway, *op. cit.*).

The main gatehouse of the priory was erected during the 13th century at the entrance from Micklegate to Priory Street. The last survivor of the monastic buildings, it was demolished in 1854, but scale drawings were made (Fig. 27).

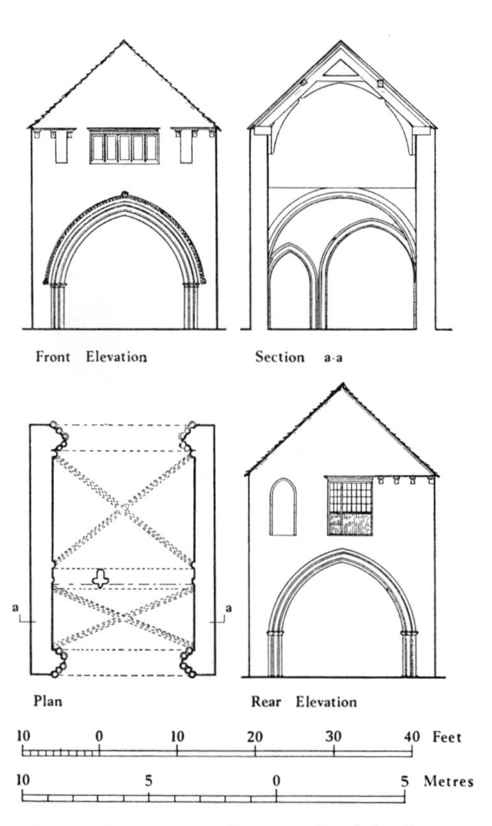

Fig. 27. (5) Gatehouse (now destroyed) of the former Priory of Holy Trinity.

Architectural Description—*The Eastern Arm* of the conventual church (91 ft. by $57\frac{1}{2}$ ft.) was aisled, of five bays with a square E. end, and was paved with small red tiles. The S. arcade was excavated by W. H. Brook in 1899, and pier bases consisting of diagonally placed squares with chamfered angles were disclosed about $4\frac{1}{2}$ ft. below present ground level. The piers were probably square with a large half-shaft on each face, for sections of a respond of this type were found, the shaft having a simple roll necking. Between the second and third bases from the E. was a mass of stone, perhaps the base of the sedilia.

The bottom courses of the S. wall of the S. aisle exist for much of the length and include good magnesian limestone ashlar in large blocks. There was an original doorway in the third bay. The wall is probably of the second half of the 12th century and has fine diagonal tooling. There is little evidence of buttresses.

The *Transepts* have disappeared, but excavations by W. H. Brook in 1905 proved that the N. transept was of two bays with E. and W. aisles. The S. respond of the W. arcade (Plate 120) is semi-octagonal, with moulded cap and base like those of the nave piers; the white magnesian limestone contrasts with the older work in buff limestone. There is no evidence for the plan of the S. transept, but it is unlikely to have had a

Fig. 28. (5) Church of the Holy Trinity. North arcade with reconstruction of Triforium and Clearstorey (*cf.* Plate 118). Masonry with joints shown by broken lines no longer exists.

W. aisle since the opening from the S. aisle of the nave has a straight face against the S.W. pier of the crossing. The cloister lay on this side of the church.

The two W. piers of the *Crossing* (24 ft. square) remain intact up to the springing of the arches; the similar E. piers were destroyed when the present chancel was built in 1887. The piers have twin half-shafts to the transept arches; the original bases had small spurs at the angles. The piers have plain surfaces to the W. arch of the crossing and semi-octagonal responds to the nave arcades. The N.W. pier is of good limestone ashlar with fine diagonal tooling where not rechiselled. The S.W. pier is similar but almost entirely rechiselled; on the S. side the late 12th-century respond is bonded and shares the same chamfered plinth. This last runs into the straight face of the opening from the S. aisle into the transept. Some stones just above the level of the pavement may belong to the footings of the early aisleless church.

The *Nave* (Plates 12, 118, 119) (82 ft. by 27 ft.) is of five bays, with heavy Transitional arcades, octagonal piers and two-centred arches of three chamfered orders. The piers have simple hollow-chamfered abaci, bell caps with roll necking, water-holding bases mostly renewed and chamfered plinths. On the N. side of the third pier is a moulded corbel, contemporary with the cap. The N.W. respond leans outwards, but is coursed through into the W. wall; a jagged line to the S. may indicate insertion. Above each pier of both arcades, internally, is an attached triple vaulting shaft; horizontally above the apices of the arches runs a moulded string which returns round the vaulting shafts.

The only part of the nave standing to its original height is the westernmost bay on the N. side. Here the triforium of the early 13th century shows within the church, having an arcade of three blind lancets with chamfered heads and round shafts with moulded caps and bases (Plate 118). This stage has externally seven similar niches, forming the wall arcading of a room built above the aisle. Above the triforium stage internally is a moulded string carried round the vaulting shafts. The clerestory, visible above the roof, has three arches with two-centred heads and chamfered reveals, the central arch opening to a window, the other two arches blind. There was a clerestory passage internally, and on the outer face a section of moulded water-table survives. The contemporary relieving arches in the two W. bays of the N. side of the nave probably formed part of the construction of the room over the N. aisle. (Fig. 28).

The N. wall has battlements of post-Dissolution date above a string, and at the E. end, an original external string at abacus level, returning on to the transept. In the blocking of each of the three E. bays is a modern three-light window.

The W. wall of the nave is mostly modern, but retains the N. jamb of the W. doorway and the corner pilaster buttresses of the N.W. angle. The N. buttress displays a trefoiled gabled niche, the moulded caps having nail-head ornament. The W. buttress has, above a string, two gabled niches divided by a shaft; on the gables and finials is dog-tooth enrichment; the top of the buttress was heightened at a later period, perhaps when the tower was built in 1453.

The masonry of the *Porch* (14½ ft. by 13 ft.) is modern, with a window of *c.* 1829 from the vestry reset in the E. wall. The N. doorway (Plate 15), on the site of an original opening, incorporates some old features reset. The two-centred head has a label with carved stops and is of three orders of which the innermost is original. Two filleted shafts with moulded caps and bases flank a band of nail-head ornament. The outer orders have round shafts with moulded caps and bases supporting a deeply moulded arch with dog-tooth ornament at the angle between the two orders. An original cap survives on the E. side, and a base on the W.

St. Nicholas Chapel (13½ ft. by 12½ ft.) (Plate 118), is contained in the western bay of the N. aisle, beneath the tower; the E. wall is of 1453. The early 13th-century N. wall, of large blocks of ashlar with claw tooling, has an original lancet with moulded label bearing flower stops externally. The window head has an inner chamfered order, which is continuous, and an outer order chamfered and supported on round shafts with moulded caps and bases similar to those of the N. doorway. The aisle wall has an external chamfered plinth and a string course below the lancet. On the inside of the W. wall a straight joint indicates the corner of the original pilaster buttress. The wall, which is of good ashlar up to the sloping line representing the former aisle roof, has a chamfered plinth and a double-chamfered string at sill level. The whole of this wall is set back some 3 ft. behind the W. front of the nave.

The *Tower* (14 ft. by 12½ ft.) (Plate 117), built above the W. bay of the N. aisle (St. Nicholas Chapel), incorporates the W. bay of the N. wall of the nave as its S. wall. The N. wall is built upon the early 13th-century aisle wall. Similarly the buttress at the N.E. angle is based upon the chamfered plinth of the early 13th-century buttress, but with a greater projection to N., with reuse of original facing stones (Plate 120); it is of five stages, with stop-chamfered angles in the first two stages. The W. buttress is of six stages and is 15th-century throughout. The W. wall, also constructed over the 13th-century aisle wall, has a buttress of three stages against the N. end. In the top stage of the tower, in each of the N., E. and W. walls, above a moulded string, is a round-headed window of two orders; the inner order is continuous with a chamfered reveal; the outer order is chamfered and springs from columns with moulded caps, necking and moulded bases of the 12th century, reused. There is a plain battlemented parapet.

The *South Aisle* (10 ft. wide) is of 1850 and modern as indicated on the plan; the 19th-century part is built of broad red brick on ashlar footings.

The *Roof* of the nave is of seven bays; the westernmost bay is modern, but the rest is of the 16th century incorporating 15th-century timbers. Each truss has a tie-beam, a high collar, no ridge, three purlins on either side and kerb principals up to the collar. The first and second tie-beams from the E. are moulded, with recesses for three bosses; the seventh has a central boss with a demi-angel; they belonged to a low-pitched cambered roof of the 15th century.

Fittings—*Altar Stone*: mediaeval, from St. John's, Micklegate, given to Holy Trinity *c.* 1958. *Architectural Fragments*: in St. Nicholas Chapel, (1) carved fragment (Plate 26), perhaps part of tympanum, with dragon in Scandinavian tradition,

11th-century (*YAJ*, xx, 209). In porch, (2) section of carving, perhaps from collar of composite cross, with coarse pelleted interlace, 11th-century, pre-Conquest (*YAJ*, xx, 208); (3) three voussoirs from a doorway, each with circular medallion of conventional acanthus, and (4), set in pier, capital with hollow-chamfered abacus and acanthus decoration, all mid 12th-century (Plate 18). Loose in church, (5) capitals, moulded stones, etc. found on Priory site, 12th to 14th-century. *Bells*: two; treble inscribed '1731 SS EBOR', for Samuel Smith II, founder, Chamberlain of York 1713, Sheriff 1723–4, buried in Holy Trinity 1731; tenor, undated, inscribed '+IHC+ campana : Beate : Marie : Iohannes : Potter : me fecit', 14th-century. *Bell-frames*, supporting tenor, part, probably 1453. *Benefactors' Tables*: on S. wall of vestry, (1) large wooden table in moulded frame, dated 1793; (2) similar to above, 1699. *Brass*: on S. aisle wall, oblong plate inscribed 'Alderman Micklethwait 1632' (Elias Micklethwait, Lord Mayor 1615, 1627), originally fixed to coffin lid (*q.v.* (1) below).

Chairs: in St. Nicholas Chapel, (1) oak, with turned front legs and shaped top with volutes to straight back having moulded uprights, *c.* 1700, (2) from St. John's, Micklegate, *q.v. Chests*: near pulpit, (1) of moulded panelling with some fluted enrichment, early 17th-century; in S. aisle, (2) of fielded panels with two drawers, 18th-century, inscribed 'The Gift of Lawrence and Elsie Dunphy 1951'; in St. Nicholas Chapel, (3) small, with sides narrowing to base and with rounded lid, covered with leather and with enriched metal straps, 16th to 17th-century. *Coffin Lids*: against W. side of N. crossing pier, (1) part only, with moulded edges and shallow roll at centre and with indents for shield-of-arms and plate (reused for Alderman Micklethwait, see *Brass* above), 13th-century; in E. side of porch, (2) with foliated cross, 13th-century, found in nave 1902–5; on E. wall of St. Nicholas Chapel, (3) upper part of coped slab bearing on one side a sword and on other a hafted cross and beginning of inscription 'H(i)c iace [t]', late 13th-century; in S. wall of S. aisle, (4) part only, inscribed in black letter 'Hic iacet Walterus fflos'. *Door*: in main N. doorway, of one leaf with central wicket, externally with mouldings in window form of six lights with rectilinear tracery, 15th-century, extensively renewed, rediscovered 1902–5 (Plate 15).

Fonts: At W. end of nave, (1) large octagonal bowl, perhaps 18th-century, set on modern shaft with cap, and 15th-century base, brought from St. Saviour's, 1953; loose at E. end of S. aisle, (2) small octagonal font fitted with modern drain, found on site of Beech House on The Mount. Font cover (Plate 28), top inscribed 'Anno Domini 1717 Richard Booth William Atkinson Church Wardens', on base 'Anno Domini 1794 Francis Hunt and Marmaduke Buckle Church Wardens', brought from St. Saviour's.

Glass: in chancel, in window at E. end of N. wall, in tracery IHS, in main lights grisaille, in side lights quatrefoils with medallions surrounded by foliage, by 'Barnett late of York' (Sheahan and Whellan, I, 541), given by Crompton family, 1850, moved 1893. *Hatchment*: on S. aisle wall, under central dormer, for Joshua Crompton of York (d. 1832). *Image*: in St. Nicholas Chapel, of stone, set in window ledge, the Holy Trinity (Plate 9), 15th-century, bought in 1952 in Delft, Holland, and said to have come from Holy Trinity, York. *Inscriptions and Scratchings*: in porch, on arch in E. wall of St. Nicholas Chapel, masons' marks (Fig. 7, p. lv). *Lectern*: of wood, with eagle, turned stem and moulded base on eight claw feet, mid 19th-century. *Lord Mayors' Table*, in vestry, wooden plaque with arms of York and seven ovals containing names and dates: 1772 Charles Turner; 1773 Henry Jubb; 1802 and 1819 William Hotham; 1810 and 1820 George Peacock; 1816 John Dales.

Monuments and *Floor Slabs. Monuments*: on N. crossing pier, (1) Thomas Condon, 1759, and Maria, grand-daughter, daughter of Charles Mellish, wife of 14th Lord Semphill, 1806, with impaling arms of Condon. On S. crossing pier, (2) John Burton, M.D., and Mary, wife, 1771, white marble, at top two books, one inscribed 'Mon. Ebor. Vol. 1' (John Burton, 1697–1771, antiquary and physician, published *Monasticon Eboracense*, vol. 1 in 1758), and, pendant from a scroll, a seal formerly bearing the arms of Burton with shield of pretence of Henson (Plate 34). In S. aisle, on E. wall, (3) Anastasia, eldest daughter of Thomas Strickland Standish, 1807, with lozenge-of-arms of Standish quartering Strickland; on S. wall, from E. to W., (4) Ann, wife of Christopher Danby, 1615, cartouche; (5) William Fryer, solicitor, 1838, children Thomas and William, and Elizabeth, wife, 1842; (6) Margaret, daughter of John Peers, wife of John Stanhope, 1637; (7) Jane, widow of Thomas Yorke of Halton Place, 1840, white marble slab with pediment and arms of Yorke impaling Reay, signed Skelton; (8) Mrs. Elizabeth Richardson, 1854, oblong white marble tablet set on black marble background with cambered sides and shaped head, signed Essex; (9) Margaret Anne, only daughter of Thomas Yorke of Halton Place, 1847, plain white marble tablet, signed Skelton; (10) William Duffin, 1839, white marble tablet, signed Skelton; (11) Henry Jubb, 1792 (Sheriff 1754, Lord Mayor 1773), Elizabeth, wife, 1793, white marble oval tablet suspended from cornice bearing urn, on shaped grey marble slab, signed Wm. Stead, York; (12) Elizabeth, wife of John Steward, merchant, 1847, John Steward, 1855, tapered white marble slab with moulded cornice and base, signed Skelton; (13) William Crumack, 1847, Martha, wife, 1854, white marble monument, signed Skelton; (14) Mary Swinburne, widow of Sir John Swinburne Bart. of Capheaton, Northumberland, 1761 (Plate 34); (15) Joshua Ingham, late of Stillingfleet House, East Riding, 1836, Elizabeth, widow, 1848, simple marble monument, signed Skelton; (16) Joshua Crompton, of Esholt Hall and Micklegate, third son of Samuel Crompton, of Derby, 1832, his wife Anna Maria, daughter and co-heiress of Anne Stansfield of Esholt Hall who married William Rookes, 1819, oblong white marble tablet with moulded cornice and draped urn, against black marble shaped slab, signed M. Taylor; (17) Elizabeth, daughter of George Ann, 1760 (Plate 34); (18) Thomas Swann, 1832, Harriet Ann, first wife, daughter of Thomas Clark of Ellinthorp, 1812, Anne Swann, second wife, widow of Joseph Bilton, 1831; (19) Elizabeth Scarisbrick, 1797, half-round white marble tablet with border panel of brown marble set on black marble beneath enriched cornice, pediment and urn, bearing lozenge-of-arms of Scarisbrick, signed Thos. Atkinson (Plate 34); (20) John Greene of Horsford, 1728, cartouche with arms of Greene. In Churchyard:

E. of main path, (21) Robert Wood, 1780, William, son, 1785; (22) Luke Graves, builder, 1792, Susannah, wife, 1826; (23) Henry Cassons, 1781, Ann, wife, 1803, Ann Ombler, granddaughter to Ann Casson, 1786; N. of church to N.W., (24) Elianor, wife of George Waud, 1784, George Waud, 21 years Clerk of Parish, 1799; to E. of chancel, (25) William Abercrombie, M.D., 1791, Sarah, wife, 1798. *Floor Slabs* (all of limestone): (1) Frances Olive, widow of Stephen Walter Tempest of Broughton Hall, nr. Skipton in Craven, 1795; (2) Jane, widow of Thomas Yorke of Halton Place, West Riding, 1810, Margaret Anne, daughter, 1847; (3) John Allanson, twice Lord Mayor, 1783, Elizabeth, wife, 1766; (4) [Jonathan Benson, chamberlain, 1725], William, son, [1741], [Mary, daughter, 1739], Ann, wife, 1746, and others; (5) William Green, 1764, wife, 1770; (6) Kezia Raper, 1797, John Horner, wine merchant, 1791, Jane Raper, widow of Leonard Raper, of Kirkby Malzeard, Yorks., aunt to John Horner, 1792, Mary, wife of John Green, fourth daughter of Jane Raper, 1802, Ann Horner, 1818; (7) Walter Richmond, merchant, Kingston, Jamaica, 1803, Jane Richmond, wife, 1808, Ann, daughter, 1798.

Piscina: now loose by W. door, octagonal bowl supported on bell-shaped foliate capital, 13th-century, found during excavations for chancel, 1887. *Plate*: includes cup (Plate 37) with baluster stem with London date-letter for 1611/12 and inscribed 'Christopher Maude, George Chapman, churchwardens of St. Trinitys in Micklegate 1666'; paten consisting of salver on three shaped feet with London date-letter for 1796/7 and inscribed 'Holy Trinity Micklegate York 1800, Roger Glover, John Gibson, Church Wardens'; paten made of secular salver with London date-letter for 1843/4 and inscribed 'Holy Trinity Micklegate York 1848', given to Church of Holy Redeemer, Boroughbridge Road; silver flagon of tankard shape with London date-letter for 1739/40. *Royal Arms*: on W. wall of vestry, wood, mid 18th-century. *Stoup*: in porch E. of entry, found during restorations of 1902–5. *Tiles*: a number, perhaps found in 1856 or 1871, now in Yorkshire Museum (Cook Collection). *Miscellanea* (see also *Architectural Fragments* above): In nave, set on first pier of N. arcade, stone shield-of-arms of Micklethwait, originally part of tomb of Alderman Elias Micklethwait, 1632 (see *Brass* above). In S. aisle, at E. end, part of Roman figure. In porch, two oak bosses said to have come from St. Crux or St. Martin's, Coney Street. Near path to church from Micklegate, stocks, of two large planks with five holes, post-mediaeval.

CHURCH OF ST. CLEMENT, Nunthorpe Road. This church of 1872–4, built as a chapel-of-ease to St. Mary Bishophill Senior, became the parish church in 1876. It includes fittings and monuments from St. Mary Bishophill Senior, *q.v.*

(6) PARISH CHURCH OF ST. JOHN THE EVANGELIST (Plate 121), formerly stood in a small churchyard in the angle between Micklegate and North Street; this has now disappeared in the widening of Micklegate. The church has walls of gritstone and magnesian limestone, and modern dressings of Whitby sandstone, tiled roofs with some slate over the central and S. aisles, and Welsh slates over the N. slope of the N. aisle.

The tower is notable as the only 12th-century example surviving in York; the upper stage is a rare example of work carried out in the period of Parliamentarian control after the victory of Marston Moor.

The earliest church, of which traces remain, was a simple rectangular cell of the early 12th century; only the western angles remain. On the N. a square plinth 9½ in. wide was found by excavation in 1955; on the W. a similar plinth, 5 in. wide, was seen passing under the tower added in *c.* 1150. This church is mentioned in 1194 (*CPL*, I, 462) and in a charter of 1189–1200 (*EYC*, I, 176). A S. aisle was added in the 13th century; in 1319 a chantry was founded by John Shupton at the altar of St. John the Baptist in this aisle (*CPR*, *1317–21*, 312; SS, XCI, 79*n.*); it was augmented in 1338 by his son-in-law Richard de Briggenhall (Skaife MS.). The early 14th-century arch leading from the chancel to the chapel at the E. end of the N. Aisle is to be associated with the foundation of a chantry in 1320 by Richard de Toller (*CPR*, *1317–21*, 420; SS, XCI, 79–90; see glass in E. window of N. aisle). The rest of the *North Aisle* dates from rather later in the same century. The *South Aisle* and arcade, with the upper part of the tower, were rebuilt late in the 15th century. Soon thereafter the N. aisle was extensively remodelled and the W. side extended in connection with the chantry founded by Sir Richard Yorke at the altar of Our Lady (see glass in E. window of N. aisle) and with money from benefactions of 1492–1506 (SS, XCI, 78–9; *TE*, IV, 135*n.*; *AASRP*, XI, pt. ii (1872), 252; Raine, 249–50). In 1519 the chancel was said to be in bad repair.

The steeple was blown down in 1551–2; repairs in narrow red brick of the later 16th century indicate that the tower fell towards the N.E. The present timber-framed *Belfry* was added in 1646, when the bell cast in 1633 was hung; three more bells, saved by Lord Fairfax from St. Nicholas Hospital without Walmgate Bar in 1644, were hung in 1653 (*AASRP*, XXVIII, pt. i, 439–40). The works of 1646 also included the making of a *Vestry*, removed in 1955, in the W. end of the N. aisle. In *c.* 1763 new steep-pitched roofs were built above the mediaeval timbers. The floor, raised by 1½ ft. after the great flood of 31 December 1763, was again raised in 1819 after Ouse Bridge was rebuilt (*ibid.*, 436; J. W. Knowles MSS.).

In 1850 the E. wall was rebuilt further W. to widen North Street, a *Porch* was added, and buttresses and windows were renewed on the S. side. The N. side also was restored with Whitby stone, and the whole church repaired and refurnished under George Fowler Jones as architect (Sheahan and Whellan, I, 517). At this time

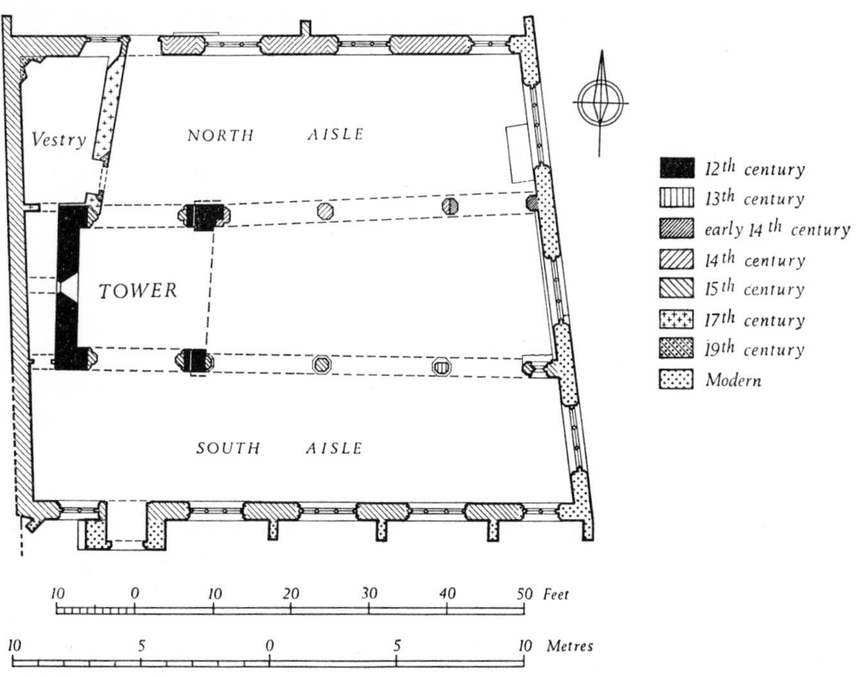

Fig. 29. (6) Church of St. John the Evangelist.

the Yorke window was restored by J. W. Knowles (J. W. Knowles MSS.). The church was reopened for worship in 1851 (*Yorkshire Gazette*, 8 March 1851). In 1866 the church was again repaired under Messrs. J. B. and W. Atkinson; in the nave an open timber roof was substituted for the flat panelled ceiling; windows were opened in the W. wall of the tower, and a new organ was made by Postill of York (*AASRP*, VIII, pt. ii for 1866, cii). The church was closed in 1934 and the fittings were dispersed in 1938–9 when the Corporation took over the fabric (*Yorkshire Gazette*, 25 Feb. and 7 Oct. 1938; *YAJ*, XXXIV (1939), 113–14), which was later given to the York Civic Trust and restored to form the Institute of Architecture of the York Academic Trust. (T. W. Brode, 'Notes on the History of St. John the Evangelist, York', in *AASRP*, XXVIII, pt. i for 1905, 435–50; also 'The Old Parish Account Books of St. John the Evangelist, York', in *ibid.*, XXIX, pt. i for 1907, 304–22; E. A. Gee 'An Architectural Account of St. John's Church, Micklegate', in YAYAS *Procs.* (1953–4), 65–82). *Now (1971) The Arts Centre.*

Architectural description—The *Church* (67 ft. by 56 ft.) is a trapezoid without any structural chancel (Plate 124): the tower is built within the W. end of: he nave.

The E. window of the *Central Aisle* (60 ft. by 17 ft.), two-centred with three lights, has intersecting tracery reproducing that of the early 14th-century original. At the W. end of the N. arcade, on the N. side, can be seen the line where the tower wall abutted on the original aisleless nave; the rest of this wall was rebuilt when the aisle was added in the 14th century. A chamfered water-table shows on the N. side cut into by the 15th-century roof timbers. Above the arches this wall was rebuilt in narrow red brick after the fall of the tower in 1552. The early 14th-century E. arch of the N. arcade is two-centred; it has two chamfered orders and responds with moulded caps; moulded bases exist below the present pavement. To the W. are two bays of rather late 14th-century work, with an octagonal pier and responds without capitals. The arches are two-centred, with two chamfered orders of which the inner merges into the pier and the outer terminates on both sides in simple corbels. The bases have a hollow and roll, partly concealed by the present floor. To E. of the S. arcade is a small square-headed opening with chamfered reveals but no glazing groove; it was probably a squint to the High Altar. The arcade is of three bays with octagonal piers and responds, without capitals. The two-centred arches have two chamfered orders, the outer resting on square corbels. The first pier from the E. has a water-holding base; this and the nine courses above it are probably 13th-century work *in situ*; the rest of the arcade and the rubble walling above are of the late 15th century, the second pier and the W. respond having bases formed of inverted 13th-century capitals.

The E. window of the *North Aisle* (65 ft. by 18 ft.) has a four-centred head, and moulded label with weathered head-stops; it is of four cinquefoiled lights with vertical tracery and reproduces the E. window of the Yorke chantry, of *c.* 1500. The N. wall is of magnesian limestone, with three buttresses. Sections of 14th-century masonry remain, but to W. of the doorway the wall is later and includes pieces of stone coffins and lids. The window at the E. end, of 1850, cuts into the head of a 14th-century canopy, perhaps the tomb of Richard Toller (d. *c.* 1335). The remaining windows, of three trefoiled lights with four-centred heads and vertical tracery of *c.* 1500, have hollow-chamfered reveals. The doorway has a two-centred head with chamfered reveals and externally shows the weathering of a former porch; inside it has a segmental rear-arch, to E. of which is the edge of the splay of an early window, now plastered over. Further W. is a late 15th-century window of two cusped lights without tracery, repaired in brick after the fall of the tower. A *Vestry* (11 ft. by 18 ft.) was formed at the W. end of the aisle in the 17th century; at the N.W. angle was a 19th-century brick chimney-breast.

The E. wall of the *South Aisle* (68 ft. by 16 ft.) has a three-light window of 1850 reproducing that of the 15th century. The S. wall (Plate 121) is of magnesian limestone and mainly of the late 15th century, with dressings of *c.* 1850. The four-stage mediaeval buttresses were renewed in 1850. The original windows have four-centred heads and are of three cinquefoiled lights with vertical tracery and hollow-chamfered reveals. To W. of the modern porch is a 15th-century window with flattened head, two trefoiled lights and vertical tracery.

The *Tower* (14½ ft. by 15 ft.), built against the W. face of an aisleless nave in *c.* 1150, was later incorporated within the church. There is a stepped plinth (excavated in 1955) and the lowest courses are of large oblong pieces of gritstone, probably reused Roman material. Above is late Norman masonry of good quality, of magnesian limestone with fine diagonal tooling. The N. and S. walls are pierced by low two-centred archways of the later 14th or 15th century with continuous chamfered orders (Plate 124). Over each arch is part of the rear-arch of a 12th-century window with well cut voussoirs, set W. of the centre of the tower. The N. wall was much damaged by the fall of 1552; at the top of the wall within the aisle is a blocked oblong window with chamfered reveals. The outer face of the S. wall shows the scar of a wall running E., now gone; rubble filling links the tower to the 15th-century arcade wall. Good Norman ashlar extends as high as the aisle roof. The W. wall contains a 12th-century window (Plate 15) with chamfered reveals and an internal splay; rear-arch and jambs have finely tooled voussoirs. The blocking removed in 1955 included an illegible 15th-century black-letter inscription. Above the aisle roofs are small square lights with internal splays in the 15th-century masonry of the N. and S. walls, and in the W. wall is a two-light window of the 19th century. The top stage, of 1646, is timber-framed with studs about a foot apart and with later brick infilling.

The W. wall of the church is rough and incorporates walls of adjacent houses. Internally, two half-arches from the W. wall act as flying buttresses to the leaning tower.

The external *Roofs* have three parallel ridges. The roof of the N. aisle has an oak ceiling of *c.* 1500 (Plate 17) at the level of the cambered tie-beams. The moulded tie-beams and purlins form panels, with bosses at the intersections, coloured in 1956. The bosses included (1) arms of Yorke (see Glass), (2) arms of Yorke impaling Mauleverer (both removed in 1850 and placed within panels at E. end) and, (3) *in situ* merchant's mark on a shield, probably for Sir Richard Yorke; (4) arms of the Merchants' Staple of Calais. There are two trusses together above the third arch of the N. arcade and the roof to W. is of the same form, with similar mouldings, possibly of *c.* 1510. The ceiling of the S. aisle is similar. At the E. end, 15th-century square bosses set lozenge-wise include: (1) head wearing bishop's mitre; (2) face under liripipe headdress; (3) eagle set on leaves (for St. John Baptist, whose altar was below).

Fittings—*Altar* stone with small incised crosses from blocking of canopy in N. wall, mediaeval, removed to Holy Trinity, Micklegate, in December 1955. *Bells*: six ((4) (5) (6) from St. Nicholas Hospital) and sanctus; (1) treble probably by William Oldfield, 17th-century; (2) inscribed '+Sancte (rest indecipherable) and with shield with three helmets placed two and one, 14th-century; (3) inscribed 'Jesvs be ovr speed 1633', by William Oldfield; (4) inscribed in black letter '[sanc]te [G]eor[gi] ora pro nobis', and with two crowned shields, one bearing a cross and one ihc, 15th-century; (5) inscribed '+ad : loca : sancta : trahe : Betris : Ros : tu : *Nicholaue*'; and above, on the crown, 'Nicholauus', 15th-century; (6) tenor, inscribed, '+Tome : propicia : sis : UUalleuuorch : Uirgo : Maria : Aº : Dⁱ : Mº : ccccº : uiiiº' and, above, 'Maria'. (5) and (6), by the same founder, have inscriptions in the same Lombardic alphabet; the donors commemorated were Lady Beatrix de Roos (d. 1414) and Thomas de Walleworth, Master of St. Nicholas Hospital (d. 1409). (7) small prayer or sanctus bell, of uncertain date, given to Dean and Chapter of the Minster and removed 1956. (G. Benson, *The Bells of the Ancient Churches of York* (1885), 4; *id.*, 'York Bellfounders' in YPS *Report* (1898), 7, 8; *YAJ*, xxxiv, 114; Terrier.) *Bell-frame*: oak, for four bells, almost identical with one at St. Martin-cum-Gregory dating from 1681, members numbered and pegged throughout, consisting of two main N.–S. frames and four E.–W. cross-frames, all of same construction and with some ovolo-moulding on top rails of cross-frames, 1646, parts only now preserved in bell chamber. The bells have been rehung in a steel frame. *Benefactors' Tables*: three, removed in 1955; on W. wall of S. aisle, (1) (Plate 22) large panel with moulded architrave and broken pediment, containing figure of Charity with two children, dated 1725, flanked by tables of Commandments, mid 19th-century; on N. wall of S. aisle, (2) large bolection-moulded panel listing benefactions, 19th-century; on S. wall of S. aisle, (3) table, *c.* 1804.

Brasses and *Indents*. *Brass*: on Yorke table tomb (see *Monuments* (1) below), at E. end of N. aisle, modern, recording restoration of 1851. *Indents*: in N. aisle, (1) oblong, in large grey marble slab (for brass to Thomas Mosley, 1624); to S. of E. arch of N. arcade, (2) small, on broken slab (for brass to John Mosley, 1624); to W. of last, (3) large oblong, in decaying slab of yellow sandstone (for brass to Mrs. Elizabeth Mosley, 1640); in centre aisle, (4) for shield and small oblong plate; to W. end of N. aisle, (5) for Evangelists' symbols at corners of large grey marble slab, and for oblong strip, slab reused by Brearey family (see *Floor Slabs* (13) below, removed in 1956 from S. of (4)); at E. end of S. aisle, (6) for broad fillet on three sides and for figures of man, wife and children and two small shields-of-arms in large slab of blue-grey marble (reused for William Brearey (see *Floor Slabs* (22) below). *Chair*: of oak, with shaped arms and enriched panelled back, 17th-century, now in Holy Trinity, Micklegate. *Doors*: in N. doorway, of planks, perhaps 17th-century; in S. doorway, *c.* 1850. *Font*: of limestone, octagonal, with quatrefoils on cardinal faces, moulded bases to bowl and octagonal shaft with moulded and battlemented cap and moulded base, *c.* 1850. Font-cover: (Plate 28) of oak, 1638, when 'made anew', much restored. Both font and cover now in church of St. Hilda, Tang Hall.

Glass: the important glass was given in 1939 to the Dean and Chapter and placed in the W. aisle of the N. transept of the Minster; other glass was given to the Chapel of Clifton Hospital (*YAJ*, XXXIV (1939), 113–14). The following summary records the position before 1939, shown in photographs by F. H. Crossley (YPS (1914/15), 144–7; Harrison (1927), 185–9).

N. Aisle, E. window, I. Designed as a memorial to Sir Richard Yorke, Lord Mayor, 1469, 1482 (ob. 1498), his two wives, Joan daughter of Richard Mauleverer and Joan, widow of John Dalton and John Whitfield, both of Hull, his six (or seven) sons and his four daughters. The late 15th-century design incorporated in the bottom range four panels of donors, on a background of leaf sprays, from a 14th-century window, presumably in the same position. Notes of Roger Dodsworth and Henry Johnston (Bodleian MS. Dodsworth 161, f. 36; MS. Top Yorks, C.14, f. 102v) identify the donors as Richard Briggenhall (M.P. for York, 1333–7, d. 1362) and Katherine (Shupton), on oak sprays; John Randeman (Bailiff 1339–40) and Joan (Settrington) on hawthorn; Richard Toller (Bailiff 1316–17, d. *c.* 1335) and Isabella (d. 1336) with priest officiating, on hops (?); William Grafton (Bailiff 1333–4) and Agnes, on vine.

Above, a two-line inscription, now partly defective, may be restored from Dodsworth and Johnston: 'Orate pro anima Ricardi Yorke militis bis maioris Civitatis Ebor. ac per [... annos maioris stapule Calicie et pro duabus dominabus Johanne ac Johanne] uxoribus suis ac eciam pro omnibus libe[ris et] benefactoribus suis. Qui [obiit .... die mensis Aprilis Anno domini M°CCCC^{mo} LXXXX° VIII°].' Above are kneeling figures of the six (seven in Johnston's sketch) sons; of Sir Richard Yorke in plate armour with arms of Yorke on surcoat; of (lost) his two wives, one with arms of Yorke impaling Mauleverer, and of his four daughters. The lost panel of the two wives is replaced by a small seated Trinity in silver stain (*YAJ*, XXXVII (1951), 228). The main panels above this inscription, of *c.* 1498, represent the Trinity, St. George and the dragon (only lower part remains), the Crucifixion (lost but recorded by Gent (p. 170) and confirmed by lead lines), and St. Christopher (Plate 123). In the tracery, angels display shields with arms of Merchants' Staple of Calais, Foster, Stapleton impaling Gascoigne, Yorke, Yorke impaling Mauleverer, Yorke impaling Darcy, Yorke impaling (unidentified 3) and City of York.

The missing parts of the main panels are partly replaced with figures and fragments from the adjacent N. windows, where they were recorded by Dodsworth in 1619 (MS. Dodsworth 161, f. 36). They commemorated [William] Stockton, mercer, (d. 1471; *see* All Saints', North Street, brasses (1)) and Alice, his first wife, widow of Roger Selby 'spycer' (d. 1425) and Elizabeth, his wife.

S. Aisle, E. Window, II. The three main lights contain 14th-century panels which Browne noted in 1846 as 'mutilated to get them into the Perpendicular window'. In the lower range are three pairs of donors. The inscriptions were lacking in 1670, though the name Richard Orinshead is recorded in 1730 (Gent, 170). The main figures and scenes record the life of St. John the Baptist, whose altar stood beneath: (a) St. Elizabeth holding figure of infant St. John, above Baptism of Christ; (b) St. John the Baptist holding the Lamb, above fragments; (c) Herod's Feast, above the beheading of St. John the Baptist (Plate 31). Tracery contains 15th-century glass including St. George, Coronation of the Virgin (Plate 123), St. Christopher and St. Michael, together with arms of the city of York and of Neville, for Ralf Neville, earl of Westmorland (d. 1425) who held the advowson of a chantry at this altar (SS, CXXV, 130).

S. Aisle, E. window of S. wall, III. Tracery contains some 15th-century glass, including two archbishops and two other figures (Plate 122).

S. Aisle, second window of S. wall, IV. Three panels of 14th-century glass from E. window of sanctuary (J. H. Parker and J. Browne in Archaeological Institute at York, 1846, *Procs.* (1848), 13). Two have medallions on stems amid oak sprays, one kneeling figure of cleric before altar.

Lectern: (Plate 44), of single desk type, of oak panelling with late Perpendicular blind tracery and shield bearing complicated merchant's mark, late 15th-century, greatly restored, now in Upper Poppleton Church, Yorks., W.R. *Lord Mayors' Table*: (Plate 22), pedimented panel, with enriched and dentilled cornice surmounted by flame between two urns, bearing arms of York between letters 'A R' (Anna Regina); below, panel inscribed: 'Richd. Thompson Lord Mayor 1708', 'Rich^{d}. Thompson Lord Mayor 1721', 'J. Wakefield Lord Mayor 1765', and fitted with rests for sword and mace, now in York Minster, at E. end of nave, on S. side.

Monuments and *Floor Slabs. Monuments*: In N. Aisle, against E. wall, (1) altar tomb (Plate 19), said to be of Sir Richard Yorke, merchant, Chamberlain of York 1460, Sheriff 1465–6, Lord Mayor 1469 and 1482, M.P. at various dates from 1472 to 1490, knighted 1483, died 1498,[1] N. and S. ends with shields set in quatrefoils and three similar panels to W., all shields having matrices for brasses, at N.W. angle two long, round-headed panels, on top grey marble slab, perhaps modern, with moulded edges and brass fillet; an inscription records restoration in 1851; the tomb has been shortened on S.; on N. wall, (2) Nathaniel Wilson, 'East countrey merchant', 1726, Catherine Wilson, widow, 1736, in pediment shield-of-arms of Wilson impaling Reynolds (Plate 33). In central aisle, over E. pier of N. arcade, (3) John Scott, 1775; over second pier of N. arcade, (4) Christopher Benson, 1801, Margaret Benson, wife, 1795, five infant children, Christopher Benson, eldest son, 1796, of white marble, signed Stead. Over first pier of S. arcade, (5) Anne, wife of John Haynes, 1747, cartouche, over second pier, (6) Elizabeth Potter, servant,

[1] Yorke in his will (*TE*, IV, 134–7) asked to be buried 'coram imagine Trinitatis, in mea propria tumba ibidem fabricata', indicating that it had already been prepared before 1498. (See *Glass*; *cf.* Raine, 250.)

1766, cartouche. In S. aisle, E. wall, (7) Thomas Bennett, 1773, Elizabeth, wife, 1825 (Plate 35), black slab with pedimental head and moulded cornice, with inscription tablet in form of scroll with sarcophagus behind and weeping willow tree of plaster applied to freestone, signed 'Bennett S.Y.'; on S. wall, (8) Luke Thompson, 1743, Grace, wife, 1776, with arms of Thompson with inescutcheon of Bawtry. *Floor Slabs*: records of twenty-nine noted in 1951 are in the RCHM archives.

Plate: the fine plate includes two cups similar in shape, with plain bowls, both inscribed 'In usum Ecclesiae Sancti Johannis Evangelistae in Civ: Ebor: A.D. 1824', and with the York mark, one for 1807/8, the other for 1824/5; stand paten with gadrooning round edge, with London letter for 1697 and inscription recording its acquisition in 1699 with names of churchwardens, I. Ibbetson and R. Greenupp (vestry minutes of 1699 record 'a compleat silver salver (or some decent patten) be bought to lye the sacrament bread upon'); flagon with straight tapering sides and flat lid with grid-iron thumb-piece, with York mark for 1790/1, given by Dorothy Bowes, 1791; two pewter flagons, one *c.* 1725, the other with seven-sided body and shaped panels, engraved throughout on exposed surfaces, *c.* 1620 on evidence of costume of engraved figures (Plate 37). All now kept at Holy Trinity, Micklegate. *Miscellanea*: In S. aisle, under third window, piece of cusping set in wall; at N.W. corner, moulded springer of arch, 13th-century, reused as corbel.

(7) PARISH CHURCH OF ST. MARTIN-CUM-GREGORY (Plates 126, 128), stands in a large churchyard S. of Micklegate. The church has walls of dressed stone and brick. The masonry is mostly magnesian limestone with a small amount of millstone grit; the roofs are of lead, tiles and pantiles.

The first church, a simple cell 33 ft. by 18 ft. coterminous with the present nave, had walls of random rubble of pre-Conquest character; an 11th-century date is borne out by two cross fragments of this date reused in the W. wall of the tower. The church is mentioned in Domesday Book (1086); it was then held with four houses by Erneis de Burun. Most of the lands of Erneis de Burun had formerly belonged to Gospatrick, a wealthy Saxon landowner. The property became a part of the Trussebut fee and the family became patrons of the church. (*EYC*, X, 23–30; *YAJ*, XL (1962), 496–505; VCH, *Yorkshire*, II, 150, 192 and 278–80.) Early in the 13th century the N. and S. walls were pierced with arcades which still survive, that on the S. being slightly earlier. An early chancel may also be assumed perhaps dating from *c.* 1230, when John Trussebut was instituted to the living of 'Sancti Martini ultra Usam in Ebor'. In the second quarter of the 14th century the *North Aisle* was enlarged to its present width, the 13th-century doorway being reused. The position of the W. wall of this new aisle shows that a tower already existed at the W. end of the nave. The enlargement was connected

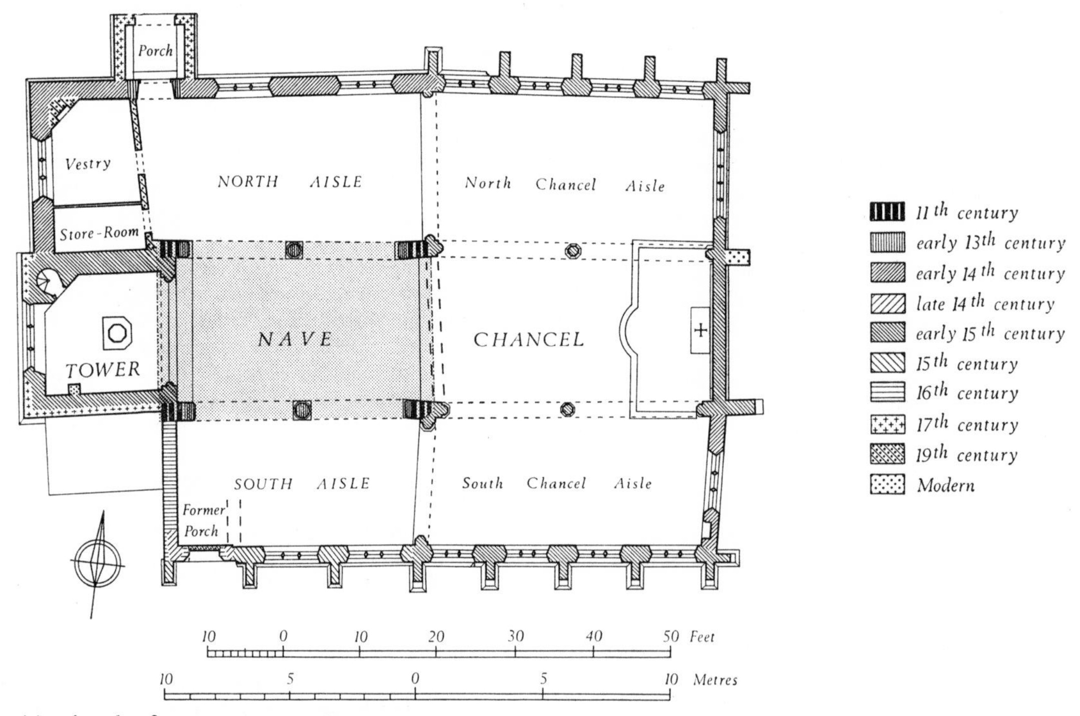

Fig. 30. (7) Church of St. Martin-cum-Gregory.

with the foundation of a chantry at the altar of SS. John the Baptist and Katherine by Richard le Toller (d. *c.* 1335) in or shortly after 1332 (*CPR*, *1330–4*, 370–1). Part of the original glazing scheme for this chapel remains and in 1670 Henry Johnston recorded a slab with an inscription beginning: 'Ricard Toller yci gist / par le grace de dieu cest chapele (f)isst . . .' (Bodleian, MS. Top. Yorks. c.14, f. 102). At about the same time a Lady Chapel was built on the S. side of the chancel; the E. wall with its window survives. This work was probably carried out with the help of Nicholas Fouke, mayor of York 1342, who is commemorated in the original glazing; in 1367 he founded a chantry at the altar of St. Mary (*CPR*, *1364–7*, 383). On the N. side of the chancel a chapel of St. Nicholas was built in *c.* 1370, above an undercroft which may have been used as a charnel house for bones unearthed in digging the new foundations. The windows retain remains of a contemporary glazing scheme with scenes from the Old Testament. The chapel was built in the lifetime of John de Gysburn, who was mayor in 1369–71 and in 1379; in his will made in 1385, he asks for burial between the high altar and that of St. Nicholas (Wills, vol. I, f. 15); his chantry was licensed in 1392 (*CPR*, *1391–6*, 145). Late in the 14th century a *South Porch* with an upper room was built against the W. end of the narrow 13th-century aisle.

In the second quarter of the 15th century an extensive rebuilding of the *Chancel* and the *North* and *South Chancel Aisles* took place, probably with the support of the patron of the living, John le Scrope, 4th Baron Scrope of Masham, who recovered the family's forfeited estates in 1425, and of William Fythian, the rector presented by Scrope in 1426. The N. aisle is of high quality; in the S. aisle the details are simplified, though the side windows are identical and (as shown by masons' marks) provided by the same shop. This aisle incorporates the E. wall of the earlier chapel. Fythian was buried in the chancel in October 1429 (Wills, vol. II, f. 569). John de Moreton in his will dated 20 July 1434 desired to be buried in St. Nicholas's choir, beside his wife Margaret already interred there (Wills, vol. II, f. 605, vol. III, f. 400). That the work was already complete by this period is borne out by the fact that Nicholas Blackburn, senior, who owned the advowson of one of the chantries in the Lady Chapel, left no bequest to the rebuilding in his will dated 20 February 1431/2.

The *West Tower* was rebuilt in the 15th century and provided with a spire of timber. The window formerly contained glass with the arms of Gascoigne and Hastings (Bodleian, MS. Dodsworth 161, f. 43; Bodleian, MS. Top. Yorks. c.14, f. 102) probably commemorating Sir Hugh Hastings, high sheriff of Yorkshire in 1480 (ob. 1489), who married Anne, daughter of Sir William Gascoigne of Gawthorpe.

The *South Aisle* was widened in *c.* 1450 to incorporate the older porch, which probably remained separate. The shields on the buttresses, now blank, formerly bore the arms of Gascoigne and Vavasour (Gent, 182), probably commemorating Sir Henry Vavasour of Haslewood (ob. 1500) who married Joan, the sister of Anne Gascoigne. A bequest of 26*s.* 8*d.* in 1477 (Raine, 237) by John de Benyngton, chaplain, to the *edificationem* of the church should probably be connected with the new chancel arch and the addition of a clerestory to the nave, the roof of which dates from *c.* 1500.

In 1548 the church roof was stripped of lead, preparatory to demolition (*YCR*, IV, 179); it was saved by Alderman John Beane and the parishioners, who, in the same year, reroofed it with tiles. In 1565 Beane, then Lord Mayor for the second time, gave 100 marks to buy three bells (Drake, 272), which were cast in 1579; one still remains. In 1570 the Rood Screen and two Roods were taken down; the screen was sold to Richard Whittington for £1 13*s.* 4*d.* According to the Visitation of 1575, the chancel was very ruinous and like to fall. In 1586, the benefice and parish of St. Gregory (in Barker Lane) were united with St. Martin (VCH, *York*, 382, 388). During the late 16th or 17th century the W. end of the S. aisle was repaired and the S. door renewed. This may perhaps be connected with the repair of the S. porch recorded in 1594 (Borthwick Inst., Y/MG.19, 9).

A new pulpit and screen were made in 1636 by Henry Harland, joiner, who was paid £31 12*s.* 0*d.*; £1 12*s.* 0*d.* was spent on Thomas Hodgson painting them; 10 years later the screen was taken down again. In 1648, a porch was demolished (Borthwick Inst., Y/MG.19, 11); the present *North Porch* was built in 1655 (*ibid.*, 9). In June 1677[1] Matthew Rayson was paid £10 for taking down the old spire; the tower was refaced in brick with stone quoins and a Classical balustraded parapet set above a heightening. 'Mr Ettie' (doubtless John Etty) was paid for a model of the battlement, and the searchers of the Carpenters and Bricklayers viewed the new steeple. In 1679 John Hindle and William Smith, masons, repaired two of the clerestory windows. New communion rails were provided in 1678 and a clock for the tower in 1680, and in 1681 work was done on the belfry by the carpenters John and Robert Rayson, under the direction of John Etty.

In 1700 the roof of the S. nave aisle was renewed and

[1] The date is established by the Churchwardens' Accounts Book for 1670–1754 (Borthwick Inst., Y/MG.20).

in 1715 a new wooden roof was erected on the N. aisle; in 1729 ceilings were inserted in the N. and middle aisles. In 1749–51 a new altar-piece was made by Bernard Dickinson, joiner, who died in November 1751, and in 1753 Matthias Butler, joiner, undertook to make new communion rails for £8 0*s*. 0*d*.

Repairs to windows were carried out in 1794 and 1829. The floor at the altar was raised in 1835; in 1836 a new organ was built by Mr. (John) Ward, and installed in the W. tower in a gallery built by Mr. (George) Lockey. Extensive repairs to the W. tower were done in 1844–5, when the Classical balustrade was removed and the top stage largely rebuilt (architects J. B. & W. Atkinson, builder apparently John Shaftoe, stonemason, freeman of York 1839). The vestry and storeroom inserted in the W. end of the N. aisle were formed between 1840 and 1846, and by 1849 the S. doorway had been blocked. Restoration, at a cost of £1,000 was carried out under William Atkinson in 1875; the old pews and the W. gallery with the organ were removed; the nave roof was renovated, the pulpit cleaned, the columns, arcades and walls were scraped, and the bells rehung (Minster Library, Hornby MS. II). In 1894 the chancel was restored at a cost of over £500, mainly spent on a new roof. In 1896 the W. window in the N. wall of the N. chancel aisle was repaired and the font was scraped and cleaned. Further repairs were done to the stone parapet of the tower in 1899, when the glazing in the windows of the N. aisle was restored.

In 1903 the doorways for the rood-loft were uncovered, and at about the same time the three N. clerestory windows were restored. The church was closed for banns and marriages in 1947 and was united with Holy Trinity, Micklegate in 1953; it has now (1971) been restored for use as Diocesan Youth Office and for secular meetings.

The church is valuable as one of the surviving early foundations, probably of pre-Conquest date. The eastern extension of the 15th century is a fine example of mediaeval architecture, of excellent quality. There is notable 14th-century glass, and the glass painted by Peckitt in the 18th century is of historical value. Of the fittings, the 17th-century pulpit and the 18th-century breadshelves and reredos are the most interesting. (G. Benson: *Notes on the Church and Parish of St. Martin-cum-Gregory in Micklegate within the City of York*, 2 parts, 1901–6; YPS *Report* for 1904 (1905); *The Inscribed Memorials in the Church and Churchyard of St. Martin-cum-Gregory in Micklegate, York* (1910); *AASRP*, XXXI (1911–12), 303–18, 613–28. D. D. Haw, *Saint Martin-cum-Gregory Church on Micklegate Hill, York* (1948). T. W. French, 'The Advowson of St. Martin's Church . . .' in *YAJ*, XL (1962), 496–505).

Architectural description—The E. wall of the *Chancel* (37 ft. by 18 ft.) to a height of about 14 ft. is of large ashlar of the 15th century; above is early 19th-century brick with the gable stepped back slightly. A chamfered plinth, some 6 ft. above ground level, is stopped against the S. buttress but continues behind the N. buttress along the E. wall of the N. aisle. The buttresses have five weathered offsets. Internally the wall is masked by the reredos. The 15th-century N. wall is of ashlar above the arcade. The slightly segmental arches of two chamfered orders spring from semi-octagonal responds and an octagonal pier, all with moulded caps and bases. The S. wall is similar but at the W. end a chamfered setback appears to mark the insertion of the 15th-century arcade into an earlier structure.

The E. wall of the *North Chancel Aisle* (38 ft. by 18½ ft.) is built of large limestone ashlar for about 9½ ft., with early 19th-century brickwork above. At the N. end is a buttress with three weathered offsets and a very decayed gargoyle. The early 15th-century E. window (Plate 16) has five cinquefoil-headed lights with two tiers of trefoil-headed grid-iron tracery. The head is two-centred and the mouldings continue down the jambs. The N. wall (Plates 13, 128), of four bays of large ashlar, incorporates 14th-century work low down in the E. bay. The moulded plinth is now visible only at the W. end. There is a moulded string course at window-sill level, now much worn and not carried around the buttresses. The five buttresses each have three weathered offsets crowned by decayed gargoyles. The projecting parapet is carried on a hollow cornice with five or six carved bosses in each bay. Above the cornice are three courses of stonework and three courses of 17th-century brickwork, capped by a moulded stone coping. In each bay is a three-light window similar in detail to that in the E. wall. Internally, the N. and S. walls have a projecting stone cornice at eaves level; it is hollow-moulded with a series of heads and flowers carved in the hollow and stops about 8 in. short of the E. wall. The W. arch into the N. nave aisle, of the 15th century with two chamfered orders to the two-centred head, is carried on semi-octagonal responds with moulded caps and bases, the latter masked by a stone step. The S. respond is bonded with the W. respond of the chancel arcade, forming a single build.

In the E. bay of the N. wall the partly buried window of a charnel vault is of mid to late 14th-century date, with two trefoiled ogee-headed cusped lights in a chamfered square head with moulded label externally.

The E. wall of the *South Chancel Aisle* (37 ft. by 16 ft.) is built of coursed ashlar, mostly of *c.* 1340. At eaves level there are two brick courses; above this the gable is set back and rebuilt in thin brick of the late 17th century. At the S. end is a 15th-century buttress with four weathered offsets and a moulded plinth which stops against the E. wall. The window has three ogee trefoiled lights and reticulated quatrefoiled tracery in a chamfered two-centred head with a moulded label. The S. wall (Plate 126) is built of squared limestone blocks with a moulded plinth. The buttresses have four moulded weathered offsets and differ in design from those on the N. side. Above the topmost weathering is a projecting

(4) CHURCH OF ALL SAINTS, NORTH STREET, from E. 13th century and later.

(4) CHURCH OF ALL SAINTS, NORTH STREET, from S.W. 12th century and later.

W. end of nave. 15th century.

(4) CHURCH OF ALL SAINTS, NORTH STREET.

From N.E. 12th century and later.

North aisle from W.

(4) CHURCH OF ALL SAINTS, NORTH STREET. 13th century and later.

Chancel from W.

North chancel aisle.

(4) CHURCH OF ALL SAINTS, NORTH STREET. *c.* 1470.

South chancel aisle.

Window III. The Prick of Conscience.

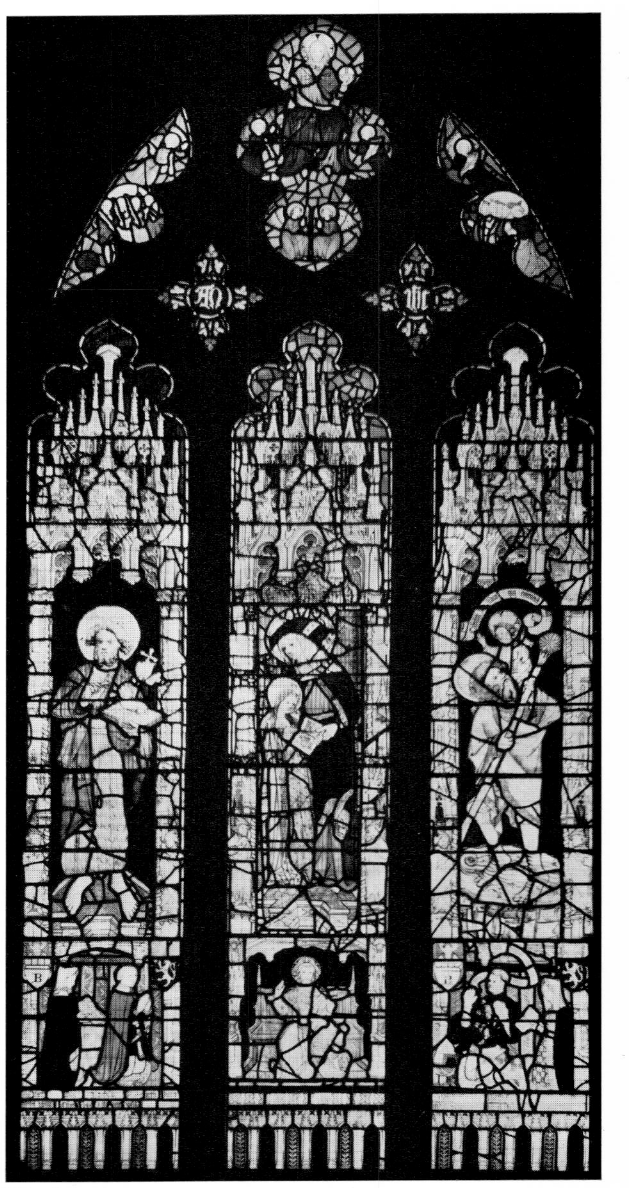

Window I.

(4) CHURCH OF ALL SAINTS, NORTH STREET. Early 15th century.

Window II. *c.* 1320–30.

(4) CHURCH OF ALL SAINTS, NORTH STREET.

Window VIII. *c.* 1340.

STAINED GLASS

Window V. *c.* 1429.

Window IX. 1425–40 (?).

(4) CHURCH OF ALL SAINTS, NORTH STREET. Before restoration of 1966.

Drawing by Henry Johnston, 1670.

After restoration in 1965.

(4) CHURCH OF ALL SAINTS, NORTH STREET. Window XI. The Nine Orders of Angels. 15th century.

STAINED GLASS

Coronation of the Virgin.

Annunciation.

(4) CHURCH OF ALL SAINTS, NORTH STREET. Window II. *c.* 1320–40.

'Stars fall from Heaven.'

'Death of all Living.'

(4) CHURCH OF ALL SAINTS, NORTH STREET. Window III. The Prick of Conscience. Early 15th century.

STAINED GLASS

St. John the Baptist.

St. Christopher.

(4) CHURCH OF ALL SAINTS, NORTH STREET. Window I. Early 15th century.

Window XII. St. Gregory. 15th century (?).

Window IX. St. John the Evangelist. 1425–40 (?).

(4) CHURCH OF ALL SAINTS, NORTH STREET.

Panel vii.

Panel iii.

Panel iv.

Panel vi.

(4) CHURCH OF ALL SAINTS, NORTH STREET. Window XI. The Nine Orders of Angels. 15th century.

Window II. N. light. *c.* 1320–30.

Window IX. E. light. 1425–40 (?).

Window IV. Centre light. 1410–35 (?).

Window XII. Centre light. *c.* 1440.

(4) CHURCH OF ALL SAINTS, NORTH STREET.

'Earthquakes.'

'Fishes rise out of the Sea.'

(4) CHURCH OF ALL SAINTS, NORTH STREET. Window III. The Prick of Conscience. Early 15th century.

chamfered stone cornice, carrying a parapet with moulded coping. In each bay is a 15th-century window of three cinquefoiled lights (Plate 16). To the W. the 15th-century archway (Plate 125) to the S. nave aisle has two chamfered orders. The springing level of this arch is higher than that of the chancel arcade, to which it is not bonded.

The E. wall of the *Nave* (33 ft. by 18 ft.), of ashlar, is carried by a 15th-century chancel arch with two hollow chamfered orders in a two-centred head; the orders die into the side walls at springing level. The early 13th-century N. arcade (Plates 12, 125) springs from plain responds and a circular pier, mostly of gritstone, with a moulded base on a square plinth and a moulded capital with nail-head ornament. The two-centred arches are of two chamfered orders. Above the arches the wall is of random rubble; at the E. end the wall is pierced with a rectangular opening (Plate 18), running slantwise from S.E. to N.W.; this led down from the aisle to the rood-loft at a level just below the springing of the chancel arch. Higher again, the wall is of ashlar with a shallow relieving arch over each bay of the arcade. Above this the wall is thickened out on a chamfered offset forming a continuous sill to the three late 15th-century clerestory windows; each of these has three cinquefoil-headed lights in a square frame. The S. wall is similar in character and layout to the foregoing. The top moulding of the responds is continued across the piers and returned round the N. and S. walls. The outer order of the arches ends in broach stops above the central cap and the responds. The jambs of the E. and W. piers below the respond moulding are chamfered, ending in mutilated bulbous stops above mutilated plinths. At the E. end there is an opening to the rood-loft, as in the N. wall; a worm-eaten wooden frame, with two iron pins on which to hang a door, has recently been removed from the opening. The W. wall is built of random rubble with a distinctive change to more regular coursed rubble about clearstorey level. The two-centred, 15th-century tower arch is of two chamfered orders, the inner carried on moulded semi-octagonal capitals and the outer terminating in broach stops at the same level. The capitals are carried on semi-octagonal shafts which run into the floor of the tower; the floor is raised three steps above the level of the nave. On each side of the head of the arch a cut in the rough stonework indicates the position of the earlier roofline.

The mid 14th-century N. wall of the *North Aisle* (35 ft. by 18½ ft.) is built of large, squared masonry with a chamfered plinth and a chamfered string immediately below the window-sills. In the E. part of the wall are two windows of *c.* 1335, each of three trefoiled lights and reticulated quatrefoiled tracery in a two-centred head. To the W. of these windows is the reset doorway of the early 13th century: it has a two-centred head and splayed jambs; externally the moulded label is enriched with nail-head ornament. On the S. the set-back to the N. wall of the tower marks the end of the surviving section of the early nave. The W. wall, built *c.* 1845, contains two doorways giving access to the vestry and storeroom formed in the W. end of the aisle.

The *Vestry* (10 to 11½ ft. by 13 ft.) has plastered walls with a N.W. angle fireplace; the window, which is the W. window of the 14th-century aisle, is identical with those in the N. aisle wall. The *Storeroom* (12 ft. by 5½ ft.) is lit by a mid 18th-century window, with ovolo-moulded glazing bars, reset in the brick partition wall to the vestry. The S. wall of the *South Aisle* (31 to 32 ft. by 16 ft.) (Plate 126), of mid to late 15th century, is built of large ashlar externally; internally it is of coursed rubble with two base courses of much larger stones. A moulded plinth runs round the buttresses and ends against the W. buttress of the S. chancel aisle. The aisle is divided by three buttresses, each with three moulded and weathered offsets; under the middle weathering of the two E. buttresses are defaced panels formerly carved with shields-of-arms; above the topmost weathering is a very worn gargoyle of a bat-like creature. The chamfered cornice, parapet and moulded coping are continued from the S. chancel aisle. The two E. bays have each a mid 15th-century window of three cinquefoil-headed lights with vertical tracery (Plate 16). The W. bay, representing the late 14th-century porch, has been considerably altered; in the lower half, rebuilt in the 16th century, is a blocked round-headed doorway with chamfered head and worn jambs. The upper part contains a mid to late 14th-century two-light window with two trefoiled ogee-headed lights in a chamfered square head with moulded label; internally the opening has been reduced by an inserted segmental head and splayed jambs of brick. The greater part of the W. wall, constructed of large rubble blocks, is not bonded into the W. buttress of the S. wall. The upper part of the wall and gable have been rebuilt in late 17th-century brickwork with a brick string, two courses deep, at eaves level.

The *West Tower* (15 ft. square) (Plate 10) is mainly mediaeval. In 1677 it was externally refaced in brick and considerably altered internally. The quoins and battlemented parapet date from 1844. The external brickwork is carried out in a variety of English Garden Wall bond, with bricks 2 in. to 2¼ in. thick. The tower is of three stages: the external face of the N. wall, to be seen in the storeroom, has no plinth and is built of large, squared rubble, mostly plastered; at about eaves level the rubble gives place to brick. The W. wall has squared stone quoins and stands on a chamfered stone plinth, above which there is one course of large blocks. The 15th-century W. window has three cinquefoiled lights with vertical tracery in a two-centred head. The N.W. angle has been built up to form a very narrow stone newel stair, entered by a low doorway and lit by two rectangular slits in the W. wall. The second stage houses the mechanism for the clock; the walls of large, squared rubble blocks have occasional brick patchings. The doorway to the stair has stone dressings with a chamfered head cut from a single stone. Centrally and near each end of the N. wall are rectangular slit windows, blocked externally by the brick refacing; there are similar blocked windows in the E. wall, beneath the present nave roof level, and in the S. wall. In the W. wall is a two-light window. In each of the four walls of the third stage is a 19th-century stone window of two cinquefoil-headed lights with vertical tracery in a two-centred head; the walls are of brick, with a scattering of rubble. A battlemented stone parapet projects on a chamfered course and has moulded merlons and embrasures. The slate-covered roof, gabled from E. to W., has a stone ridge and a large central stone cap. The *North Porch* is [illegible] squared stone and has a chamfered two-centred [illegible] archway.

Roofs: the S. chancel aisle [illegible] 15th-century roof, slightly cambered and divided [illegible] twenty compartments by four tie-beams [illegible] longitudinal beams, all moulded, and originally with bosses at the intersections. The S. wall-plate is supported on two later wall posts on moulded stone corbels; the posts are moulded, with ogee bar stops. The nave roof is of *c.* 1500, divided into twenty compartments by six moulded and cambered tie-beams, and three moulded longitudinal beams with bosses at the intersections. On the central beam the bosses have heraldic shields (from E. to W.): (a) *argent a cross gules*; (b) *argent a bordure azure*; (c) *per pale azure and argent*; (d) *argent a bend sinister azure*, all recently repainted. The N. nave aisle had a new wooden roof in 1715, underdrawn with a ceiling in 1729; the present flat and plastered ceiling may represent this. The roof of the S. nave aisle dates from 1700 and has four cased tie-beams carrying principal rafters. It is ceiled at collar level and externally is roofed continuously with the S. chancel aisle.

Pre-Conquest Stones: In vestry in W. wall (1) small *grave-cover*, 4 ft. by 1 ft. 1 in., now plastered over; in storeroom, in N. wall of tower (2) two pieces of tapered *cross-shaft*, now plastered over; in tower, in W. wall, set in course above plinth, two fragments each displaying one decorated face (3) *cross-shaft* (Plate 26), fragment 31 in. by 16 in., tapering to 15 in. by 11 in. with scroll in relief and showing 'pecked' dressing of typical Saxon character, probably 11th-century (*cf.* fragment at Haile, Cumberland, CWAAS, Extra Series XI (1899), 182). (4) *cross-shaft*, fragment 18 in. by 15 in. by 12 in. from base of shaft, displaying crudely drawn upright human figure in shallow relief, probably 11th-century; the crude drawing and shallow relief compare with those of crosses from the Chapter House, Durham (Haverfield and Greenwell, 79–91).

Fittings—*Bells*: three; (1) with bands of decoration, upper band inscribed 'Gloria in altissimis Deo 1697', lower with stamp 'SS Ebor' (for Samuel Smith), damaged; (2) (Plate 21) inscribed 'Iohn + Beane + alderman + gave + theis + three + bells', and, below, 'Robert + mot + made + me + MCCCCC + LXXIX' with very worn shield, apparently bearing a crown between three bells; (3) sanctus bell, uninscribed, perhaps 16th-century. The other two bells of 1579 were sold in 1792 and 1904 (*AASRP*, XXVII, 631). *Bell-frame* (Plate 21), of oak, with members of E.–W. cross-frames numbered I, II, III, IIII, V, on third frame carpenter's mark and on upper rail of fifth frame circular scratching, 1681, with some reused timbers. *Benefactors' Table*: on N. wall of N. nave aisle, board (Plate 22) with moulded surrounds and shaped top with finials, late 18th-century. *Brackets*: in N. Chancel aisle, five of stone, two flanking E. window and three between windows of N. wall, with enriched mouldings; in S. chancel aisle, N. of E. window, of stone, moulded; all probably 15th-century. *Breadshelves*: on W. wall of N. aisle, of oak (Plate 23), 18th-century.

Chandelier: hanging in chancel, of brass (Plate 24), *c.* 1715 (Haw, 14), globe inscribed 'Hoc Candelabrum Dedit Robertus Fairfax Armiger Ecclesiae Parochiali S^{ti} Martini Micklegate Eborac'. *Chairs*: in chancel, two with high backs, not a pair, one 19th-century, the other incorporating 17th-century carving. *Coffin Lid*: built into outer face of N. nave aisle, fragment, probably late 13th-century. *Communion Rails*: of oak (Plate 127), centre opening part projecting in semicircle, made by Matthew Butler, joiner, in 1753 at a cost of £8 (*AASRP*, XXXI, pt. ii (1912), 620). *Communion Tables*: in Sanctuary, (1) with bulbous turned legs, late 17th-century; in S. aisle, (2) with turned legs, moulded rails and modern deal top, 18th-century. *Doors*: in doorways to N. aisle, Vestry and Storeroom, all early 19th-century. *Font*: in W. tower, of stone, with octagonal bowl, stem and base, 15th-century. Font-cover of oak (Plate 28), late 17th or early 18th-century.

Glass. E. window of N. aisle, I: fragmentary panel, probably a 'Noli me tangere' scene in a border with a merchant's mark above 'R' (perhaps from W. window of N. aisle) and fragmentary panel of Christ and St. Thomas both of *c.* 1335; canopy and many quarries of same date together with inserted fragments of 15th and 16th centuries including a shield of arms with a *bend azure*, possibly for Scrope of Masham; in the lower panels were formerly donors, probably the family of John de Gysburn (Gent, 182). Windows in N. chancel aisle, II–IV, contained remains of a glazing scheme of the third quarter of 14th century, perhaps from the original Gysburn chapel. These panels are stored in the workshop of the York Glaziers Trust. Window II has three Old Testament scenes, possibly Creation of Eve, Sacrifice of Isaac and Worship of Golden Calf. Scratched on quarry: 'William Nichols, Plumb and Glazier, Novm. 3 1810'. Window III had, possibly Expulsion from Paradise and Creation, and many later fragments, including a figure possibly of St. Jude. Scratched on quarries; 'William Wilson 1771'; 'Thomas Peacock Dr. 4th 1844'; 'William (Hill vued ?) this window November 19th 1844'. Window IV had Tree of Knowledge in Paradise and many later fragments. Scratched on quarries: 'Richd. Stead Glazor October 20th 1780'; 'Richd. Stead'; 'William Lonsdale Novembr. 30 1745 in time of Reb'; 'Richd. Richard'; 'Thos. Simpson April 16 1742'. 'These windows bigun to be Repared 8^{br} ye 28th 1747 By Wm. Lonsdale & John Durham'; broken '. . . foot (t)hese Windows 1830' 1884; 'Edwin Lynell cleaned this Window July 9th 1855', (*See also* under *Inscriptions* and *Scratchings*). The westernmost window in N. chancel aisle, V (Plate 115), has in main lights late 18th-century glass from Peckitt workshop including in central light a narrow urn on a plinth, a memorial to William Peckitt, 'glass painter and stainer' (ob. 1795) 'designed and erected' by his widow in 1796; in tracery four 15th-century figures, possibly St. Eadmund, archbishop, St. Nicholas, St. Egbert and St. Albert. The eastern window in N. aisle of nave, VI (Plates 29, 115), has in central light female figure symbolising Resurrection, pointing upwards and holding scroll with Job xix, 25, signed 'Peckitt, Ebor, 1792', in E. light St. Catherine, in W. light St. John the Baptist (Plate 31), both of *c.* 1335 with borders showing many repeats of a merchant's mark above 'R' presumably for Richard Toller.[1] N. window of N. nave Aisle, VII,

---

[1] This and the mark in E. window of S. Chancel aisle are possibly the earliest surviving merchant's marks in stained glass in England (C. Woodforde, *English Stained and Painted Glass* (1954), 27).

fragments and quarries, including parts of a Resurrection scene, borders with presumed Toller mark and monogram 'RR' perhaps Richard Roundell of Hutton Wansley (ob. 1718); scratched on quarry: 'John Cussons Glazed this window In 1844'. E. window of S. chancel aisle, VIII, retains much glazing of *c.* 1340 (Plates 29, 114). N. light has Blessed Virgin, inserted shield-of-arms, perhaps Staveley impaling Plesyngton, and start of a Lombardic inscription '+Priet pur Nicho[las Fouke]'. Central light, probably a Crucifixion, had disappeared by 1730 and was replaced in 1846 by St. Martin of *c.* 1335 from W. window of N. Aisle (Plate 116). S. light has male figure, probably St. John. The canopies display an elaborate merchant's mark, probably of Nicholas Fouke. The border of central light has groups of three leopards alternating with three fleurs-de-lis, in allusion to Edward III's assumption of the French arms in 1340. The donors, inserted at the base of the side lights may represent Edmund Grey Earl of Kent and Katherine (Percy) his wife; the female figure wears a cloak with arms, probably for Grey, formerly impaling *argent a lion.* The S. windows have collected fragments of old glass and include two Flemish 15th-century panels of Betrayal of Christ and of David and Goliath (Plate 30). In third S. window, scratched on quarries: 'John Pick Plumber & Glazier York 1844'; 'John Cussins Plumber & Glazier 1844'; 'T. Harper Plumber & Glazier Sheriff Hutton near York'; 'John Hewso[n]'; 'John (Cussins) Glazie[r] Re[par]ed this Window in 1844'; 'John Thompson. York', etc.

Inscriptions and *Scratchings*: in N. chancel aisle, in third N. window, on various glass quarries, 'I hope this may be a plase for true protestants to resort to & never to be ruled by Papists God Bless King George ye 2^d^ & Billy off Cumberland Whome God long preserve'; 'Our Noble Duke Great Georges Son who Beat ye Rebels near Collodon the 16th Day of Aprill 1746'. For masons' marks *see* Fig. 8, p. lv.

Monuments and *Floor Slabs. Monuments*: In N. chancel aisle, on E. wall, (1) Jane, widow of Thomas Boulby of Whitby, 1803, Adam, son, 1819, white marble slab with moulded top and base, on dark veined marble, with shield-of-arms of Boulby, by Taylor; (2) Rev. Robert Benson, M.A., Vicar of Heckington, Lincolnshire, 1822, white marble slab with moulded top, carrying gadrooned casket, all on black ground with shaped base, with impaling arms of Benson, by Stead, York; on N. wall, (3) William Gage, 1819, Margaret, wife, white marble slab with patera at each lower corner, probably originally carrying a casket, against black marble, by C. Fisher, York; (4) Sir William Stephenson Clark, Lord Mayor 1839, 1851. In S. chancel aisle, on E. wall, (5) Thomas Carter, Alderman and Lord Mayor, 1686, Sarah, wife, daughter of John Pierson of Lowthorp, 1708, nine children, erected by daughter Frances, wife of Richard Colvile of Newton, Isle of Ely (Plate 32), with arms, now nearly illegible; (6) Jarrard, second son of Walter Strickland of Sizergh, Westmorland, 1791, Mary, wife, second daughter of Walter Bagenall of Bagenall, Co. Carlow, 1744, Cecilia and Mary, 1821, daughters, rectangular white marble slab with moulded top and draped urn, against a black ground with arched head and shaped base, by Taylor; on S. wall, (7) Joseph Volans, 1826, Elizabeth, wife, 1834, Harriet, daughter, 1850; (8) Alicia, daughter of Henry Iveson of Black Bank and Alicia his wife, 1729. In Nave, on S. wall, (9) Susanna, widow of Henry Wray, M.A., Rector of Newton Kyme, daughter of George Lloyd of Hulme Hall, Lancashire, 1830, rectangular white marble slab with scrolled pedimental-like enrichment at top and two feet at base, with lozenge-of-arms of Wray impaling Lloyd of Co. Waterford, all against shaped black marble, signed 'Fisher, Sculp^tr^'. In N. nave aisle, on N. wall, (10) Mary, widow of William Peckitt, Glass Painter and Stainer, 1826, Mary Rowntree, grand-daughter, 1846, square white marble slab with stylised acanthus leaves of plaster in each corner, against square black marble, both set lozengewise, by Plows; on W. wall, (11) William Garforth, 1828, white marble slab, framed in draperies, against shaped black marble, by Bennett, York. In S. nave aisle, on S. wall, (12) Frances, widow of Doctor Walker, physician at Newark, 1788; (13) John Atkinson, Captain, 68th Regiment, 1808, white marble slab with shaped top and moulded and reeded base, against shaped black marble, signed 'Fishers, York'; (14) Ann Collett, 1829, Elizabeth Fletcher, sister, 1833, Sarah Collett, 1838, Humphrey Fletcher, 1838, white marble slab against shaped grey-mottled marble, by Fisher, York; (15) Andrew Perrott, Alderman, Lord Mayor, 1701, cartouche (Plate 32); (16) Dame Martha, widow of Andrew Perrott, 1721, cartouche (Plate 32); on W. wall, (17) Lucinda, widow of Rev. Robert Benson, M.A., 1830, white marble slab with fluted side pieces and pediment, and urn (now missing), all against shaped black marble, signed 'Stead'; (18) Samuel Dawson, 'late Merchant, Son and Grandson of two worthy Gentlemen who were (in their turns) Lord Mayors of this ancient city; which honour he himself modestly declined', 1731, erected by widow, Sarah, daughter of Robert Watson of Whitby (Plate 33), with arms of Dawson quartering Hutton with an inescutcheon of Watson. In churchyard, N. of chancel, (19) Thomas Hurworth, 1830, Elizabeth, wife, 1839, Cha[rles] John Armstrong, grandson, 1841, slab with shaped head, by Shaftoe. S. of nave, (20) Grace Cave, 1779, Thomas, 1779, five grandchildren, William Cave, 1812, slab with shaped head; (21) Henry Cave, artist, 1836, Elizabeth, widow of William Cave and mother of Henry, 1843, table-top on brick base with shaped head.

Floor Slabs: in chancel, (1) [Hannah wife of Charles] Perrott, Alderman, daughter of E[dward] Trotter of Skelton Castle, Cleveland, and Ma[ry], daughter of Sir John Lowther [of] Lowther, [Westmor]land, Baronet, [1713], and infant children, with arms of Perrott impaling Trotter; (2) [Richard Perrott], 1670, Dorothea, mother, 1680, John, son of Dorothea, 1691, Alderman Perrott; (3) Thomas Bawtry, Lord Mayor, 1673; (4) [Samuel] Coyne, Fellow of Sidney Sussex College, Cambridge, Rector, 1690; (5) Mary, wife of Samuel Dawson, Alderman, 1692, Thomas Dawson, Mayor, 1703; (6) Thomas Perrott, Rector, Prebendary of Ripon, 1728; (7) Sir Gilbert [Metcalfe, Lord Mayor 1695], Alderman, 1698; (8) Mary Garforth, 1725, Isabella, daughter, 1726, Ann, daughter, 1731, William, husband, 1746, Isabel Dring, niece, 1754, Rev. Edmund Garforth, nephew and heir, 1761, Elizabeth, wife of Edmund, 1799, William, 1828; (9) Thomas Carter, Lord Mayor [1681], 1686, Mrs. John Peirson, eldest daughter of John Peirson of Raysthorpe, niece of Sarah, wife and widow of Thomas Carter, Alderman, 1746; (10) Susanna, wife of William Beilby of 'Miclethwait Graing', 1664, A. Iveson, 1729 (Plate 36), below first inscription worn arms of Beilby impaling Sunderland; (11) Ann, daughter of William Peckitt, 1765; (12) James Mayson [1733], Elizabeth, [1745], George Wright [1746], Ann Malton, 1754, Elizabeth Wright, 1770, Thomas Mayson, 1772, John Malton, 177[3]; (13) I.A., 1808; (14) Henry Augustus, fourth son of William and Mary Sarah Hargrove, 1830; (15) Davies Toplady, 1785, Dorothy, 1796, William Gimber, 1823, Elizabeth, 1828. In N. chancel aisle, (16) Thomas Garland, 1777, Ann, 179[2], Frances, 1809, Eleanor, 1814, Elizabeth, 1824, Jane Kidd, 18[39]; (17) Jane Heath, 1778, John, 1784; (18) Christopher Yates; (19) Mary, wife of John Swann, 1756, John, 1766, Elizabeth, wife of Thomas, son of John Swann, and daughter of Mr. Marm^{a} Prickett of Kilham, 1771; (20) Mary, widow of William Peckitt, Glass Painter and Stainer, 1826, Mary Rowntree, grand-daughter, 1846, reused stone (Plate 36) with original marginal inscription in black letter '+hic iacet d(omi)n(u)s henricus Cattall quondam capell(anu)s hui(us) Cantarie Qui obiit VII° die Febr̄ An(no) d(omi)ni m^{o} IIII^{mo} LX° cui(us) a(n)i(m)e p(ro)picietur de(us) Ame(n)' (in his will of 1460 (Wills, vol. II, f. 439), Henry Cattall asks to be buried 'in choro sancti Nicolai'); (21) Philemon Marsh, Rector, 1788; (22) A.B./J.B., probably early 19th-century; (23) Robert, son of George Benson, Mayor, 1765, Thomas, youngest son, 1794, Mary, widow, 1816 (Plate 36); (this slab is mentioned in accounts rendered in the Court of Exchequer by Benson's widow, 'Paid Mr. Carr for a Marble Stone laid ... £10 and for the Inscription cutting £2. 7. 4. and for laying down at 11s. 6d.—£12. 18. 10' (PRO, E.134/14 Geo. III, Hil.10; *cf.* E.112/2059/75 and E.134/7 Geo. III, Hil.5)); (24) Thomas Varle[y], 1771, Ann, daughter of Richard and Ann Bealby, 177[4]; (25) Jane Beckley, 1837, Nathaniel, 1843; (26) William Sawrey of Plumpton Hall, Lancs., 1727; (27) Eleanor, wife of John Blower, Rector, daughter of Francis Billingsby of Astby Abbotts, Salop., 1719, John Blower, 1723; (28) Frances, wife of Charles Bathurst of Clints, daughter and heir of Thomas Potter, grand-daughter of Edward Langsdale, M.D., 1724 (Plate 36), with arms of Bathurst with inescutcheon of Potter, impaling Potter; (29) traces of black-letter inscription [John Burton] B.A., Rector, 1475; (30) Susanna, relict of Rev. Henry Wray, 1830; (31) Nathaniel Robinson, 1770; (32) Elizabeth, wife of Henry Henwood, 1820, Henry, 1825, Elizabeth, widow of J. Holroyde of Rochdale and mother of Elizabeth, 1827. In S. chancel aisle, (33) Elizabeth, daughter of William Some[rs] of Bampton in Oxfordshire, 1726; (34) Joshua Earnshaw, Lord Mayor [1692], 1693; (35) William Ramsden, Lord Mayor, 1699; (36) Eleanor [Armst]rong, 1781; (37) Henry Stainton, 1764, Elizabeth, wife, 1794; (38) William Sharp, 1703, Elizabeth and Dorothy, daughters, William, son, 1718; (39) Captain William Rousby, 1761; (40) John Bradley, 1775, Antonia, wife, 1777, Catherine Marshall, sister to Antonia, 1779; (41) William Berry, 1835; (42) John Telford junior, 1770, John Telford senior, 1771, Isabella, daughter of John junior, 1775, [Hannah], widow of John junior, 1803; (43) Francis [R]amsden Hawksworth, 1825, Elizabeth Ann Mary, widow, 1835; (44) members of Iveson family; (45) John Brooke, V.D.M., 1735, Ann, daughter, 1735; (46) Joseph Volans, 1826, Elizabeth, wife, 1834, Harriet, daughter, 1850; (47) Mary Batty, 1795; (48) Elizabeth Ann, daughter of the Rev. John and Mary Richardson, of Cleaves, near Thirsk, 1809, Mary, 1830.

Plate: now kept at Holy Trinity, Micklegate, includes (1) cup inscribed 1636, presented by Henry Barker; (2) cup, copy of last, inscribed 1819, with York mark for 1818, made by Barber & Whitewell of York; (3) deckle-edged paten, originally domestic, with London mark for 1737, made by George Hindmarsh; (4) jug-shaped flagon with no date letter, but apparently 1720–9, and two makers' marks 'R.B.' and 'H.P.' (Humphrey Payne); (5) jug-shaped flagon with Newcastle mark for 1740, made by Stephen Buckles; (6) pair of pewter candlesticks inscribed '1754'; (7) silver-mounted rod inscribed '1678' (*YAJ*, VIII (1884), 327–9, and T. M. Fallow and H. B. McCall, 'Yorkshire Church Plate' in YAS *Extra Series* III (1912), i, 16–17; G. Benson, *Notes on the Church ...* (1901), 20). *Poor Box*: (Plate 23) probably late 19th-century, with 18th-century oak back board, inscribed with quotation from Acts XX, 35. *Pulpit* (Plate 38): of oak, hexagonal and panelled, made 1636, standing on late Victorian carved wooden base and stone plinth; between panels and sloping top, painted inscription 'Preach the Word in season and out of season'.

Recesses: in S. chancel aisle, at S. end of E. wall, (1) tall recess with moulded sides, incomplete, upper part masked by Strickland monument, at base of which two flat stones have been inserted to roof recess, 15th-century. In N. nave aisle, in N. wall between two windows, at floor level, (2) low tomb recess with flattened two-centred head of two moulded orders, 14th-century. *Reredos*: (Plate 127) of oak, in three sections, with Ten Commandments flanked by the Lord's Prayer and Apostles' Creed, 1749–51, made by Bernard Dickinson, joiner, apparently at a cost of £33 7*s*. 6*d*. (G. Benson in *AASRP*, XXXI, pt. ii (1912), 620; Borthwick Inst., Y/MG.20). *Royal Arms*: over tower arch, on E. face, of William and Mary, with Garter, mottoes, supporters, etc. *Stoup*: in S. nave aisle, reset in rebuilt section at W. end of S. wall, immediately E. of blocked doorway, with bowl semi-octangular outside and circular inside, base tapering to corbel carved as human head, late mediaeval.

Miscellanea: (1) in 19th-century chancel ceiling, circular panel with septfoil decorations, probably reused from ceiling of 1729; (2) in N. nave aisle, against S.E. pier, Roman tombstone, formerly outside W. wall of tower (RCHM, *Eburacum*, No. 97, 128a); (3) on steps N. of tower arch, very worn stone figure, possibly Roman; (4) next stoup in S. wall, column drum built into wall; (5) hanging on S. wall of tower, four leather fire buckets (Plate 24), each with painted inscription 'St. Martin cum Gregory 1794'; there are said to have been twelve buckets, bearing date 1699 (Hargrove in *Yorkshire Herald*, 1907); (6) on W. side of churchyard, loose pieces of tracery, apparently the original stone dressings of the N. clerestory windows renewed in 1903.

(8) PARISH CHURCH OF ST. MARY BISHOPHILL JUNIOR (Plate 129), is in the street now called Bishophill Junior.

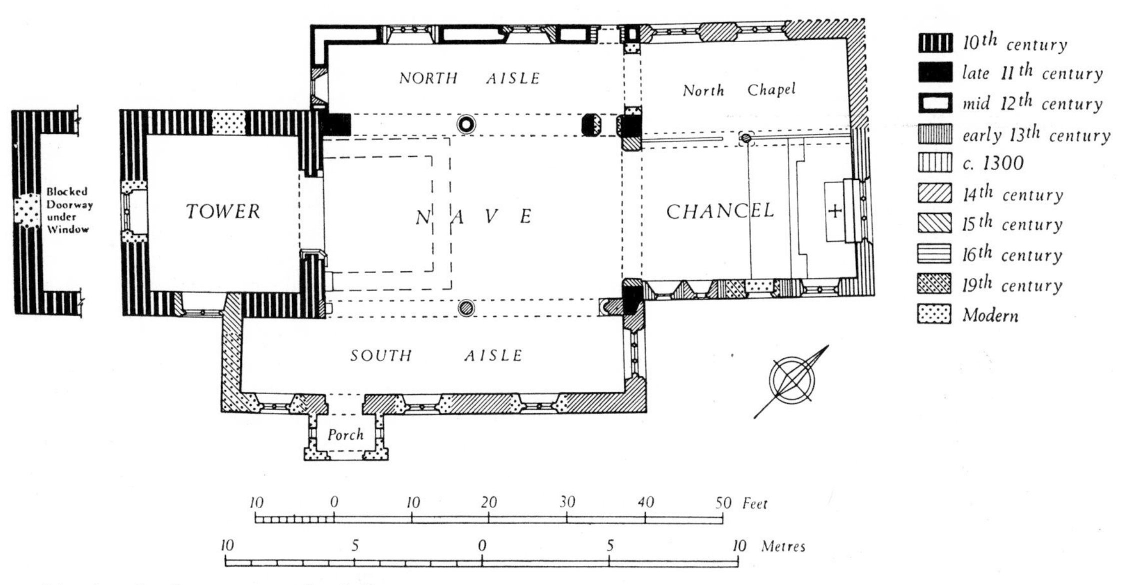

Fig. 31. (8) Church of St. Mary Bishophill Junior.

It stands in a churchyard of some size and is built of random rubble comprising gritstone, Roman *saxa quadrata*, tile, and some magnesian limestone, with limestone dressings. The roofs are covered with tiles and slates.

The oldest carved cross fragments from this church may be as old as *c.* 900, and burials of the early 10th century were found N. of the church in 1961 (excavation by Mr. L. P. Wenham). The base of the tower is of the 10th century; the stubs of two walls running E. suggest that the church of this date had a tower-nave with a small narrower E. cell, like Barton-on-Humber (cf. *Arch. J.*, CXVIII, 171–2). An architectural fragment of this date, probably part of a sculptured tympanum, was found reused in the belfry. The tower was raised in height in the 11th century, probably before 1066. A small *Nave*, of the same width as the tower and about 40 ft. long, replaced the earlier E. cell soon after the Conquest. A *North Aisle* was added in the mid 12th century; the *Chancel* was added early in the 13th century and extended *c.* 1300. The *North Chapel* and the *South Aisle* were built in the 14th century. In 1411 the W. side of the tower and the timber roof were said to be in great need of repair (*YFR*, 250); it was at this time that the battlements were added. About this period windows were inserted in the S. wall of the tower and in the N. aisle, and the nave roof and the lower floor of the tower were renewed. By 1481 the church was again in a poor state (*YFR*, 258).

In the 16th century a doorway was inserted in the N. wall of the N. aisle, and in the late 17th century square-headed windows of secular character were inserted in the S. aisle. A doorway in the S. wall of the chancel was made in the early 19th century, the W. end of the S. aisle was extended to form a vestry, and a brick S. porch was built. The church was restored in 1860 by J. B. and W. Atkinson ('an unintelligent and destructive restoration' *Ecclesiologist*, N.S., XIX (1861), 207), most of the S. aisle wall being rebuilt and the present windows inserted. The *Porch* and much of the E. end of the chancel were rebuilt, and the S. side of the roof was renewed with Welsh slate.

Architectural Description—The *Chancel* (28 ft. by 17 ft.), largely rebuilt in the 19th century, retains the E. window of *c.* 1300, of three uncusped lights with geometrical tracery, masked by a modern reredos, and an original early 13th-century lancet at the W. end of the S. wall. The 14th-century N. arcade of two bays has chamfered arches springing from the end walls and from an octagonal pier with a moulded capital. The partly renewed window near the E. end of the S. wall has two two-centred lights without cusps under a round head. The doorway, now blocked, replaces an earlier doorway recorded in 1840. The inserted trefoil-headed lancet retains some original 14th-century stones. The asymetrical chancel arch of *c.* 1400 is of two chamfered orders, the inner corbelled, the outer merging into the responds.

The N. wall of the mid 14th-century *North Chapel* (27 ft. by 11½ ft.) has two segmental-headed three-light windows with trefoiled ogee heads and inverted tracery, largely renewed. The wall is not bonded into the earlier nave aisle. The tiled roof is hipped.

The E. and N. walls of the *Nave* (38½ ft. by 20½ ft.) (Plate 130) are early Norman. A deep respond at the E. end of the

N. arcade now contains an archway (in 1843 a 'confessional window near the chancel arch', *AASRP*, XXVIII, i (1905), 424). At the W. end the rubble is exposed and the facing of the respond, with coarse diagonal tooling, is an obvious addition. The 12th-century arcade has round arches of two orders, the inner double-chamfered and the outer chamfered on the S. side only. The small limestone voussoirs have fine diagonal tooling, showing remains of colourwash. The pier is round, with a simple hollow-chamfered cap and a plain round chamfered base of gritstone. The S. arcade has two arches with two-centred heads and deeply cut mouldings of 14th-century character. The E. respond is half-round with a hollow chamfered capital and roll necking and a base of similar, but reversed, profile on a square plinth. The pier is round, with a hollow-chamfered octagonal capital and similar necking and base. The similar W. respond, let into the tower wall, has been largely cut away.

The *North Aisle* (38 ft. by 8½ ft.) has a tiled roof continuous with that of the nave; the archway at the E. end is of 1860. The N. wall is of the 12th century, retaining some of the original quoins of the N.E. angle. The badly coursed masonry includes large blocks of magnesian limestone. The 16th-century doorway at the E. end has a four-centred head with chamfered reveals and a round rear arch of brick. The eastern window of mid 15th-century character is square-headed with a segmental rear arch; it is of two lights with trefoiled heads and inverted ogee cusping in the tracery. The W. jamb of the rear arch has stones with diagonal tooling and belongs to an original 12th-century window. The western window is square-headed with a segmental rear arch and is of three lancet lights, trefoiled with pierced cusps of soffit type, perhaps of *c.* 1280. The early 14th-century lancet in the W. wall has an ogee trefoiled head and renewed jambs.

The 14th-century *South Aisle* (49 ft. by 9½ ft.) now incorporates a later vestry at the W. end. The renewed E. window of *c.* 1330 (Plate 16) has reticulated tracery and a moulded label. The S. windows, all of 'Decorated' type, are modern. The original doorway between the two W. windows has hollow-chamfered reveals and a four-centred rear arch with square jambs.

The *Tower* (20 ft. square) (Plate 10) is 73 ft. high, with walls 3 ft. thick. The lowest stage, of roughly coursed *saxa quadrata* of magnesian limestone, has large gritstone quoins. From about 20 ft. to about 52 ft. above ground level the construction is similar with some herringbone courses defined by bands of limestone and gritstone and quoins of gritstone set alternately. Above a square string the top stage is set in slightly with thinner walls, almost entirely of greyish gritstone; it is crowned by a battlemented parapet and pinnacles of magnesian limestone added in the early 15th century. The E. wall contains the contemporary tower arch, 10 ft. wide and nearly 16 ft. high (Plate 14), of two square orders with a label of square section resting on large imposts of two slightly corbelled courses. The jambs are of two square orders with bases each formed of one large chamfered block of gritstone with bold axing of haphazard broken lines. To S. of the arch a large projecting piece of gritstone represents a wall running to the E. with the inner face set against the label of the arch. On the N. a corresponding wall is probably represented by two courses near the floor level. Above the roof of the nave a steeper roof line can be seen; the next two stages are lit by small oblong lights. In the belfry is a window surrounded by strip work, with a round arch formed by a square-section label on square imposts; below these last are simple pilasters on square bases. Beneath the arch a turned baluster supports a square dosseret on which rest two small round-headed arches; the lower part of the window, which was of exceptional size, has been blocked. The N. wall is similar; there is said to have been a doorway on this side, but the slight evidence indicates that it was later. The similar S. wall was pierced in the 15th century at ground stage for a square-headed window of two lights, now largely a modern reconstruction. In the W. wall at the ground stage is a window inserted in 1908; the wall beneath it is much patched, indicating the former presence of an original doorway. The belfry window has original pilasters, but the head has been renewed in magnesian limestone and now has two two-centred lights inserted after 1846, when the window still preserved the same form as the others. At the restoration of 1908 the tower was found to stand on good foundations of rubble composed of Roman tiles and bricks and broken stone.

The 15th-century *Roof* of the nave (Plate 17) is of five bays with moulded tie-beams, wall-plates, three longitudinal beams, forming panels, and modern bosses. The moulded E. beam, now hidden from below, is 1 ft. in advance of the wall and probably formed the head of a boarded and painted tympanum against which the Rood was set. The roof of the N. nave aisle, of 17th or 18th-century date, has plain square rafters, purlins and wall-plates, with provision for a dormer removed in 1949. The ceiling of the ground stage of the tower contains a chamfered beam supported at the N. end by a large wall-post on a stone corbel, and with a brace from the post to the beam; this work is of 15th-century character.

Pre-Conquest Stones. Seven carved stones have been found in this church (Plate 26). Nos. 1 and 3 are now in the Yorkshire Museum; the rest are in or built into the church. No. 2 was formerly in the S. wall of the chancel; Nos. 4 and 5 were in the tower walls, inside, at the level of the bells; Nos. 1 and 3 were in the walls. No. 7 (two fragments) is in the E. face of the tower, S. of the ridge of the nave roof. All are of gritstone and, except where otherwise stated, of the 10th or 11th century. (1) *Cross-shaft*, 27½ in. by 12 in. to 13 in. by 11 in., front with two male moustachioed figures in secular costume facing one another and half turned towards the onlooker; below their feet traces of animal in interlace; on left side a dragon in narrow single-strand interlace; on right side a bold double scroll with large boss-like ends to tendrils; probably early 10th-century. (*YMH* (1891), 76; *YAJ*, XX (1908), 176–7; *cf.* early 10th-century Cross of the Scriptures at Clonmacnoise (F. Henry, *Irish Art 800–1020* (1967), pl. 92).) (2) *Cross-shaft*, 34 in. by 9½ in. to 14½ in. by 7 in. to 8¼ in., front with debased vine scroll; on sides a basket plait; on back a serpent with coiled tail within a basket plait (*YAJ*, XXIII (1915), 260; *cf.* also Crathorne (e) *YAJ*, XIX (1907), 305, and T. D. Kendrick, *Late Saxon and Viking Art* (1949), 65). (3) *Hog-back*, fragment, 27½ in. by 20 in. by 7 in. to 9¾ in., below a tile-roof pattern a debased scroll with berries above a coarse plait of two straps, 10th-century (*YMH* (1891), 76 no. 8; *YAJ*, xx, 170; *cf.* two hog-backs from Crathorne (*YAJ*, xix (1907), 305)). (4) *Cross-arm*, 13 in. by 9 in. to $10\frac{1}{4}$ in. by 6 in. to $6\frac{1}{4}$ in., plain with pellet border and rounded arrises on all faces; with dowel-hole in base for fixing to the shaft (*YAJ*, xx (1908), 207). (5) perhaps part of *Tympanum*, 16 in. high with top partly rounded and originally 32 in. long, decorated with incised circle with six rays, late 10th or 11th-century (*ibid.*); not now visible. (6) *Fragment*, 15 in. by 12 in. by 4 in., with worn interlace within a border, on three sides, of two incised lines; possibly architectural, or the upper part of a small headstone. (7) *Cross-shaft* (?), two fragments, about $4\frac{1}{2}$ ft. by $1\frac{2}{3}$ ft. tapering to $1\frac{1}{3}$ ft.; the lower with worn remains of close interlace, the upper, tapering, with a large vine scroll, possibly late 9th or early 10th-century.

Fittings—*Bells*: two; with identical technical details but differing widely in inscriptions (1) with stamp of St. John Baptist and Lombardic inscription '+Fac. tibi. Baptista. fit, ut. acceptabilis ista' 14-century; (2) almost certainly cast by John Hoton of York (*fl.* 1455–73), with inset shields (Fig. 32) of (a) three crowns in pale (St. Edmund), (b) three crescents (Ryther), (c) St. Edmund, (d) a chevron with a chief indented (Thornton), (e) Thornton, and Virgin with another figure (Fig. 33), inscription in black letter spaced between the shields: '+.Mater Dia me sana Virgo Maria'. *Bell-frame* of oak, for two bells, late 15th-century, seriously decayed (destroyed 1968, when bells were placed at W. end of N. aisle to await rehanging). *Brass* and *Indents*. *Brass*: on S. wall of chancel between two E. windows, of Mary, wife of William Spence, 1811, inscribed plate set in white marble with reeded frame. *Indents*: in N. aisle, at W. end, grey marble slab with indents of shield and figure, perhaps a woman; in 1908 there were other indents in S. aisle floor (Hargrove in *Yorkshire Herald*, 20 September 1908). *Chairs*: inside altar rails, (1) armchair with fielded panel to back, turned front legs, rails and cross-piece, plain back legs, *c.* 1700; (2) armchair (Plate 44), of oak, perhaps 17th-century; both given by Frances Eliza Cobb, 1875. *Coffin Lids*: in N. aisle, reused in head of N.E. window, (1) part only, 13th-century; S. of tower arch, (2) part only, of gritstone, with incised lines, perhaps Saxon; in the Yorkshire Museum, (3) two parts bearing a cross flory (Plate 15), 14th-century, found under floor in 1861 (*YMH* (1891), 90; nos. 942–3). *Communion Table* (Plate 44): in N. chapel, as altar, of oak, 17th-century, top modern, brought from St. Sampson's in 1929 (notes kept at Vicarage). *Font* and *Cover*: font, under tower arch, round cylinder on octagon, with octagonal stem on modern base, mediaeval. Cover (Plate 28): of oak, late 17th or early 18th-century. In 1909 a font from this church was transferred to Whitwood Mere (Borthwick Inst., Faculty Papers 1909/29). *Glass*: in chancel, in second S. window, four panels (a) St. Michael, (b) St. Mary the Virgin in Glory, (c) archbishop holding pastoral cross, (d) archbishop with pallium, all late 15th-century.

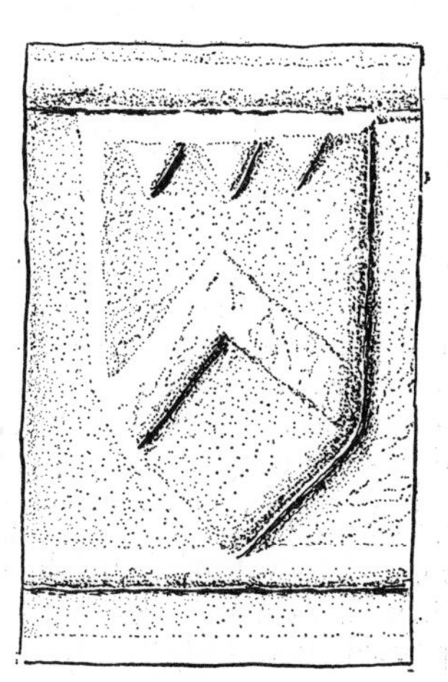

Fig. 32. (8) St. Mary Bishophill Junior. Bell no. 2. Armorial stamps. Arms of St. Edmund, Ryther and Thornton.

Monuments and *Floor Slabs*. *Monuments*: in chancel, on S. wall, (1) Robert Stockdale, Vicar, 1780, white marble tablet with cornice and brackets on black marble slab, signed 'Stead'; (2) Frances, widow of John Nicolson, Doctor in Physick, 1721. In N. chapel, on N. wall, (3) George Hotham, 1823; (4) John Burgess, 1837, Mary, wife, 1829, white slab on black marble, signed 'Skelton, York'; on W. wall, (5) Maria Dorothy, third daughter of Henry Smales, 1849, white marble slab with shaped head and brackets on grey veined marble, signed 'Skelton'; (6) Ann, wife of Henry Smales, 1835, Francis, infant son, 1834, white marble slab with cornice and pediment on grey veined marble, signed 'Skelton'. In S. aisle, (7) George Steward, comb manufacturer, 1820, Elizabeth, wife, 1833, Edward, son, 1839, Elizabeth, daughter, 1851; (8) Elizabeth, daughter of William and Deborah Stead, 1818, set up by husband, John Thompson of Higher Ardwick, Manchester. *Floor Slabs*: all of freestone unless otherwise stated. In chancel, (1) Maria, wife of Thomas Procter, druggist, 1698, Francis, son of Thomas and Mary Procter, Jane, wife, 1733; (2) [Mary] Burgess, 1829, John, husband, 1837; (3) William Bulmer, grandson of John Allanson, 1806, John Allanson of Holgate, 1812, Ann, widow of John Bulmer, daughter of John Allanson,

1813; (4) Rev. John Fuller, subchanter of the Cathedral and Vicar, 1747; (5) Ann, wife of [Richard] Dawson, 1758, [Ric]hard Dawson, [1762]; (6) Richard, son of Rev. Richard Forrest, Vicar, 1793, John Allenby, grandfather, 1811, Mary, wife of Rev. Richard Forrest, 1821, Rev. Richard Forrest, 1829. In nave, (7) Catherine, wife of Rev. [William] W[illiamson], Vicar, 1753, George, Elizabeth, and ... children, [1751–5]; (8) partly covered, [Michae]l Hansby and [Mary] Russell, 1762; (9) Mary Merry, 1829, Phillip Knapton, 1833; (10) Elizabeth, [wife of D]aniell Awtie, 16[9]1, black marble; (11) [Robert Beal], 1763, Ann [Beal, 1775], Thomas [Beal], father, [17]90, Ann [Beal, 1795]; (12) Ann Knowles, 1746, John Taylor, 1774, Ann, wife of John Taylor, 1780; (13) Isabell, wife of Thomas Beal, 1813, Thomas Beal, son, 1829.

Plate: includes (1) cup of 1570, by Robert Beckwith, with (2) paten, as cover; (3) two pewter patens, (4) two pewter flagons, and (5) pewter basin, all inscribed 'St. Mary Bishop-Hill Junr Geo. Beal Jno Lawrence Churchwardens 1774'; and the plate from the demolished church of St. Mary Bishophill Senior, *q.v.* (Fallow and McCall, I, 18–19). *Pre-Conquest Stones*: *see* entry above, before 'Fittings'. *Royal Arms* (Plate 41): in S. aisle, on W. wall, in oblong moulded frame, with 'G 3 R' and date '1793'. *Stoup*: in S. aisle, octagonal stoup on rough square shaft, mediaeval. *Miscellanea*: in S. aisle, set in E. wall, twin water-holding bases and a moulded capital, probably from Holy Trinity Priory, 13th-century.

Fig. 33. (8) St. Mary Bishophill Junior. Bell no. 2. Stamp with figures of Prophet (?) and Virgin.

(9) PARISH CHURCH OF ST. MARY BISHOPHILL SENIOR (Plates 131, 132), stands in a churchyard of considerable size in the angle of Bishophill Senior and Carr's Lane; it is built of ashlar, rubble and red brick, the earliest walls being 2¼ ft. thick and having quoins of large millstone grit blocks and many reused Roman *saxa quadrata* of magnesian limestone. The roofs are of Welsh slate.

Excavation by the Commission has shown that soon after A.D. 350 the site of the church was occupied by a suite of heated rooms on the S. side of an open courtyard (*York* I (18), Fig. 40; *JRS*, LV (1955), 204). In the 10th century a quadrilateral enclosure was made facing S. to Bishophill Senior; the N. wall followed the line of the N. wall of the 4th-century suite. Disturbed burials were found, including one on the line of the E. wall, together with contemporary pre-Conquest carved stones. Early in the 11th century a rectangular single-cell church of stone was built alongside the N. wall of the enclosure. It survives to form much of the present *Nave*. One fragment of a 10th-century cross (no. 19) was incorporated in the footings, and pottery of 11th-century character was found beneath the N. wall.

In *c.* 1180 an aisle was built N. of the nave and the existing doorway inserted in the S. wall. A change in the masonry of the S. wall of the churchyard shows that the enclosure was enlarged to the E. at this date; an extension to W. is also attested by excavation, and the addition of the N. aisle would have necessitated a further extension in this direction. No evidence of a contemporary chancel was found during the excavation but its existence cannot be excluded. Early in the 13th century the enclosure was again extended to the E. and the present *Chancel* added, more than doubling the length of the pre-Conquest church. About 1300 the *North Aisle* was widened and extended one bay E. Early in the 14th century a N. chapel was added; it may be associated with the founding in 1319 of a chantry at the altar of St. Katherine for which Roger Basy had obtained a licence in 1311 (*CPR, 1307–13*, 343; SS, XCI, 68–9). In a severe thunderstorm on 6 April 1378 damage was done to the stone belfry and the timbered porch of the church (J. Raine (ed.), *Letters from the Northern Registers*, RS, LXI (1873), 419). In the 15th century the *North Chancel Aisle* was rebuilt, possibly when Basy's chantry was re-endowed in 1403 (*CPR, 1401–5*, 193). Later in the century two large windows were inserted in the S. wall of the nave, and the roof was renewed; to this period also belonged the former E. window of the chancel. In the 17th century there was an extensive restoration; the chancel walls were raised in red brick and a new roof was made, copied from that of the nave. The N.W. *Tower* was built over the W. bay of the aisle in 1659, replacing a detached bell-tower in the churchyard seen in Speed's plan of 1610. The windows of the N. chancel aisle were modernised by the Fairfax family, who had inherited rights in the Basy chapel. A brick *Porch* was added in the late 18th century and a gallery built in 1841 (Borthwick Inst., Faculty Papers, 1841/1). In 1860 the church was restored: the chancel aisle was remodelled and the E. window of the chancel

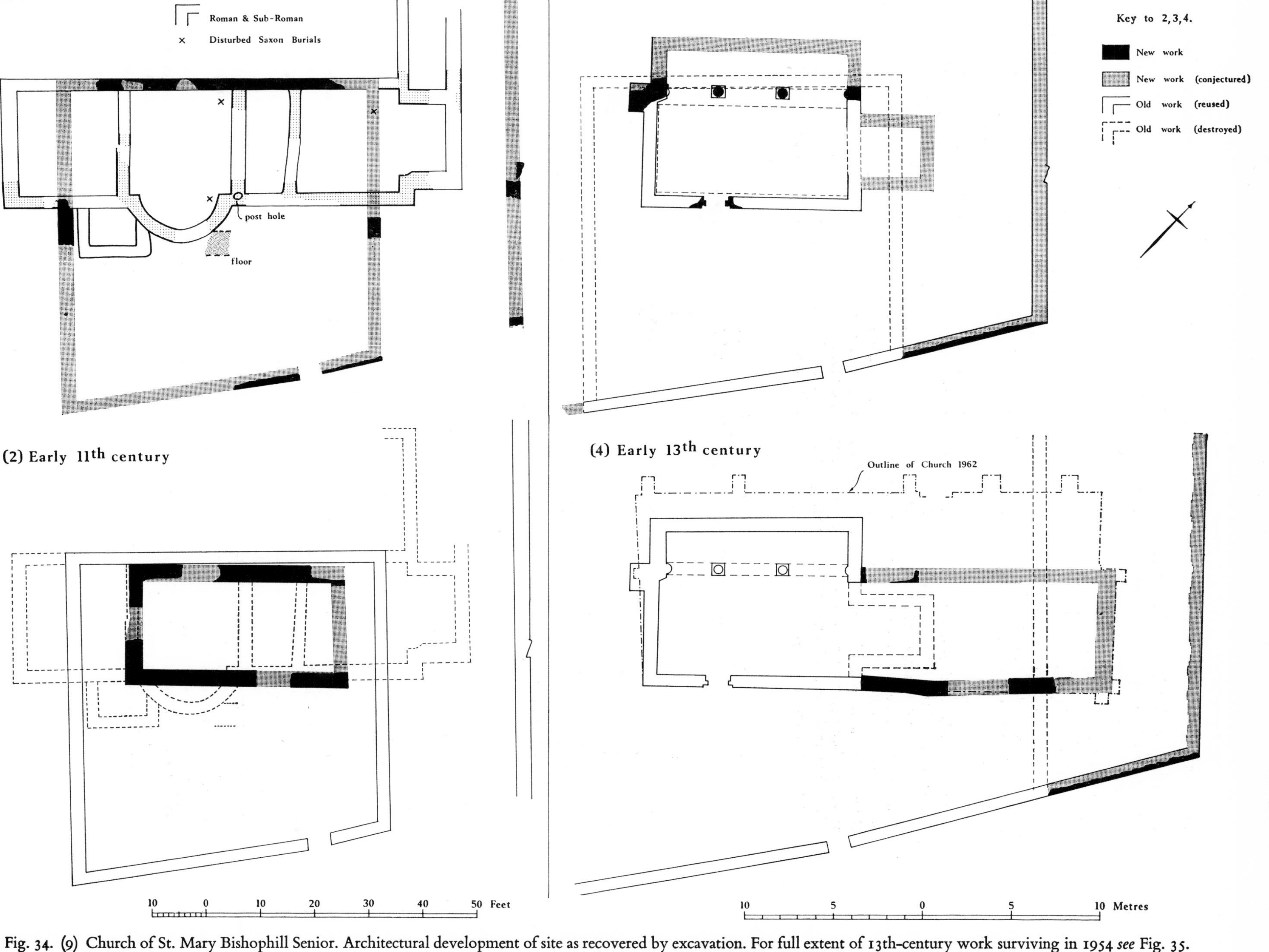

Fig. 34. (9) Church of St. Mary Bishophill Senior. Architectural development of site as recovered by excavation. For full extent of 13th-century work surviving in 1954 *see* Fig. 35.

replaced. A ceiling inserted in the nave *c.* 1810 was removed. In 1876 the church ceased to be parochial and was annexed to St. Mary Bishophill Junior. By 1930 it was completely disused, and it has since fallen into a bad state of disrepair. (YAYAS *Procs.* (1949–50), 36 ff.).

Since the recording of the church in 1951–4 and subsequent dispersal of the fittings, the fabric was demolished in 1963. The arcade, the south doorway, and some details have been re-erected in the Church of the Holy Redeemer, Boroughbridge Road.

Architectural Description—The 13th-century E. end of the combined *Chancel* and *Nave* (82 ft. by 19 ft.) has pairs of pilaster buttresses at the angles and a moulded string under the E. window; the N. buttress is only a fragment in part incorporated in the E. wall of the aisle. The 19th-century window, of four lights with geometrical tracery, has vertical lines at each side indicating the outer sides of a group of three lancets, of which traces also remain internally. To N., patching blocks a doorway to a former vestry.

Inside (Plate 133), the E. respond of the N. arcade incorporates part of the E. jamb of a lancet; outside, a stringcourse and the roof-line of the 13th-century chancel are visible. The first arch rests on a square respond to E. and upon a monolithic octagonal pier with recut capital and moulded base; it is probably reformed and is of 19th-century character. The three arches to W. (Plates 13, 133) are similar but have smaller voussoirs with diagonal tooling. The second pier from E. has a simple capital and a moulded base of millstone grit; the third pier is similar but has no base. The fourth pier is of *c.* 1300 to the E. and retains part of the earlier respond of the nave arcade to the W. The bays to W. have round arches with two square orders on the S. side, but only one order to the N.; the voussoirs are small, with fine diagonal tooling. On the W. side of the fourth pier is a corbel with scalloped capital of the late 12th century. The fifth pier has a square chamfered abacus, plain hollow-chamfered cap with round necking, round shaft, and moulded base; the sixth pier is similar, with a chamfered plinth. The S. wall (Plate 133) is of two main builds: Saxon for the three western bays and early 13th-century in the E. part. The Saxon wall is of millstone grit and magnesian limestone blocks; the later wall is of yellowish blocks of limestone. At the E. end is a narrow lancet, not originally glazed, the rear arch showing coarse diagonal tooling. A mid 13th-century round-headed window of two two-centred lights and a segmental rear arch and modern spandrel cuts the external string. Two windows of *c.* 1330 each have two trefoiled lights with curvilinear tracery under two-centred heads; between them is a round-headed doorway of large voussoirs with a segmental rear arch and probably of the 17th century. Above the narrow lancet further W. is a 17th-century square-headed light splayed internally. To E. of this opening the chancel walls have been heightened in 17th-century brick; to W. a fragment of the 13th-century roof-line of the nave can be seen, with a 15th-century heightening. Westward the wall is set back and a base-course shows the original length of the Saxon nave. Above, the wall is rough and built of reused stone and contains two large 15th-century three-light windows with vertical tracery under square heads with moulded labels. The wall above these has a 15th-century heightening in pale grey limestone ashlar.

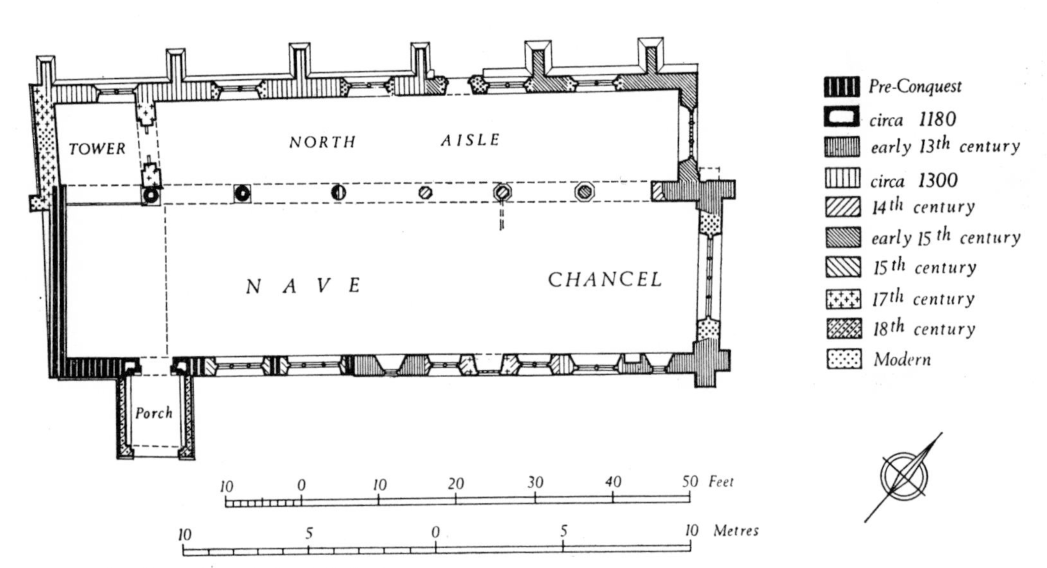

Fig. 35. (9) Church of St. Mary Bishophill Senior.

The S. doorway (Plate 14) has a round head with a chamfered inner order and a moulded outer order supported on shafts with water-leaf caps and moulded bases of *c.* 1180; the square-headed rear arch is later. The S.W. quoin of the nave is built of large pieces of gritstone and limestone alternately. In the later heightening of the wall is a small oblong light (*c.* 1841) to the gallery.

The W. wall of the nave, of reused Roman stone, is built on an earlier wall of magnesian limestone. The gable is of red brick laid in irregular facing bond, with a stone coping.

The E. wall of the *North Aisle* (67 ft. by 10½ to 11½ ft.) is of magnesian limestone in large blocks. The 15th-century E. window, now blocked, was of three lights, altered in the 17th or 18th century; the head is almost round, with mullions running straight up without tracery. The N. wall is built of good ashlar and is of two periods: to E. of the N. doorway the outer face sets in slightly and has two narrow four-stage buttresses with gabled tops and an elaborate high plinth of the 15th century. The first two windows are modern, replacing 15th-century windows similar to the E. window. The W. section of the wall is thicker and is of three bays, with the tower built over the W. bay. The W. buttresses are of four stages with sloping tops; the two westernmost originally had gabled heads. The third and fourth windows are also modern, but the western window is partly original, with soffit cusps of *c.* 1300. The W. wall of the aisle is a 17th-century patchwork and includes a coffin lid beneath the plinth and another lid forming the N. jamb of a blocked square-headed window. Above and to S. is another square-headed light. At the S. end of the wall is a modern buttress, not quite in line with the nave arcade. Inside the aisle, the E. wall of the tower, of narrow brick, has a tall two-centred archway of two square orders with the imposts carried across as a stringcourse. The *Tower* (10½ ft. by 10½ ft., not square), of red brick in irregular English bond with stone quoins, rises two stages above the aisle. Each face is lit in the lower stage by a plain opening with a flat moulded stone head and, in the upper stage, by windows of two uncusped lights beneath four-centred heads. The tower is crowned with a chamfered string and stone battlements.

The 18th-century *Porch* (8 ft. by 8½ ft.) is of red brick, well laid in Flemish bond. The doorway has a rubbed-brick round arch with a stone key-block and moulded stone imposts. Above is a gable with a stone coping. Inside there is a flat plastered ceiling with a moulded cornice.

Roofs: the nave roof, of *c.* 1500, of six bays, has a very shallow pitch with cambered, stop-chamfered tie-beams, and principal rafters, purlins and common rafters all chamfered. The 17th-century chancel roof of four bays is similar to the nave roof, though with rafters of smaller scantling. An outer high-pitched roof above the ancient nave roof was apparently added when the tower was built in 1659.

Pre-Conquest Stones: the following are from the fabric of the church. (1–16) and (21–3) were found in the walls during demolition in 1963 and are now inside the Church of the Holy Redeemer, Boroughbridge Road, (1) set in the E. wall, (2) and (3) in the lectern, (4) and (22) in the pulpit, (5–13, 21) and (23) in the S. wall and (14–16) above the arcade. (17) is in private hands, (18–19) are in the Yorkshire Museum and (20) is in St. Clement's Church, Scarcroft Road. All are of coarse sandstone or fine gritstone and of the 10th to 11th centuries unless otherwise noted. The occurrence of gritstone probably indicates the reuse of Roman building blocks and tombstones.

(1) *Cross-shaft* (Plate 25) in two joining fragments 2 ft. 4 in. in total height and tapering from 10 in. by 9 in. to 8 in. square. On the front are two figures one above the other. The larger, lower figure is shown full-face wearing a long garment draped in many folds. On the breast is a rectangular object decorated with eight pellets, perhaps a book satchel, and between the hands is an oval object, possibly a chalice. Only the legs remain of the upper figure which was in profile in a half-sitting position with a sheathed sword at the left side. On the back is an animal with head downwards interlaced, perhaps linked to another. On the left side is a sinuous serpent crossed by a two-strand strip with knots in the curves; on the right a broad undulating band with two incised lines similarly crossed by straps probably indicates another serpent. The angles of the shaft are edged with cable moulding. The figures resemble those on crosses at Nunburnholme, Yorks. (*YAJ*, XXI (1911), 267), and at Edenham, Lincs. (*Arch. J.* LXXXIII (1926), pl. V, no. 13). Late 10th to early 11th-century. Stone (13) may be the lower portion of this same shaft which could then have been about 6 ft. high.

(2) *Cross-shaft* fragment 6½ in. by 8¼ in. carved on one face with cable moulding round two sides and formerly visible around a third, enclosing part of a circle formed by two incised lines, probably remains of a two-strand interlace or possibly of the nimbus of a figure. Probably 10th-century.

(3) *Cross-shaft* fragment or headstone 20 in. by 11½ in. by 6 in. (Plate 25). The lower half is undecorated where it stood in a socket stone or in the ground, and tooling here resembles Roman dressing. On the front is a simple plait within a border of pellets surrounded on the top and sides by a cable moulding. On the left side is a narrower, more complicated, plait, but about 1 in. of the right side has been removed. The stone resembles one from Parliament Street, now in the Yorkshire Museum (*YAJ*, XX (1908), 162), and stone (18) below.

(4) *Cross-shaft base*, 17½ in. by 13 in. by 8½ in. Only two faces are visible. On the wider face is a two-strand loop tied in a Stafford knot, through which is plaited a second loop; a similar knot presumably formed the upper half of the panel which is bordered by two incised lines and on the sides by a cable moulding. The left side has two plaited loops with a rectangular ending at the bottom. Probably 10th-century.

(5) *Cross-shaft* fragment, 10¾ in. by an original width of 13¼ in., with a simple close interlace on one face, so degenerate as to give the impression at first sight of incised cross-hatching. *Cf.* cross from Gainford (Haverfield and Greenwell, 97–8, No. XXXI).

(6) *Coped tombstone*, perhaps hog-back, fragment, 16 in. by 10 in. by 5 in., decorated on a chamfered edge with a band of a narrow two-strand interlace with a double cable moulding to one side.

(7) *Cross-shaft* fragment, 8½ in. by 6 in. to 7 in. The decorated face, probably a lower corner of a larger panel, has a pattern of closely woven two-strand straps with traces of a pellet or cable-moulding border below.

(8) *Cross-shaft* fragment 15 in. by 14 in. by 10 in. with traces on two faces of a very worn interlace and of a wide plain margin on two sides of the larger face.

(9) *Hog-back* fragment, 12 in. by 19½ in. to 21 in. On one

face is part of an over-all flat in-and-out weave interlace below a very coarse moulding.

(10) *Tomb-slab* fragment, 18½ in. by 13 in. (Plate 25), but broken on all sides. On one face a cable moulding runs across a panel of animal ornament with entwined sinuous beasts, resembling that on a coped stone from St. Denys's Church, York (*YAJ*, XX (1908), 162).

(11) *Cross-shaft* (?) fragment, 10 in. by 9½ in. by 5¾ in., rounded at the top. On the face part of an interlace of two-strand straps is enclosed on three sides by a cable moulding within a plain border ending in a curve. *Cf.* the Two Dales Cross, Darley Dale, Derbyshire (*Arch. J.*, XCIV (1937), pl. XX).

(12) *Cross-shaft*, fragment 9½ in. by 6½ in. by 9 in. On two faces is an incised key pattern. (Cf. *YAJ*, XIX (1907), 285; XXI (1911), 291.)

(13) *Cross-shaft*, 23 in. by 10 in. by 8½ in. On the face is a panel of very closely woven two-strand interlace, and similar interlace appears on the right side. The left side is very worn and the back face has apparently been removed. This may be the lower portion of the shaft to which stone (1) belonged.

(14) *Fragment*, approximately 1 ft. by 8 in., decorated with traces of an interlace on the only visible face.

(15) *Cross-shaft*, approximately 17 in. by 14 in., decorated on two adjacent faces with a close interlace within a narrow plain border.

(16) *Tomb-slab* fragment, 10 in. by 13 in., with, on the only visible face, a single two-strand strap woven in and out of rings interlaced to form a chain.

(17) *Hog-back* fragment, 31 in. by 13 in. to 23½ in. by 11 in. (Plate 26). On the central ridge are two cable mouldings with a simple two-strand interlace on either side. The worn original end probably had a terminal animal as at Barmston or Lythe (*YAJ*, XXI (1911), 258, 295). Found in a car park in Burton Stone Lane paved with rubble from the demolished church and now in private hands.

(18) *Headstone* fragment, 13 in. by 15 in. by 6½ in. On three sides are interlace panels within a border, plain on the sides and of pellets on the front; the back is mortar-covered. Traces of diagonal tooling and a raised border on the bottom show that this is a reused Roman stone. Found built into the late Norman footings at the N.W. angle of the nave and now in the Yorkshire Museum.

(19) *Cross-shaft* or cross-head fragment of magnesian limestone, 4 in. by 2½ in. by 3 in., showing part of a deeply carved interlace. Found in the footings of the 11th-century church and now in the Yorkshire Museum.

(20) *Tomb-slab*, 47½ in. by 15 in. to 17 in. (Plate 25), having on one face a patriarchal cross with interlace of two-strand straps in the panels between the arms and the border, which in places is a double cable moulding and in places a narrow plait. (*YAJ*, XX (1908), 207).

(21) *Fragment* of yellow limestone 14 in. by 6 in. by 6 in. carved with a crucifix in shallow relief 4½ in. high and 3 in. wide. The figure is naked except for drapery at the waist, and the feet are separate. A hollow above the head may be the trace of a Hand of God. Reused in 13th-century footings; probably pre-Conquest.

(22–3) *Tombstone* fragments, 8½ in. by 10 in. and 13½ in. by 7½ in., each with a foot and part of a leg in relief above a wide plain border. These are probably from Roman tombstones or sarcophagi brought from the cemetery in the Baile Hill–Clementhorpe area.

Fittings—*Bells*: six; each inscribed 'Pack & Chapman of London fecit 1770' (Plate 21) (*York Courant*, 12 February 1771), removed 1954 to St. Stephen's Church, Acomb. *Benefactors' Tables*: two; (1) oblong panel with moulded surround, benefactions of 1778 and 1843; (2) bolection-moulded panel with semicircular head (Plate 22), given by Thomas Todd (d. 1703), lettering of *c.* 1771, benefactors include Henry Beckwith, £100 towards new bells, 1771 (both now on W. wall in St. Clement's, Scarcroft Road). *Brass*: in freestone slab near altar step, (1) 'Hic jacent reliquiae G.D., C.D., E.W., P.G. ut supra in marmore scriptum est' (see *Monuments* (2)). *Bread Shelves*: oblong oak cabinet (Plate 23), mid 18th-century (now in N. chapel at St. Clement's, Scarcroft Road). *Candlesticks*: two enamelled candlesticks (Plate 24), 14th-century, found in 1859 under church floor (in Yorkshire Museum (*YMH* (1891), 237)). *Chairs*: three; (1) and (2) oak, with turned legs and front rail and enriched panelled backs with shaped tops (now at St. Hilda's Church, Tang Hall); (3) oak (Plate 44), 17th-century (now in N. aisle of St. Clement's, Scarcroft Road). *Coffin Lids*: above E. window, (1) fragment (Plate 27), 14th-century; in W. wall of tower, in plinth, (2) complete lid without markings, mediaeval; in W. wall of tower and forming N. jamb of blocked window, (3) fragment (Plate 27), 14th-century (found in course of demolition, now built into S. wall of Church of the Holy Redeemer, inside); in lintel of W. window of tower, (4) part of coffin lid inscribed in Lombardic capitals 'priez : pvr : lealme :', early 14th-century (now under E. lancet of Church of the Holy Redeemer); (5) part of lid with incised stepped cross, 12th-century (now in Church of the Holy Redeemer); (6) with floriated cross in relief, with chalice (now on S. wall in Church of the Holy Redeemer). (See also *Pre-Conquest Stones*, above.)

Collecting Shovels: two; of oak (Plate 23), 18th-century (now in St. Clement's, Scarcroft Road). *Communion Table* (Plate 44): of oak, with heavy turned legs, upper rails with slightly incised Jacobean enrichment, 17th-century (now in St. Clement's, Scarcroft Road). *Door*: in blocking of S. door of chancel, of planks, possibly 18th-century. *Font*: fragment found in 1964, with Geometrical panelling, late 13th-century. *Gallery*: at W. end of nave, pitch pine, front with five panels and moulded rail, 19th-century. *Lord Mayors' Tables*: on W. wall, two, (1) (Plate 22), inscribed Elias Pawson, 1704, W.ᵐ Coates, 1753, John Carr, 1770, second time, 1785; on frame between inscribed panel and City arms a small iron sword-rest, 18th-century; (2) tall panel with cinquefoiled head under battlemented cornice, 'VR' in spandrels and arms of York City at top, inscribed James Meek, 1848, James Meek, 1850, Henry Cooper, 1851, Fred Gains, 1945–6, 1946–7, *c.* 1850 with added names (both now in St. Clement's, Scarcroft Road).

Monuments and *Floor Slabs*. *Monuments*: on E. wall, N. of window, (1) [Hester, daughter of Robert Bushell and widow of Robert] Fairfax, 1735; on S. side of arcade, above E. arch, (2) George Dawson, late of Minster Yard, 1812, Catherine, second wife, 1807, and her two sisters, Elizabeth, wife of Rev. Henry Wood, D.D., 1799, and Philadelphia Gore, 1808, erected by George Dawson, son, large white marble sarcophagus with tapering sides, enriched cornice and pedimental lid, moulded base with fluted corbels, all against black marble, said to be signed 'Fisher sculpt' (J. W. Knowles, MS. Notes in York City Library); above second arch, (3) Mary Sophia, wife of Frederick Hill of Clementhorp, daughter of N. P. Johnson of Burleigh Field, Leicestershire, 1819, and two children, Charles Frederic, 1811, and Frederica, 1813, white marble sarcophagus set on oblong pedestal, against black marble slab, by Melor?, York; in spandrel of second pier from E., (4) inscription, now defaced, to Edward Prest, 1821, and Elizabeth, wife, 1841, oblong white marble slab with cartouche above between palm leaves, with arms of Prest, sculptor's name indecipherable; over third arch, (5) Henry, fourth son of Edward and Elizabeth Prest, 1827, white marble slab on black, by Skelton; in spandrel over fourth pier, (6) Francis, son of Stephen and Martha Beckwith, 1818, and Stephen, M.D., Senior Physician, brother, 1843, oblong white marble slab on projecting oblong pedestal supported on foliated consoles, all against black marble, by Skelton; on S. wall, (7) Thomas Suttell, 1789, white marble oval with border formed by snake with head in mouth, on black marble slab, by Thomas Atkinson; (8) Joseph Harrison, merchant at Newport in Rhode Island, later private secretary to the Marquis of Rockingham, by whom he was appointed Collector of Customs at Boston in New England, Helen, wife, 1794, oval marble slab as (7); (9) Sarah, wife of Peter Atkinson, architect, 1825. In N. aisle, (10) Christopher Brearey of Middlethorpe, 1826, Jane, wife, 1835, oblong white marble slab on black, by Skelton; (11) Rev. Richard Tillard, Vicar of Wirksworth, Derbyshire, 1736, with arms of Tillard with escutcheon of pretence of Yoward. The following (12–17) are in St. Clement's, Scarcroft Road. On W. wall, (12) Catherine, widow of Henry Pawson and only daughter of Robert Fairfax of Steeton, 1767; (13) Alathea, sister of Robert Fairfax of Steeton, 1744, with lozenge-of-arms of Fairfax; (14) Thomas Rodwell, 1787, white marble slab set on another, supported by two plain corbels, by John Fisher (Morrell, *Monuments*, 84); restored by George Cressey, relative, 1844; (15) Elias Pawson, Lord Mayor 1704, 1715, and Mary, wife, daughter of William Dyneley, 1728, cartouche with impaling arms of Pawson (Plate 32); (16) Rev. John Graham, 48 years rector, 1844, white marble slab with moulded top and base, on grey marble slab with shaped head and base bearing ivy wreath and palms, by Fisher, York; (17) Henry, son of Elias, grandson of Henry Pawson, 1730, white marble slab between recessed panelled Corinthian pilasters and moulded cornice surmounting segmental pediment, supported on enriched corbel, signed 'W. Palmer fecit' (William Palmer of London). In churchyard, among some twenty-five late 18th and 19th-century tombstones *etc.*: N. of chancel, (18) Hannah and Mary, daughters of Samuel and Mary Lucas, and Henry, son, 1848, signed 'Lucas' (the father); N. of tower, (19) Peter Atkinson, York architect and partner of John Carr, 1805; (20) James, son of Peter and Magdalen Atkinson, 1791; S. of chancel, (21) John Stevenson, 1828, by Waudby, Coney Street; S. of porch, (22) Jonathan Tomlinson, 1828, and wife, Ann, 1825, by Plows; W. of church, (23) James, 1827, Henry, 1828, and Martha Bromley, 1836, by Plows. The last three are in fine-grained limestone with shaped heads and good lettering.

Floor Slabs: (1) Lydia Robinson, 1831; (2) Sarah Pawson, 1724; (3) Ralph Yoward, 1714, Sarah, widow, 1716, Richard, nephew, Receiver General of Archiepiscopal Rents of York, 1748, Elizabeth Morrice, widow, sister of Richard, 1768, Ralph, son of Richard, 1781; (4) William Coates, merchant, Lord Mayor of York 1753, 1758; (5) Edward Prest, 18[21], Elizabeth, widow, 1841, Henry, fourth son, 182[7]; (6) Thomas Suttell (*see* monument (7)); (7) Elias, son of Elias Pawson, merchant, 1700, Alice, daughter, 1702, Elias, son, 1705, Dyneley, son, Elizabeth, daughter, 1708, Thomas, son, 1710, the said Elias Pawson, 1715, Mary, wife, 1728; (8) Mr. Tho . . . 178.; (9) Francis Brown, late of Leicester, 18..; (10) Mary Bewla(y), 1752, Henry Bewlay, Common Brewer, 1762; (11) Elizabe(th), widow of John Taylor, butcher, 1759, two children, Elizabeth and Ann; (12) Mary, wife of Robert Fairland, 1749, (Eliz)abeth, daughter, (Ro)bert Fairland, 1753; (13) Elizabeth, wife of George Cressey, 1805, three children: Selina, 1799, Thos. Rodwell, 1811, John, 1813, Susanna, wife of George Cressey, 1819, George Cressey, 1846; (14) Thom(as Thackray), Ann, wife, 1806, Thomas, son, 1847; (15) . . . . Wm. Porter and grandson of above Thomas and Ann, 1772, Ann, wife of Wil. Porter, merchant, William Porter; (16) . . . . William Jackson, junior, 1751, Mary Wharton, daughter of William Jackson, senior, 1751, Elizabeth, daughter of Thos. and Ann Thackray, 1772, Lucia Elizabeth Thackray, 1781; in N. aisle, (17) Margaret, daughter of Tho. and Ann Middleton, 1720; in porch, (18) slab inscribed 'Hic jacet Richardus Cordukes A.M., Rector hujus Ecclesiae', (1796).

Organ: by John Ward, 1846 (*Yorkshire Gazette*, 23 May and 12 September 1846), moved to St. Mary Bishophill Junior in 1901. *Wall Painting*: at E. end of N. arcade, on S. side fragment of black-letter inscription in English, set in panel with round pedimental head, 16th-century. *Piscina*: in S. wall, W. of easternmost lancet, with cinquefoiled head and stop-chamfered reveals, bowl destroyed, early 13th-century. *Plate* (now at St. Mary Bishophill Junior): includes (1) cup of 1674, inscribed 'Deo et Ecclesiae St. Mariae de Bishophill Senr Ebor. Anno. Dom. 1674. John Place Geo. Smith Church Wardens'; with (2) paten as cover, York letter for 1674 and mark of Robert Williamson; (3) salver on tapered stem, inscribed 'Deo et ecclesiae St. Mariae De Bishopphill Senr. Anno Domino 1706. John Mawman and William Tesh Church wardens', with London letter for 1706–7 and mark of Andrew Raven (Fallow and McCall, I, 17–18). *Pre-Conquest Stones*: *see above* before 'Fittings'.

Miscellanea: the S. side of the churchyard is divided from Carr (formerly Kirk) Lane by a wall largely of brick (17th–18th century) and partly of 19th-century masonry. Its base incorporates remains of a much earlier stone wall, part of which may well be that defining the S. side of the Saxon burial ground, the E. and W. sides of which were marked by walls of post-Roman date found in the excavations of 1964. The eastern of these walls apparently joined the wall to the lane at a point where the character of masonry changes, a very large block indicating the external angle. There is an early blocked gateway, 4 ft. wide, towards the lane, some 20 ft. to E. of the modern gateway.

In the excavations of 1964 the following were found: in N. aisle, at the W. end, (1) pewter paten with large knop, probably 16th or 17th-century; from graveyard N. of church, (2) fragments of spreading foot of pewter chalice and segment of flat paten with raised rim; from burial disturbed by destruction of E. enclosure wall of Saxon churchyard and by subsequent burials, (3) bronze strap-end (D. M. Wilson in *Med. Arch.*, IX (1965), 154) probably 10th-century, now in Yorkshire Museum.

(10) PARISH CHURCH OF ST. PAUL (Plate 134), stands in a small churchyard on the N. side of Holgate Road. It is built of brick faced with coursed sandstone rubble and ashlar dressings; the roofs are of Bangor slate with Staffordshire ridge-tiles.

Topography of the City of York; and the North Riding of Yorkshire (1857), I, 562), when the original perpetual curacy became a rectory.

Alterations were proposed in 1874 (Borthwick Inst., Faculty Papers 1874/2) and in 1890 (Faculty Papers 1890/22), when the architects were Demaine and Brierley (descendants in practice of Messrs. Atkinson). In 1890 the chancel was extended into the nave and the organ was removed from the gallery to the N.E. corner; new screens were put on either side of the chancel and the pulpit was moved further W. In 1906 a Faculty was obtained to replace the three tall lancets in the E. wall with a Geometrical window filled with stained glass. The architect was G. H. Fowler Jones and the glass was made by Heaton, Butler and Payne of London (Faculty Papers 1906/22).

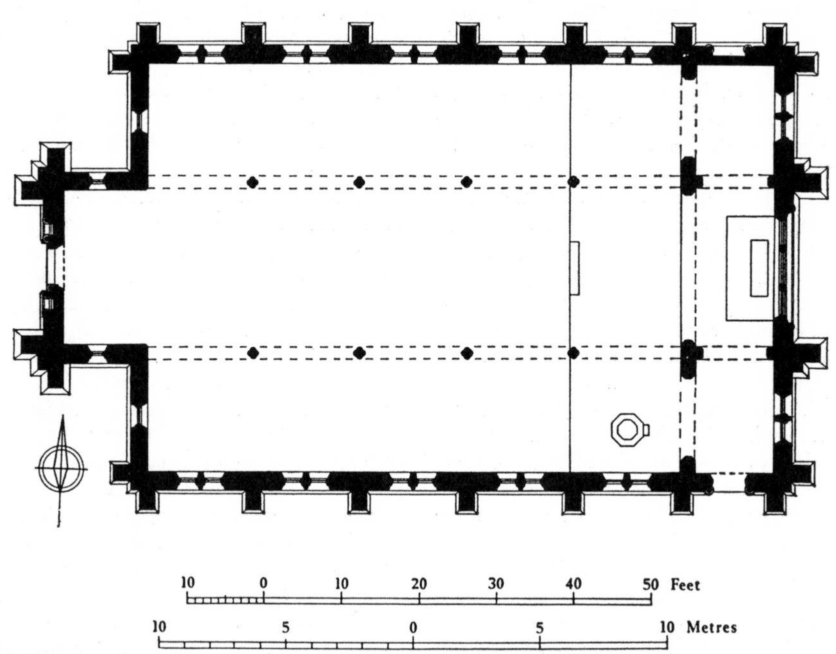

Fig. 36. (10) Church of St. Paul, Holgate Road.

This part of the old parish of St. Mary Bishophill Junior developed rapidly as a residential area for railway employees after 1840. A new District church was proposed some time before 1850 and designs were made by J. B. and W. Atkinson. Plans and elevations survive (archives of Messrs. Brierley, Leckenby & Keighley) and bear dates of October and November 1850. The contractors were John Beal and John Bacon. The church was opened on 7 October 1851; it cost about £2,400 (York City Library, YL/Gray, Letters 5(b), 8(b)). The consecration by the Archbishop of York took place on 3 January 1856 (T. Whellan and Co., *History and*

The church is an interesting example of the Gothic Revival in York, as practised by a leading firm of local architects.

Architectural Description—The church is an aisled parallelogram, with a projecting W. bay of full height forming a narthex. The *Chancel* (26 ft. by 20 ft.) comprises the original structural chancel of a single narrow bay and the E. bay of the nave. The arcades are of boldly moulded two-centred arches, and the chancel arch is similar, resting upon responds with triple shafts.

The arcaded *Nave* (54½ ft. by 51 ft.) is of four bays and, with the E. bay now forming part of the chancel, is uniformly treated, having paired lancets in the aisle walls. In the arcades, the slender piers of painted cast iron are widely spaced and support deeply moulded two-centred arches of plastered brick. The narthex projects W. one bay beyond the end walls of the aisles. The W. wall carries a bellcote capped by a gable with foliated cross. There is a large rose-window above the W. doorway, and the latter is flanked by elaborate niches.

The main *Roof* has collar and king-post trusses with large arched braces resting on moulded corbels some distance below wall-plate level; there are intermediate trusses, with collars only, between the main trusses. The roofs of the aisles also have main and subsidiary trusses, with braced collars carrying queen posts. All the panels are plastered.

Fittings—*Galleries*, at W. end, supported on beams bolted to the iron arcade piers, the intermediate beams being supported by circular iron columns; the fronts rounded, and panelled.

with spire replaced the previous bellcote. In 1850 the chancel was taken down, and in 1851 the old chancel arch was removed and the new *Chancel* built. A *Vestry* was built, over the Hale vault, in 1889, and in 1890 the E. window was inserted.

The church is an interesting example of early 19th-century Gothic.

Architectural Description—The *Chancel* (19¾ ft. by 17½ ft.) is of ashlar, with two-stage buttresses, and has a three-light window in the E. wall and a two-light one in the S. wall, both with geometrical tracery. The chancel arch is of two chamfered orders, the inner resting on corbelled shafts. The *Nave* (36¼ ft. by 29 ft.) and *Transepts* (23 ft. by 12½ ft.) form a large single compartment, T-shaped on plan. They are of white magnesian limestone, with uniform buttresses and lancet windows. Extending across the whole of the two

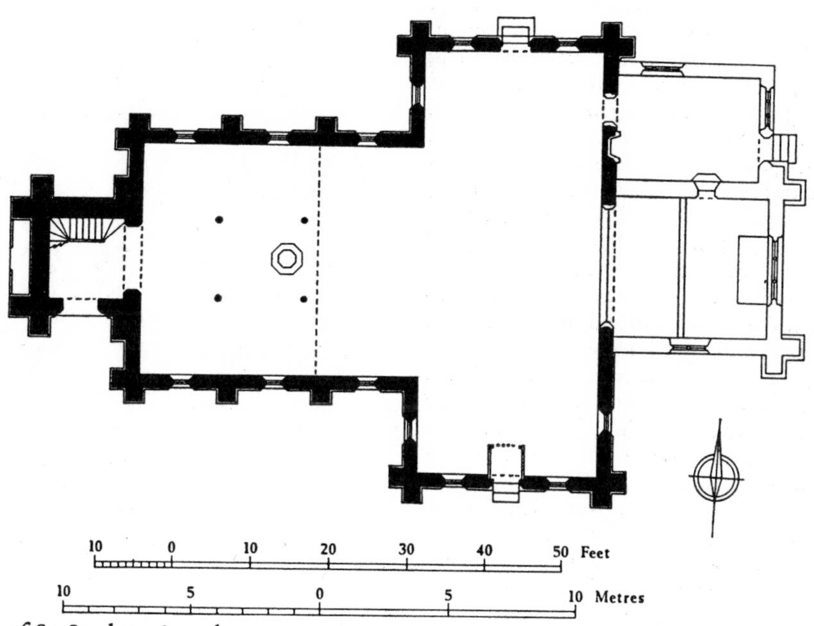

Fig. 37. (11) Church of St. Stephen, Acomb.

(11) PARISH CHURCH OF ST. STEPHEN, Acomb (Plates 135, 136), stands in a large churchyard set back from the N. side of York Road. The E. end is built of rubble ashlar; the transepts, nave and tower are of brick, ashlar-faced in Tadcaster stone. The roofs are slated.

The earlier church consisted of an aisleless nave and chancel (G. Benson, *AASRP*, XXXVIII, i (1926), 78), perhaps dating from the 12th century, but modified in the 15th century. In 1831–2 the nave was replaced by the present *Nave* and *Transepts* designed by G. T. Andrews, perhaps in association with Peter Atkinson, who had made a design approved in 1829; a new *Tower*

western bays of the nave is a gallery supported on iron pillars. The tower arch has a four-centred head with chamfered reveals, and a similar arch above gives access to the gallery. The *Tower* (10 ft. square) is of two stages, buttressed. It is surmounted by corbelling and a moulded cornice and carries a simple broach spire. The chancel *Roof* is of scissor-rafter type; the transepts and nave roofs, the latter arched, are ceiled.

Fittings—*Bells*: three; (1) inscribed 'Jesus be our speed 1660. Recast by John Warner and Sons, London'; (2) 'Venite exultemus Domino. Given by M. A. Hale 1889. R. P. T. Tennant vicar. Cast by John Warner & Sons, London 1886';

(3) inscribed 'Iesvs be ovr speed 1633' (*YAJ*, XVI, 47). *Chairs*: within altar rails, two, of Jacobean character with inset panels, Adam and Eve, and Hope (Plate 42); panels and parts of frames, early 17th-century. *Gallery*: at W. end of Nave, with pitch-pine front, supported by four octagonal iron pillars, 19th-century, before 1850. *Glass*: in lancet in E. wall of N. transept (from E. window of older church), royal arms of Charles II in Garter and with supporters, scrollwork, initials and mottoes (Plate 41), dated 1663, probably by Edmund Gyles.

Monuments and *Floor Slabs*. *Monuments*: In nave, on E. wall, (1) Frances Etridge, 1825, white against grey marble, with round pediment enriched with swag and scrolls between semicircles, and with anthemion above and moulded plinth below, signed Plows of York; (2) Rev. Joseph Pickford, 1804, son of Sir Joseph Radcliffe, Bart., Mary Pickford, 1834, daughter of Sir Archibald Grant of Monymusk, Aberdeen, white marble with moulded cornice against pedimental-headed black marble, signed Williams of Huddersfield; (3) Thomas Smith, 1810, twice Lord Mayor of York, Ann Smith, wife, 1818, John Smith, fourth son, 1805, Barnard Smith, youngest son, of Murton Hall, 1822, Rev. William Smith, M.A., eldest son, 1823, Henry Smith, fifth son, 1823, and several others including Thomas Smith (junior), second son, Alderman and twice Lord Mayor of York, 1841, freestone, with black marble inscription panels, in the Gothic style and with the arms of Smith (Plate 35) signed M. Taylor York, from chancel of older church; on N. wall, (4) Rebecca, 1761, wife of Samuel Armytage; (5) William Philips, 1827, signed M. Taylor; (6) William Kay, 1798; on S. wall, (7) Jeremiah Barstow, 1821, and Dorothy, wife, daughter of John Kilvington of Acomb, 1803, white marble slab and urn on pedestal, against grey marble pyramid, signed Fisher of York (Plate 33); (8) Elizabeth, wife of Nathaniel Wilson, 1758, and Nath. Wilson, 1765 (Plate 33); (9) Jane, 1837, widow of John Weatherill of Bootham, and John Weatherill, 1820, on pedimental-headed black marble, white marble sarcophagus (lid missing) with claw feet between swags of laurel, on plain base under ascending dove, erected by Anne Hutchinson, signed Davies of Newcastle upon Tyne. In N. transept, on W. wall, (10) Frances, 1841, wife of Richard Hale, Jemima, daughter, 1816, William Lawrence, son, 1830, Elizabeth, daughter, 1832. In Churchyard: E. of church, (11) Samuel Smith, baker in Acomb and York, 1713; N.E. of chancel, (12) Ellen Jones, 1830, signed Fisher; S.E. of chancel, (13) William Burdeux, 1690, mostly indecipherable, flat freestone slab on four plain legs, with incised scrollwork at base, with reference to benefaction; N. of church, (14) Robert Driffield, 1816, with shield-of-arms (Fig. 38); S.W. of tower, (15) Jonathan Ferrand, 1741, jeweller, son of John Ferrand, born in Normandy; W. of church, (16) Thomas Royston of Knapton, 1680, wife, 1712, Matthew, son, 1721. There are some other well-preserved headstones of the period 1774–1822 with simple shaped heads and fine lettering; one to the Clarke family is inscribed 'E.B. 1798' on the back. A headstone to George Hubback, 1810, and flat slabs to members of the Driffield family, 1806–16, and of the Smith family, 1822–52, have inscriptions on brass plates. *Floor Slabs*: of freestone, (1) Sylvester Richmond, 1793; (2) Anna, 1743, Elizabeth, 1745, children of John and Elizabeth Kilvington, John Kilvington, ...

Panelling: in gallery, flanking archway to tower, dado of fielded panelling, and section of similar forming back of seat on N. side, 18th-century. *Plate*: cup with bell-shaped bowl, conical stem and domed foot, with added inscription 'Acom Church' in lettering of late 17th-century character, York mark, no date letter, stylistically of *c*. 1570 (Fallow and McCall, II, 10–11). *Royal Arms* (Plate 41): in gallery, on wooden panel, William IV.

Miscellanea: in nave, on E. wall, inscribed limestone tablet recording rebuilding of church, 1831. On step to S. doorway to tower, part of black-letter Latin inscription, the winder under tower stair having formed part of a tombstone. In wall near S.E. entrance to churchyard, stone inscribed 'E.B. 1727', possibly the date of the wall.

Fig. 38. (11) Church of St. Stephen Acomb. Shield of arms from headstone of Robert Driffield.

(12) PARISH CHURCH OF ST. EDWARD THE CONFESSOR, Dringhouses (Plates 135, 136), stands in a small churchyard on the S.E. side of the Tadcaster Road. It has walls of limestone quarried at Clifford and roofs of Cumberland slate. A chapel had recently been built at Dringhouses in 1472 by parishioners of Holy Trinity, Micklegate, who had hitherto worshipped at Acomb (*YFR*, 254); it stood on the site of the present Library in front of the former Manor House. Another chapel, to N.E. of the present church, was built in 1725 (VCH, *York*, 376); its foundations still remain. Both of these earlier chapels were dedicated to St. Helen. The present *Church* was built in 1847–9, the architects being Vickers and Hugall of Pontefract; the contractor was Roberts, and the carpenter Mr. Coates of York. The

(4) CHURCH OF ALL SAINTS, NORTH STREET. Window IV. The Corporal Acts of Mercy. Early 15th century.

STAINED GLASS. BORDER FIGURES

Window III.

Window XII.

Window V.

(4) CHURCH OF ALL SAINTS, NORTH STREET. 15th century.

STAINED GLASS DETAILS

Feeding the Hungry.

Clothing the Naked.

(4) CHURCH OF ALL SAINTS, NORTH STREET. Window IV. The Corporal Acts of Mercy, early 15th-century.

STAINED GLASS DETAILS

Giving drink to the Thirsty.

Visiting the Sick.

(4) CHURCH OF ALL SAINTS, NORTH STREET. Window IV. The Corporal Acts of Mercy, early 15th-century.

Entertaining the Stranger.

Relieving those in Prison.

(4) CHURCH OF ALL SAINTS, NORTH STREET. Window IV. The Corporal Acts of Mercy, early 15th-century.

(7) CHURCH OF ST. MARTIN-CUM-GREGORY. Window VIII. Angel, *c.* 1340, restored.

Window V. Memorial to William Peckitt, 1796.

Window VI. Centre light, by William Peckitt, 1792.

(7) CHURCH OF ST. MARTIN-CUM-GREGORY.

St Mary (?)

St Martin.

St. John (?).

(7) CHURCH OF ST. MARTIN-CUM-GREGORY. Window VIII, early 14th century.

design resembles that of Littlemore Church, Oxfordshire, as the work was done at the expense of Mrs. Trafford Leigh, an adherent of the Oxford Movement. An organ was built in a new recess in 1868, and a chancel screen to the design of C. Hodgson Fowler of Durham was inserted in 1892 (Borthwick Inst., Faculty Papers 1868/7; 1892/22). The *Vestry* was enlarged in 1902 (Faculty Papers 1902/10). The spire was renewed in fibreglass in 1970.

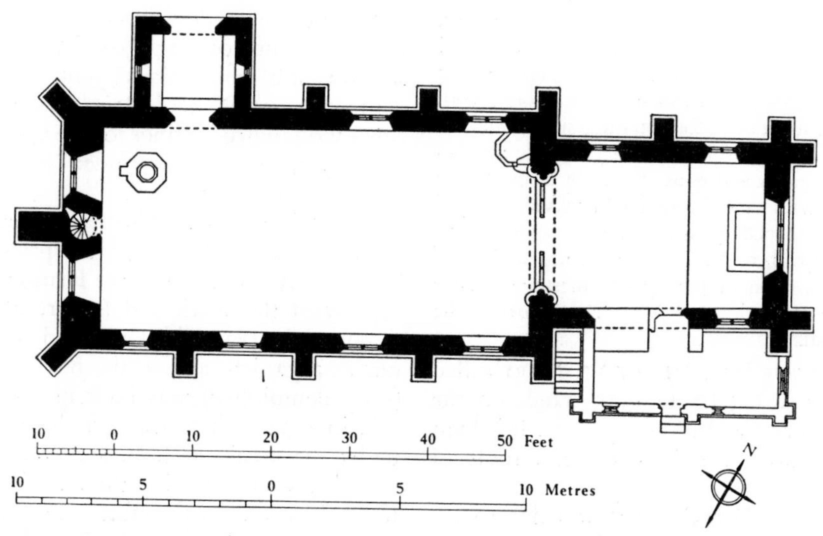

Fig. 39. (12) Church of St. Edward the Confessor, Dringhouses.

The structure is a pleasing example of scholarly work of the Gothic Revival, and the craftsmanship is good.

Architectural Description—The church consists of a chancel and aisleless nave, a vestry to the S.E., a N. porch and W. bell turret. The *Chancel* (27 ft. by 18¼ ft.) is built of ashlar with enrichments of ballflower. It is of two bays with buttresses. The E. window is two-centred with three lights and geometrical tracery; the side windows are two-light with two-centred heads and geometrical tracery. Inside, the enrichments, which include a stencilled inscription, are painted and gilded. The *Nave* (55½ ft. by 25½ ft.) is of four bays, with buttresses and two-light windows with geometrical tracery in two-centred heads. The W. wall has diagonal buttresses and a centre buttress of four stages supporting a clock turret on the gable. The turret, set diagonally, carries a small octagonal spire and is flanked by two-light windows with geometrical tracery. At various points on the internal walls of chancel and nave are inscribed scrolls in relief. The nave roof is supported by a series of corbels, each consisting of a demi-angel supporting an abacus and holding a shield bearing the emblems of the Passion. The floor is of red and grey tiles, by Minton, set diagonally.

The oak *Roof* of the chancel has three trusses each with principal rafters and a tie-beam supported by braces and carrying tracery; the purlins are moulded. The nave roof has two struts above the beam instead of tracery. The simple *Porch* (11 ft. by 9¼ ft.) has a roof of scissor type.

Fittings—(All of 1847–9 unless otherwise stated.) *Doors*: to N. porch, (1) of vertical oak planks with scrolled hinges externally; to newel stair, (2) of vertical planks. *Font*: of Clifford stone, octagonal, with decoration of window form of two trefoiled lights with geometrical tracery on each face. *Glass*: in chancel, (1) in E. window, Christ between St. Mary and St. John; at foot of N. light Christ tempted in the Garden, of centre light Christ carrying the Cross, of S. light SS. Mary and John walking away, with background of crosses; this glass won a first prize in the Great Exhibition, 1851; (2) in E. window in N. wall, SS. Luke and John, (3) in W. window in N. wall, SS. Mark and Matthew; (4) in window in S. wall, SS. Peter and Paul. In nave, (5) in E. window in N. wall, two scenes of Our Lord with children; (6) in W. Window in N. wall, Parables, (a) Good Shepherd, (b) Sower, (c) Talents, (d) Prodigal son; (7) in third window from E. in S. wall, Christ and a cripple and Christ walking on the water; (8) in lancet in S. wall, set in grisaille, round panel of Virgin and Child, the latter reading a book, perhaps 18th-century; (9) in N. window in W. wall, St. John Baptist and St. Joseph carrying the Child; (10) in S. window in W. wall, Moses and Elias. Excepting (8), all the foregoing made for the original church by William Wailes of Newcastle, whose monogram and the date 1849 are in (3).

Monuments: in vestry, three wall monuments and brass plate recording their removal in 1849 from the old chapel, namely, (1) Rev. Francis William Dealtry, M.A., 1822, Rector of Over Helmsley, with impaling arms of Dealtry; (2) Elizabeth Frances Dealtry, 1793, daughter of Francis Barlow of Middlethorpe, wife of Rev. William Dealtry, Rector of Skirpenbeck and Wigginton, Yorks., large slab of white marble, signed T. Atkinson, York; (3) Samuel Francis Barlow, 1800, son of Francis Barlow who built the second chapel, against pyramidal-shaped grey marble, white marble slab with brown veined marble strips, moulded cornice and fluted base, under draped urn on pedestal between two flaming lamps, signed Wm. Stead, York. In nave, to S. of chancel arch, (4) Edward Trafford Leigh, 1847, buried at Lymm, Cheshire, 18 years Rector of Cheadle, Cheshire, in whose memory the church was dedicated. *Plate*: includes cup with semicircular bowl, octagonal stem and large jewelled knop, with London mark 'IK' and date-letter of 1848/9; paten with same mark and, on top, lamb and flag in sexfoil; flagon with round body, shaped neck and foot, with same mark. *Processional Cross*, embodying the original brass altar cross. *Pulpit*: of stone, octagonal, with decoration of window form as on font on three sides, moulded rail, carved base of naturalistic foliage. *Seating*: simple benches of oak with panelled backs and ends. *Stalls*: in chancel, five to N. and four to S., with carved, moulded backs and shaped arms, and desks with shaped ends and carved poppy heads. *Miscellanea*: in churchyard, to N.E. of church, foundations of 18th-century chapel, section of paving of limestone squares with diagonal insets at angles and some red quarries; the Barlow vault at E. end.

(13) CONVENT OF THE INSTITUTE OF ST. MARY called THE BAR CONVENT (R.C.) (Plate 138), stands on the corner of Blossom Street and Nunnery Lane. It is built of red brick with stone dressings and has roofs of Welsh slate.

The founding of the Bar Convent was due to the establishment of an Institute of Religious Women, a body concerned with the education of girls, as the outcome of the work of Mary Ward (1584/5–1644/5). A house of the Institute, endowed in 1678 by Sir Thomas Gascoigne, was temporarily set up at Dolebank near Fountains, but was transferred to the present site in York, where a messuage and garden had been purchased, on 5 November 1686, to provide a home of Religious women and a boarding school for young ladies. Mother Frances Bedingfield, a companion of Mary Ward, became the first Superior and the Bar Convent Grammar School can claim to be one of the oldest girls' schools in England.

A prospect of York by John Haynes, dated 1731, and other early views show that the original 'Nunnery' was accommodated in a two-storey house of late 17th-century type, with a porch. Between this house and the corner of Nunnery Lane was a house, belonging to Mr. Benson in 1731, of early 18th-century character (Plate 162), which was later added to the convent; it was rebuilt as the Poor School in 1844–5.

The present buildings were begun in 1765 to designs by Thomas Atkinson, and a foundation stone was laid on 4 March 1766. The residential parts were ready for occupation early in 1768 but the Chapel was not used until 27 April 1769. The moving spirit behind the building programme was the Superior, Mother Ann Aspinall (d. 1782), who obtained from Rome a small model for the chapel. The fittings of the chapel were not completed until 1775, when a new wing containing rooms for music and drawing was built between the chapel and the old house on Blossom Street. Rebuilding of the street frontage followed between 1786 and 1789, and a total of £1091 6*s*. 6*d*. was spent. In 1790–3 a new range, with children's refectory below and dressing room and a long dormitory above, was built to N. of the main court, costing £987 13*s*. 4½*d*. Thomas Atkinson was still the architect; John Prince was the main contractor and Richard Hansom supplied the roof and other woodwork. Fittings for the Chapel were provided and payments in 1790 included 'the Glory at the Large Alter in Burnished Gold'.

A building E. of the Chapel was remodelled in 1814–16 at a cost of £1149 4*s*. 1*d*., the principal builder being Thomas Rayson. Richard Hansom, the carpenter, supervised the work and Robert Mountain provided masonry. In 1826 additional land was bought and laid out as a garden, and in the next year a new oratory (since demolished) was built in the burial ground to S.E. In 1834–5 there was a fresh campaign of building under the architects J. B. and W. Atkinson, and this included a range closing the N. side of the main court to S. of the children's refectory, and a large conventual range running N. from the E. end of the Chapel. The latter range cost £2313 9*s*. 9¼*d*.; the main builder was Richard Dalton, bricklayer; Richard Hansom was the carpenter, and Michael Taylor, a distinguished sculptor, provided stone from Elland, Harehill and Oulton and chimney pieces made of Illingley and Roche Abbey stone. The building was lit with gas and provided with flushing water closets. Minor works were also done in the Chapel, which was fitted with gas, and in 1837 the Sacristy was enlarged; two marble basins were supplied by Michael Taylor.

The house on the corner of Nunnery Lane was demolished in 1844 and a new Poor School, designed by G. T. Andrews, was built on the site. The total cost, inclusive of the fees to the architect and the clerk of works, was £1472 3*s*. 11*d*. The principal builder was John Lakin, bricklayer; Noah Akeroyd, mason, provided the stonework, and John Bacon was the joiner. The work was completed in 1847. In 1844–5 a third storey to contain a dormitory was built over the children's refectory, and in 1846 the sanctuary of the Chapel was enlarged and seats for the bishop, priests and acolytes provided. Lunettes in the nave and the window lighting the S. transept were inserted at this time. The architect throughout this period continued

Front Elevation

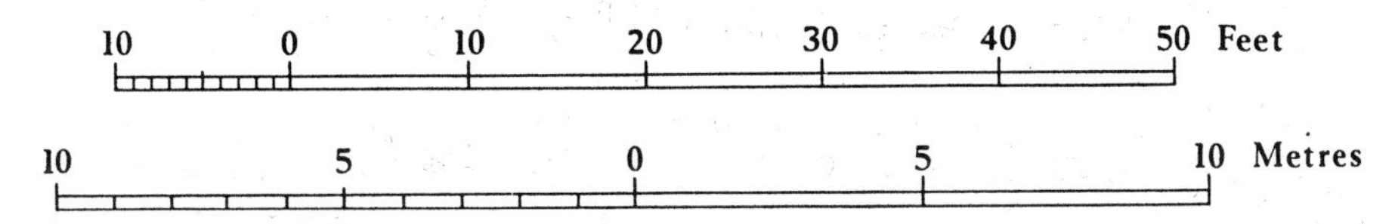

Fig. 40. (13) The Bar Convent. Building of 1789–9.

to be G. T. Andrews, who in 1850 added new kitchens adjoining Nunnery Lane, at a cost of £999 4*s.* 11*d.* Later works included the formation of a Lady Chapel and its decoration in 1853–5 and the roofing of the central court in railway-station style after 1865. The School Infirmary was destroyed by enemy action in 1942. Further accommodation was added between 1945 and 1957 (VCH, *York*, 442). The Chapel was altered and redecorated in 1968–9. (H. J. Coleridge, *St. Mary's Convent, Micklegate Bar, York* (1887); Convent archives.)

Architectural Description—The main *Front Range* of 1786–9 is of three storeys and attics, built of red brick with stone dressings; the windows have rubbed brick flat arches, stone sills and 19th-century plate-glass hung sashes (Plate 138), recently (1969) fitted with glazing bars. The pedimented central feature (Plate 137) is set within an arched recess; this design was used by the architect Thomas Atkinson also for the façade of his own office in St. Andrewgate. The ends of the range are cloaked by other buildings, and the four-storey rear face has an annexe of three storeys built against it. The annexe has a pent roof above a modillioned cornice and, projecting above the roof, a square brick clock-tower (Plate 151) rising to the height of the wall of the main range and supporting a wooden cupola. This last has an ogee lead-covered roof and moulded cornice supported on columns. The clock was made by Henry Hindley of York (d. 1771) and cost £60 in 1789, additional expenses were £39 2*s.* for erection and £3 4*s.* for the two faces and scaffolding to put them up. In 1790 Robert Dalton supplied three 'Turrett Clock Bells' for £24 15*s.*

Inside, the main rooms retain their late 18th-century character and have moulded cornices and skirtings but in general have lost the chair rails. The doorways have moulded architraves and doors with six fielded panels; the windows towards the street retain their shutters. The entrance hall has arcading with pilasters on each side with doorways opening into the Great Parlour (so called in 1834) on the S., and the portress's rooms on the N.; N. again is the Little Parlour. The *Great Parlour*, which contains original portraits of Sir Thomas Gascoigne and four of the Mothers Superior, has round-headed recesses in the E. wall; these flank the fireplace for which £56 7*s.* 10*d.* was paid in 1788. It is of white marble with inserts of veined green marble in the entablature and originally had oval wreaths on the pilaster entablatures and a large wreath set above swags and festoons on the key block. The passage in the annexe on the inner side of the main block is open to the Court.

The *Court* to the rear of the main range has arcades of round-headed arches on oblong piers to E. and W. Cast-iron columns (not shown on the plan), supporting a glass roof on iron trusses were added after 1865. An ornate floor of Minton tiles and iron benches, chairs and tables, all of mid 19th-century date show the influence of G. T. Andrews. Large benches in iron and wood, with a nasturtium pattern, are by C. B. Dale.

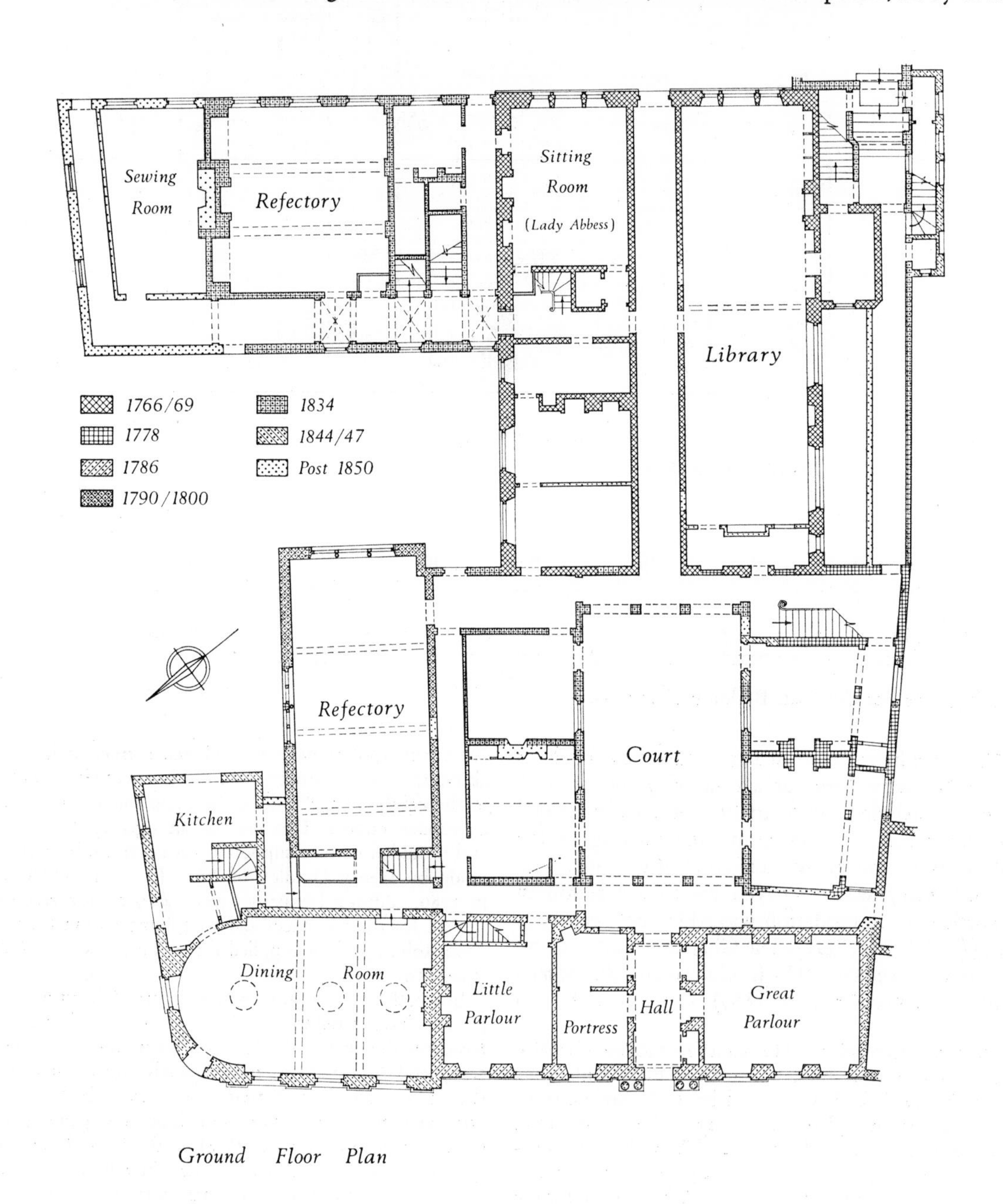

Fig. 41. (13) The Bar Convent.

The *North Range*, of 1834–5 and by J. B. and W. Atkinson, is built of dark red brick. On the S. side, facing the Court, the ground floor has two windows flanked by two doorways, all under segmental heads and with deep reveals. Above the glass roof to the Court the first floor has four large windows and one blind recess all with slightly segmental heads and stone sills; above a projecting band of four courses of brick-work the second floor has square-headed sash windows cutting into the cornice. In the E. wall there are round-headed windows above the doorway. There is a straight joint between this range and the parallel Refectory range to N. On the ground floor the W. room, formerly a scullery, has been

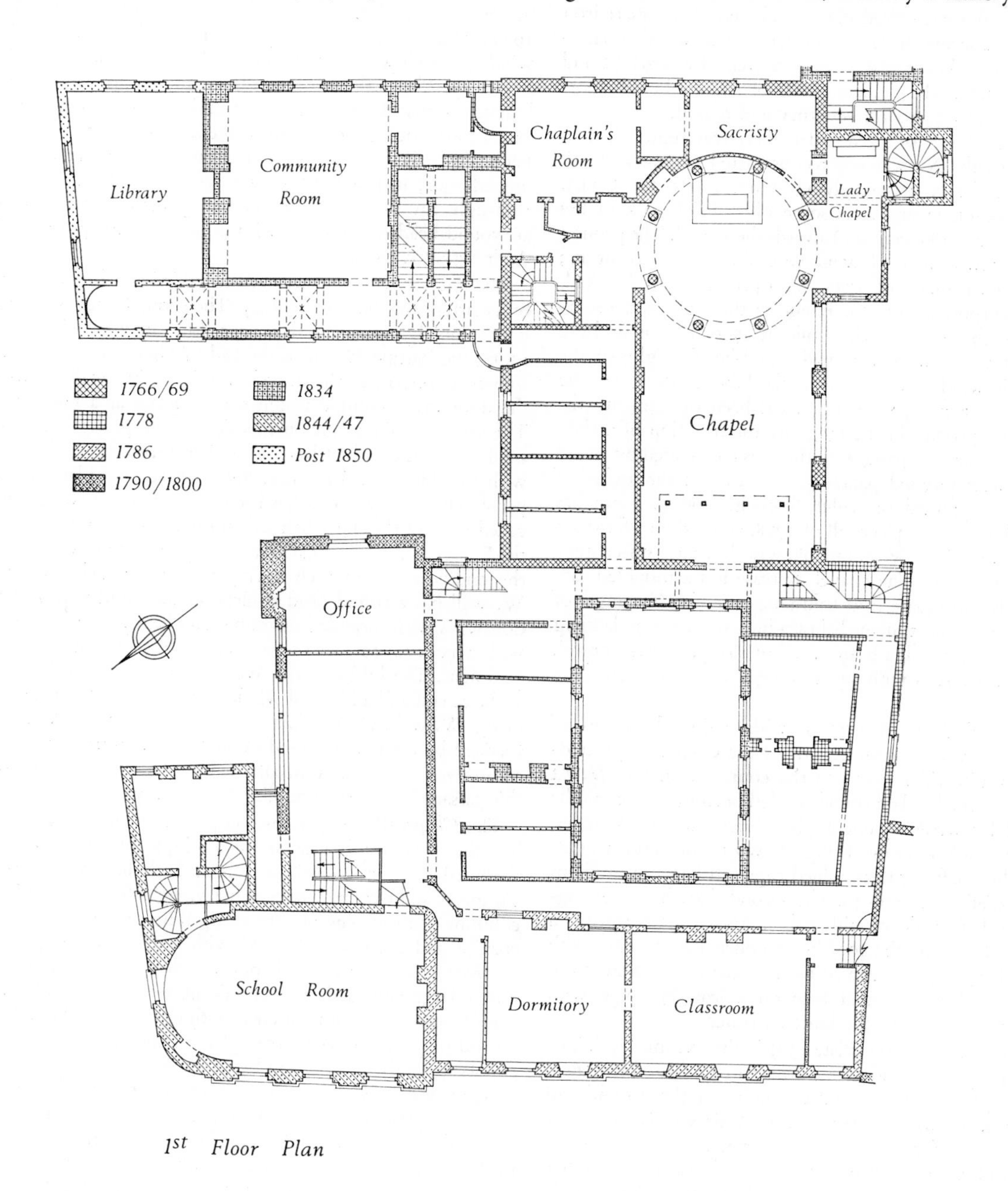

Fig. 42. (13) The Bar Convent.

remodelled and the chimney-breast has been rebuilt; there are some remains of a former chimney carried by a moulded corbel on the E. wall and the ceiling in the E. part of the room is lower than in the W. The first-floor passage on the N. has a round-headed archway leading E. to a staircase off the N. angle, rising from first-floor level. The latter, of 1834–5, has a moulded rail, cantilevered stone treads and plain square iron balusters. The second-floor landing has turned wooden balusters with square knops and a heavy newel, of *c.* 1765–70 and probably reused. The fielded panels of the cupboard doors under the stairs are also 18th-century and reused.

The *South Range*, of 1778 and earlier, was remodelled in 1834–5 when the N. wall was reconstructed to match the S. elevation of the N. range, giving symmetry to the Court. The S. elevation incorporates some older brickwork of the late 17th or early 18th century towards the E. end. The ground floor is lit by two hung-sash windows of the early 19th century; higher up two arches of 1778, for openings now blocked, remain. Incorporated into the E. end of the range is the staircase to the chapel, originally built in 1766–9; it is lit by a window of *c.* 1840 with two small round-headed lights on the S., and a big window with semicircular head on the E. On the ground floor, a passage on the S. has been opened into the two adjoining rooms in the range by the insertion of arches. On the first floor, the passage on the S. is terminated by round archways with recessed panels to the reveals; the arches of 1834–5 are supported on volute corbels. The E. room has doors flanking the fireplace, all of 1778, and a door of 1834–5 in the S. wall. The W. room likewise has an 18th-century fireplace and doors. The chapel staircase has a moulded rail, stone treads and square iron balusters, as in the N. range, of 1834–5. Again the timber balustrading to the top landing embodies a heavy rail, a bottom rail or string, a heavy newel, and turned balusters with square knops, of 1766–9, but here *in situ*.

The *Chapel Block*, of 1766–9, with major alterations of 1847–53, closes the fourth, E., side of the Court. At ground-floor level a corridor opens to the court, as on the W. At first-floor level a blocked round-headed opening in the middle is flanked by windows under plain brick segmental arches, each of the two windows being of two round-headed lights with Gothic shafts, with moulded caps and bases to mullions and jambs, all of *c.* 1830–40. The second floor has three segmental-headed windows with sashes. Above the cornice is a hipped roof, and on the roof is a wooden bell-housing with pyramidal lead top. At each corner is a tapered rainwater head with moulded cornice and base on a lead fall pipe with opposed fleurs-de-lis on the junction bands.

The rear, E., elevation (Plate 138) is also symmetrical. The window proportions and eaves cornice are unusual for 1766–9. The plate glass casements and top lights of the ground and first-floor windows may be by G. T. Andrews. The heads of the side lights of the ground-floor Venetian windows are of painted ashlar. The arched and segmental heads of all other windows on this elevation are of plain brick. The simple cornice on shaped modillions is cut into by the small-paned sash windows of the second floor.

The N. side, of buff-red brick, has four-brick bands separating the floors. The windows all have small panes and are a mixture of sash and casement types set irregularly. Widely-spaced brackets support a plain wooden cornice.

On the S. side two floors correspond to three floors elsewhere. Towards the E. end the Lady Chapel projects southward; it was lit by a round-headed two-light window with vesica at the head at first-floor level, but this was replaced in 1969 by a sash window. The ground floor is lit by two Gothic windows, each with four round-headed lights under lozenges as tracery. Above the ashlar heads is a four-brick band. Three lunettes at first-floor level are set under a brick arch, which cuts a four-brick band related to the sash windows originally here. There is a gutter supported on metal brackets to the hipped roof; in this last are two flat-topped dormers and a roof-light towards the E. In annexes against the Lady Chapel are round-headed windows, with stone sills, perhaps by G. T. Andrews, of *c.* 1850.

The interior retains many cornices, skirtings, doorways and other fittings of the 18th century. The ground floor is divided longitudinally by a central passage, N.E. of which is a large room, the 'Sitting Room of the Lady Abbess' in 1834, with a Venetian window in the E. wall and, in the W. wall, between the doorways a simple round-headed niche at a high level. The second main room on this side of the passage has been subdivided; the E. half has a good late 18th-century fireplace with moulded dentil cornice, fluted frieze, and blank centre panel. The *Library*, to S., has been formed out of two rooms, called in 1834 the 'First School' and the 'Drawing School'. In the E. part of the S. wall are two groups both of three recesses, the middle recess of each having a round-arched head. The W. wall has a round-headed niche in the middle flanked by doorways with moulded surrounds and heavy overdoors each with a pedestal to carry a bust centrally placed in a broken pediment. The lobby at the W. end has a blocked doorway at the S. end and a blocked window at the N. end respectively of the W. wall. On the first floor, a passage at the W. end is spanned by a series of round-headed archways with moulded architraves. The square-headed entrance to the chapel from this passage has a surround like the archways.

The *Chapel* (Plate 139) has complex internal arrangements, but the structure is hidden by a simple hipped roof; this disguised its function before Roman Catholic emancipation. To the E. and defining the Sanctuary, eight freestanding Ionic columns with an entablature comprising an enriched dentil cornice and a frieze decorated with festoons of grape vine between urns and ribboned sprays of flowers (Plate 139) support a large ribbed dome. The dome rises to a cupola with glazed top and a drum enriched with close-set brackets above a band of palmette. A narrow glazed zone at the base of the cupola, possibly an insertion by G. T. Andrews and removed in 1969, formerly separated the tops of the individual radial ribs; these last, eight in number, are round in section and enriched with leaves. Each bay of the vault contains an oval medallion framed by enrichments which alternate from bay to bay: some with curved swags of conventional foliage supporting crossed sprays of foliage and fruit, and the others with intersecting swags supporting palm leaves. The entablature towards the nave (as opposed to the Sanctuary) is simpler and has a heavy dentil cornice; the 19th-century painting on the frieze was obliterated in 1969. The plasterwork, probably by Thomas Atkinson, which is now white and gilt, was painted in 1847 by Mr. Bulmer, perhaps to designs of G. T. Andrews. The N. and S. bays of the rotunda open to small transepts under a curved trabeation supported by heavy foliated corbels and pilasters with a round recessed panel between two shaped panels in their height. They are probably a remodelling of 1847 by G. T. Andrews. In the N. wall of the N. transept, round-headed panelling frames both the square-headed doorways, and the window to the staircase has a round head framing two round-headed lights and a blind spandrel; the E. wall has a square-headed recess between two oblong panels. A trapdoor in the transept floor gives access to a cavity about 8 ft. square. The S. transept extends eastward beneath a small dome and the cupola to form the Lady Chapel; entry to the extension is between responds similar to those at the entry to the transepts. The cupola has a glazed top and elongated brackets and is supported on radial ribs springing from a modillioned cornice. The recess departs from G. T. Andrews's plan of 1845, but it was constructed to his design in 1853 and decorated in 1855. A niche was discovered in the E. wall in 1969 when the decoration was altered.

The *Nave* has a coved plaster ceiling, with an iron grille in the centre. The three round-headed recesses in both N. and S. walls may be by Thomas Atkinson. Those on the N. are blind, those on the S. contain glazed lunettes probably by G. T. Andrews, of 1847. The W. gallery may have been fitted between 1847 and 1855 by G. T. Andrews, for the W. end had assumed its present shape by 1855 (Sheahan and Whellan, I, 556). The gallery, which contains the organ, is carried by four round columns of painted timber with caps of Early English type, square abaci, moulded bases, and bands at mid height. The columns support round-headed arches with recessed spandrels under a bold coved cornice. The organ gallery has an elaborate scrolled iron railing. The cornice returns on either side of the organ recess at gallery level.

The *Sacristy* on the E. has a curved W. wall to accommodate the rotunda of the chapel. A blocked square light in this wall formerly provided indirect light for the High Altar. The doorway at the S. end of the curved W. wall which opens to the Sanctuary has an 18th-century architrave and a door of *c.* 1845. The doorway in the S. wall is of 1847–55. An S-shaped staircase of 1847–55 by G. T. Andrews, of similar design to his staircase in the Schoolroom block (*see below*), is entered through a doorway in the S. wall of the Lady Chapel. A staircase E. of the Lady Chapel is not accessible from the Chapel block at this level.

The Chapel *Roof* is original, but where the cupola has been heightened one of the principals has been trimmed back. The truss above the cupola has a king-post and struts above a collar; all other trusses have tie-beams, queen-posts and struts. The 'garet over the chapel' was underdrawn in 1791 and partitions were put in an attic storey.

Fittings—*Altars*: in Sanctuary, table incorporating two mahogany scrolled legs with winged cherub heads and pellet enrichment and, on floor in front, a pelican in her piety, of oak, from the 18th-century High Altar shown in a lithograph of 1840, and now reinstated in the same relative positions (Plate 139). In Lady Chapel, table of *c.* 1855 by a Mr. Hayball, possibly designed by G. T. Andrews. *Images*: in nave, in niches on N. and S. walls respectively, St. Joseph with infant Jesus, and Virgin of Assumption with crescent moon, clouds and cherubs' heads, two white marble statues bought in Florence in 1823 for £20 each; St. Sebastian; St. Michael; St. Margaret, three 18th-century Baroque alabaster statues (Plate 140), probably Spanish, brought from the Dominican Priory of Bornheim in the Netherlands by Fr. Anthony Plunkett O.P., its last Prior, then in Stourton Lodge Chapel for some time, and placed in the Bar Convent in 1805 when Plunkett was chaplain. At foot of staircase to S.W. of chapel, seated Virgin and Child, of limestone painted buff with some gilding, early 17th-century, Flemish, on 19th-century wooden restored base, from St. Martin's, Coney Street, removed by Mr. John Leadbitter, after whose death it was in the Spital at Hexham from 1867 to 1893, then given to the Convent by M. Elizabeth, John Leadbitter's grand-daughter. *Reredos*: in modern setting, carved oak figures of the four Latin doctors (Plate 40), N. to S., St. Jerome with lion, St. Gregory as pope holding book, St. Augustine as archbishop holding flaming heart, St. Ambrose as archbishop holding book, 18th-century, now painted white and gilded. *Paintings*: at top of staircase to S.W. of Chapel, triptych in oils, depicting, *l.*, kneeling man, bald and bearded, with eight sons behind, *c.*, Lamentation, with Virgin pierced by dagger flanked by angels bearing instruments of the Passion, *r.*, kneeling woman with two grown-up daughters behind, perhaps 17th-century; the family of Sir Thomas Gascoigne (*c.* 1596–1686) has been suggested for donor, but the number of children is incorrect. *Panelling*: on staircase to S.W. of chapel, eight carved wooden panels (Plate 42), four with enriched round-headed arches, cherubs' heads in spandrels, and fluted pilasters, four with figures in niches and strapwork surrounds, with Netherlandish inscriptions, (1) Resurrection, inscribed 'CHRISTUS IS MIN LEV-. NDT VND SAE EVEN IS MIN GEWIN [illegible]', (2) Crucifixion, inscribed 'CHRISTVS IS IMME VI. SER SUNDE WILN GEOF. FERT Aō 1597', (3) Adoration of the Shepherds (?), inscribed 'EHRE SI GOT IN DER HOGE. FREDE VP ERDE VND DEMN. GHE- EIN WOL (blank) ALLEN', (4) Annunciation, inscribed 'DE ENGEL GABRIEL GENA. NT VAN GODT WERT. TO MARIEN GESANT' and Virtues, (5) Prudence, as double-headed man, inscribed '. . HOFAIOT'; (6) Justice, with sword and scales, inscribed '. . GE. DH. DH'; (7) Fortitude, woman carrying a column, inscription illegible; (8) Charity, woman with child on left arm and another by right side, inscribed 'D . . LE TE', 16th-century. Carved wood panel of Entombment, with dead Christ supported by an angel, probably 18th-century and French. *Miscellanea*: patriarchal cross, relic, of silver-gilt chased with scroll work and inscribed, with seal of Chapter of St. Omer 1657–62, presented by Rev. Thomas Lawson, S.J., 1792, enclosed in silver shrine of 1860, perhaps from the Bromholm Rood in existence by the 13th century.

The *School Rooms* in the angle between Blossom Street and Nunnery Lane were built to the designs of G. T. Andrews in 1844–45. The Blossom Street, W., front (Plate 138), is set slightly forward from the adjoining main 18th-century front and has a two-storey giant order supporting a heavy pedimented entablature, with a moulded dentil cornice above a deep frieze of good quality ashlar. The brick podium has a chamfered stone plinth. The windows have rubbed brick flat arches; the plate glass sashes were changed to small panes in 1969. The upper sills extend from pilaster to pilaster, the lower sills form a continuous moulded stone top member to the podium. A curved section of brickwork at the angle is pierced at street level by an inserted doorway of *c.* 1920. The N. front to Nunnery Lane, though of coarser brick, has the stone plinth and heavy entablature continued from the W. front, but the cornice is without dentils. Windows and doors have segmental heads. The windows are at varying levels since the S. part is of three storeys, the central part fronts a staircase, and the N. part is of two storeys. The E. wall is of similar brick, of three storeys, and the heavy stone entablature continues across it. Inside, on the ground floor a large *Dining Room* with apsidal N. end has cornices running round cased beams and a circular surround to a gaslight point in the centre of each ceiling bay. A curved door on the N.E. leads to the kitchen. On the first floor is a lofty *School Room*, also with an apsidal end. The S-shaped staircase has a round handrail and cast-iron balusters and a heavy turned wooden newel; each baluster is decorated with an oval medallion at mid-height with a four-lobed leaf on each side and foliation at head and foot.

The *North Range* containing the Refectory was built to the designs of Thomas Atkinson in 1790–3 and formed the original N. side of the Court. The Long Dormitory on the second floor was added by G. T. Andrews in 1844. All three storeys are of reddish brick. On the N. side, to W. of a large chimney-breast, the ground floor is lit by an 18th-century four-light Gothic window set high; it has four round-headed lights divided inside by a round shaft with moulded cap and base and subdivided by lesser round shafts. Three windows on the first floor are grouped within a concrete surround, with a single window adjacent to the chimney-breast. On the second floor two large sash windows with small panes are placed together under a concrete head. A blocked window to the E. has a brick dentil cornice as the sill, showing that the roof has been raised. There is a plain plank cornice.

In the E. end, on the ground floor is a Venetian window with a central arched brick head, but the intermediate jambs and the heads of the side lights are of concrete or painted ashlar; inside, the central opening is flanked by freestanding pillars with moulded caps and bases. The first-floor sash window has a slightly segmental head and retains the original sashes. Immediately above is a statue of the Virgin and Child in a round-headed niche. The second floor has a similar window, but with twelve panes instead of six.

Inside, the *Refectory* on the ground floor, remodelled in the 19th century, has in the W. wall a simple round-headed entry between two slightly lower shallow niches also with round heads; over the entry is a moulded corbel of *c.* 1830–40. In the S. wall is a blocked central doorway and a second doorway of *c.* 1855. Three cased girders have been inserted beneath the ceiling. The 18th-century window in the N. wall contains 19th-century glass in the spandrels, with gold foliated patterns set in blue and red. On the first floor, a room at the E. end has a fireplace and other features of *c.* 1800.

The *Range* extending N.E. from the Chapel block, designed by J. B. and W. Atkinson, was built in 1834–5. The top storey was added in the late 19th century, and the two N. bays were rebuilt after damage during the 1939–45 War. It is three storeys high. The E. elevation has six sash windows under slightly segmental arches of ordinary brick and with stone sills on each floor. The 19th-century cornice has reeded wooden brackets set widely apart. The N. elevation is modern. The W. elevation has similar sash windows with flat arches and stone sills on the ground floor, some original round-headed sash windows on the first floor and modern windows above. The S. part of the elevation is now masked by a modern annexe but when the Commission's plan (Fig. 42) was made a curved passage projected at mezzanine level. It had round-headed lights and resembled an oriel window. The cornice is a wooden gutter supported on widely-spaced brackets.

Inside, the *Sewing Room* on the ground floor has been rebuilt in the style of 1834–5. The *Refectory* has a fireplace of Illingley stone in the N. wall set in a round-headed recess and flanked by round-headed recesses; it is by Michael Taylor and has a mantelshelf supported by shaped brackets and, above, a horizontal panel set under a moulded cornice. Three round-headed niches with simple moulded architraves are set high up in the W. wall. The passage on the W. is somewhat altered; it was similar to the passage on the first floor. The latter has a series of transverse arches on pilasters separating plaster groined vaults and intermediate barrel vaults; at the N. end is a large round-headed niche. The plasterwork is by David Jennings. The first-floor *Library* has been largely rebuilt following bomb damage, eliminating a slightly curved apsidal N. end which existed before. The *Community Room* has round-headed recesses containing original chests of drawers in the E. wall and another fireplace by Michael Taylor. The *Staircase* has unpierced walls between flights; it has broad stone steps, stone landings, and plaster barrel vaults. Above the landing between the ground and first floors are two shallow domical vaults and in the corresponding position above are groined vaults.

The *Range* (Plate 138) extending S.E. from the Chapel block (not shown on the plans) and on the site of an earlier building, was drastically remodelled in 1814–16 and its interior retains nothing of interest. It is three storeys high, seven bays long, and one room thick, with a single-storey prolongation of four bays to the E. The N. elevation is stucco-faced. A chimney projects between the third and fourth bays. The ground floor has seven round-headed windows with sills joined to form a continuous band, moulded imposts, and moulded archivolts. The first floor has similar fenestration, but with shorter windows, and the two W. bays are blind. The second floor has square-headed windows; the two W. bays are again blind. The simple wooden cornice is supported on widely spaced modillions. The E. elevation is also stucco-faced and has the first-floor impost moulding and second-floor sill band returned from the N. elevation. The S. elevation is of brick. There is a central three-sided projecting bay rising through all three storeys with three sash windows with stone surrounds in each storey. Various annexes have been put against the W. end of the elevation, and there is a modern range against the other end. A single-storey *Range* to the E. of the last block has four round-headed openings on the N. side with moulded archivolts and imposts, and roundels with raised mouldings in stucco above the piers. There is a roof light in the slate roof. In the stucco-faced E. end is a single round-headed opening.

(14) Former ALBION CHAPEL (Wesleyan Methodist), Skeldergate (Plate 57, Fig. 43), was designed by the Rev. John Nelson junior and opened for worship by the Rev. Robert Newton on 11 October 1816 (J. Lyth, *Glimpses of Early Methodism in York* (1885), 228; Hargrove, II, 172, gives the date as 16 October). It is of plain red brick and cost £2,250. After the erection of a new chapel in Priory Street, it was sold in 1856–7. The building has been converted to a warehouse for the storage of heavy electrical gear with much loss of original detail. It has a rectangular ground plan and originally had galleries down both sides and across the N.E. end; the galleries were approached by staircases at the entrance. (R. Willis, *Nonconformist Chapels of York* (YGS, 1964), 23–4; VCH, *York*, 409.)

The street front (Plate 57), in good quality brick, is symmetrical, with the centre three bays slightly projecting. The pedimental gable has ashlar coping and a central circular window with plain ashlar surround. In the centre projection the ground floor originally had three openings; of these, the main entrance has been removed and blocked with modern brick though the semicircular arch of brick headers remains *in situ*; to either side are secondary entrances, possibly originally giving access to the gallery staircases; the doors have been removed and the entrances blocked, but the simple doorcases remain and have shallow cornice-hoods supported on simple brackets. Over the three entrances are rectangular stone panels; no inscriptions remain. In both recessed flanking bays are round-headed sash windows, with the tympana or fanlights now filled with boards. On the first floor are five windows with ashlar sills and rounded arches of brick headers. The N.W. side is in stock brick and has three segmental-headed recesses and, at the S. end, a simple doorway. There are or were three round-headed openings to the first floor. The three centre bays of the S.W. elevation project 4½ in.; in the projection are two tall round-headed windows and, to E. and W., blocked openings in each storey. The S.E. side has been altered by the opening of a large doorway, and some stucco-rendering.

All fittings have been removed from the inside, but scars on the side and end walls denote the former positions of the galleries, which had tiered seating. The two windows in the S.W. wall have simple moulded architraves.

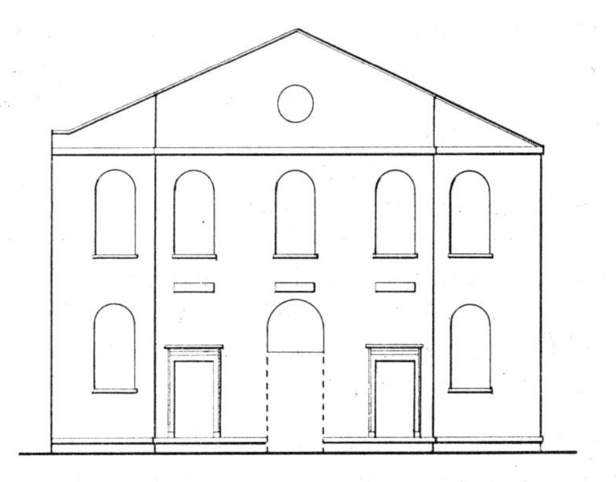

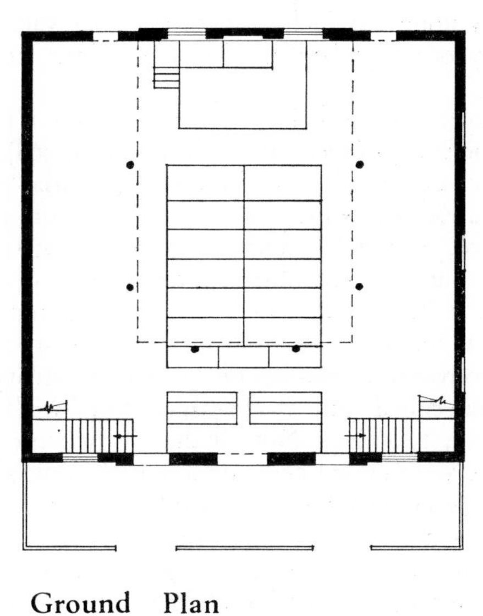

Fig. 43. (14) Albion Chapel, Skeldergate, before alteration.

(15) Former PRIMITIVE METHODIST CHAPEL, known as Benson's Chapel, Acomb Green, was built *c.* 1846. It was bought by the Society of Friends in 1912 and converted to form their Meeting House. No original features of interest survive. (H. Richardson, *History of Acomb* (1963), 29; VCH, *York*, 406.)

(16) Former WESLEYAN CHAPEL, now Nos. 91, 93 Front Street, Acomb, was built in 1821 and opened on 20 November in that year. It ceased to be used when Trinity Chapel was built in 1879, and the building was sold in 1881. After that date the structure was so greatly altered in conversion into two dwelling-houses that little can be seen of the original, apart from traces of a gallery at first-floor level and a circular plaster ornament on the ceiling where a chandelier was formerly suspended (Allen, I, 469; H. Richardson, *History of Acomb* (1963), 29; VCH, *York*, 412.)

(17) CHOLERA BURIAL GROUND (59755178) lies between Station Road and a footpath which runs parallel with the City wall. It was originally about 200 ft. by 45 ft., with the main axis lying S.W. to N.E. The ends have since been rounded, and the area now bounded by a low stone kerb includes part of a stone yard which formerly occupied land to the N.E. of the site. (OS 1852.)

A cholera outbreak in York lasted from 3 June to 22 October 1832 (Sheahan and Whellan, I, 368). Consideration of possible sites for a special burial ground had occurred before the epidemic reached York (House Book B50, 273) and permission to use the present site for the purpose was obtained on 8 June (*Yorkshire Gazette*, 9 June 1832). As the Archbishop refused to consecrate the land if only leased from the Corporation, it was conveyed to trustees on 21 and 22 January 1833 and consecrated on 23 January (Borthwick Inst., R.4K.180.A, B, and C). Not all the 185 victims of the epidemic were buried here; some burials took place in St. George's old churchyard (*Yorkshire Gazette*, 23 June 1832: report of 18 June). An attempt to secularise the ground for road widening in 1925 was successfully resisted, for the Chancellor of York Diocese decided against the grant of a faculty for the purpose on 25 January 1927 (*AASRP*, XXXVIII, ii (1927), lxxiii).

There are twenty surviving memorial stones. They are all of sandstone, and the date of death is generally 1832.

(18) FRIENDS' BURIAL GROUND lies in Bishophill. The ground was bought in 1667 and closed in 1855 by an Order in Council of 1854 (YCL, Cooper's MSS., 7.121); it is marked as 'Quakers' Burying Place' on Cossins's plan of York of *c.* 1727 and mentioned in 1736 (Drake, 269).

All the headstones are to one simple pattern with a rounded top and are of the 19th century, despite the dates upon some of them; those recording deaths in the 18th century bear the inscription 'within these precincts' and were erected after 1818 (Hargrove, II, 160). The lettering is of uniform style. The stones are arranged in seven rows, with two outliers. Four of the rows are marked by small stones, A, B, C and D. (Record of the names and dates is in RCHM archives.)

(19) ST. CLEMENT'S PRIORY, better known as Clementhorpe Nunnery, was a small house of Benedictine nuns founded by Archbishop Thurstan *c.* 1130 (*EYC*, I, 278–81). It incorporated an earlier parochial church dedicated to St. Clement, which had already given its name to the locality. Though ruins of the priory church (Plate 6) were still standing in the early 18th century (Drake, 247), the last remains of the buildings were swept away in 1873 excepting a small section of the *Precinct Wall* on the S. side of the street of Clementhorpe (60305108), with traces of the wall further E. in the footings of a brick wall on the same alignment; the remains are of magnesian limestone (for the ruins visible in 1825 see drawings by George Nicholson in York City Art Gallery, box B5, PD410; B7, PD353, ff. 49, 50). (VCH, *Yorkshire*, III, 129; *York*, 360, 377.)

The Yorkshire Museum contains a few worked stones from the site (W. H. Brook, MS. Catalogue, II, nos. 419–21, 424–5) and a fine though mutilated carved figure of the Virgin and Child seated in a niche (Plate 141), 51½ in. by 23 in. by 9 in., 14th-century. This can be identified with a woodcut reproduced by William Lawton, who gives details of rediscovery (William Lawton, in W. Bowman, *Reliquiae Antiquae Eboracenses* (1855), 59–61).

SECULAR

(20) OUSE BRIDGE (Plates 142, 143), crossing the River Ouse from Bridge Street to Low Ousegate, is an early 19th-century structure of brown limestone. It was probably at this crossing that a timber bridge collapsed under the weight of the multitude that welcomed St. William in 1154, precipitating 200 people into the river (W. H. Dixon and J. Raine, *Fasti Eboracenses* (1863), 225–6). A stone bridge must have been built in the second half of the 12th century as on it there was built a chapel of which there are 12th-century remains in the Yorkshire Museum. The chapel on Ouse Bridge is mentioned in 1223 (*York Memorandum Book*, SS, CXXV, 68) and by 1228 it had been rededicated to St. William (*CPR, 1225–32*, 175) who had been canonised in the previous year. Archbishop Walter de Gray appealed for money to repair the bridge in 1233 (*Archbishop Gray's Register*, SS, LVI, 60–1) and further sums of money for repairs are documented in 1307, 1407 and 1493 (YCA, G.1; *CCR, 1405–9*, 214; TE IV, 39).

St. William's Chapel stood on the N. side of the bridge, at the Bridge Street end; W. of it was the City Council Chamber, to which there was access from the chapel, and associated gaols known as the 'kidcotes'. Houses and shops, for which several 15th-century leases survive, were built on the bridge, as well as a stone cross (YCA, Memo. Book B/Y, f. 39v.) and an almshouse or Maison Dieu for women (*York Memorandum Book*, SS, CXXV, 291–2; Raine, 207–25). In the winter of 1564–5 two arches of the bridge and twelve of the houses upon it were destroyed by floods. Leonard Craven, carpenter, undertook in 1565 to complete a 'gytty' or caisson to allow masons to put down new foundations (YCR, VI, 100). Martin Bowes, Lord Mayor of London, sent Thomas Harper, who had worked on London Bridge, to advise the Corporation (YCR, VI, 103) and a single arch spanning 81 ft. was designed to replace the two spans destroyed. Christopher Walmesley was appointed mason for the rebuilding. John Todd, carpenter, made the centring (YCR, VI, 113, 121) and Walmesley was assisted by 'Walmesley Junior' and Whithead (*ibid.*, 120), masons; squared stone was delivered by William Oldred, mason (*ibid.*, 113). Further stone was taken from Foss Bridge Chapel, Holy Trinity Priory, St. George's Chapel by Castle Mills and the 'Bichedoughter Tower' of the walls by the Old Baile, and was requested from the Archbishop either from St. Mary's Abbey or the former Archbishop's Palace (YCR, VI, 73, 114, 115). The main work was probably completed in 1566, but further repairs were executed by Walmesley until 1572 (YCR, VI, 120, 139; VII, 8, 49).

In 1745 it was agreed 'to pull down the little houses on the top of Ouse Bridge'; houses were being removed until 1793 (YCA, M.17), and balustrades were erected. In 1795 a committee reported on the state of the bridge, recommending that it should be rebuilt, if possible in iron, and that it should be 35 ft. wide within the parapets (YCA, M.17). A competition for widening the bridge, utilising the existing arches, was held in 1809 and the results announced in the *York Courant* of 25 September. The first premium of £100 was awarded to Peter Atkinson (II), architect and City Steward, the second of £60 to John Rawsthorne, and the third of £40 to Charles Watson. Subscription lists were drawn up, and an Act of Parliament passed to authorise the work. Thomas Harrison (1744–1829) of Chester, in a 'Report on the Present State of Ouse Bridge' in the *York Courant* (5 March 1810), recommended the complete rebuilding of the bridge, as the existing structure was in such a state of disrepair. Atkinson produced new designs, with three spans instead of five (Plates 143, 144), and on 10 December 1810, the first stone was laid. Debts of £30,000 were soon incurred, and the scheme was suspended. On 7 June 1815 a second Act, for an enlarged and amended scheme, became law, and provided that the County should help with the cost. Atkinson remained architect, David Russell was Clerk of Works, and the contract was let to Messrs. (Abraham) Craven. The bridge was built in two halves, longitudinally, to permit the retention of the old bridge until free passage was obtainable across the new. The centre arch was to have a span of 75 ft., and the two side arches, spans of 64 ft. each. The first half of the bridge was opened on 1 January 1818, and the second half completed in summer 1820; tolls were abolished on 18 June 1829 (VCH, *York*, 515–18).

The bridge (Plate 142) is in three spans of segmental arches with heavily rusticated masonry. The rustication is continued on the under side of the arches and a straight joint shows the junction of the two builds. On either side of the river on the S. side are stairways to the tow-paths; a small building has been removed from the S.E. angle of the bridge. The fabric has been much patched, and the lamp standards are modern. A contemporary newspaper cutting records the inscription on a brass plate deposited in a cavity of the new bridge:
The FIRST STONE of This BRIDGE was laid December 10th in the Year MDCCCX And in the Reign of GEORGE the THIRD by The R^{t}. Hon. GEO. PEACOCK Lord Mayor. PETER ATKINSON Architect.

St. William's Chapel (Plate 145) was demolished in 1810 but some of its masonry survives in the Yorkshire Museum, and the clock from it is now in Scunthorpe Museum. A chapel was standing in 1223 and was rebuilt, incorporating parts of the earlier structure, after 1228 when letters of protection were granted for David, collecting money for the Ouse Bridge and the Chapel of St. William of York (*CPR*, *1225–32*, 175). The chapel was built above a cellar, which in 1376 was leased to John de Shirburn (*York Memorandum Book*, SS, CXX, 5). The cellar had one round or segmental-headed window flanked by two rectangular windows in its E. wall (Halfpenny, Plate 22). A watercolour of 1776 (Gott Collection, Wakefield Museum, III.12) and a view by Cave (Plate 23) show a small door fronting the street to E. of the main door to the chapel, though this is not shown on the plan of *c.*1800 (Evelyn Photographic Collection, No. 51). It may have led to a staircase to the lower level. In 1547 it was agreed to take down the steeple and all the lead from the chapel, and re-roof it with stone. All the fittings were to be sold, with the exception of the clock and bells (YCR, V, 39) but the Marian reaction led to refurnishing. In 1554 repairs to the glass windows and altars were ordered, and payments made to Edmond Walkyngton and (William?) Fornes, glaziers, and John Wedderell, locksmith, for mending the broken glass windows in the chapel; to Thomas Yaits, tiler, for repairs to the walls about the windows; and to (Richard) Graves and (Thomas) Grethede, carvers, for eight new pillars to the 'parclose' (YCR, V, 100; Raine, 216). In 1578, Queen Elizabeth's arms were set above the chapel door to show that the building was being used as a law court (YCR, VII, 175). Alterations were ordered in 1585 to chambers in the chapel used for the detention of prisoners (YCR, VIII, 99–100). The chapel was demolished in 1810 during the rebuilding of the bridge but drawings were made during its demolition (Cave, Plates 23–7; original drawings in York City Art Galley, Evelyn Collection, PD 1325–8).

Monumental fragments survive in the Yorkshire Museum:

(i) Two voussoirs (No. 482). A chain moulding with two rows of small pellets around each link, with three-leafed plants in triangular panels. This formed part of the middle of three orders of a round-headed porch to the chapel, known from 1749 'Drawings of Saxon Churches' in the Society of Antiquaries, F14 and F17, Halfpenny, Plate 23, and Cave, Plate 25.

(ii) Two voussoirs (No. 504). Similar to 482.

(iii) Four voussoirs (Nos. 430, 431, 432, and 433).

(iv) Remains of a round-headed arcade (No. 434, sixteen stones forming two and a half arches), and a moulded string-course (Nos. 435–8) (Plate 145). The arches are decorated with chevron ornament on the face and soffit. No. 435 is the returned end of the string. The source was in the W. wall of the chapel, to right of the pointed-arch doorway which gave access to the Common Hall (Cave, Plates 26 and 27).

(v) A carved stone, apparently a reworked voussoir (Plate 141). It exhibits both diagonal tooling and claw tooling, and so was probably part of a 12th-century arch reused in the 13th-century rebuilding. The carving, in low relief, does not seem to relate to its earlier use as a voussoir, and is incomplete. The upper register represents the Annunciation and the lower the Flight into Egypt. The latter includes a building with piercings possibly representing tracery, in which case the carving is unlikely to be earlier than the 13th century; this is borne out by the claw tooling on the stone. Identification rests on a detailed sketch with full notes made in 1825 by the Rev. J. Skinner (BM, Add. MS. 33684, ff. 30v., 33).

Clock (now in Scunthorpe Museum) (Plate 146). A clock at the chapel on Ouse Bridge is mentioned between entries of 1390 and 1399 (*York Memorandum Book*, SS, CXX, 223). This is presumably the same as the clock of the Corporation on Ouse Bridge of 1428 (*op. cit.*, p. 183). The earliest specific mention of a clock with a dial is in 1666 (YCA, Chamberlains' Accounts 26, f. 17). This is the surviving clock, made in 1658 by William Edwards of London (YCA, House Book B. 37, f. 111v.; Chamberlains' Accounts 1658, f. 17). The clock tower appears in many views. The clock was altered 'unto a long pendulum' in 1703 (House Book B. 40, f. 156v.). It was taken down in 1809 (House Book B. 47, f. 356) and sold for 25 guineas to the parish of St. Michael Spurriergate. The bell was sold to George Thomas Richardson, brazier, for a shilling per pound (*op. cit.*, f. 353). The clock was acquired from Barrow-on-Humber church by Scunthorpe Museum and Art Gallery in *c.* 1954. The bell mechanism does not form part of the 1658 clock. On the clock is a metal plate (Plate 146) inscribed:—

1658
Robert Horner} Maior
Sir Thomas Dickinson }
Henry Tomson } Aldermen
John Geldart }
Gulielmus Edwardus Cambriae Britaniae me fecit

(21) HOLGATE BRIDGE (58795133) crosses Holgate Beck, on Holgate Road. The N. side is probably part of a bridge built in 1824 (Sheahan and Whellan, I, 663); the S. side is modern. The N. side (*c.* 1824), pierced by a round arch of ashlar with a projecting keystone, has moulded architraves and a heavy ashlar parapet. The remainder of the elevation, and the soffit of the arch, is of brick. At the W. end of the N. parapet is a piece of magnesian limestone, reused, inscribed 'H. M.' probably for Henry Masterman, Lord of the Manor of Acomb and Holgate (d. 1769).

(22) SCARBOROUGH BRIDGE (59615205) is a combined rail and pedestrian bridge of two spans across the River Ouse to W. of Lendal Bridge. It was built in 1845 for the York to Scarborough railway line (Blythe and Moore's *Stranger's Guide* (1846), 56); a contemporary description occurs in an account of the opening (*Yorks. Gazette*, 12 July 1845). It was reconstructed in 1874, when all the original ironwork was replaced by new girders and the footway was moved from the centre to the E. side of the bridge. The earlier arrangement is on record (OS 1852). The main alterations can be deduced by comparison with an early photograph (NMR, CC61/12) and a water-colour in York Railway Museum (No. 60).

The two spans are carried on a central pier, battered on all four sides, and abutments, all of gritstone ashlar. Both abutments were pierced by barrel-vaulted arch-ways, with moulded architrave, key-block, and a string running across the piers at impost level. The S. archway is now blocked, and appears as a round-headed recess. The curved retaining wall to the embankment has been cut back, and is now free-standing in front of the pedestrian way. A square-headed opening to the S. of the S. pier, roofed with flat slabs, dates from 1874. The N. archway is intact, although the passage floor has been raised. Straight joints in the N. wall of the passage indicate the blocked entrances to the two flights of steps, which rose to a common landing, from which a single flight returned to the pedestrian way along the centre of the bridge. The pier and abutments have moulded parapets above a bold cornice. This upper portion of masonry has been heightened by the insertion of an additional string and a deep flat band of masonry above the level of the keystones. The original ironwork has been replaced but the seatings for the main struts, four to each side of a span, remain.

(23) KNAVESMIRE RACECOURSE GRANDSTAND, built in 1755–6 and originally of two storeys, was removed from its original site before 1925 (Benson, III, fig. 21), and the lower storey only has been rebuilt in the Paddock. The first recorded race at York was in 1530 (VCH, *York*, 159). In 1708–9 regular racing began on land provided by Sir William Robinson at Clifton and Rawcliffe

Ings. In 1730 the Wardens of Micklegate Stray were ordered to chain the Knavesmire, and the following spring the Pasture Masters were told to spend £100 on levelling, spreading and rolling the ground. The first meeting there was in summer 1731 (YCA, B.42, f. 136). The prime mover in the project for the grandstand on the Knavesmire was the Marquess of Rockingham (1730–82), and there is a survey of York racecourse in his estate papers (R. B. Wragg, 'The Stand House on the Knavesmire', York Georgian Soc. *Report*, *1965–6*, 4). On 7 December 1753 an application to erect a stand was made to the City, a lease of the ground was granted, and in 1754 subscriptions were requested by advertisement. Designs were submitted by Sir Thomas Robinson, James Paine and John Carr. Carr's design was chosen, and work began on the site in 1755. The architect's fees were £160 10*s*., and the Clerk of Works, Thomas Terry, received £20 for 2 years' supervision. John Carr provided 20 tons of Elland slate for £10, and most of the stone came from Hooton Roberts near Rotherham, a quarry owned by Rockingham, who himself paid £10 to Joseph Wood for clearing the quarry head, and probably gave a large amount of the stone. Richard Raisin was the carpenter and the plasterer was Richard Ward. The Grandstand cost £1,896 and in addition the Marquess of Rockingham paid £415 out of his own pocket, to cover outstanding tradesmen's bills, to George Thompson, who had administered the finances.

The front elevation forms an arcade of nine bays: the central and end bays are faced with rusticated stonework and project slightly to form three pavilions, the central one surmounted by a pediment; the intermediate bays are of brick with stone bases and imposts. A stone cornice runs the length of the building, above which is a balustrade interrupted by solid stonework over the piers and over the pediment. The rusticated stone arches are repeated in the end elevations. The whole of the brickwork is modern.

An engraving, dated 12 August 1755, by Fourdrinier, and a perspective of 1759 by Carr's assistant William Lindley, show above the surviving ground floor a symmetrical lofty first floor of seven bays. Round-headed windows with small panes set in round-headed recesses, of a design used also in Castlegate House, flank a central Venetian window. At the top is a bold balustrade like that below, and in the centre over the Venetian window is a large rectangular block of masonry decorated with swags and festoons beneath a rococo cartouche. Its condition in 1818 has been described (Hargrove, III, 515):—'The Ground Floor of the stand comprises several convenient rooms and offices for a resident, and for the entertainment of company, who may be accommodated with any kind of refreshment. On the Second Floor is a very commodious and handsome room, with a balustrade projection in front, more than 200 feet in length and supported by a rustic arcade 15 feet high, and commanding a fine view of the whole course. The top or roof of the building is leaded and constructed peculiarly for the accommodation of spectators.'

(24) MIDDLETON'S HOSPITAL, Skeldergate, was founded by Mrs. Ann Middleton, widow of Peter Middleton (Sheriff in 1618), in 1659 (Drake, 266). This building, which stood on ground leased from the Vicars Choral (York Minster Library, Vicars Choral Plans), was taken down, and the present hospital built by the Corporation of York (as Trustees for the Charity) in 1827–9 (Allen, I, 331; YCA, M.17/A, 8 June 1827) on the freehold portion of the site, further back. The architect was Peter Atkinson, the builder Mr. Dalton and the joiner Henry Hansom.

The building has two storeys built of stock brick in Flemish bond with stone dressings to the front. (Plate 151; Fig. 44). It is in seven bays of which the central three break forward under a pediment. At the eaves is a moulded gutter carried on modillions which are continued across the pediment. The central entrance has a stone architrave and a cornice supported by brackets; above is a round-headed niche with moulded architrave and stone blocks flanking the base, carved with honeysuckle issuing from volutes in bas-relief. Within the niche is a painted stone figure of the foundress in mid 17th-century Puritan costume, probably from the earlier almshouse. To each side are hung-sash windows with rubbed brick arches. At the back is a small central porch which has been extended.

The accommodation consists of eleven rooms on each floor, six along the front and five rooms and a staircase at the back.

(25) ST. CATHERINE'S HOSPITAL, No. 45 Holgate Road, replaces an earlier almshouse which fronted The Mount in the centre of the site now occupied by Nos. 116–28. This older building is prominent in all the earlier prospects of York from The Mount (Plate 2) and had itself been built in 1652 on the site of a mediaeval foundation variously described as a lazar-house and a *xenodochium* or place of hospitality for poor travellers (Drake, 246; Hargrove, II, 508–10; Davies, 105). The close of land in which the old almshouse stood was acquired by Leonard and John (afterwards Sir John) Simpson, who in 1833 petitioned the Corporation to be allowed to transfer the hospital to a new site on Holgate Lane, now Holgate Road, at the rear of the property, in order that they might develop the valuable frontage. They had obtained designs for the new building from the architect George T. Andrews. After taking legal advice the City agreed to the proposal and the hospital was built in 1834–5; on 7 May 1835 it was stated that 'the new building . . . is in all respects more commodious

than the Old Hospital—the inmates shall be peremptorily required to remove within fourteen days from this time' (YCA, K.82; M.17/A, B). In fact completion seems to have been delayed until 22 August 1835.

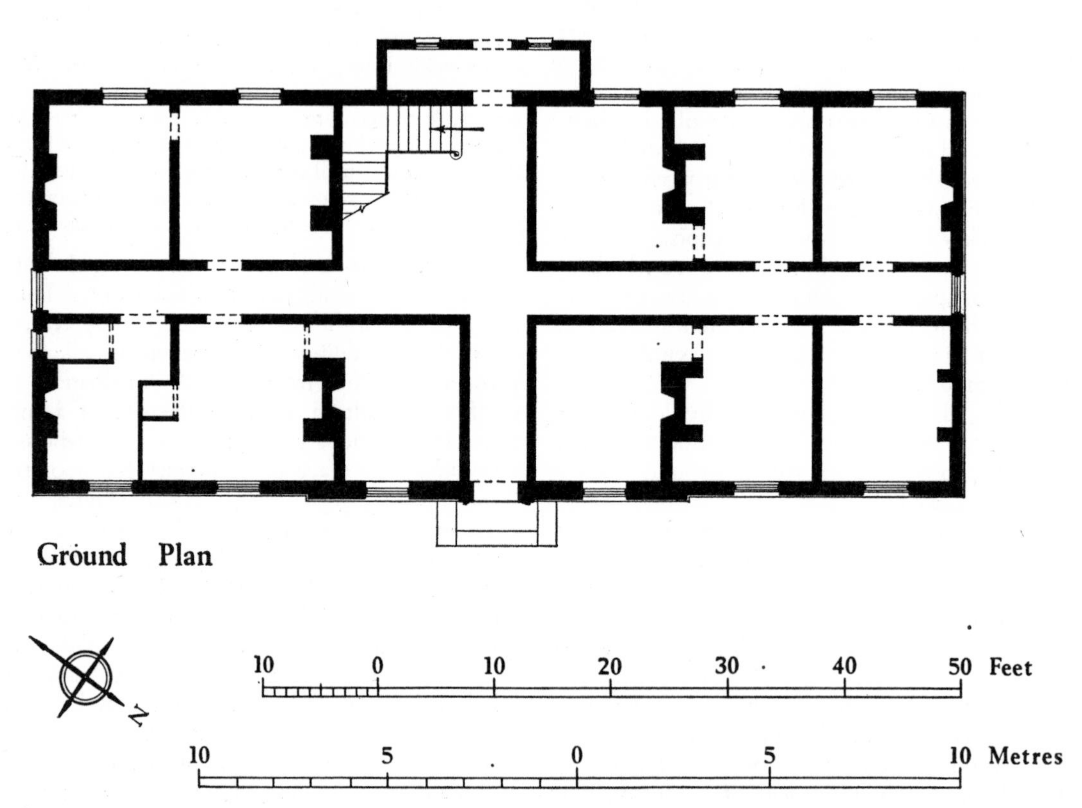

Fig. 44. (24) Middleton's Hospital, Skeldergate.

The building is of a single storey, in brick with ashlar dressings in the Tudor style, with square-headed two-light mullioned windows; the roof is of Welsh slate. The centre of the almshouse is recessed at front and rear, between two side wings each containing a front and a back room. *Demolished 1962.*

(26) MOSLEY'S SCHOOL, Cambridge Street, now a fireplace factory, was founded in 1844, and its buildings erected shortly afterwards; at first it was called Holgate Classical and Commercial Seminary, but was usually known as Mosley's School. 'It provided an efficient commercial education for the sons of business and professional men' and closed soon after 1901 (Knight, 696).

The building consists of a large hall on the S.W. side and a low annexe to N.E. divided into two rooms. The walls are of broad red brick, and the windows have stone heads; the roofs are covered with slates. The N.E. front is of four bays defined by buttresses with sloping stone heads and above them is a moulded stone string and plain parapet. The S.E. bay is blank, the next two have large recesses with four-centred heads and chamfered reveals, and the end bay has a blocked doorway with similar features. There are gables at both ends with stone finials and copings prolonged to a lower slope above the annexe.

The S.E. end has a later opening cut into the bottom of a three-light window with a square stone head, chamfered reveals and a high transom, flanked by similar two-light windows, beneath a single-light window at the head. The N.W. end has similar windows to the S.E. gable. The S.W. wall is plain except for brick pilaster buttresses and has no openings. The main hall is of four bays: in the N.E. wall are four two-light clearstory windows from the annexe, and below are a series of plastered recesses with chamfered reveals. Under the second roof truss from S.E. is a doorway with a four-centred head. Each truss has principals and collar, with braces to wall pieces below, and king post and struts above, strengthened by iron ties. Purlins, braces and collars are all chamfered, and wall pieces and braces rest on stone corbels. The walls are plastered in imitation of stonework. The hall formerly had a basement. Some huge turned legs and heavy deal rails remain of the supports for the tiered seating, which sloped down from S.W. to N.E. *Demolished 1968.*

(27) THE QUEEN'S STAITH forms the W. bank of the Ouse to S. of Ouse Bridge. It was built in 1660 by Christopher Topham, Lord Mayor of York, was repaired in 1676 and enlarged in 1678 (Drake, 282); it was rebuilt early in the 19th century. The revetting wall is of yellowish limestone ashlar and the pavement is of cobbles.

(28) OLD RAILWAY STATION stands within the City Walls to N.W. of Toft Green and Tanner Row and was opened on 4 January 1841. The site had held important Roman buildings (*York* I, (34), pp. 54–7) and is known to have been that of the King's House and Royal Free Chapel of St. Mary Magdalene *c.* 1133 (*Rolls of Eyre for Yorks.*, Selden Soc., LVI, 1937, nos. 1142, 1143, 1147 and pp. xxxvii–xxxviii). In 1227 Henry III granted the chapel and a plot of land to the Dominican Friars for their house (C. F. R. Palmer, in *YAJ*, VI (1881), 396 ff.), and subsequent grants extended their precinct to upwards of 3 acres.

No remains of the Friary survive and accounts of the excavations of 1839–40 and 1846 do not describe many post-Roman finds. Some of the skeletons found (*York* I, 80) may belong to a burial ground associated with the Friary or the chapel that preceded it, particularly twenty-seven without coffins towards Tanner Row (Hargrove, Yorks. Museum Misc. MSS. (typescript), 2). A cross-head and base found under the mediaeval walls should, like the Roman material found with them (*York* I, (34g), p. 57), derive from the Friary site and are therefore to be associated with the chapel or its predecessor, which on this evidence had probably been founded in the 8th century.

The *Cross-head* (Plate 26), of light-red sandstone 18½ in. by 11 in. by 6½ in., with a flat central boss within a circle of pellets, and a double bead around the edges of the arms now reduced to stubs, is well carved and may be 8th-century (*YAJ*, XX, pt. 78 (1908), 179; Yorks. Museum). The base is not now identifiable in the Museum but was described by W. Collingwood (*YAJ*, *loc. cit.*) as not of a pre-Norman type.

After the surrender the house was sold by the crown to William Blitheman on 24 April 1540 (LPH, XV, 296, no. 613 (16)), and eventually passed into the hands of Lady Sarah Hewley, to form part of the endowment of her charity (Hargrove, II, 182). The site had always contained much garden; in 1380 the garden of the Friars Preachers was mentioned and at the dissolution in 1538 1 acre was in garden and orchard. In the 17th century it had become a nursery garden which under the management of the Telford family became the most important in the North of England (J. H. Harvey in *YAJ*, XLII, pt. 167 (1969), 352–7). The business was sold in 1816 to Thomas and James Backhouse (*York Courant*, 6 May), who had to vacate their lease when the freehold was sold by the Trustees to the York and North Midland Railway in 1839. Along with the gardens the railway acquired the house, which had been occupied successively by members of the Telford and Backhouse families, at the N.E. end of the precinct. From the York Corporation they obtained the House of Correction which had been built in 1814 on the open land of Toft Green, which lay beyond the precinct to the S.W., in the angle of the City Walls.

A railway between York and London had been con-

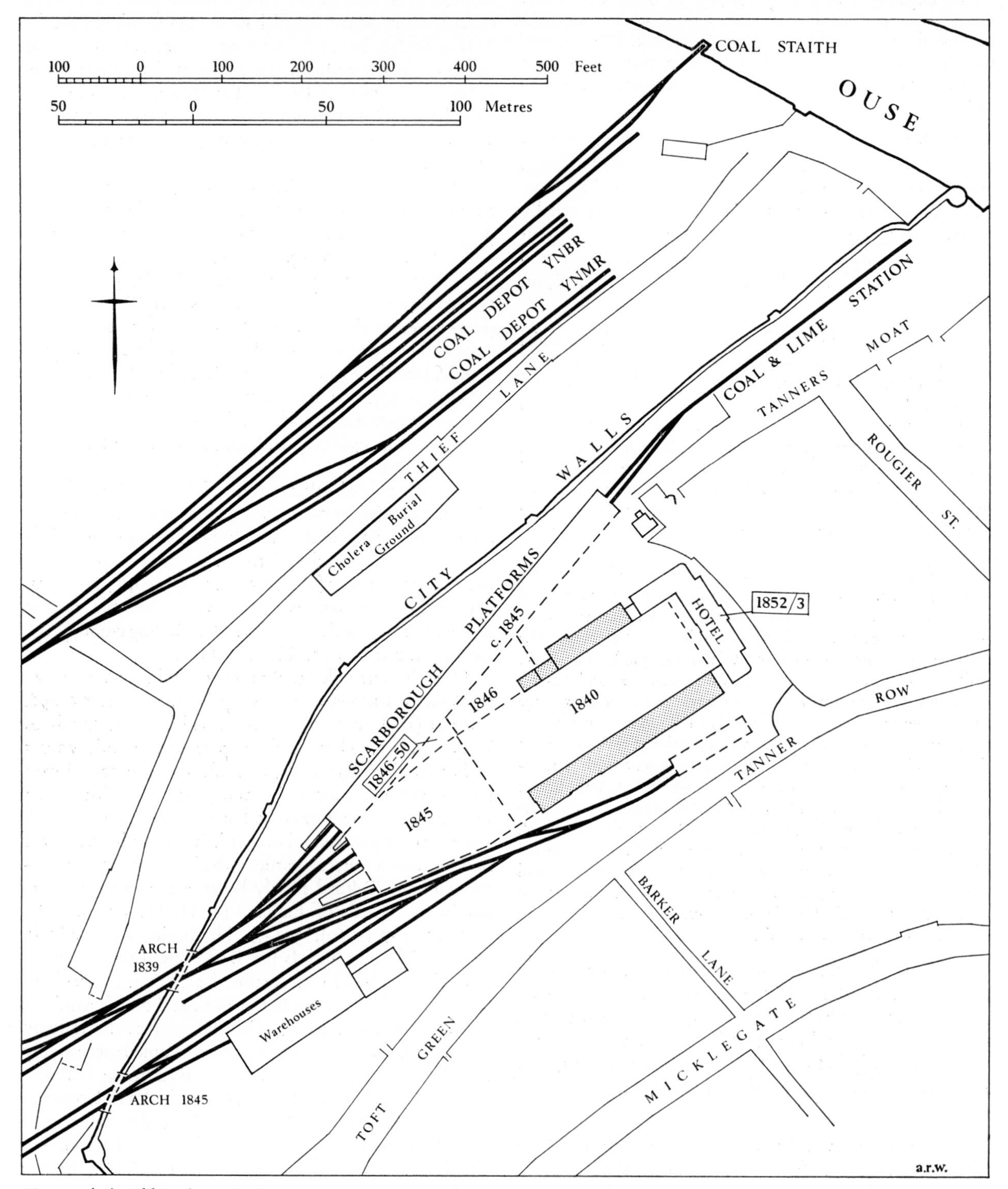

Fig. 45. (28) Old Railway Station.

From N.E.

From S.W.

(5) CHURCH OF THE HOLY TRINITY. Tower, 1453.

(5) CHURCH OF THE HOLY TRINITY. W. end of nave and St. Nicholas' chapel, early 13th century.

(5) CHURCH OF THE HOLY TRINITY. Nave from N.W., *c.* 1200.

Buttress at N.E. angle of tower, 15th century.

(5) CHURCH OF THE HOLY TRINITY.

Respond of N. transept arcade, *c.* 1200.

(6) CHURCH OF ST. JOHN THE EVANGELIST, from S.E., 15th century and later.

STAINED GLASS

(6) CHURCH OF ST. JOHN THE EVANGELIST. Window III. Tracery, 15th century. © Dean and Chapter.

Window I. The Trinity, and St. Christopher, *c.* 1498.

Window II. Coronation of the Virgin, 15th century.

(6) CHURCH OF ST. JOHN THE EVANGELIST. © Dean and Chapter.

W. end from S. aisle.

From S.W., before restoration.

Chancel, from W., after restoration.

(6) CHURCH OF ST. JOHN THE EVANGELIST. 12th century and later.

From S.E.

From S.W.

(7) CHURCH OF ST. MARTIN-CUM-GREGORY. 11th century and later.

(7) CHURCH OF ST. MARTIN-CUM-GREGORY, from S.E., 15th century.

(7) CHURCH OF ST. MARTIN-CUM-GREGORY. Reredos, *c.* 1750, and communion rails, 1753.

(7) CHURCH OF ST. MARTIN-CUM-GREGORY, from N.W., early 14th century and later.

(8) CHURCH OF ST. MARY BISHOPHILL JUNIOR, from S.E., 10th century and later.

(8) CHURCH OF ST. MARY BISHOPHILL JUNIOR, from E., 10th century and later.

(9) CHURCH OF ST. MARY BISHOPHILL SENIOR, from N.W., by F. Bedford, *c.* 1843 (York City Art Gallery).

(9) CHURCH OF ST. MARY BISHOPHILL SENIOR, from S.W., Pre-Conquest and later.

Exterior, from S., Pre-Conquest and later.

Interior, from S.W., *c.* 1180 and later.
(9) CHURCH OF ST. MARY BISHOPHILL SENIOR.

Exterior, from S.W.

Interior, from W.

(10) CHURCH OF ST. PAUL, HOLGATE ROAD. 1850.

(11) CHURCH OF ST. STEPHEN, ACOMB, from N.E., 1831 and later.

(12) CHURCH OF ST. EDWARD THE CONFESSOR, DRINGHOUSES, from N.W., 1847–9.

(11) CHURCH OF ST. STEPHEN, ACOMB, from E., 1831–2.

(12) CHURCH OF ST. EDWARD THE CONFESSOR, DRINGHOUSES, from W., 1847–9.

sidered as early as 1835 (YCA, M.17/B) and in 1836 the City Corporation agreed to proposals for a 'York and North Midland Railway' and to a 'Great Northern Railway' (YCA, Council Minutes I). In 1837 the Finance Committee reported on land bought by the York and North Midland Railway in the Holgate Road area and by 1838 the Y.N.M. wished to purchase the House of Correction and land within Toft Green for their station. Agreement was reached as to the necessary breach through the city walls (YCA, Council Minutes II, 63) and in 1839 George Stephenson, representing the Great North of England Company, was invited to York to plan a joint station (*Yorkshireman*, 13 July 1839). The station was built by the two companies jointly after the G.N.E. had agreed to pay £5,000 for their interest. The local company, led by their chairman George Hudson, played the chief role in negotiations (*see* YCA, Council Minutes II, 63, above). The (N.) archway was built in the summer of 1839. The architect was G. T. Andrews of York, with Thomas Cabrey, engineer to the Y.N.M., as consultant. By May 1840 the contract for building the station was let to Messrs. Holroyd and Walker of Sheffield (*York Courant*, 14 May 1840; *Yorkshireman*, 16 May 1840). The first line to be completed, in July 1840, was that constructed by the Y.N.M. from Normanton, where it made connections with lines to London and Leeds. A temporary station in Queen Street was used until the completion of the Old Station (VCH, *York*, 478). Plans for a booking-office block facing Tanner Row, to cost about £7,900, a refreshment room, and a train shed were approved in 1840. Although the station was to be used for passenger traffic only, the Y.N.M. was to use the joint line for other purposes and to build a goods depot within the walls; the G.N.E.'s depot would be outside. The station was to have been finished by 30 Aug. 1840 (*York Courant*, 14 May 1840), was reported on as nearing completion on 10 September (*York Courant*), but was not opened until 4 January 1841, owing to difficulties in building the train shed of cast iron and glass. The iron roof was by Mr. Bingley of Leeds (*Yorks. Gazette*, 9 Jan. 1841), with iron columns made by Thompson, the York firm of iron-founders (Plates 149, 150). The platforms were extended and covered-in shortly after the opening of the station. The G.N.E.'s line from Darlington to York, which joined the Y.N.M. outside the walls, was opened in 1841. In 1844 the Y.N.M. was authorised to construct a line to Scarborough, and obtained Corporation approval to pass under Bootham and to cross the Ouse (YCA, Council Minutes III for 5 Aug. 1844). A request to make a second breach in the city walls to enable the tracks from the warehouses situated between Toft Green and the passenger tracks to unite directly with the main line outside the walls was granted in 1845 (*Yorks. Gazette*, 15 Nov.). An extension to the station was under construction by 6 December 1845 (*Yorks. Gazette*), probably the extension and roofing of the platforms to the S.W. The additional passenger traffic was provided for by converting the G.N.E.'s coal depot to the N. of the station into another arrival platform (*Yorks. Gazette*, 12 Sept. 1846), and constructing a canopy. The arrangements at York, already complicated because through traffic was using a station built as a terminus, became worse when the Scarborough line was built, because the G.N.E. insisted that it should join their Newcastle line, and not cross it independently to reach the station. The necessary manœuvres were only possible because traffic was light; only eighteen trains a day ran from York in 1845. During the first two decades the expansion of traffic meant continuous alteration and extension. A hotel was erected to the designs of G. T. Andrews in 1852 across the end of the tracks, and was opened in February 1853. The open area behind the colonnade at the front of the hotel has been closed with brickwork in recent years to provide additional accommodation in the offices for which the old hotel now serves. The large and elaborate block of offices N.E. of the hotel was completed in 1906 to the designs of H. Field and W. Bell.

The new station outside the City Walls was designed by Thomas Prosser in 1867 but its construction was delayed for some years and it was not opened till 1877. The very impressive ironwork carrying the glass canopy over the platforms is part of Prosser's design with some modification by Benjamin Burley who succeeded him as architect to the N.E.R.

Architectural Description—The station (Figs. 45, 46) is aligned N.E. to S.W., with the main station building and departure platform on the S.E. side, facing Tanner Row–Toft Green. The main building was symmetrical except for the two-storey end blocks, which were of three and five bays respectively (Plate 148), and 250 ft. long. The central block of five bays, and the flanking blocks both of six bays, are three storeys in height, of West Riding carboniferous sandstone at ground-floor level, and of white brick with stone dressings and cornices above. The centre block containing the main entrance and booking office is of rusticated ashlar at ground-floor level, with five round-headed openings. The two outer openings were originally entrances, but have been replaced by copies of the adjacent windows (Plate 147). At either end of the block are pilasters at every level, and two added pilaster chimney-flues flank the central round-headed opening. The top member of a deep masonry band above a heavy moulded ashlar cornice on modillions acts as a continuous sill for the second-floor windows, which have segmental arched heads and eared architraves of ashlar. There is an ashlar cornice and parapet, and a hipped roof of slate. The adjacent three-storey blocks are lower and have Tuscan colonnades at ground-floor level each of five bays carrying a simple entablature and terminated by a projecting bay, rusticated at ground-floor level, with a round-headed recess containing a window; the colonnades have been altered by rebuilding the back walls to abut the column bases, and closing the entrances from the Booking Hall. There are moulded ashlar strings at both first and second-floor levels, and a moulded cornice and parapet. At the N.E. end, a third storey has been added to the two-storey end block shown in Andrews's drawing of 1858. Both outer blocks have windows set within round-headed recesses of fine ashlar at ground-floor level, the centre window of the N.E. end block replacing an entrance. On the N.W. side despite alterations earlier this century and in recent years, and the removal of most of the train shed canopy, part of the ground floor remains unaltered. The five bays of the centre block have round-headed arches of gauged brick, and plain ashlar sills. All the other original windows at ground-floor level have flat arches of gauged brick apart from four at the S.W. end. Where exposed beneath later accretions, the first and second floors are in Flemish-bonded white brickwork, and have sash windows with flat arches of gauged brick, and ashlar sills. Cast-iron fixtures on a simple ashlar band at first-floor level secure the roof members of the train shed. There is a moulded ashlar band at second-floor level, and a moulded ashlar cornice capping.

The plan (Fig. 46) shows the arrangement after additions and alterations in the 1850s. The earlier arrangement is traceable on the Ordnance Survey plan of 1852. On either side of the Booking Office were the 1st and 2nd Class Waiting Rooms, with toilet facilities. To the N.E. end of the range was the Parcel Office, and at the other end a Post Office; railway company offices occupied the upper storeys. Little of interest survives internally apart from a staircase with cast-iron balusters and a wooden handrail (Plate 89). The buildings on the Arrival (N.W.) side of the station have been much altered. Least altered is the original First Class Refreshment Room (Plate 147) with round-headed windows, and retaining its elegant decoration in Regency style. The entrance to the bar at the S.W. end of the room, though blocked, has simple pilasters and entablature framing the original opening. The Second Class Refreshment Room, the First Class Ladies Waiting Room, the Bar and Tearoom adjoining the Refreshment Room, have all been altered internally. The elevation to the platform is intact, with round-headed windows in recesses, but many of the openings have been modified. The original colonnade at the S.W. end of the range has been reduced by 2½ bays, by the erection of a later building. The N.W. elevation towards the City Walls has a centre block of two storeys and seven bays in red brick, with gauged red brick round-headed arches to the lower windows and flat arches to the upper, all with ashlar sills, and a simple moulded ashlar band to the first floor. A single-storeyed building of three bays in red brick to the S.W. is probably part of the original build, but has had a storey added and has been extended to the S.W. The three-storeyed range in white brick at the N.E. end, of seven bays, was built 1852–3 as part of the hotel. Only one of the eight original turn-tables survives, at the S.W. end of the arrival platform.

The Train Shed. Much of this remained intact until 1965. The width between the arrival and departure platform buildings is 100 ft., and the roof of iron and glass, contrived in two spans, covered the platforms and four tracks. In the middle there were originally pairs of cast-iron columns 16 ft. apart spaced at 20 ft. intervals along the entire length of the train shed and connected longitudinally by pierced arched members. The train shed, when completed in 1841, measured 300 ft. in length, and ended at a point just S.W. of the four turntables, still marked by a masonry pier at the S.W. end of the arrival platform. The N.E. end was originally enclosed by a pierced wall but when the hotel was built its ground floor was open to the train shed, with the upper floors supported on iron girders spanning between four columns. The longitudinal arched members were altered and abut the capitals awkwardly. A later brick wall fills in the spaces between the columns. To S.W. of the arrival platform, where the train shed projected beyond the station buildings, cast-iron lintels with a raised pattern on the underside replace the arched members connecting the columns.

Four main phases can be seen in the train shed and its extensions. The cast-iron columns of phases 1 and 3 are known to be by the York firm of ironfounders, Thompson. No nameplates could be found on the columns of the other phases. (1) *Original Train Shed.* The columns have enriched capitals and square bases standing on high plinth blocks (Plate 149). In the spandrels of the arched connecting members are a series of diminishing circles. The surviving column to S.W. of the arrival platform is an exception, with a plain column surmounted by a double bracket-shaped block (Plate 149). (2) *South-West Extension of Train Shed.* An addition about 185 ft. long probably of 1845, necessitated by the extension of the platforms. The caps are plain (Plate 149), and bases round, again on high plinths. The column bases on the platform are no longer visible, and were probably encased when the platform level was raised. The arched spandrel panels are similar to those of the main train shed. (3) *Scarborough Platform—First Period.* Probably dating from 1845/6, covers a length of about 130 ft., adjacent to the first stage of the train shed. The caps are plain bell-shaped, with an upper ovolo moulding instead of the cyma used elsewhere. The bases are more bulbous than those of the other columns, and stand on high plinths (Plate 149). The roof of this part differs from the rest in being constructed of timber. The iron columns were originally not connected by arched members, though these were later inserted. (4) *Scarborough Platform—Second Period.* Built before 1851, covered a long, narrow triangular space at the S.W. end of the platform, adjacent to the second stage of the main train shed. The enriched caps are similar to those of the original train shed (Plate 149). The bases resemble those of the earlier columns of this platform, but have low plinths (Plate 149). The arched members have short vertical struts in the spandrels. A fifth section of canopy, in 1851, covered the whole length of the Scarborough platform. This was entirely demolished before the building was recorded; it was probably of the same date as the work of the second period of the Scarborough platform. The ironwork of the roof (Plate 150) consists of light trusses spanning between the arcades; the finish is of lead-covered boarding except at the ridge where there is a raised clearstorey, glazed on top and with open louvres at the sides.

The Station Approach. In 1850 a lodge and gates, seen in a drawing by Andrews (Plate 150), at the N.E. corner of the site gave access to Tanner Row. At a later date the cast-iron work (Plate 150) and ashlar piers were moved to their present position in the boundary wall of the area before the Old Station.

(29) THE OLD WAREHOUSE, Skeldergate, is of the 17th century, built of brick, and has two storeys; the roofs are tiled. It was built probably for wine merchants, the ground floor being in the nature of a cellar. The plan is a long rectangle, with a small wing near the W. end of the S. side. A modern warehouse has been built against the N. side. A late 19th-century bonded warehouse adjoins the E. half of the S. elevation. At both stages of the E. end are large openings for access of goods from boats in the river below. Two parallel ranges, apparently of similar design, adjoined the building on the N. until early in this century (Fig. 47).

Fig. 47. (29) Old Warehouse, Skeldergate. Reconstruction.

The original wall of the *West Elevation* (Plate 152) has been altered by the insertion of various openings at the ground and first floors, but the curved Dutch gable with pedimental apex and the S. brick kneeler remain. In the upper opening, of 19th-century date, is the swivel post of a hoist or derrick; this has led to some ill-founded references to the building as the 'Old Crane', which was in fact further down the river. The small projecting wing (Plate 152) has a modern opening to the ground floor, a band of two courses at first floor, a small window with brick segmental head at second floor, and a Dutch gable with coping of two brick courses; the E. kneeler remains. To E. of the wing on the ground floor is a blocked original opening with segmental arch, and above it a blocked reconstructed opening. Internally the *Ground Floor* is brick-vaulted throughout, with blocked openings at intervals on both side walls; the *First Floor* has stone flooring, brick walls, and a simple trussed-rafter roof. *Demolished 1970.*

(30) HOLGATE WINDMILL (58425148) was rebuilt in its present form (Plate 151) between 1770 and 1792 by George Waud senior, miller, who was stated in a surrender of the property on 26 May 1792 to have 'lately erected' the adjacent house and 'brick built wind-mill' (YCL, Court Rolls of Acomb and Holgate, 25 April 1793); Waud had been living in the manor from 1770. A mill has been on the site since the 15th century; in 1432 occurs a mention of 'the windmill standing near the hill of Holgate in the common field of the Archbishop' (*York Memorandum Book*, SS, CXX for 1911 (1912), 216). The mill had five sails, an unusual arrangement first introduced at Newcastle-upon-Tyne by the engineer John Smeaton. Although provided in the 19th century with a steam auxiliary engine, the mill continued to make use of the sails until they were seriously damaged in the heavy gales of January 1930; they were then taken down leaving only the struts, but milling continued with electric power (*Yorks. Gazette*, 1 Feb. 1930). The mill ceased to work, and the building was taken over in 1938 by York Corporation. It was repaired as a landmark in 1939 by Messrs. Thompson and Son, millwrights of Alford, Lincolnshire, but the intended refitting of sails was deferred by the outbreak of war. In 1940 the mill was sealed up and the warehouse and outbuildings demolished (N. M. Mennim in *Yorks. Gazette*, 22 July 1949; report by Rex Wailes dated 1 August 1954).

The tower mill is five storeys high, round in plan, of narrowish red brick, with broad mortar joints, and is painted with black bitumen. It batters up to a waist at the level of the third stage, and then finishes vertically with a six-course corbelled section to carry the cap. The bricks of the battered base are set horizontally and the wall is of constant thickness, so that both inner and outer faces slope inwards. At ground-floor level are two doorways, that to W. blocked with brick on the inside. Some of the eight staggered windows have segmental heads; the two bottom windows, which have sills, are blocked. The round cap is covered with steel plates on wooden cap spars or rafters, with a ball finial. Projecting main timbers hold the surviving sail stocks which retain some iron ties in front, and formerly carried a fantail behind. The walls are plastered inside; the ground floor is of concrete. The boards have been removed from the four upper floors: each floor has two main parallel square-sectioned beams, and stop-chamfered joists, all of pine, and a heavy softwood ladder. The first three ladders ascend the E. side. The machinery is largely intact, but some iron supports carrying the lower bearings have been broken off, and there are no bins nor means of feeding-in the corn. The main drive is of iron, but the stone-nuts have a metal frame carrying wooden cogs. One pair of millstones is of gritstone, but the other is of French Burr stones bound with iron.

(31) Bound Stones (59235075), by N. entrance to Knavesmire, two: (a) of brown limestone, 21 in. by 12½ in. by 5½ in., inscribed 'The Boundary of Micklegate Stray', with round head (18th-century) (Plate 92). The boundary marked is that between Micklegate Stray and ancient enclosures to N., coinciding with the former parish boundary between St. Mary Bishophill Junior and Holy Trinity Micklegate; (b) of magnesian limestone, 29½ in. by 12 in. by 9 in., uninscribed; both are in badly weathered condition.

(32) Bound Stone (58085022), of magnesian limestone with segmental head, bears the inscription 'Bounds of Bishophill' (18th-century). This stone marked the point where the ancient parish of St. Mary Bishophill Junior met the parish of Acomb and of Holy Trinity Micklegate (Dringhouses detached portion); it is in badly weathered condition.

(33) Bound Stone (59065037) of weathered limestone, 11 in. high by 7 in. by 17 in., on S.E. side of the Tadcaster Road opposite to Hob Lane, marking the ancient city boundary. It has been reused by the Ordnance Survey for a benchmark.

(34) Hob's Stone (58915042), on the N. side of Hob Moor Lane 140 yards W. of the Tadcaster Road, 39 in. by 21 in. by 15 in., consists of a heavy coffin-lid bearing a much weathered effigy of a knight, now set upright. On the left arm is a shield-of-arms of three water bougets, presumably for the family of Ros (probably early 14th-century). The original edge of the lid remains on the right side but a recess has been cut into it, and there are three dowel-holes in front. On the back was an 18th-century inscription already nearly defaced by 1818 (Hargrove, III, 513, and Drake 398): 'This Statue long Hob's name has bore, / Who was a knight in time of yore, / And gave this Common to the poor', with the names of the Pasture Masters who erected it in 1717, as well as the later date '1757'. At the back of the lid is a separate flat stone, 25 in. by 22 in., with a shallow basin cut in it, probably used for the disinfection of money when the plague was in York; and in the surface of the lane to S. are two blocks of stone which may have formed an 18th-century pedestal for Hob's Stone. No evidence has been discovered to support the view (Davies, 98) that the effigy came from St. Martin's, Micklegate, or that the name Hob commemorates an historical Robert Ros; occurrences of the place-name elsewhere imply that it contains the element *hob*, a goblin (EPNS, XIV (1937), lx, 290). The two blocks of stone in the lane had been removed, or covered with tarmacadam, by 1969.

(35) Mounting Block (59145058), on the footpath on S.E. side of the Tadcaster Road, with three steps, was formerly used as the first milestone from York, measured from Ouse Bridge, and on the side towards the road an iron pin, set in lead, still remains to hold a metal inscription-plate. A similar pin set in the face towards the city would seem to have carried an Ordnance Survey benchmark added at the survey of 1850. The stone also served as a bound stone, marking a re-entrant angle of the City Boundary of 1832.

(36) Memorial Gates, Rowntree's Park (60465063), were brought, it is said, from 'Ritchland Park, near Windsor, Berks.', in 1954–5, by Messrs. Rowntree & Co. Ltd. They were set up at the main entrance to Rowntree's Park on its river frontage, as a memorial to the War of 1939–45. They are said to date from 1715 and to have been made by Jean Tijou; they were restored by W. Dowson of Kirbymoorside and erected under the supervision of the York City Engineer. It seems probable that 'Ritchland' is a mistake for Ritchings Park, near Iver, Bucks., where in 1960 only garden features of the former mansion survived (N. Pevsner, *The Buildings of England—Buckinghamshire* (1960), 177).

The wrought iron gates (Plate 45) are set between square piers of Portland stone bearing stone cherubs and buttressed by oblong projections surmounted by volutes. There are curving sections of railing on each side set on brick dwarf walls coped with stone, leading inwards to the main gates which stand between two smaller gates for pedestrians. The ironwork is painted black with gilt enrichments. There are no features of identification such as monograms or heraldry.

(37) Knavesmire Wood (around 59204880) contains an avenue of tall lime trees, nearly a ¼-mile long, aligned between the Archbishop's Palace at Bishopthorpe and Dringhouses. The avenue appears on the engraved maps of Francis White and Robert Cooper, published in 1785 and 1832, but not on the atlas of Thomas Jefferys issued in 1772. It is probable that the planting was connected with the improvements at Bishopthorpe evidenced in 1773–4, when the archbishop's head gardener, Thomas Halfpenny, was paid for extensive work in the gardens and for 'clearing prospeck to Minster', which indicates an interest in such vistas (Borthwick Inst., CC 67885). York Corporation in 1965–6 removed decayed timber and planted new lime saplings as eventual replacements for the old trees.

ACOMB ROAD runs W. from Holgate Bridge across the township of Holgate to Acomb. Development did not take place until 1828 (*see* p. 123), apart from the few houses which constituted the hamlet of Holgate.

ALBION STREET was one of the earliest redevelopments of York within the walls, projected and in part built in 1815 (*see* p. 123). The ground had formed the gardens behind John Carr's own house in Skeldergate, left to his nephew William Carr, who sold the land in 1815 in two lots. It seems that the main developers were George Willoughby of Old Malton, builder, and Leonard Overend of York, slater, but one lot was acquired by Ralph Peacock, raff merchant (YCA, E.96, f. 243v, 249, 249v).

BAR LANE leads from Micklegate, immediately within the Bar, to Toft Green. The 'Jolly Bacchus' public house and a few other small houses, which formerly fronted on its W. side, have all been demolished; standing on the city rampart, they were Corporation properties (YCA, M.10D).

BARKER LANE, formerly known as Gregory Lane from the small parish church of St. Gregory which, until *c.* 1585, stood on its E. side, now contains no monuments. The lane follows the line of a Roman street and is evidenced in documents from the early 13th century. It led from Micklegate to the main gateway of the Dominican Friary, built on the site of the earlier King's House and Chapel (*see* TANNER ROW, with TOFT GREEN).

BISHOPGATE STREET, the first section of the Bishopthorpe Road, S.E. of the Old Baile, was so named by 1830, when there were four houses in it. Little further development took place until after 1850 (*see* p. 123).

BISHOPHILL (including Victor Street). The name Bishophill was formerly used to include the three streets now known as Bishophill Junior, Bishophill Senior, and Victor Street. Of these the last is certainly identical with the mediaeval *Lounlithgate*, evidenced in documents from the 12th century. Bishophill Senior was probably *Besingate*, mentioned from the 13th to the 15th centuries. The name Bishophill was originally (from 1344) that of a district, known earlier as *Bichill* and probably a possession of the pre-Conquest church of York (*YAJ*, XLI (1966), 377–93).

In 1282 this was not a populous part of the city, as husgable was paid only upon twelve tofts in *Besingate*. In *Lounlithgate* were forty-one tofts but these were in the hands of only twelve persons (YCA, C.60). Later the postern (*Lounlith*), on the site of the modern Victoria Bar, was blocked and the whole area became a backwater. In 1632 the parish of St. Mary Bishophill Junior was the poorest in the whole city, with twenty-six persons receiving relief against only four paying the Poor Rate (YCA, E.70). In the 17th century only a small part of Bishophill was built up, but several houses were of considerable standing, notably the great mansion of Lord Fairfax and later of the Duke of Buckingham to N. of the churchyard of St. Mary Bishophill Senior. The last remains of this house, known as Duke's Hall, were cleared away in the 18th century.

From 1756 onwards the City Corporation granted leases of plots along the N. side of Bishophill Junior W. of the church (YCA, B.44, f. 28 etc.), and a ribbon of small houses was built during the next 50 years (YCA, M.10D). Serious redevelopment of the area on a speculative basis began in 1811 with the building of the first houses of St. Mary's Row (in Victor Street), opposite to the Rectory (42) of Bishophill Senior, probably the earliest small 19th-century terrace houses in York (*see* p. 130). The builder was probably Thomas Rayson (YCA, E.96, f. 169). Near Bishophill Junior church a group of properties was bought up *c.* 1810–20 (YCA, E.97, f. 208) by John Tuke (1759–1841), surveyor and land agent. He rebuilt on some sites and in other instances resold to builders such as Ambrose Gray (E.97, f. 242v.), who *c.* 1825 put up Gray's Buildings, now demolished. The building-up of the whole area with streets of small houses at a high density did not take place until after 1850.

(38) BISHOPHILL HOUSE, Nos. 11, 13, was built in the early 18th century on an L-shaped plan with a front four bays wide (Fig. 48). In 1740 the house was acquired by Richard Dawson (1696–1762), a prominent merchant and the wealthiest parishioner, who on 6 May 1740 advertised his house in Trinity Lane (128) as to let from Michaelmas (*York Courant*). Dawson enlarged the house by the addition of two further bays to the N.W. and built the present staircase in the re-entrant angle of the original house. He remodelled the front elevation, framing the present entrance, and refitted much of the interior. The house was subsequently tenanted by Lady Gascoign and was sold in 1764 on the death of Dawson's eldest son Thomas, a Portugal merchant of London; it was advertised as including 'a handsome large Drawing-Room, hung with India Paper, two Parlours fronting a pleasant Garden, belonging to the House . . .', (*York Courant*, 10 April 1764). The property passed to James Fermor, esq., who in 1771 married Mrs. Henrietta Standish, a widow, upon whom he made a large settlement including Bishophill House (Borthwick Inst., York Wills Reg. 128, f. 10). It was probably when Fermor took over the house that Dawson's addition was extended N.E. to allow the formation of a large Saloon with a semicircular bay at the N.E. end, and the fine plaster ceiling, so close in style to Francesco Cortese's work of 1764–5 at Newburgh Hall, was inserted. After Fermor's death (1783) his widow married in 1785 William Carr, nephew of John Carr the architect (*York Courant*, 18 Jan. 1785); William Carr lived in the

house during his uncle's life, but about 1811 sold the property to John Tuke, who by 1825 had converted it into three tenements (YCA, E.97, ff. 96, 208). This remodelling is evidenced by the refitting of several rooms and the alteration of windows, including the two N.W. windows of the main front. Further alterations were made when the house was bought by Mrs. Sarah Preston, who was living there in 1828–30 (Directories), but soon afterwards leased it to the Misses Lucy and Eleanor Walker, who used it as a girls' boarding school from 1834 or earlier until *c.* 1850 (Directories; Tithe Map of 1847). In the course of the 19th century plate glass was put in all windows; it is probable that the cornice and roof are also of the second half of the century.

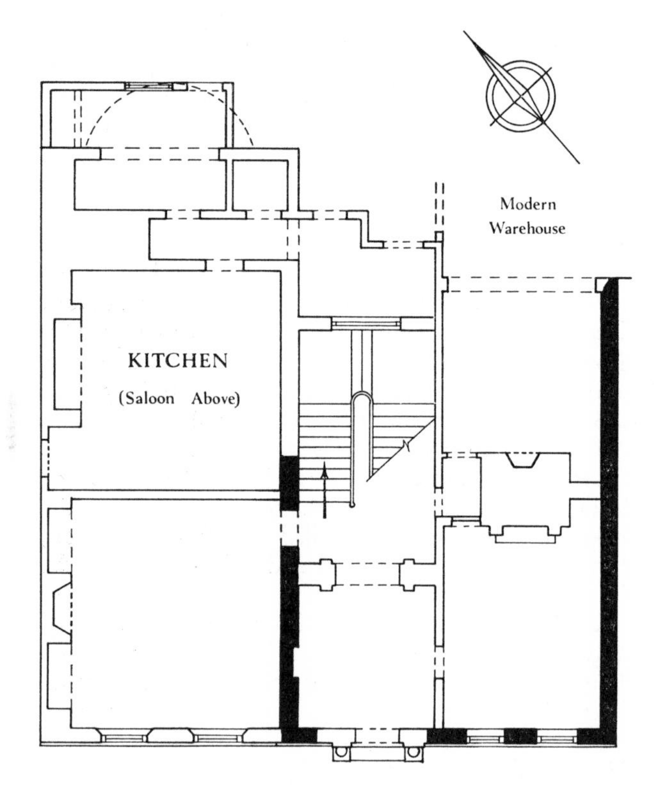

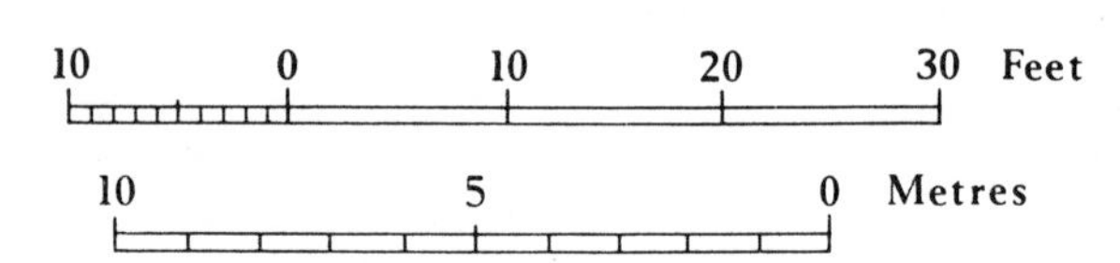

Fig. 48. (38) Bishophill House.

The *Front Elevation* (Plate 57), facing S.W. on Bishophill, is of two builds, the original S.E. part being in good stock brick, Flemish bonded, with fine brick dressings to openings, band and quoins. The three-course plinth, with chamfered weathering was originally returned round both ends, and so was the string course. The imposing entrance, roughly central, has round Ionic columns to the jambs, a pulvinated frieze to the entablature; a moulded, modillioned cornice and pediment; and a semicircular fanlight over a heavily moulded and fielded eight-panelled door (Plate 62). Beneath a four-course band, at ground floor, are two sash windows with ashlar sills and flat rubbed-brick arches. The first floor has two original window openings with ashlar sills and stuccoed flat arches with keys; two windows have been removed and a window has been inserted in the blocking immediately over the entrance; dressings of the former openings are visible. To the second floor are four sash windows, almost square, with stone sills and flat arches of rubbed bricks. About half-way up these windows, the character of the brickwork changes at the level of a timber plate one course deep. It is likely that the upper part of the wall was rebuilt when the structure was reroofed and the pre-existing cornice was replaced by the present one. The front of the second build to the N.W., in pale red stock brick in Flemish bond, with good quality red brick dressings, has to the ground floor two large early 19th-century sash windows, with stone sills, carried up to the brick band and without arches. On the first floor are two sash windows with narrow stone sills and stuccoed arches with key-blocks, and on the second floor smaller sash windows matching those further S.E. The same change of brickwork and timber occurs half-way up the second-floor windows. The eaves are supported on shaped brackets, those to the S.E. build being in pairs, the six to the N.W. being almost evenly spaced. The *Rear Elevation* has a projection on the S.E. 3–4 ft. deep, and formerly provided with a shallow segmental bay (OS 1852); the original wall remains above a modern warehouse extension. In the middle of the house, above a single-storey addition, large semicircular arches with rubbed brick voussoirs remain over the openings for two windows lighting the staircase, one above the other. To the N.W. a wing projects approximately 20 ft. and has on the first floor a large semicircular bay window now underbuilt; it has stucco dressings and bronze frames and glazing bars; a similar bay on the second floor has been removed and the wall built up flush.

Inside, the *Entrance Hall* has a moulded and enriched cornice and skirting. In the S.E. wall is a doorway with moulded entablature with dentil cornice and pulvinated frieze. To the N.E., opening to the stair hall, is a large archway with panelled reveals between Ionic pilasters. The room to the S., with moulded cornice and skirting, has a reeded surround to the doorway in the N.W. wall; in the N.E. wall is a chimney-breast between a segmental-headed recess (to S.E.) and a blocked doorway retaining its door with six fielded panels; the windows have reeded surrounds with plain angle pieces.

In the stair hall, which has a moulded and enriched cornice and skirting and a floor laid with diagonally-set limestone flags, the *Staircase* (Plate 82) rises to the second floor in five flights with two landings and three half-landings, with a solid mahogany moulded rail curving round the angles, but no string; it has strongly cantilevered treads with recessed panels under them and moulding on the edges, all of soft wood. The heavy balusters have graduated bases stepped like those at Micklegate House (81) and Nos. 134, 136 Micklegate (98). The newel and spiral rail at the foot have been removed. The side wall has a rising boarded dado and a moulded dado rail and skirting. The window in the N.E. wall lighting the stair consists of two sashes placed together which have heavy ovolo-moulded glazing bars and a moulded surround. Doorways in the N.W. and S.E. walls of the stair hall have or had moulded entablatures with dentil cornice and pulvinated frieze.

A large room to the W. has a moulded cornice and skirting of *c.* 1820–30. On either side of the chimney-breast is a deep segmental-headed recess. The doorway has moulded jambs and lintel, square angle-pieces with handsome foliated paterae (*c.* 1820–30), and a reused door with six panels, fielded on the outside. The two S.W. windows have handsome moulded surrounds and the reveals have small elegant panels with applied moulding; there is similar panelling under each window.

To the N. the *Kitchen* is entered from a rear passage. On the N.W. is a large chimney-breast containing an open fireplace with a great segmental-headed arch, like the kitchen fireplace at Micklegate House (81). In a recess to S.W., above a doorway, is a window with six panes and heavy ovolo-moulded glazing bars. Between the S.W. wall and the chimney breast is the springer of an arch, probably cut away to insert this window.

The central *Cellar*, under the entrance and stair halls, has barrel vaults of brick rising from walls with two large attached piers. Two compartments have vaults at right angles to the rest and two others, belonging to the earlier build, have plastered ceilings.

On the *First Floor*, the landing, with plain walls and a moulded plaster cornice enriched with egg and dart and dentils, has four doorways serving the rooms, the one in the N.W. wall being like that on the ground floor with eared surround and pulvinated frieze. The E. room, redecorated with Regency fittings, has a reeded plaster cornice, with formalised flower paterae to the angles; in the N.E. wall is a 19th-century sash window with narrow lateral sashes. In the completely panelled N.E. wall of the S. room is a fireplace with an overmantel with moulded and eared surround of *c.* 1740. The W. room has in the N.E. wall a doorway opening to a small landing leading off the main landing and giving access to the main saloon to the N.E. The main feature of the *Saloon* is the plaster ceiling (Plates 61, 154, 155), possibly the finest example of rococo plasterwork in York. The bay window has bronze bars to the sashes. On the *Second Floor*, over the stair well, is a moulded plaster ceiling, and the landing has a moulded dado, skirting and doorways with simple doorcases. One of the rooms has a simple late 19th-century cornice. (*Damaged by fire. Staircase and many fittings destroyed.*)

(39) House, No. 15, was 'new erected' for John Tuke in 1825 (YCA, E.97, f. 208). The site was that of two old cottages of interest as the homes of the prominent York stonemasons Andrew Kilvington (d. 1774) and, next door, his son George Kilvington (1760–89) (YCA, E.94, f. 149v.; E.95, f. 11).

The street front, three-storeyed, is in Flemish bond brickwork, the bricks being of a pale colour and rough texture, and has a timber cornice with brackets and large dentils. On the ground floor are two windows, that to the N. somewhat broader than the other, with simple box-framed sashes, stone sills and slightly segmental arches of stock brick. The entrance to S. of centre, has a plain surround with a crude cornice. In the upper storeys are two sash windows with moulded flush frames, segmental brick arches and plain, painted stone sills, those to the first floor having twelve panes and those above being shorter and nine-paned.

(40) House, No. 17, was described as a 'cottage newly erected where an old house was' in 1755, when it was sold to Benjamin Grosvenor, gent., by William Carr, carpenter, and his wife Diana (YCA, E.94, f. 2; *cf.* ff. 6, 7v., 139v.); Carr was probably his own architect and builder. It has two rooms to each floor, a small yard to rear, and buildings adjoining both sides.

The street front, in Flemish bonded brickwork with red brick quoins, window dressings and bands, has a gable with ashlar coping. The ground floor has been considerably altered, the original window being replaced by a casement; the entrance is of *c.* 1835. At both first and second-floor levels are three-course bands. The first-floor and second-floor windows have segmental red brick arches, flush frames and plain timber sills.

Inside, the staircase, which is original, is the only feature of note; it has a closed string, moulded handrail, square newel posts and turned balusters with square knops.

(41) House, No. 19, is of the early 18th century and retains some original features. Built on part of the great Fairfax estate, it was a town house of standing, with a very large garden, extending from Bishophill almost to Skeldergate. The home for many years of the famous benefactor of York, Dr. Stephen Beckwith, M.D., who died there on 26 December 1843, the property was sold (*Yorks. Gazette*, 16 March 1844) to become the York Female Penitentiary. In this century the Penitentiary moved to Clifton and the site was acquired by Messrs. Cooke as an extension to their adjacent Buckingham works. In the 19th century an extension to N.E. and a two-storey porch in front of the entrance were added; the entire front elevation was probably stucco-dressed at this time. In this century a range of industrial buildings was constructed along the Bishophill frontage, abutting on the S.E. front.

The stucco-dressed *South-East Elevation* towards the garden was originally symmetrical, of three bays width, the centre bay projecting by some 4 in. To N.E., on both first and second floors, are original windows having flat arches with key-blocks. At the first floor there is a window in each of the N.E. and S.E. sides of the porch, matching the original openings, as do those in the N.E. extension, all windows being reglazed. The original entrance has been brought forward to the face of the porch. Above, is a timber modillioned cornice; the roof, hipped to N.E. and S.W., has a central dormer. The *Interior* has two rooms to each floor with a central stair well. The *Staircase* (Plate 88), rising in two flights with a half-landing between floors, has two turned balusters (Fig. 18n) to each tread and a swept moulded handrail. Apart from the fireplace in the first floor S.W. room, nothing of note remains. The kitchen was originally to S.W. in the basement-cellar, having a fireplace in the N.W. wall. Two attics within the roof space have part of the roof construction exposed.

(42) THE OLD RECTORY, Nos. 3, 5 Victor Street (Plate 46), was built in the late 17th century. The first surviving terrier (Borthwick Inst. R.III.A, xvi.1), undated (?1684) but signed by William Stainforth, rector 1668–1705, probably refers to an earlier building on the site as 'a House and Garden . . . worth about five pounds per annum'; the next, of 1716, clearly describes the present building, the particulars remaining substantially unchanged until 1865 (*ibid.*, xvi.2–18). It had been declared unfit for residence by 1818 (Lawton, 26), and was sold by the Ecclesiastical Commissioners soon after 1876. In the late 19th century, the ground floor was converted to shop premises and a carriageway passage inserted.

The front and back elevations are each divided into four bays by brick pilasters with a projecting string-course at first-floor level. The original eaves cornice has been replaced by hardboard. The roof is covered with pantiles. Hung-sash windows to the front are all later than 1818 when the building was described as having 'small windows'. Some of the segmental arches for the original windows remain at the rear. One original window, now blocked, retains the original timber frame and mullion.

The interior has been much altered. Terriers mention two upper rooms only until 1764 but from 1770 refer to three; there was also a garret in the roof which is not now accessible. A fireplace on the first floor, probably of the late 18th century, has a frieze enriched with applied composition of unusually low relief.

BLOSSOM STREET (Plates 7, 158, 162) shares with THE MOUNT and MOUNT VALE a single continuous numbering and together they represent the suburban stretch of the main Tadcaster Road without Micklegate Bar. (For the continuation of this road see DRINGHOUSES, p. 116.) The name Blossom Street is derived from *Ploxswaingate*, the street of ploughmen, a name traceable to the early 13th century. Its great width allowed a horse and cattle market to be held along it. Beyond the turning of the Holgate Road the street continues as The Mount, and subsequently as Mount Vale, names derived from the Civil War fortification on the summit of the hill. It was the presence of this royalist outwork which enabled a number of old houses to survive in Blossom Street, the only suburb without the Bars to escape destruction in the siege of 1644. The road was already built up in the 13th century, as husgable was paid on twenty-nine tofts without Micklegate Bar in 1282 (YCA, C.60). By 1639 there were sixty-eight houses (Bodleian, MS. Rawlinson C.886, pp. 51–2). Some of these were probably destroyed in the siege, and there is evidence that in the early 18th century there were many empty plots between inhabited houses. By the middle of the century these plots were being filled with new houses, sometimes detached, but often abutting on the side walls of the earlier buildings on each side.

From an early date the street contained a large number of inns and hostelries and it would seem that this was a main centre of accommodation for merchants of the lesser sort. In the second half of the 18th century the street began to be fashionable (a little later than Bootham, at the northern exit from the city) and several houses were built for letting to tenants of the gentle class. Interspaced with these were small farmhouses belonging to cowkeepers who kept cattle on their tofts and put them out to common on the Knavesmire during the daytime. By 1830 Blossom Street had several shops and many of its houses were the private residences of those in trade. The Mount was largely residential and included the homes of several of the minor gentry. New residential development began to fill vacant land off the street soon after 1824 (South Parade) and both terrace houses and detached villas were built on The Mount from 1824 onwards. This new development was of a high standard and catered largely for the growing professional class (*see* pp. 123–4, 127–9).

(43) WINDMILL HOTEL, Nos. 14, 16 (Plate 157), consists of a large U-shaped complex of four or five separate builds, from the late 17th to the late 19th centuries. The earlier works have been considerably altered and considerable internal rebuilding and refitting was carried out in 1965. In the early 19th century the Windmill Inn was 'a noted house' (YAYAS coll. in York City Library, letter M. Johnson to W. A. Evelyn, 6 Nov. 1913).

Stage 1, the block at the N.E. corner, with two adjacent gables to Blossom Street and steep-pitched roofs running back, is of the late 17th century and consists of brick outer walls with timber-framed internal partitions. It probably represents an early stage of rebuilding after the damage caused in the Civil War. The bay windows to the front, though largely renewed, were in existence by *c.* 1785; the timber barge-boards to the gables are modern. Surviving internal features include chamfered beams, running N.–S., and rafters of the S. gabled roof, halved and pegged together at the apex. There are two large internal chimney-breasts, but all the fireplaces are of the late 19th or 20th century. The staircase is of mid to late 18th-century date with turned balusters to the bottom flight. It was probably only the structure constituted by these two parallel ranges that formed the 'two messuages cottages or tenements called the Windmill Inn', described in a deed of 1735, by which the children of the late Henry Lee conveyed the premises to the occupier, George Benson, innholder (YCA, E.93, f. 86). Henry Lee (1665–1727) belonged to the fifth generation of a family of millers who, from 1621 to 1690, had been lessees of a windmill belonging to the City near the top of The Mount (YCA, I, ff. 102, 103), so that the inn must have taken its name from the mill worked by the family.

Stage 2, immediately to S., with a frontage to Blossom Street of about 21 ft., was probably originally a separate building, perhaps erected soon after the purchase of 1735. The roof, which is modern, has two ridges with a valley between, parallel to Blossom Street; the narrower span, to W., possibly represents a later addition, but no internal brick wall is thicker than 4½ in. A segmental bow window was put into the front elevation in the early 19th century.

Stage 3, a square three-storey block (No. 16) adjoining and S. of Stage 2, includes a carriageway to the hotel yard. It retains most of its original fittings. That it was built as an addition to the hotel is proved by the position of the staircase, accessible only from within Stage 2. It has a closed string, square balusters and turned newels. A long stable range behind, together with this block and carriageway, are shown on Baine's map of 1822, when they had probably just been completed, since No. 16 was first assessed to rates in 1823 (Borthwick Inst., Rate Books of Holy Trinity, Micklegate).

Stage 4, a long range running W. of Stage 1, is of *c.* 1890 but replaces an older range (OS 1852). There were intermediate stages of internal alteration, one of the mid 18th century including the main staircase and dado panelling in two ground-floor rooms; several doors are of *c.* 1840.

(44) HOUSE, No. 19, is on the site of an older house rebuilt in *c.* 1760, and in 1761 occupied by William Thornton, clockmaker (YCA, E.94, f. 34v.); of this house some external walling at the rear, the staircase and some doors, remain. It was later tenanted by William Green, esq.; in 1781 it was sold to Mrs. Ann Aspinall, Superior of the Bar Convent (E.94, f. 235); and William Hotham (Alderman from 1792; Lord Mayor, 1802 and 1819) was tenant from 1791 until his death on 8 August 1836 (Skaife MS.).

There was work on the house in 1791–3, but details are not available. The architect was Thomas Atkinson; John Prince was paid for bricks, plaster and work; Richard Hansom was responsible for carpentry, Mr. Croft for lead and glass, Mr. Haxby for ironwork, Mr. Smith for painting, Mr. Rusby for slates, and Mr. Fothergill for fixtures (Bar Convent Archives, 7 B 2(4)). A bill presented by Richard Hansom, specifically for this house, mentions work on the staircase, including a centre for a Venetian window, cutting a way for the stairs and hipping a roof over it, and various cornices (7 B 2 (8)). In *c.* 1815, the front of the house was taken down and rebuilt by Thomas Rayson (receipt dated 17 May 1821, 7 B 3(11)). A plan of the house by J. B. and W. Atkinson, dated August 1834, was doubtless a prelude to the alterations of 1837, to produce a residence for the chaplain (7 B 9 and 7 B 10). Richard Hansom provided staircase wainscotting and repaired bannisters (7 B 9(2)); took out front windows and refitted the sashes and shutters, removed the door-case, and provided a new front and a back staircase (7 B 10(1) and (3)). Richard Dalton provided bricks, lime and cement (7 B 9(5)); Matthew Walker did plumbing and glazing (7 B 9(6)) and in particular was paid 'for 10 windows in front glazing Best, for glass 12 squares each containing 213 feet', a description of the present windows in the lower two storeys. James Haxby provided ironwork (7 B 9(8)); Michael Taylor stonework, such as thresholds, sills and slips for fireplaces, and also four fireplaces (7 B 9(15)). Perhaps the most interesting payments are to Judith Jennings for plasterwork (7 B 9(7)), the details describing many of the cornices and features still existing. A lithograph of Blossom Street by Monkhouse of 1846 shows the house as still of two storeys, and indicates that a pair of windows to N, shown in the 1834 plan, had been replaced by one (probably in 1837) and that the doorway to S. must have been moved in 1847, when the staircase was inserted at that end. In that year G. T. Andrews added a third storey: work was carried out by William Shaw, joiner (7 B 14(3)), and John Ellis, bricklayer (7 B 14(5) and (6)); Henry Buckley provided window sills, moulded string, chimney pieces and hearth stones (7 B 14(7)); Matthew Walker did roof work in lead; Richard Knowlson plastered; and John Henry Cattley put best Bangor slates on the roof with copper nails. The house may have been divided at this time and the S. end combined with No. 21, newly built (1845), the new staircase being provided to give access to this complex; its iron balusters are characteristic of G. T. Andrews's work.

The front to Blossom Street (Plate 158), of six irregularly spaced bays, is of good quality red brick with a stone plinth, moulded and modillioned cornice, and a Welsh slate roof, hipped to S. At ground floor, the doorways each have two engaged, fluted columns with moulded caps and bases, supporting an entablature with plain frieze and moulded cornice; over the doors are radial fanlights. The doorway to S. was reset in 1847 and is not aligned with the windows above. The sash windows have flat rubbed brick arches and stone sills, and six windows at first floor are similar. Although the second storey was an addition of 1847, the brick and windows match up well with those below, but the windows are not quite so tall. No. 21 has a single window to each storey and is entered from No. 19.

The back of No. 19 shows the different builds clearly, the

18th-century work being in red brick and the additions of 1847 in large buff brick. To N., running through two storeys, are two red brick pilasters (*c.* 1760), between which is an infill of later brick with sash windows (1847). Against the second bay is a modern annexe, blocking a round-headed stair light at first floor. A projecting third bay, of late 18th-century brick, is lit at each floor by a large sash window with slightly segmental arch and thin stone sill. At the top of the first floor is a coping of stone flags on projecting blocks, representing the top of the house of *c.* 1760; in the recessed part, a lower band produces the effect of a parapet. The second floor is all of large brick of 1847. A fourth bay, slightly recessed and refaced from top to bottom in 1847, has at ground floor a doorway, cloaked by a one-storey annexe, with a sash window to S., both of *c.* 1760, reused. To S., again, a fifth bay, brought forward in 1847 to align with the back of No. 21, contains a large round-headed stair window with hung sashes and small marginal panes.

The interior fittings include doors and doorcases of *c.* 1760 and a staircase of the same date with cut string and turned balusters with plain umbrella-shaped knops spiralling at the bottom over a heavy newel similar in form to the balusters. Many of the fittings are of 1837, exemplified by a doorway in the stair hall (Plate 68). The S. staircase, of 1847, has cast-iron balusters.

(45) HOUSE, Nos. 22, 24, 26, was built in 1789 by John Horner, a wine merchant from Liverpool, as a pair of dwellings of unequal size pierced by a central carriageway leading to a warehouse (No. 24) behind. Horner occupied the smaller house (No. 22) and advertised the other for letting (*York Herald*, 27 Feb. 1790); it was taken by Joseph Newmarch, wine and spirit merchant. Horner died in 1791 (*ibid.*, 12 Feb. 1791). In 1795 the property was described as 'a large, genteel, well-built Freehold Dwelling-house, with spacious cellars and convenient out-buildings (No. 26) . . . with a commodious warehouse and wine-vaults under the same (No. 24), a yard, stabling for three horses, and a very good garden, well stored with a variety of choice fruit-trees'. No. 22, the smaller house, was similarly described except for the omission of the word 'large' (*York Herald*, 14 March 1795). Mrs. Horner remained in No. 22 and Newmarch in No. 26 until 1798. Later occupiers of No. 22 included, in 1808–26, the widowed Lady Mary Stapleton, daughter of the 3rd Earl of Abingdon; and of No. 26, the architect Charles Watson. Watson moved from Wakefield to York at the end of 1807 (*York Courant*, 18 Jan. 1808) and resided and carried on practice in the house until 1821, when he was succeeded by James Pigott Pritchett (1789–1868), taken into partnership on 1 January 1813 (*ibid.*, 4 Jan. 1813). The practice was carried on from Blossom Street until the partnership was dissolved in 1831, when Pritchett moved his office to Lendal (*Yorks. Gazette*, 1 Jan. 1831). Thomas Cabry, engineer of the York and North Midland Railway, lived in No. 26 in 1841–4 and was succeeded from 1845 to 1848 by Joseph Rowntree (1801–59), founder of the famous firm; later occupiers were the Rev. Robert Whytehead, rector of All Saints' North Street 1854–63, author of *A Key to the Prayer Book*, and his widow (Borthwick Inst., Rate Books of Holy Trinity, Micklegate). The whole property was conveyed in 1888 to the North-Eastern Railway (British Railways, York, Estate & Rating Dept., Survey Vol. 17, p. 6, No. 32) and in 1895 a rent-charge of 10*s.* a year payable to Holy Trinity Micklegate by charity of Christopher Waide, Sheriff in 1619 (d. 1623), was redeemed (*ibid.*, No. 32A). During the period of ownership by the N.E.R. it was usual for No. 26 to be the residence of the York stationmaster and No. 22 that of a railway inspector. In 1934 the London & North-Eastern Railway sold the freehold to the York Railwaymen's Club, and extensive alterations were made: the ground floor of No. 26 was formed into a single large room, and the first floor of the whole property thrown into one. Many of the internal fittings are, however, in Regency style and presumably the work of either Watson or Pritchett.

The building is of special interest, both on account of its plan with central carriageway, very unusual in York, and because it is one of the earliest three-storey houses outside the city wall.

Nos. 22 and 26 form a simple rectangular building of three storeys. The E. elevation to Blossom Street is in five bays with a timber cornice at the eaves. The windows to the upper floor are regularly spaced, but on the ground floor those to the N. are offset to allow for the carriageway. The timber pilasters flanking the carriageway and the entablature above are all modern. At the back each house has a lofty round-headed window lighting the staircase.

(46) HOUSES, Nos. 32, 36 (Plates 64, 156), were built on part of a large plot which belonged to an old house to N. (site of the modern Nos. 28, 30), sold by Thomas Lupton to William Smith in 1747 (YCA, E.93, f. 195); less than 6 months later, in January 1748, Smith mortgaged the property, which then included 'three messuages two of which have been lately new erected by the said William Smith' (*ibid.*, f. 199). The building was subsequently united in a single occupation and later redivided on several occasions, belonging for a considerable time to the family of Healey, merchant tobacconists. George Healey (1734–1824), Sheriff of York in 1789–90, and his younger brother John Healey (1751–1823) were occupiers of the two houses for nearly 50 years until their deaths (Borthwick Inst., Rate Books of Holy Trinity, Micklegate). From the middle of the 19th century onwards the occupiers were shopkeepers and various alterations were made. When No. 38 was built, to S., about 1822, an entrance passage to No. 36 was

provided through the later building which stood upon part of the ancient curtilage.

The houses are roofed in two ranges: that over the front rooms has a central gable towards the rear; the back rooms are covered by a low-pitched lead roof belonging to the original build. At first floor the back elevation has a central round-headed stair light with rubbed brick arch. There is a rain-water head on this side dated 1777, probably the period when the Healeys took over the property.

Internally a staircase rises in two flights with a full landing and half landing, with a heavy moulded rail, a turned newel with spiral fluting, and turned balusters with alternately plain and spirally fluted stems; other fittings include original doors with two large fielded panels and angle hinges, and a marble fireplace of *c.* 1800. *Demolished 1964.*

(47) HOUSE, No. 40 (Plate 156), closely resembles Nos. 32, 36, built in 1747, and was presumably by the same designer. The original work of this period was an L-shaped building consisting of the front block to the street and a projecting wing at the back. The house belonged to Ann, widow of William Collingwood, gent., before her remarriage to Henry Casson in 1773, and it may have been built for Collingwood. From 1773 it was let to William Phillips Lee, esq., a wealthy bachelor of distinguished family and friend of Laurence Sterne, who put up £100 for the original publication of *Tristram Shandy* (*YAJ*, XLII (1967), 103–7). After Lee's death in 1778 the house was held for several short terms until, in 1792, the freehold was acquired for £700 by Thomas Swann, a prominent York banker. The property was described as a messuage 'with Coachouse, Stables, Outbuildings, Garden and Yard' (YCA, E.95, f. 131v.). It was probably Thomas Swann (d. 1832) who extended the range behind the main building and added a third storey to it. Further extensions at the rear, and other alterations, were carried out soon after 1850. Members of the Swann family continued to live in the house until 1846; later occupants were Joseph Crawshaw, the rail-way contractor, from 1847 until his death in 1856, and during the 1870s the Rev. John Metcalfe, rector of Holy Trinity, Micklegate (Borthwick Inst., Rate Books of Holy Trinity, Micklegate; Directories). In the present century the ground floor was altered to form a shop.

The street front, in six bays, has a projecting brick band at first-floor level and timber eaves cornice. The doorway (Plate 64) is similar to that to No. 32. Some of the windows retain the original sashes. The staircase, opening off the central passage in the N. corner of the main block, has turned newel and turned balusters, three to a tread. *Demolished 1964–5.*

(48) BAY HORSE INN, No. 55, contains a nucleus which goes back at least to the 17th century, possibly to the period of reconstruction after the Siege of York, when the property belonged to Joseph Denton (free of York 1677). It was then perhaps a small farmhouse, with a croft running back to Scarcroft, and by 1726 was described as having a Kiln, Barn and Stable (YCA, E.93, f. 30). In 1748 it belonged to Matthew Spence (1700–65), inn-holder (*ibid.*, f. 204) and became an inn, though there is no evidence of its sign until 1798, when it was already 'The Bay Horse', very likely in reference to the famous Bay Malton, which won the Gimcrack 500 guineas at York in 1765 and even greater prizes at Newmarket in the two subsequent years (W. Pick, *The York Racing Calendar*). Among the later landlords of the house, during the 1860s and 1870s, was George Benson, father of George Benson (1856–1935), the York historian, most of whose childhood was spent there (York City Library, T. P. Cooper MSS.).

The earliest build is evidenced by a group of heavy ceiling joists in the ground-floor bar. The original house may have been L-shaped and it had only two floors with attics, but was later converted to three storeys. Enlarge-ment to the W. and the addition of a second floor, with the existing staircase, doors, etc., at first floor, probably belong to the conversion of the house to an inn by Matthew Spence between 1748 and 1765. Some of the windows were altered in the 19th century, and in the back wall 18th-century work in 2½ in. bricks contrasts with very large bricks of the mid 19th century. Internally, the staircase is of the third quarter of the 18th century and there are other fittings of the Regency period, but much of the ground floor is modern.

(49) HOUSE, Nos. 82, 84, 86 The Mount (Plate 159), was built late in the 17th century on an L-shaped plan. It appears clearly on the view of York by John Haynes, engraved in 1731, but not on any of the prospects taken from the same point by Gregory King, William Lodge or Francis Place between 1666 and 1678. It was probably the messuage with an orchard, garden, yard etc., which descended from William Pemberton, merchant grocer of York (free of the City 1695 and Chamberlain 1699), to his relatives the Geldart family, and was sold to William Smith in 1750. It was then said to have been lately in the occupation of Mr. Abercromby (YCA, E.93, f. 244). By 1778 (York Minster Library, Terriers, K.2) the property was owned and occupied by Mr. Ikin (?William Aitken) and *c.* 1835 was taken by Robert Davies, the Town Clerk, who in 1851 built his own large house, Nos. 88, 90, on the garden to S. Later in the century the house became the home of Davies's sister, Mrs. Skaife, and her son Robert Hardisty Skaife, the antiquary. About 1895 it was divided into separate occupancies. Earlier in the century there had been sub-stantial additions and alterations, including the provision

of two staircases, probably to fit the house for occupation by members of the Davies and Skaife families. By 1847 it was described as 'two comfortable dwelling houses, with spacious garden'.

The street front is an early 19th-century symmetrical façade in yellow-red brick. In 1963 a modern shop front was removed from No. 82 and the sash windows replaced with modern casements. A second front door, to No. 86, was inserted in 1963. The steep-pitched pantiled roof has two small square dormers, now modern reconstructions. The rear elevation includes the stuccoed end of a 17th-century range with a Dutch gable having a pediment at the head and curved sides. The S. elevation has a stuccoed ground floor with three modern windows, and a first floor of rather narrow red brick in mixed bond, with some lines of headers, having two 19th-century sash windows in old openings to E. and a blocked window to W. The S. gable of the front range, now masked, has an attic window set in the blocking of a 17th-century window with ovolo-moulded brick cornice. The gable has a coping with, at the bottom of the W. side, a badly weathered Classical woman's head, the hair arranged in cable fashion, of dark gritstone and most probably Roman.

Internally the fittings are mostly of the late 18th and early 19th centuries. In the attics some late 17th-century oak rafters and purlins of broad flat section are visible.

(50) NUNROYD, No. 109 The Mount, and No. 1 Mill Mount, form one building with an unusual front with bay windows in three storeys (Plate 46). The N.E. third of this front is modern. The building was originally one house, of which the nucleus may go back to the late 17th century and first appears on the prospect of York published in 1731 by John Haynes. This building may be represented by the present entrance hall to No. 1 Mill Mount (to S.W.), with stone footings and a small isolated cellar. An addition to S.E. along Mill Mount, in slightly different brick, may have been built soon afterwards. The extent of the early work is defined by the extra brick string-course appearing in the S.E. part of the elevation to Mill Mount. The main part of the building was erected in the first half of the 18th century with a symmetrical front to The Mount. By 1740 the property was described (YCA, E.93, f. 121) as a house with a barn, stable and cowhouses, occupied by Richard Middleton, yeoman, who with others conveyed it to Thomas Hungate (1710–77), the eccentric herald-painter, occupier until 1776. Hungate took up the freedom of York in 1736–7 and was Chamberlain in 1751; in 1749, on the death of Sir Charles Hungate, bart., of Saxton, Thomas was considered the next heir to the title but did not take it up, 'being a man of penurious habits and of reserved and singular manners. His friends, however, usually styled him Sir Thomas' (Skaife MS.). Later occupants were the Rev. Robert Stockdale, (d. 1786), vicar of St. Mary Bishophill Junior, and the Rev. John Walker, rector of St. Denys, who lived here in 1786–92. The rate assessment was raised from £5 to £7 in 1798, probably on completion of the alterations which included the building of the polygonal bay windows which give the house its marked individuality. The bays were in existence by 1802, the date of a watercolour of The Mount by Thomas White (Plate 7). In 1803–9 the tenant was the widow of Edward Bedingfield, Mrs. Mary Bedingfield, who moved here from No. 114 Micklegate (94) after her husband's death. Another phase of work is associated with the division of the house into two moieties in 1815 (Borthwick Inst., Rate Books of Holy Trinity, Micklegate). The first occupant of No. 109, the N.E. moiety, from 1816 to 1821, was Richard Allanson (Chamberlain of York, 1797), whose initials appear on a Georgian teaspoon found wedged into a lintel. Among later occupants of No. 109 were Leonard Simpson, J.P. (d. 1868), brother of Sir John Simpson; and from 1904 until his death in 1924 Alderman Norman Green, Lord Mayor in 1911–12, who added the block to N.E. of the older house, with a third bay window. When a passage was driven between two cellars a heavy rubble foundation was encountered, probably part of the foundations of St. James's Chapel, known to have stood near this spot.

The staircase to Nunroyd is of the 18th century, with square newels and turned balusters. That to No. 1 Mill Mount has a lower part of 1815 but is of the 18th century above. Some of the other fittings are 18th-century; several of the fireplaces are of *c.* 1815.

(51) HOUSE, Nos. 110, 110A The Mount, was built early in the 18th century and appears on John Haynes's view of York of 1731 (Plate 2) as the next house downhill from the old hospital of St. Catherine. In the last quarter of the century it was occupied by John Simpson, a farmer (York Minister Library, Terriers, K.2; parish registers). Early in the 19th century the house was altered and most of the fireplaces inserted; for about a century it remained a residence and from 1896 to 1906 was the home of William Angus Clarke, manager of the alpine department of James Backhouse & Co., the nurserymen; later the ground floor was converted to form a shop (Directories; Voters' Lists).

The colour-washed front elevation has 19th-century and modern features but retains a three-course band beneath the attic storey, heightened *c.* 1860. The rear elevation shows the original reddish-yellow brick. Original internal fittings include the staircase with square newel and turned balusters, and doors with two large fielded panels and angle hinges. *Demolished 1962.*

BRIDGE STREET. Before the rebuilding of Ouse Bridge in 1810–20 this short stretch of street, sometimes called Briggate, was regarded as part of the approach to the bridge. It was entirely redeveloped (in 1815–22) with the new bridge and was known in the 19th century as New Bridge Street (*see* p. 124).

CAMBRIDGE STREET, laid out in 1846 and completed in 1851, consisted of terrace housing for railway employees (*see* p. 124).

CARR'S LANE was formerly Kirk Lane or *Kirkgail* (13th century), and is a steep and narrow passage from Bishophill Senior leading down to Skeldergate. Its modern name goes back to the early 19th century and appears to commemorate the famous architect John Carr, who owned the large property at the foot of the lane on its N. side (*see* ALBION STREET).

CHERRY HILL, which presumably got its name from the adjacent Cherry Orchard referred to in a deed of 1780 (YCA, E.94, f. 220), is a narrow lane leading from Bishopgate Street to Clementhorpe, and was undeveloped until *c.* 1830 (*see* p. 124).

CLEMENTHORPE, originally the main street of a suburban village in the fee of the Archbishop of York, and leading to a staith, declined greatly in importance from the dissolution of the Nunnery (19) in 1536. By the 18th century it contained one or two small houses, and redevelopment, on a small scale, began only in 1823 (*see* p. 124).

CYGNET STREET was formerly Union Street, laid out in 1846 as small-scale terrace housing (*see* p. 124).

DALE STREET was built, as small-scale terrace housing, in 1823–8; it was occupied largely by railway employees and by minor artisans (*see* p. 124).

DOVE STREET was built in 1827–30 as small-scale terrace housing and was occupied by minor artisans and by railway employees (*see* p. 124).

FETTER LANE was originally *Feltergail* (13th century), the lane of the felt-workers, and in 1282 it comprised ten tofts on which husgable was paid (YCA, C.60). On the N. side were the backs of the Micklegate tofts, so that the houses of Fetter Lane were mostly S. of the street. Properties in the lane fell into decay and orders were given to rebuild two of them in 1587 (YCA, B.29, f. 108v.). By the early 19th century Fetter Lane contained a few small houses and the workshops of minor craftsmen.

HOLGATE ROAD was formerly Holgate Lane, leading W. from Blossom Street to the hamlet of Holgate and there dividing to become the roads to Acomb and Wetherby, and to Poppleton and Knaresborough. The hamlet of Holgate, beyond the Bridge (21), contained only about half-a-dozen houses until the 19th century. Apart from a few small cottages on Corporation land at the entrance to the lane, no buildings seem to have been erected until after 1823. Lindley Murray, the famous grammarian who occupied Holgate House (52) from 1786 to 1826, 'being unable to walk himself... contributed largely towards forming and keeping up a walk by the side of the road' and 'a seat, on which to rest the weary traveller, was put up by the side of this walk, entirely at Mr. Murray's own expense' (*Memoirs of the Life and Writings of Lindley Murray*, ed. Elizabeth Frank (1826), 221 n.). By 1850 a considerable amount of development had taken place to E. of the railway and on the N. side of Holgate Hill (*see* p. 125–6).

(52) HOLGATE HOUSE, No. 163 (Plate 160), was built, apparently as a speculation, by Edward Matterson, plumber and glazier, who had acquired the site in 1770. He disposed of the 'new erected messuage... with two gardens and stables and outbuildings with the back room called the Garden House' to John Iveson a dealer, who went bankrupt in 1783 and the property was sold to George Dawson, R.N. The latter intended to retire there, but on receiving the command of the frigate *Phaeton* in 1785 he sailed for the Mediterranean and the house was sold to William Tuke, acting on behalf of his fellow Quaker, the American lawyer and grammarian Lindley Murray, who had been recommended to settle near York for the sake of his health. (YCA, Acomb Court Rolls; *Memoirs of... Lindley Murray*, ed. E. Frank). Murray lived in the house for over 39 years and died there on 16 January 1826. In 1859 the Backhouse family, proprietors of the famous nurseries, moved here from No. 92 Micklegate (83); James Backhouse (1794–1869), founder of the firm, died here. Later occupiers were W. W. Morrell from 1882 and the Pressly family in 1912–22. Finally the property was acquired by the Railway and it is now (1970) British Transport Police Headquarters.

The original house, consisting of the central block, by the 19th century already had single storey additions to E. and W.; there was a portico but no porch (engraving by T. Sutherland after H. Cave). Later in the first half of the 19th century, the E. annexe was replaced by a two-storey wing, of larger bricks but still utilising part of the E. wall of the earlier addition; the W. addition was extended to W., a second storey added, and a small one-storey annexe built against its W. side. In the late 19th century the N. porch was built, the hall paved, and one of the bay windows on the S. front very much enlarged. A modern storey has been added to the W. annexe, and there have been internal changes.

The N. front has an 18th-century doorcase reset at the

entrance to the later porch and there are bay windows to the ground floor only. On the S. side bay windows are carried up through three storeys on each side of a modern porch and of upper windows of one large light flanked by narrow side lights.

Inside, original fittings include the staircase, with turned balusters and fluted newel, and on the first floor two fireplaces with pilastered surrounds (Plate 75). An 18th-century doorway (Plate 67) is reused in a later partition. Some renovation was carried out in the early 19th century and most of the other fittings on the first floor belong to this period.

An original *Stable*, W. of the house, is of two storeys under a pantiled roof and has bull's-eye windows in the S. front. A *Summer House* in the Classical style, presumably the 'Garden Room' of 1774, formerly stood in the garden but has been removed to the Mount School in Dalton Terrace (Plate 56).

(53) HOUSE, Nos. 167, 167A, was built in the second half of the 18th century on a nearly symmetrical plan. Minor changes were made early in the 19th century and later in that century two large bay windows were added to the front. In modern times the house has been divided into two.

The front, symmetrically designed in red brick, has two large bay windows, and a porch behind which the original entrance has fluted pilasters supporting a frieze with delicate swags and festoons beneath a moulded and dentilled pediment. Above are five original sash windows with flat arches, and a moulded and modillioned eaves cornice.

Internally the fittings of the entrance hall and the staircase are original. Most of the rooms have been refitted with fireplaces and other details of the early 19th century.

MICKLEGATE (Plates 153, 162; Figs. opp. pp. 69, 94), the great main street of York S.W. of the river, existed on its present alignment before the Conquest, and its name is evidenced from the mid 12th century. Micklegate was of unique importance since it formed the only way between the Bar and Ouse Bridge and was thus an integral part of the old main route between London and Scotland. Its mercantile importance, close to the dockside formed by North Street and Skeldergate, was great, and from the Middle Ages until modern times it has housed a substantial proportion of the greater citizens of York. In it too were placed the town mansions of noble and gentle county families, mostly those from the Ainsty and the West Riding.

In 1282 husgable was paid on at least one hundred and eighteen tofts along the street (YCA, C.60; part of the entry is missing), implying that the whole of the frontages had been built up by then. From the 13th century until the opening of the 19th the mixed character of this great thoroughfare continued. Mansions of the nobility stood next to the houses of merchants and citizens and to the shops of artisans. In the Georgian century, 1720–1820, most of the properties were either rebuilt or refronted in the prevailing style, and this major capital outlay applied not only to the great mansions but also to the quite small buildings which comprised a good part of the frontages. In spite of the stagnation of York trade, much complained of throughout the 18th century, Micklegate was able to retain its prosperity. In it, in front of St. Martin's Church, stood the Butter Market, the staple for the north of England. In the same neighbourhood vintners and bacon-factors congregated, while higher up the hill lived the members of distinguished families who visited the city for its annual 'season'.

Soon after 1800 the proportions of professional premises and of shops increased, as the residential aspect of the street declined. At this period a number of occupiers can be traced as moving out from Micklegate to suburban houses in Blossom Street and The Mount, without the Bar. The mixture of professional, commercial and official buildings, along with a number of hotels and inns, has ever since remained characteristic; characteristic too is the flanking position of its three churches, Holy Trinity, St. Martin, and St. John the Evangelist, which have occupied the same sites since the Conquest and perhaps earlier.

(54) HOUSE, Nos. 2, 4, 6, occupies the plot of land given before 1189 by Erneis de Mykelgate to St. Peter's Hospital (*EYC*, I, 176–7). The present building was originally a large timber-framed, L-shaped structure, jettied towards Micklegate and with a N.–S. range at the back, against the E. end (late 16th-century). In the early 18th century the S. part of the E. wall was cased in brick, and in Nos. 4, 6 an impressive staircase was inserted and some good doorcases and bolection-moulded panelling were fitted. The main front to Micklegate was rebuilt *c*. 1840, the jetties removed and the house heightened; in the later 19th century the W. side of the back range was cased in brick. In 1759 the property was acquired by Alderman John Wakefield, Lord Mayor in 1766, and was later known as Wakefield Court (YCA, E.94, f. 25v.). In 1831 the property was bought by Joseph Shilleto, a butcher, and No. 2 became his shop, while the entry between Nos. 2 and 4 took the name of 'Shilleto's Yard', though officially described as St. John's Court.

The main front, of *c*. 1840, is of large red bricks, containing modern shop fronts on the ground floor. The window sills are of stone. The first floor is of six bays with peculiar spacing: to W. of the first sash window is a window recess and then four more windows, unevenly spaced, each of the foregoing having a very deep and markedly splayed flat arch; the sills are continuous above a band. The second floor has a similar disposition, but the windows are not so high and have separate sills. The third floor has or had a range of six small windows, with heads formed by the moulded cornice above, which has heavy rectangular modillions. In the slate roof above it are three pedimental-headed dormers. The wall facing E. on St. John's churchyard is of 18th-century brick but is still jettied above the ground floor. It has a 19th-century heightening. At the N. end of the range behind No. 2 is a patched timber-framed gable. The main block, in general of 18th to 19th-century brick, consists of two sections, each with a gable and of four storeys. In the E. section is a through passage and above the second floor is one exposed timber, probably a wall plate. The W. section bears the scar of a three-storey wing, now demolished. There is an 18th-century doorway on the ground floor, and blocked doorways occur on the first and second floors.

Inside No. 2 the staircase of *c.* 1840 has a simple moulded rail, turned newels and square balusters. The S. room on the first floor has early 17th-century oak panelling with a carved fluted frieze on three walls, though that on the E. wall is partly modern; the S. wall is plastered and has windows of the early 19th century, when this wall was rebuilt. The central fireplace in the N. wall has an early 18th-century bolection-moulded surround and an arcaded overmantel of *c.* 1600 with carved decoration and simple fluted demi-columns and pilasters; the fluted frieze may well be a modern reproduction. The room behind in the projecting wing has an exposed transverse beam, supported by a post; a spine beam joins the N. wall plate and the transverse beam. On the second floor, a large room to the N. has cased posts with enlarged heads at each end of the E. wall. In the S.W. room at the N. end of the W. wall is a post with enlarged head and, at the top of the wall, a wall plate. There are some simple original roof trusses with collar and principals, all cased; the jettied front to Micklegate has been cut back and hipped, and elsewhere the roof has been altered and heightened in the mid 19th century.

The house Nos. 4 and 6, of rectangular plan, has a very thick internal wall. In general there are two rooms to the front on each floor, and behind the chord wall are the staircase and landing. The ground floor has been gutted but the N.W. corner-post of the timber-framed building remains in battered condition. A doorway to the rear has a bolection-moulded and eared surround and heavily moulded cornice. The door has eight fielded panels and opens outwards; to E. of it is a round-headed window with heavy ovolo-moulded glazing bars. The cellar has stone walls. The early 18th-century bolection-moulded panelling from the front first-floor rooms has been taken to Skelton Manor, near York.

The staircase (Plate 161, Fig. 17e) rises from ground floor to attics and has closed strings, with mouldings top and bottom, turned balusters with square knops and a swept handrail and matching dado. Up to the second-floor landing, the baluster stems are twisted; above, they are plain; all have similar turned elements below the knop. On the first-floor landing is a large surround to a doorway, now blocked, which led to a vanished rear wing, and paired round-headed doorways with keyblocks under a cornice leading to the front rooms.

On the second floor, in the E. room, some timber framing remains visible. Under a wall plate, which disappears into the later front wall, is a series of studs, and at the N. end a curved strut from a corner-post down to the lower rail; in two places on the wall plate is a carpenter's mark. In the middle of the wall is another strut, from a stud, dropping below floor level; it bears the numbers v and iv, corresponding to studs abutting it. There are two marks on the strut. The numbering suggests that the jetty may have been about 4 ft. beyond the present front. On removal of panelling in the first-floor room below, various studs were visible, tenoned into a heavy rail. The N. strut and a stud which ran into it were both numbered iii. The next stud S. was numbered iiii. S. of this were two pegs, probably for a brace and, after a long gap, studs marked respectively ii, vii, iiiv. The last stud was cut by a strut, numbered iiiv, which had a stud coming down to its end. Only the struts and stud ii were pegged. *Demolished 1964, some parts of the fittings were removed to the Castle Museum, York.*

(55) Houses, Nos. 3, 5 and the Queen's Hotel (7, 9), were built by Mr. Henry Thompson (1677–1760) of Kirby Hall and Alderman Richard Thompson (d. 1753), the latter being Lord Mayor in 1708 and 1721; the family were wine merchants (Davies, 187). The building was completed by 1727, when it is shown on Cossins's plan (Fig. 49), and is described as 'new built' in 1736 (Drake,

Fig. 49. (55) From John Cossins's Map of York, *c.* 1727.

280). It is probably on the site of the 'stone house' of Roger de Cnarresburg which abutted on the corner property given to Fountains Abbey early in the 13th century (W. T. Lancaster (ed.), *Abstracts of the Charters . . . in the Chartulary of . . . Fountains* (Leeds, 1915), i, 277), for mediaeval masonry is still incorporated in a wing behind Nos. 3, 5. Evidence exists of another earlier structure (possibly late 16th-century) incorporated at the back of the Queen's Hotel, and a wing at the rear of Nos. 3, 5 is late 17th-century. Nos. 3, 5 remained in the Thompson family until 1788, when this part was sold to William Wallis, grocer, already in occupation (YCA, E.95, f. 65). After being a grocer's residence, it became a grocer's shop by 1834 (Directories). In 1830–45 Nos. 7, 9 were occupied by Miss Alicia Rawdon (YCA, E.98, f. 114; Rate Books; Directories). In the E. house (Nos. 3, 5), two shops were inserted in the ground floor, the first and second floors were turned into flats, and the main staircase was removed (Fig. 50). The W. house (Nos. 7, 9) was converted into the Queen's Hotel in *c.* 1845; considerable alterations were made on the ground floor and the staircase was removed. Later a wing was added to the rear. Despite alterations leaving nothing of the original plan of either house on the ground floor, the first-floor suite of rooms of the Queen's Hotel is an outstanding example of early 18th-century craftsmanship of high quality.

The *North Front* to the street (Plate 163) retains on the ground floor only the two original doorways, not in their proper positions; above, it is in good quality original brickwork. The bold brick string below the cornice is moulded and has a plain band above it. The W. half of the S. elevation is partly covered by a modern block. The three-storey wing behind Nos. 3, 5, of the late 17th century with some 19th-century alterations and rebuilding, incorporates a mediaeval stone wall (possibly Norman) in the W. wall; the E. wall is in good quality brick and, architecturally, of four bays with a heavy oak modillioned cornice. Between the windows are sunk panels, and all openings have flat arches of single gauged rubbed bricks; the four ground-floor openings have been more or less altered and in part blocked, but straight joints below the northernmost

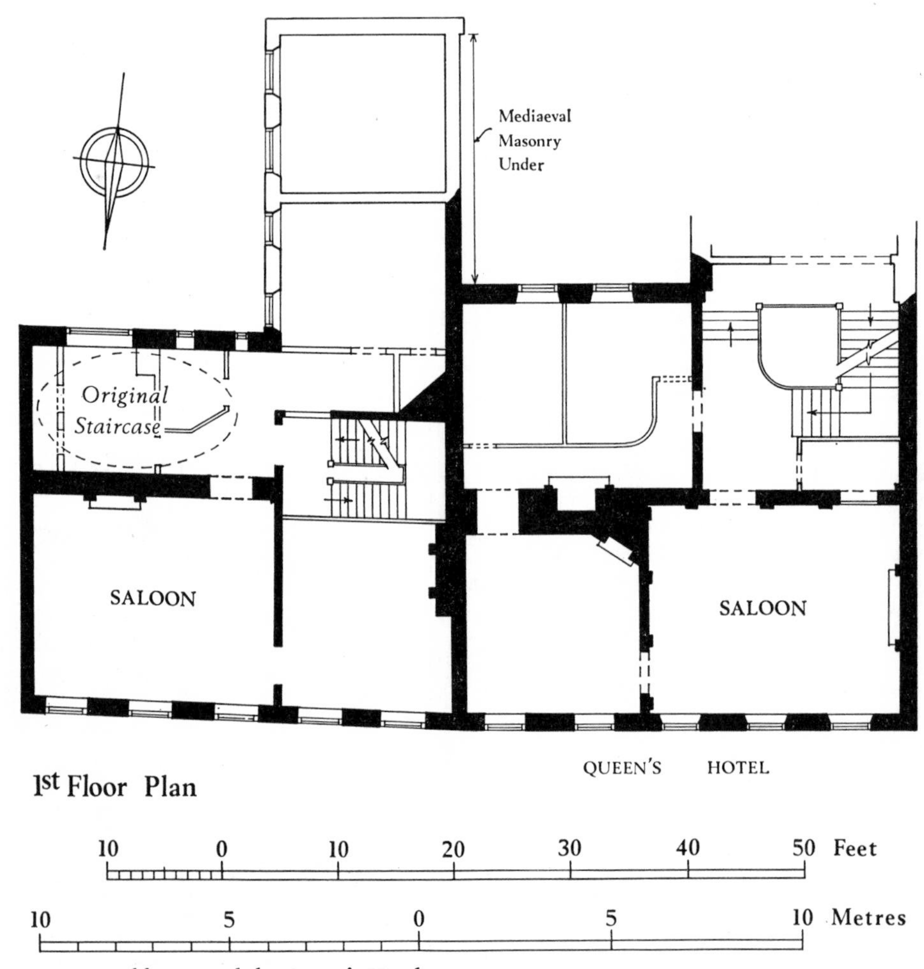

Fig. 50. (55) Nos. 3, 5 Micklegate and the Queen's Hotel.

(13) THE BAR CONVENT, BLOSSOM STREET. Central feature of main front, 1786–9.

Main front, 1786–9, and School Rooms, 1844–5.

From E.

(13) THE BAR CONVENT.

Dome.

From W.
(13) THE BAR CONVENT. Chapel, 1766–9.

St. Margaret.

St. Michael.

St. Sebastian.

(13) THE BAR CONVENT. Figures in Chapel, 18th century.

(20) OUSE BRIDGE. St. William's Chapel. 'Flight into Egypt', 13th century.

19) ST. CLEMENT'S PRIORY. Virgin and Child, 14th century.

(20) OUSE BRIDGE. 1810–20.

Design by Peter Atkinson junior, 1810 (Messrs. Brierley, Leckenby and Keighley).

From S.W., 1810–20.

(20) OUSE BRIDGE.

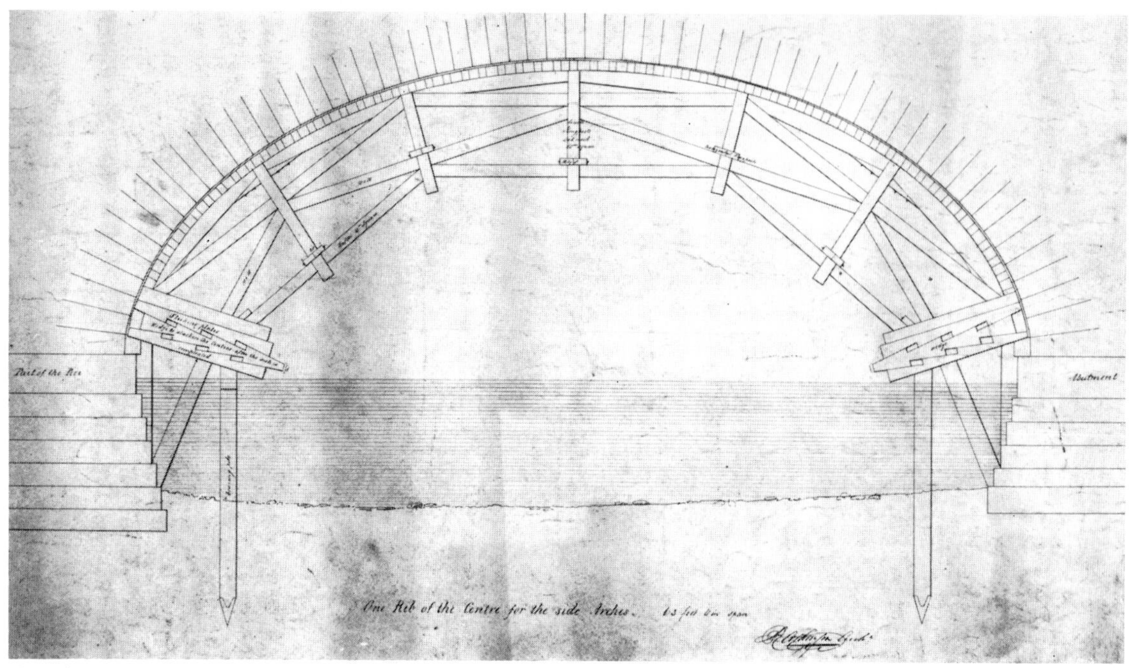

Centring.

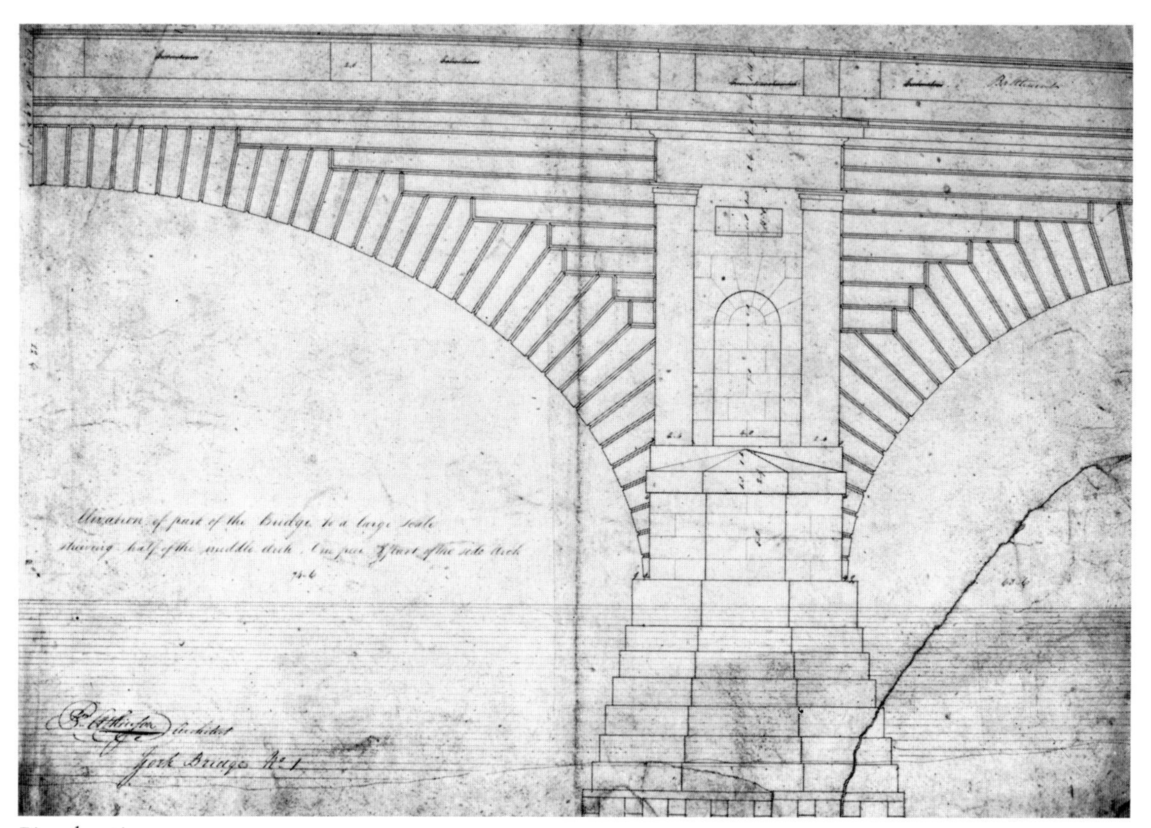

Pier elevation.

(20) OUSE BRIDGE. Drawings by Peter Atkinson, junior, 1810 (Messrs. Brierley, Leckenby and Keighley).

From W.

From E.

(20) OUSE BRIDGE. St. William's Chapel. Drawings by Henry Cave, during demolition, 1810 (York City Art Gallery).

(20) OUSE BRIDGE. St. William's Chapel. Clock and inscription, 1658.

First Class Refreshment Room from W.

Design for main front by G. T. Andrews.

(28) OLD RAILWAY STATION. 1840.

28) OLD RAILWAY STATION, from E., 1840 and later.

Train Shed, S.W. extension, 1845.

S.W. of Arrival Platform, 1841.

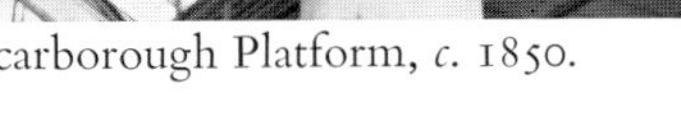

Scarborough Platform, *c.* 1850.

Scarborough Platform, 1845/6.

Original Train Shed, 1841.

Scarborough Platform, *c.* 1850.

(28) OLD RAILWAY STATION. Details of cast-iron columns.

Scarborough Platform roof, *c.* 1850.

Entrance gate, *c.* 1850.

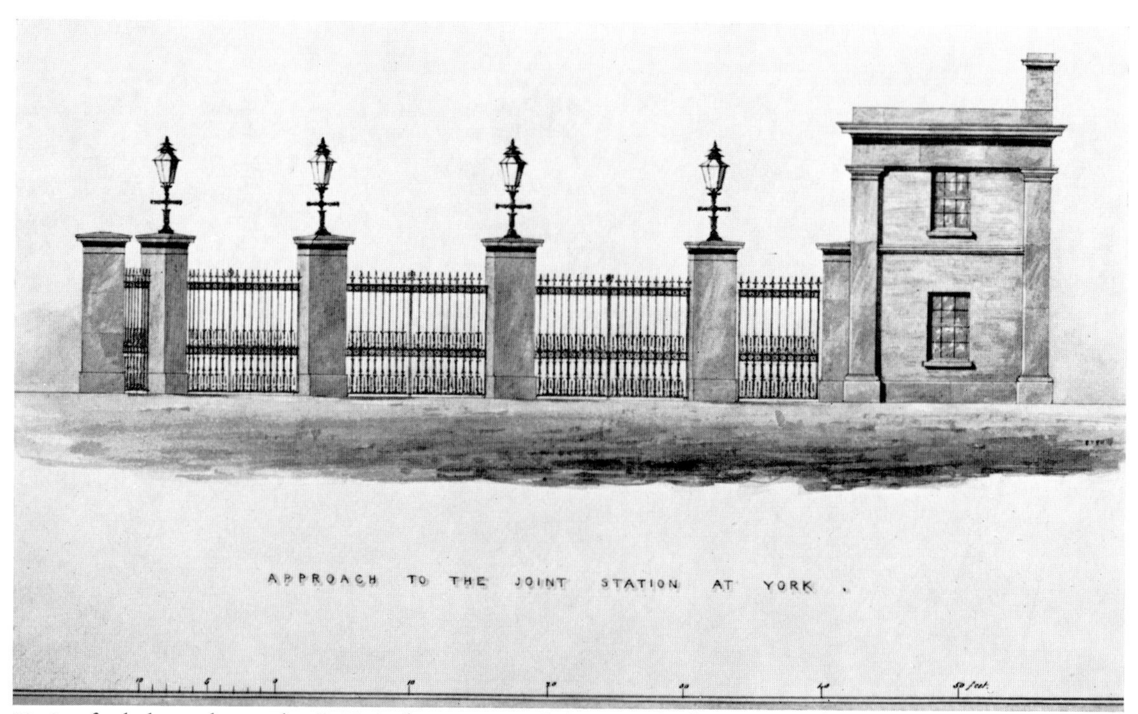

Design for lodge and gates, by G. T. Andrews, *c.* 1850 (Victoria and Albert Museum).

(28) OLD RAILWAY STATION.

(13) THE BAR CONVENT. Inner face of clock tower, 1786–9.

(30) HOLGATE WINDMILL, before removal of sails. *c.* 1790.

(63) MICKLEGATE. House, No. 37. Entrance to rear wing, early 19th century.

(24) MIDDLETON'S HOSPITAL, SKELDERGATE. Detail of front elevation, 1827–9.

S. elevation.

From S.W.

(29) THE OLD WAREHOUSE, SKELDERGATE. 17th century.

suggest that it was originally an entrance. On the first floor are four flush-framed sash windows and on the second floor three sash windows and a casement to S. of uncertain date.

Inside, the original plan of the *East House* can be traced on the first floor: to the E. was the saloon with a great staircase behind it which has been removed; to the W. were two smaller rooms with the earlier wing behind. The front room to the W. has since been shortened to make space for a modern staircase. The saloon has in the W. wall a bolection-moulded doorcase and, in the S. wall, a mid 19th-century white marble fireplace. The plaster ceiling has a heavily moulded and enriched cornice and a central cast-iron decorative roundel for a gas fitting. In the truncated N.W. room is a pine fireplace of *c.* 1780 with pewter and cast-lead enrichments (Plate 61), moulded, dark-grey veined marble slips and a contemporary cast-iron hob grate by Carron (Plate 74). The late 18th-century plaster cornice is enriched with motifs similar in style to those on the fireplace. The plaster ceiling which survives over the site of the former staircase is domical above an elliptical moulded plaster cornice contained within a rectangle (Plate 165). The enrichments of the cornice include, at the cardinal points, the Thompson crest of a mailed arm holding a dagger. In the S. wing, oak beams of square section are exposed in the ground-floor S. room.

In the *Queen's Hotel*, an early 18th-century corner cupboard (Plate 69) survives on the ground floor, now reset in the modern S. wing. The first floor of the hotel comprises a fine suite of three rooms, with most of the original fittings intact. The N.W. room, the *Saloon* (Plate 165), probably the best example of the period 1700–20 to be found in York, has an architectural decorative treatment comprising the Corinthian order with a boldly enriched modillion cornice and framing bolection-moulded panels (Fig. 16a). In the N. wall are three tall sash windows with late 18th-century glazing and bolection-moulded architraves with carved acanthus leaf enrichment, terminating at sill level and resting on fluted pilasters. The E. wall is divided into three bays by pilasters and has a doorway with a bolection-moulded architrave. The S. wall, in four bays, has the principal doorway flanked by pilasters; the door is similar to that in the E. wall; the W. panel was originally hinged to open into a butler's service closet. In the W. wall is a central chimney-breast with the fireplace flanked by pilasters (Plate 164). The ceiling of the room is coved above the cornice. The *N.E. Room* is almost square and in part lined with bolection-moulded panelling. There is a heavy moulded dado rail, panelled dado and moulded skirting, and the windows have bolection-moulded architraves with pedestal bases. In the W. wall is a doorway (Plate 166) with an overdoor containing an oil painting of *c.* 1620 depicting a pastoral scene. In the S.W. angle is a fireplace (Plate 166) surmounted by a rectangular panel intended to house a painting. The *S.E. Room* has been sub-divided but the original cornice, panelling and other fittings remain. The fireplace in the N. wall has over the blocked opening a horizontal foliated panel with raised moulded surround, a moulded cornice breaking out over the fireplace, and a rectangular panel above.

The *Staircase* has been rebuilt, but the two doorways on the first-floor landing are original (Plate 166); they are remarkable for their width of moulding and the acanthus carvings. On the second floor, the N.W. room contains reused 17th-century panelling with a frieze of carved formalised foliage. Among other early 18th-century fittings the fireplace in the W. wall is flanked by plain pilasters with superimposed small pilasters with sunk panels; the breast above the fireplace opening is simply panelled. The N.E. room is panelled throughout (Plate 69); the panelling on the E. wall is probably modern. A room to the S.E., eight steps lower than the front rooms, has a plain plaster cornice and contains a simple late 18th-century Carron fire-grate.

(56) HOUSE, Nos. 8, 10, 12, was possibly the new house built by Alderman William Brearey, who died 1637 (NRCRO, ZBM 183, 627, 629). About 1700 it was extended to the N. on the evidence of fittings of that period within the extension. In the early 19th century the house was occupied by the Rev. Thomas Lund, son of John Lund the surveyor, and rector of Barton-le-Street 1783–1832, and later by Henry Cave (1779–1836), the artist, who died here after moving across from No. 13 (57). At this time the owner was Mrs. Custobadie. The property had been divided by 1830 and a second staircase built in the W. part.

The main front, originally set back, was rebuilt on the street frontage *c.* 1860 (*cf.* OS 1852; views by Whittock 1858–63). The ground floor has a through passage, with a contemporary shop to the E.; the shop window to the W. is modern. Above, the wall is of large red precision-cut bricks; the roof is of slate. The N. elevation, of stuccoed brick, is two-storeyed with basements; a small closet wing projects at either side of the elevation. At ground-floor level is a stucco-dressed band of three courses; there are two tall sash windows with flush frames and late 18th-century glazing. At first floor is a band with oversailing course, and two similar sash windows at second floor. The lead rainwater head is bowl-shaped with flutings, and the lead fall pipe has holdfasts bearing opposed fleurs-de-lis. There is a simple timber cornice, probably of 18th-century date. Each wing has an entrance to a porch at ground floor, and one late 18th-century sash at first floor, replacing an earlier one. The basement has two 19th-century three-light sashes to the main elevation, and to each wing, a door and sliding sash opening to the area.

Inside No. 8, a ground-floor room to the N.E. with a very high ceiling contains an angle fireplace with plain surround and stone mantelshelf fitted with an early 19th-century range by Bowsfield of York. A stair hall to S. of this room has a tall round-headed opening to the shop in the S. wall. The *Staircase* of *c.* 1700 (Plate 86) has a moulded rail, closed string, square newels with attached half-balusters on the faces, and turned bulbous balusters. On the first floor, a room to the N. has a moulded cornice of *c.* 1700 and panelling from floor to ceiling, in seven heights, on the E. and part of the S. walls; on the W. wall is a dado of similar panelling. A fireplace in the S.E. angle has a bolection-moulded surround with square stone slips and a round-headed grate of *c.* 1830–40. In the N. wall, a doorway with moulded architrave leads to a small room at the N.E.

corner which has a moulded cornice like that of the main room and, in the N. wall, an early 18th-century window with heavy ovolo-moulded glazing bars. The roof, which runs E. to W., is of common rafter construction with collars and single purlins to each side.

Inside No. 12, the ground-floor room to the N.W. has an original doorway in the S. wall opening to the staircase passage, a similar doorway in the N. wall to the porch entrance, a large sash window with bolection-moulded architrave in the centre of the N. wall, and a simple plaster cornice. The room at the front has been converted to a shop. Between these rooms are a stair well and a passage. The *Staircase*, of *c.* 1830, has a swept handrail and simple turned softwood balusters. The flight to the basement has stone steps, with cast-iron square-section balusters. On the second floor, the rear room contains reset run-through panelling, in part ill-assorted, and a timber cornice. In the centre of the N. wall is a sash window with a refitted raised architrave of *c.* 1700. The central fireplace in the W. wall is original, but contains an early Victorian grate and surround; over the fireplace is a bolection-moulded panel. *Demolished 1964*.

(57) HOUSES, Nos. 11, 13, were built as a single structure *c.* 1740. The building was in two moieties, of which that to E. (No. 11) was bought in 1830, from the heirs of John Green, plane-maker, by Robert Gray, builder, for £745; it had recently been occupied by Thomas Rayson senior, builder, who had moved to No. 16 South Parade (*see* p. 129). The adjacent No. 13, which had earlier been occupied by John Batty, drawing master, was for many years the home of Henry Cave the artist, who soon after 1830 moved across to No. 12 (*see* (56). YCA, E.98, f. 114; Directories).

The brick front (Plate 53) has 19th-century shopfronts inserted at ground floor and two angular bay windows, probably of *c.* 1830–40 added above. On the second floor two windows retain original sashes. The two houses were identical in plan, with open staircases placed transversely between front and back rooms.

(58) HOUSES, Nos. 16, 18, have been formed from an important timber-framed structure erected in the late 16th or early 17th century with three full storeys and semi-attics above, fully jettied to N. and S. There were few contemporary buildings in York of such large size. An abstract of title (York Co-operative Society, held by the Co-operative Bank, Leeds) shows that in 1565 the property comprised 'messuages and a garden' occupied by Margaret Catton, widow, when the freehold was sold by William Harrynton to William Winterburne, armourer. Winterburne died in 1586/7 and by 1602 his widow, who had remarried, sold the house to William Cowper of York, innholder, and his wife Rosamond. Cowper, free of York in 1561, was already the occupier, and when the freehold passed to Thomas Herbert, in 1629, another innholder Lancelot Geldart (free, 1621) was in occupation. Presumably the house had been built as a major inn under a building lease, perhaps *c.* 1590. The freehold descended in the Herbert family until 1711, when it passed to Richard Reynolds, who in 1727 sold it to James Robinson, an apothecary. The property was for sale in 1764 (*York Courant*, 13 March), when part was occupied by Mrs. Robinson and under-tenants, and part vacant. At this time the upper part of the front was cut back and a brick façade built, the property divided into two parts, each fitted with its own staircase, and new chimney-breasts inserted. This work is dated to 1764 by a rain-water head bearing the crest of Walker, *a greyhound sejant, collared*, referring to the family of the Rev. John Walker, rector of St. Denys 1797–1813, who died in No. 16 on 25 August 1813 at an advanced age (*York Courant*, 30 August). The freehold in 1801 passed to William Cobb (YCA, E.95, f. 247v.) who had already acquired No. 18, which was occupied for a time early in the 19th century by Peter Atkinson junior, the architect (YCA, E.96, f. 201v.; E.97, f. 185v.). No. 16 was bought in 1814 by William Price, grocer, who occupied it for a time; later occupiers were Amos Coates, surgeon, Sheriff of York 1833–4 (YCA, E.98, f. 169v.) and Henry Keyworth, surgeon (YCL, St. John's Rate Books; Directories). In 1824 the adjacent Pack Horse Inn, to W., took over No. 18 to extend its premises (YCA, E.97, f. 185v.). During the 19th century the front of No. 18 was stuccoed and the upper parts of both houses covered with Roman cement.

On the front to Micklegate (Plate 52) No. 16 is faced with 18th-century brickwork above a later shopfront. No. 18 has been covered with early 19th-century stucco. The back wall (Plate 52) retains the jetties of the second and attic floors, and has hung-sash windows mostly of the 18th century. The remodelling of *c.* 1764 formed two houses each with a staircase (Plate 86) placed transversely between front and back rooms, the staircase probably occupying the site of original chimneys. One of the first-floor rooms has an original enriched plaster ceiling (Plate 51) curtailed when the front was cut back. Other fittings are of the 18th and early 19th centuries. In the roofs original collar-beam trusses remain with principals rising from short sole-pieces like the stub ends of tie-beams (*see* p. lxxiv and Fig. 13j). *Demolished 1964*.

(59) HOUSE, Nos. 17, 19, 21, was originally timber-framed, of three storeys with two jetties to the street and none to the back. It is of the late 15th century, and may have been of double width with a second gabled roof to the back. Parts of the timber-framed structure remain *in situ*, some being exposed on the second floor with the infilling between the timbers placed against pegs. The roof, despite later alterations, gives enough evidence for a reconstruction of the mediaeval framing.

The floors are all original and have rough-chamfered joists.

In the 16th century an annexe was added at the back, the panels of the framing being filled with brick on edge with the mortar let into grooves cut in the timbers. Behind No. 21 a wing was added *c.* 1600, now of four bays but originally longer; it has semi-attics with principal rafters rising from short sole-pieces instead of tie-beams to allow uninterrupted access to the whole floor (*cf.* Nos. 16, 18, 111 Micklegate, Fig. 13i p. lxxiii). Remodelling of the original house in the same period included the construction of an enriched plaster ceiling on the first floor of No. 17, with panels containing fleurs-de-lis. There are also remains of early 17th-century panelling. In the first half of the 18th century a staircase (Plate 85; Fig. 18k) was inserted in the 16th-century annexe and the houses were given a new front in one vertical plane, some windows retaining the sashes of this date; at the same time the roof pitch was lowered. Early in the 19th century, No. 21 was refaced in stucco and refenestrated, moulded stucco architraves being applied and a bay window added. Shop fronts were inserted and re-arrangements made at ground floor in the late 19th century. In 1792 No. 17 belonged to James Hilton, limner (d. 1814), and Thomas Hilton, painter (d. 1793) (YCA, E.95, f. 119), and passed to Thomas's widow, Mrs. Mary Hilton (d. 1833); the painters' business was carried on as Messrs. Beadle and Perfect. By 1828 the house was occupied by Joseph Perfect & Sons, painters; and by Henry Perfect, painter and paperhanger, from 1837 until 1889, when it became a butcher's shop (YCL, St. John's Rate Books; Directories). No. 21 became a grocer's shop early in the 19th century, and later a restaurant.

The ground floor of No. 17 has been greatly altered by conversion to business premises. A small room to E. is painted throughout (Plate 167), probably the work of the craftsmen painters who occupied the premises *c.* 1779–1889; the work is probably of the late 19th century, but is of two dates, the panels being outlined with *papier-mâché* borders overlying painted arabesques and the end panels having painted paper over earlier bolder painting.

The late 15th-century roof, of oak, has been modified and its slope reduced (Plate 50). The crown posts, originally carrying a collar-purlin and collar, have been converted to king posts by removing the collars and utilising the collar-purlin as a ridge. The main cross braces are retained, though one of each pair has been shortened to carry the side-purlins. The raking struts affixed to the purlins and the foot of the brace (supporting the purlins) are reused, being secured by nails, but the mortices for their original housings are still visible in the braces (Fig. 13d). *No. 17 demolished 1966.*

(60) CROWN HOTEL and HOUSE, Nos. 23 and 25, stand on a plot held in 1282 by Roger Basy of the Master of St. Robert (of Knaresborough), paying 1*d.* husgable (YCA, C.60, m.5/23; *YASRS*, LXXXIII (1932), 181–94). The property in the 14th and 15th centuries descended to the Knottyngley and Wenteworth families (YCA, B/Y, ff. 62–62v.). There was an inn on the site by 1733 (Benson, III, 164) and early in the 19th century this had become The Grapes. Among former owners was Thomas Varley (1693–1771), Sheriff of York 1766–7. The W. end of the original building, No. 25, had been divided from the rest shortly before 1830, when it was mortgaged by Richard Dent, a miller, who sold to Charles Robinson, druggist, owner of the adjacent property, for £512 in 1832 (YCA, E.95, f. 84v.; E.98, ff. 107, 143, 165v.).

The building, of three storeys, is of the early 19th century and incorporates part of a late 17th-century brick structure at the rear. The larger section, to E., appears to be of *c.* 1825, and includes the earlier structure; the smaller lower part to the W. is slightly later in date. About 1850–60, in an attempt to give a symmetrical appearance, the whole front elevation was redesigned with a stucco rendering, a central decorative feature rising the full height, and remodelled windows. Later, shop fronts were inserted to the ground floor. The interior has been much altered but some of the early 19th-century fittings survive.

(61) ADELPHI HOTEL, Nos. 26, 28, comprises two separate buildings: one facing Micklegate, No. 28, known as 'The Micklegate', and the main structure along Railway Street. They have little historical or architectural merit and, in recent years, have been considerably altered. Behind No. 28 is a separate building (No. 32) of earlier date, now derelict.

The buildings formed part of a large block of property owned by the Benson family and extending N. to Tanner Row. The main block of the Adelphi Hotel represents a house occupied by George Cook, butcher, later by Mrs. Storre or Torre, in 1793 by Mrs. Henrietta Leedes (YCA, E.95, f. 132), and *c.* 1800 by Mr. Ralph Robinson (E.97, f. 87). In 1819 it was sold to Thomas Cattley, raff merchant (YCA, E.97, f. 87). No. 28, formerly purchased from George Benson, stapler (Lord Mayor, 1738) was in 1749 in the possession of John Malton, goldsmith, and his wife Ann, who had it from her parents, James and Elizabeth Mason. A moiety of the property was conveyed to trustees and the whole described (E.93, ff. 217, 231) as 'a Fore Part late in the occupation of B. Bradley, surgeon and Richard Wilkinson; and a Back Part with two low rooms or kitchens with Chamber over them and a Turf Chamber on the right hand side of entry, formerly enjoyed by Mr. Christopher Easby'. By 1810 the whole property belonged to the Rev. Robert Benson, grandson of George Benson, who sold it to John Nicholson, yeoman, and Robert and Ambrose Gray, bricklayers (YCA, E.96, f. 131). They immediately rebuilt the front part, as an advertisement of 13 August 1810 (*York Courant*) describes the house as 'modern built'; and by February 1811 (E.96, f. 154) the Grays conveyed their two-thirds of the property to Nicholson, who in 1815 sold the whole to Henry Henwood, occupier of the 'lately rebuilt' front house (E.96, f. 248v.).

The main hotel block appears to have been reconstructed after 1850, but part of the Micklegate elevation makes use of a mid 18th-century moulded and modillioned cornice. No. 28 retains the upper part of its front of 1810 and incorporates in the walls some timbers, possibly remains of a 17th-century structure engulfed in the rebuilding. The front is of painted brick with two sash windows on each floor and a shallow modillioned cornice.

Behind the hotel lies the 'Back Part', of two storeys with attics; it has walls of late 17th-century brick encasing part of a mid 16th-century timber frame and retains parts of original roof trusses. Early in the 17th century a large chimney-stack was built near the middle of the house with a staircase at the side of it (Plate 83); the decoration of roses and thistles on the strings may indicate a date soon after 1603. There are also moulded beams and joists of the early 17th century. New windows were put in in the 18th century and new fireplaces in the early 19th century.

(62) CROMWELL HOUSE, Nos. 27, 29, 31, occupies a site of considerable historical interest. In the 13th century the house belonged successively to William and Thomas Fairfax, and to the latter's son Bego who, in 1280, sold the property to Master William de Muro, called 'de Skeldergate'. By 1290 it belonged to the Staveley family and in 1310 a life interest was sold to John de Hothum, afterwards Bishop of Ely. Hothum and the Staveleys conveyed their interests to Sir Geoffrey le Scrope in 1317–22, when the property consisted of a messuage, three cellars, four cottages and gardens extending back to Fetter Lane. The Scrope family held the freehold for several generations (*YASRS*, LXXXIII (1932), Nos. 526–9, 531, 533–5, 537, 540–4, 546, 552, 554–5). By the 18th century there were two houses on the site, in 1790 occupied by Thomas Prince, merchant, in the W. moiety, and by Thomas Cave, copperplate printer. Between 1800 and 1805 the two tenements had been reunited, and in 1823 the rating assessment was raised (YCA, E.95, f. 84v.; Rate Books; Directories). For a long period the property belonged to Messrs. Theakston, Robinson & Co., druggists.

The present house seems to incorporate remains of an older building, notably a bold moulded dentil cornice and small-paned sash windows to the top stage, but otherwise all visible features belong to a general remodelling of *c.* 1860.

(63) HOUSE, Nos. 35, 37, stands on the site of a mediaeval stone house on which, by 1282, husgable of 2*d.* was paid, indicating that it occupied a double plot (YCA, C.60, m.5/23). The freehold of Hugh de Selby, Mayor of York in 1230, the property passed in 1274 to the Clerevaus (Clervaux) family and in 1336 to Sir Geoffrey le Scrope. In 1275 there were two cellars beneath the hall and a part only, on the W. side, let to the owner of the adjacent property, had a frontage of 25 ft. and a depth of 80 ft. Behind the principal house were subsidiary houses and a garden. (*YASRS*, LXXXIII, Nos. 526–9, 531, 533–7, 540–2, 552, 554–5; YCA, E.20A, ff. 62–62v.). Early in the 16th century the owner was John Beane, Sheriff 1538–9 and Lord Mayor 1545 and 1565, who saved the nearby church of St. Martin from demolition. By the marriage of Beane's daughter Mary in 1554 the property ultimately descended to the family of Wharton and Anthony Wharton (*c.* 1653–1703) probably built part of the existing building. His daughters Mary (d. 1776) and Margaret (d. 1791 aged 94), known as 'Peg Pennyworth', lived there. The front part of the house was built in the early 18th century. By 1812 the house had been acquired by Peter Atkinson junior, the architect, who divided it and himself lived for some 15 years in the larger portion, No. 37, letting No. 35 to John Bellerby. By 1829 the larger part was occupied by Thomas Hands, cabinet-maker, and Atkinson had leased No. 37 to William Hargrove, proprietor and editor of the *York Herald*. This W. moiety was from 1843 the home of the famous surgeon, Sir William Stephenson Clark, Lord Mayor 1839 (d. 1851).

The 17th-century part of the premises forms an L-shaped block set some way back from the street, the wings extending E. and S. In the early 19th century Atkinson added a third storey to part of the S. wing, rearranged the accommodation and constructed a new staircase hall in Regency style to the rear of the 18th-century house.

The street front (Plate 52) is of two storeys and five bays in width. The whole ground floor has been converted to shop premises with modern fronts. At first floor are five tall sash windows under arches of rubbed gauged brick. The eaves have been remodelled with a modern cornice.

At the back the S. elevation of the E. wing of the 17th-century building (Plate 54) had three openings to each of the main floors, and in a gable an attic window framed by brick pilasters. In the E. wall of the S. wing a new entrance (Plate 151) was made in the early 19th century and most of the windows were altered but a small oval window is original (Plate 51). There are moulded brick bands and part of an original brick cornice and parapet.

Inside, two first-floor rooms are lined with panelling, of the 17th and 18th centuries respectively, the latter with bold bolection mouldings. The roof to the 18th-century building is carried on trusses with angled principals of a type usually associated with very boldly projecting eaves.

(64) HOUSES, Nos. 42, 44, 46, 48, standing opposite St. Martin's Church, include a back wing of *c.* 1710 but the main part of the building (Plate 53) was erected as two messuages in 1747, when a Mithraic altar stone was discovered whilst digging a cellar (Stukeley, III, 358; Wellbeloved's *Eboracum*, 79–85; *Gentleman's Magazine*, May 1751). The site had been acquired from Thomas Mell, merchant, by Thruscross Topham (d. 1757), who married Ann Sanderson on 16 May 1747 (YCA, E.94, ff. 1, 2v.); R. H. Skaife, *The Register of Marriages in York Minster* (1874), 115), and the rebuilding was probably in connection with the marriage settlement. In 1774 the property was sold to Thomas England, butter factor, who, until his bankruptcy in 1781, lived in one house; the other had been occupied to his death in 1770 by George Eskrick, haberdasher, Lord Mayor in 1739 and 1747 (YCA, E.94, f. 153). In 1791 the whole property belonged to George Beal, butter factor, including a third house in the yard behind (E.95, ff. 111v., 113v.). For about 10 years from 1823 a girls' boarding school, kept by Miss Patience Nicholson, occupied No. 48.

Above modern shop fronts the street elevation is in 18th-century brick with a 19th-century cornice at the eaves. Only one of the windows retains original sashes with heavy glazing bars.

The back wing of *c.* 1710 retains some original fittings and contains a mid 18th-century staircase (Plate 88). In the main range No. 44, in the middle, has a staircase of the middle of the 19th century. In the W. house, of which the upper part forms No. 48, a room on the first floor is lined with panelling of 1747 (Plate 69; Fig. 16d), and other original fittings remain including a staircase with alternate turned and twisted balusters; the top part of the staircase was altered at the end of the 18th

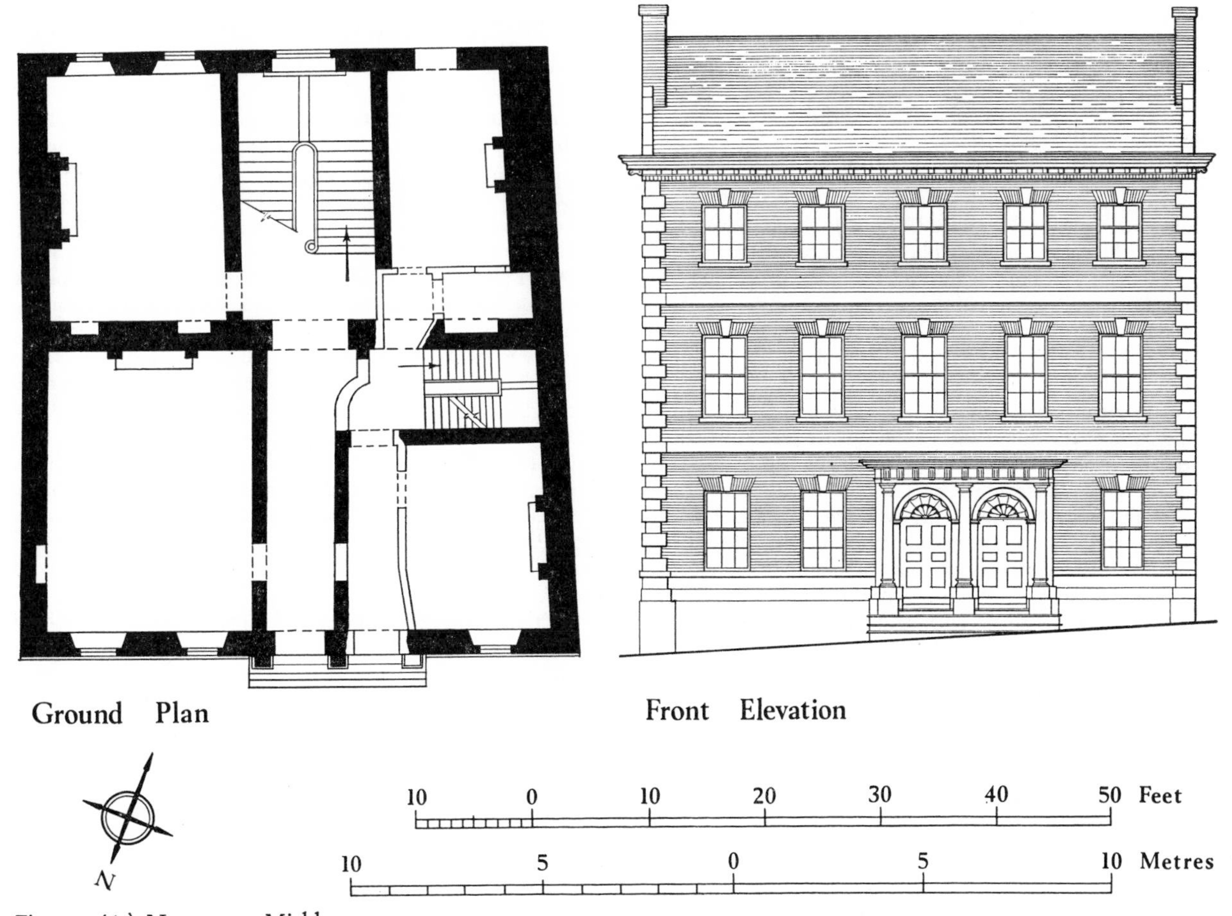

Fig. 51. (65) Nos. 53, 55 Micklegate.

century. Stop-chamfered ceiling beams are of the 17th century reused.

(65) House, Nos. 53, 55, had probably been built shortly before 29 May 1755, when the Corporation required that 'the wall lately built before the house and steps' should be removed as an encroachment (YCA, M.17). It may have been an early design of John Carr, having details similar to those of Micklegate House and Garforth House (also attributed to Carr), as well as Fairfax House, Castlegate, and Castlegate House, known to have been designed by him. Details given by Davies (*Walks*, 168–9) of the home of Lady Dawes and her second husband Paul Beilby-Thompson, refer to this house, which can be identified from abuttals in a deed of 1791 (YCA, E.95, f. 108), but it was not the house 'newly-built' in 1736 (Drake, 280). In 1806 the house 'the late residence of the Countess of Conyngham' (widow of the 1st Earl) was advertised to be sold or let, with stabling for eighteen horses (*Yorks. Gazette*, 20 Jan.). About 1813 (Borthwick Inst., Rate Books of St. Martin-cum-Gregory) the house was divided into two parts and numerous alterations made. Since the middle of the 19th century, No. 53 has been occupied by someone connected with the adjoining wine and spirit merchants' establishment; No. 55 was occupied by offices of the Inland Revenue and other Government bodies from 1853 onwards; after 1905 it became a girls school.

The front elevation (Plates 170, 78; Fig. 51) was symmetrical until the entrance was doubled (Plate 62) when the house was divided. The back, in buff-red brick, with red brick dressings, has later additions built against it. There are projecting brick bands above the windows of the two lower storeys and a heavy moulded cornice at the eaves. The central stair window has a semicircular arch of rubbed brick, key-block, stone imposts and original sashes.

The original central entrance hall leads to the main staircase at the back of the house and to the former servants' staircase placed transversely to the W. and now reached by a passage taken out of the W. front room. The E. front room has decoration of the highest quality; the woodwork is moulded and enriched, the walls are panelled under an elaborate cornice and the fireplace surround has flanking columns and enriched overmantel (Plate 72). The W. front room was refitted in the 19th century with a simple cornice and cupboards with applied composition roundels of female heads. The secondary staircase has closed strings, square newels and turned balusters. The sumptuous main staircase (Plate 171) rises in a stair hall in which the walls have enriched panels and over the window are floral swags and pendants (Plate 90); the ceiling is decorated with rococo plaster-work (Plate 180). On the first floor the door-cases are surmounted by enriched pediments, and pilasters flank the arched opening to a recessed lobby leading to the front rooms (Plate 168). The front rooms were completely refitted at the beginning of the 19th century (Plate 68). One of the back rooms is lined with panelling (Fig. 16e). The attics are floored with gypsum plaster. *Stair hall ceiling renewed 1970.*

(66) Garforth House, No. 54 (Figs. 52, 53; *see* p. xcii), the town house of the Garforth family of Wiganthorpe, was nearing completion in 1757. It is said to have been designed by John Carr (YCL, Knowles, 'York Artists' MS., I, 104), which is probable since Carr remodelled Wiganthorpe Hall for the family (H. M. Colvin, *Dictionary*, 125), and it resembles Fairfax House, Castlegate, designed by Carr. The site in 1732 had been divided into two tenements, occupied by Matthew Rayson, carpenter, free of York 1708 (No. 52, to E.), and Nathaniel Earby junior (No. 54); the freehold was then sold by Benjamin Barstow to William Tesh, a wine-cooper (YCA, E.93, f. 65). By 1736 William Garforth had acquired No. 54, and No. 52 came into the hands of his nephew, the Rev. Edmund Garforth (born Dring), by 1755 (E.93, ff. 90, 197; E.94, ff. 1, 2v.). Building of the house may have started about the time of Edmund Garforth's marriage, in 1750, to Elizabeth, daughter of the Hon. Thomas Willoughby of Birdsall; their son William, High Sheriff of Yorkshire in 1815, died in 1828 without issue. In 1831 the house was for sale (*Yorks. Gazette*, 27 August), and it became the residence of Barnard Hague (Davies, 179). It had at times been let to tenants, notably to Walter Fawkes of Farnley Hall, the occupier in 1788 (E.95, f. 67v.). The house retains an exceptional number of original fireplace surrounds.

The street front (Plates 172, 78) is of red brick with stone dressings and original railings protect the area in front of the basement. The entrance (Plate 63), flanked by original wrought-iron lamp brackets, is in the W. bay and the corresponding window in the E. bay has been altered from a doorway which would have completed the symmetry of the elevation. Wrought-iron balconies to second-floor windows were added in the early 19th century.

The back (Plate 55) is of light-coloured brick with red brick dressings, and is assymmetrically arranged to allow for a large Venetian window lighting the main staircase. A lead rainwater pipe (Plate 81) bears the date 1757 with the initials of Edmund and Elizabeth Garforth and the Garforth crest, a goat's head *couped*.

On the ground floor the entrance hall is paved with white stone and black marble, and has an enriched cornice with modillions; opening to N. is an archway flanked by pilasters, with a glazed fanlight. A large richly decorated room occupying the central part of the front of the house was probably the *Dining Room*; it is lined with panelling in two heights with dado rail and enriched modillion cornice above a frieze decorated with arabesques, shells and a female head. The doorways are pedimented (Plate 66), one to each side of the fireplace (Plate 71) which has a surround of white marble. A small room to E., balancing the entrance hall, was formerly the servants' entrance hall, with a doorway to the street; it has

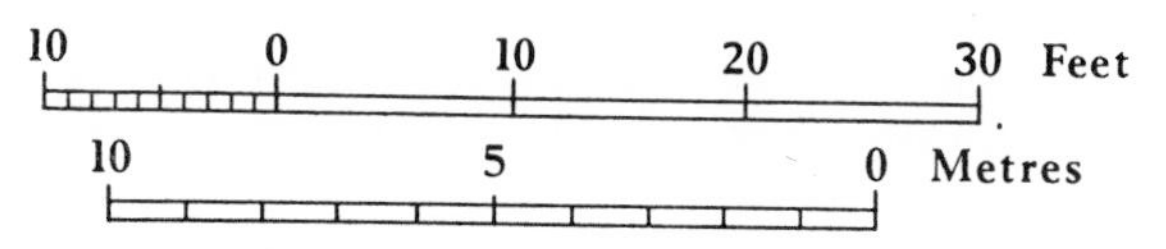

Fig. 52. (66) Garforth House, No. 54 Micklegate.

a late 19th-century fireplace. The larger back room has a shallow cornice and an early 19th-century fireplace with a marble surround; the chimney-breast is flanked by segmental-headed recesses. The main *Staircase* (Plate 88) has cantilevered treads and pine balusters (Fig. 18p), three to a tread, and a mahogany handrail finishing in a volute at the bottom. The *Servants' Staircase* (Plate 88) has turned balusters, two to a tread, and a pine handrail. The doorways have carved and enriched doorcases and doors of six fielded panels.

The *Basement* has vaulted store-rooms flanking the entrance passage from the front area. At the back are three vaulted rooms, the largest of which was formerly the kitchen.

On the *First Floor* the main stair hall has an enriched cornice with dentils and modillions and a decorated panelled ceiling (Plate 169) extending over the landing to the S. The staircase is lit by a Venetian window (Plate 90) set in an arched recess with rococo plasterwork and a cartouche bearing a shield-of-arms of Garforth.

The central passage of the ground floor is repeated giving access to three front rooms with moulded cornices with dentils or modillions, entablatures over the doorways (Plate 67), and original fireplaces, one (to W.) decorated with fine

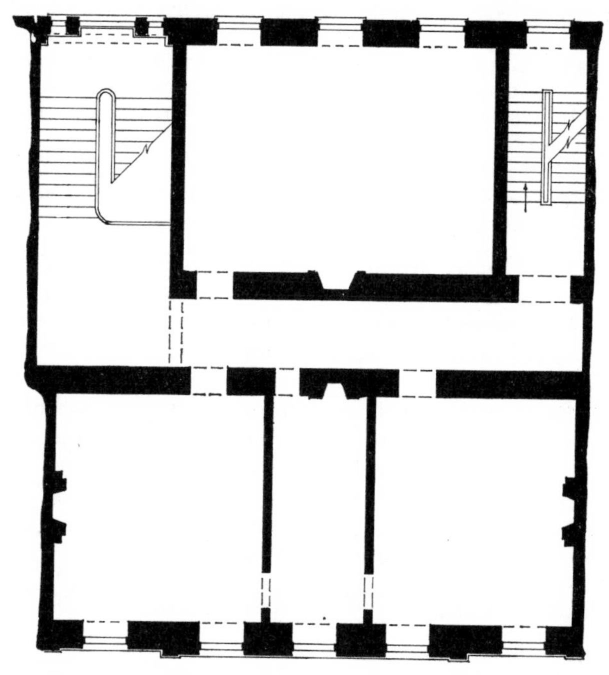

1st Floor Plan

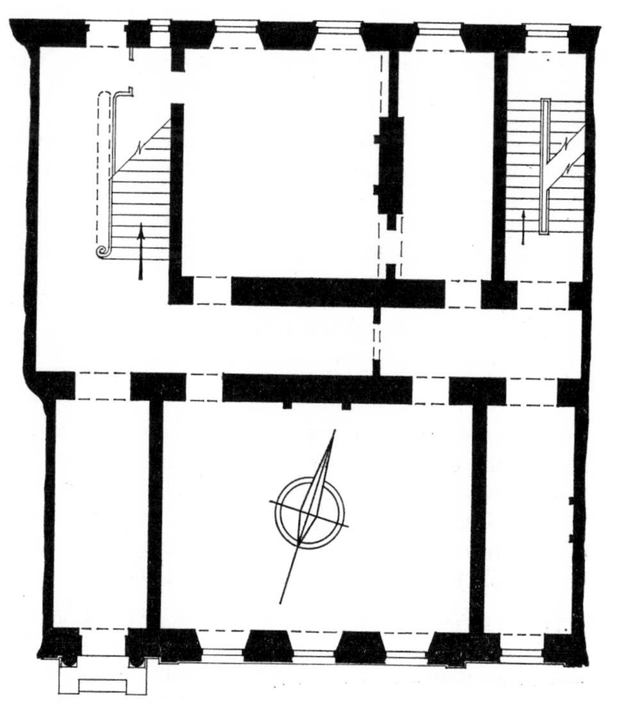

Ground Plan

Fig. 53. (66) Garforth House, No. 54 Micklegate.

carving (Plate 73). At the back, to N., is a fine *Saloon* entered by a doorway pedimented on both sides (Plate 67); it has a dado of fielded panels, an enriched dentilled cornice and a rococo ceiling decorated with fruit, flowers and musical instruments (Plate 169); the fireplace has a moulded surround of white marble.

On the *Second Floor* most of the rooms have moulded cornices and in the S.E. room is a fireplace with moulded eared surround, set against plain pilasters (Plate 71).

(67) HOUSE, No. 56, is mainly of the second half of the 18th century, but there are remains of an earlier structure, probably of the 17th century; a shop front was inserted in the late 19th century. It is of three storeys and attics, built in brick with modern pantiled roof. The property was bought from Christopher Rawdon in 1747 (E.93, f. 197) by John Bradley, apothecary (d. 1775), who probably carried out extensive rebuilding before his term as Sheriff in 1755–6. After his death it was the home of his widow Antonia (d. 1777) and her sister Catherine Marshall (d. 1779). Later the house was occupied by tenants, including the Misses Mary and Ann Brickland, who carried on a girls' boarding school here from 1823 for some 10 years.

The front elevation is of two bays with sash windows under arches of gauged brick voussoirs, and with a continuous band at first-floor sill level; the second floor has two sash windows with flat arches lower than those below. At the eaves the fascia board and rainwater gutter, supported on simple timber brackets, are not original. The E. gable adjoins No. 54. The back is partly plastered and has 19th-century windows. At the wall-head is a gable with flush coping two courses wide, and a chimney stack to W. Behind this gable, and at a higher level, is another gable, of 17th-century origin but altered to accommodate the mid 18th-century staircase.

Internally most of the rooms have original plaster cornices and two are lined with panelling. On the first floor is an original fireplace (Plate 71). In the E. end of the attic storey is a blocked window, showing that the brickwork here is earlier than 1757, when Garforth House (66) was completed.

(68) HOUSE, Nos. 57, 59 (Fig. 54), has a lead rainwater head to the rear elevation bearing the date 1783, agreeing with the stylistic appearance. The house was occupied at the end of the century by Robert Swann the banker, who in 1801 moved to No. 128 (97); and later by Mrs. Susanna Wray (d. 1830 aged 83), widow of the Rev. Henry Wray of Newton Kyme (Borthwick Inst., Rate Books of St. Martin-cum-Gregory; Directories).

The street front (Plate 173) is of three storeys and built in brick with stone bands and a wooden cornice (Plate 78) at the eaves. The ground floor was reconstructed in 1946–7 (*Report* of York Civic Trust) but both original doorways remain: that to E. (Plate 62) is the main entrance while the other originally gave access to a passage leading to the rear. At the back is a lead rainwater head with initials and date, W A 1783 (Plate 81).

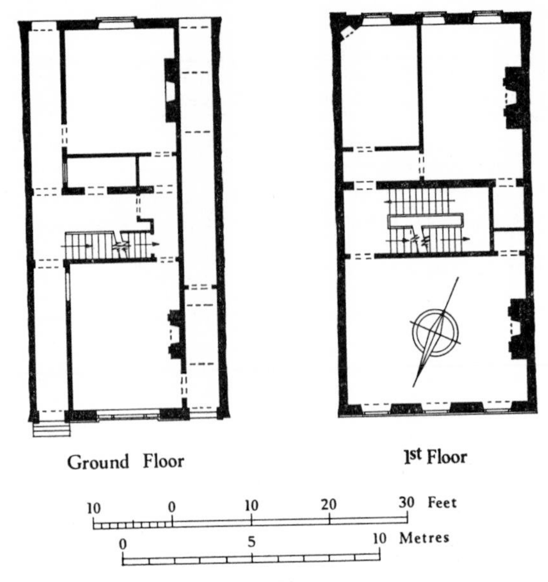

Fig. 54. (68) Nos. 57, 59 Micklegate.

To the E. the main entrance opens into a through passage, giving access to front and back rooms and the staircase between them. There were formerly stairs leading down to the kitchen in the rear of the basement, with service to both ground-floor rooms from the head of these stairs. To the W. a second passage leads directly to the rear of the house. The front room has an original fireplace (Plate 73). The staircase has two slender turned balusters (Fig. 18q) to each tread, shaped cheek pieces and a swept mahogany handrail.

The front saloon on the first floor is a typical room of the period, with moulded skirting, dado rail with milled enrichment, and a moulded and enriched cornice. The doors and windows have moulded and milled decoration to their architraves; the window reveals have sunk panels with foliated decoration in applied composition; in the W. wall is a fireplace with applied composition decoration (Plate 72). There is also an original fireplace in one of the back rooms.

(69) HOUSE, Nos. 58, 60, includes a large house built in the late 18th century on the E. two-thirds of the site; this was refronted *c.* 1830–40 when an older house to W. was rebuilt, giving a uniform elevation to Micklegate. The dentilled brick cornice at the back suggests that the earlier part was designed by the firm of Carr.

The front elevation to Micklegate, of three storeys, is in large pinkish-white bricks; the roof is of Welsh slate. Two shop fronts are probably of *c.* 1830–40, each flanked by fluted pilasters, and each with a house doorway to the W. Each of the upper floors has four hung-sash windows and above is a moulded cornice on coupled gutter-brackets. At the back the earlier building (No. 58) is in pale brick with red brick dressings. The two lower floors have wide three-light sash windows with segmental arches.

In No. 58 the staircase has an open string to the first flight, the remainder having a closed string and very slim turned balusters with square knops and bases. In No. 60 the staircase has a closed string and, up to the second half-landing, slender turned balusters with large turned newels, and beyond, plain square balusters.

(70) HOUSE, No. 61, must have been built late in the 18th century; the three-bay front (Plate 173) and the plan, with staircase placed transversely between front and back rooms, are typical (*cf.* adjoining house, Nos. 57, 59, dated 1783). An unusual feature is the placing of the service passage adjacent and parallel to the entrance passage. Later alterations include a mid 19th-century surround to the front door, and a large single-storey extension to S. The property was occupied by George Telford (1749–1834), of the family of York nurserymen, from 1786 until 1809 (*York Courant*, 12 Dec. 1786; 13 March 1809), and the freehold descended to his grandson Colonel Charles Telford, J.P., who lived in the house from 1876 to his death in 1894, when it was bought by Dr. W. A. Evelyn, physician and antiquary.

The staircase has cantilevered stone treads, shaped on the underside, square balusters with hollow faces, and is lit by a lantern in the roof. In the front room on the ground floor is an original fireplace with applied decoration (Plate 74).

(71) HOUSE, No. 67 (Plate 158), was originally a three-storey timber-framed structure of two bays, probably of the 16th or 17th century, with two gables towards the street. Early in the 18th century it was rebuilt in brick with a new roof parallel to the street, but some of the original structural timbers remain inside. The fenestration of the top floor was altered later, and the front of the ground floor has been removed to insert a new shop front, which forms a unified composition with Nos. 69 and 71 adjoining. The interior has been much altered, and there are later additions to the rear of early 19th-century date. This was probably a butcher's shop for a long period: the successive occupiers William Hill (from before 1798), Chamberlain of York 1807, and Peter Armistead, Chamberlain 1823, belonged to that trade (YCA, E.96, f. 105; E.97, f. 143v.; Rate Books; Directories).

On the street front the windows have exposed box frames and heavy glazing bars. A rainwater head is dated 1763 (Plate 81).

(72) HOUSE, No. 68, was built in the mid 17th century; an entrance hall archway, the fine staircase, and possibly the cellar doorway are of this period. In the early 19th century the upper storey was added and most of the house remodelled. Edmund Gyles (1611–76), the glass-painter, and his more famous son, Henry Gyles (1645–1709), lived here (Davies, 171–2); later occupants were William Stead junior, stonemason (d. 1823), Thomas and William Kirby, druggists (Rate Books, Directories), and George Hornby, surgeon (Davies, 175).

The street front, three-storeyed, is in stucco-rendered brick; it has a modern shop front and, to W., a doorway of *c.* 1800, with Roman Doric pilasters, glazed fanlight, and door of four fielded panels. Above the entrance is a four-light casement with raised moulded architrave, and over the shop front a Victorian bay window. In the upper storey are two casements, with raised moulded architraves and plain sills. There are moulded bands to both first and second floors, and a moulded cornice. The back elevation, originally two-storeyed, of the late 17th century, has an added early 19th-century storey with a pediment-like gable containing a central bull's-eye window. The original windows were replaced at this time.

Inside, on the ground floor, a large room to N.E. has two cased ceiling beams running E. to W., supported by large posts of *c.* 1650, stop-chamfered on the N. side. The fireplace in the E. wall is modern, but the doorcase and door and a segmental-headed recess and cupboard are of *c.* 1800. A room to the N.W. has in the N. wall a Regency three-light sash window, with panelled half doors below the centre sash; the raised bead moulding to the panels suggests an early 19th-century date. A doorway of *c.* 1820 in the S. wall has a fluted architrave with formalised floriate paterae, and in the W. wall is an early Victorian fireplace. The central stair hall leading off the entrance hall has two round-headed archways with moulded architraves and key-blocks by the S.W. corner; one gives access to the staircase. To the E., in a large well, the oak *Staircase* (Plates 82, 176; Fig. 17a) of *c.* 1650 has members of large scantling; the treads have been renewed in soft wood. Under the staircase half-landing, the cellar doorway has an ovolo-moulded case of *c.* 1650, containing a door of planks with a frame forming two large panels.

The *Cellar* is in two main parts. The N. part has a chamfered ceiling beam running E.–W. In the E. wall are two segmental-headed lamp niches, and in the W. wall is a kitchen range with an early 19th-century grate decorated with raised bars with foliate ends. To E. is a wine cellar with barrel-vaulted roof. The other part has large ceiling joists of the 16th or 17th century running E.–W. There is another cellar, of narrow red brick, to the S.

On the first and second floors the rooms contain various fittings of the early 19th century including windows and doorcases and a white marble fireplace with reeding and carved sprays of flowers. The staircase of *c.* 1820 from the first floor has a moulded mahogany rail, turned newels, plank string and square balusters.[1]

(73) HOUSE, Nos. 69, 71, was originally timber-framed of two bays, probably of the late 16th or early 17th century, and was later the home of Samuel Dawson (1691–1731), Sheriff in 1718–19 (YCA, E.94, f. 50). In

[1] Painted glass from a window in this house, by members of the Gyles family dated 1665 and consisting of a family register and shields of arms (Davies 175–6) is said to have been acquired by the Yorkshire Philosophical Society (J. A. Knowles, *Essays in the History of the York School of Glass-painting* (1936), 17). The various accounts of this glass are very confused.

the second quarter of the 18th century extensive alterations were made, including the building of a new brick front and the insertion of a fine staircase with a lantern above and other internal fittings. Considerable later alterations include additions at the rear and the conversion of the ground floor to shops. Despite the 18th-century alterations, the original roof survives, with the gables to the street hidden by the later brick front. The W. half of the property (No. 71) was the Minster Inn, mentioned in 1736 as 'of good resort' (Drake, 280), which survived until 1830 (Directories). The house was advertised for sale in 1794 (*York Herald*, 6 Sept.), when it was occupied by the widow of the Rev. Philemon Marsh (d. 1788), rector of S. Martin-cum-Gregory for 43 years. The crest of Marsh on a rainwater head confirms the stylistic evidence suggesting that the major alterations belong to *c.* 1745–50, soon after Marsh came to the parish. Since the private dwelling house was in 1794 stated to have 'four rooms on a floor, good garrets, and a garden, stable, etc.', it is uncertain where the accommodation of the Minster Inn was placed.

The brickwork of the street front (Plate 158), though thickly painted over, appears to be in Flemish bond. All the openings on the upper floors to the front are original though the first-floor windows and some of the second-floor windows have sashes of later date with narrow glazing bars. The parapet is finished with a moulded stone coping, now much decayed. At the E. end is a rainwater pipe with ornamental head bearing the crest of Marsh (Plate 81).

Inside, the ground floor has been converted into shops and only in the front room of No. 71 are there remains of a moulded and dentilled plaster cornice. On the first floor are cased axial and transverse beams supported at their intersections by a Doric column, perhaps of the 18th century. On the second floor it is evident that the roof is gabled toward the street and that the roof-slopes cut across two of the upper windows, which are blocked accordingly behind the glazing.

The *Staircase* of No. 69 has no visible string (Plate 87); the balusters, two to a tread, have square knops, and at the foot the handrail finishes in a scroll supported on a turned newel. There is a panelled dado about the same height as the stair balustrade, consisting of large fielded panels separated by panelled pilaster-strips below a swept dado-rail. The stair well rises through the whole height of the building, the second floor being carried across at one side on a narrow gallery with a balustrade similar to that of the secondary staircase. The well has a bold enriched and moulded plaster dentil cornice, above which the ceiling is coved toward a modern lantern light. The secondary staircase, from first to second floor, has turned balusters of Doric type, square newel posts and closed strings.

(74) HOUSE, Nos. 70, 72, includes in the front range remnants of a two-storey, late 15th to early 16th-century timber-framed house; a third storey and attics were added in the 17th century. A middle range is of the 16th century, and a block to the N. is of the early 19th century. The property was refronted to Micklegate *c.* 1823, when the house was empty for part of the year and the assessment was raised (Rate Books of St. Martin-cum-Gregory). Over a long period from 1802 the premises were occupied by Christopher Simpson, a saddler, but parts of the property were sublet. In 1825, when Simpson mortgaged the freehold, he was stated to have recently converted the former tenements into one house (YCA, E.95, f. 262; E.97, f. 213v.).

The street front (Plate 60) is of fine dull red brick with markedly thin mortar joints; the two shops on the ground floor are modern. The two first-floor bow windows have moulded framing with moulded paterae, plain friezes and simple cornices. At the W. end is a contemporary moulded and fluted rainwater head. At the back, the original range has been engulfed in later heightening and early 19th-century additions. A three-storey block standing forward from the rest, in very poor condition, is in stucco-dressed brickwork and has sash windows to all three storeys; the four-storey block to the E. is in large common brickwork of the early 19th century. All the roofs are tiled.

No. 70 has no internal features of interest at the ground floor. The S. first-floor room was refitted in the Regency period, only the boxed transverse beam remaining of the earlier building. The *Staircase* of *c.* 1800–10 has a moulded rail, plank string, turned newels and square balusters. The common-rafter roof includes a large number of reused timbers.

Inside No. 72, the *Staircase* has a moulded rail, plank string and square balusters. On the first floor, in the room in the N. range is a fireplace with a plain surround with moulded architrave containing a basket grate, signed Carron, which has on each jamb an oval containing a pair of doves between musical instruments, and, on the cheek pieces, scrolls and rabbits. To S. is the staircase in which the stair landing has a minute cornice of palmettes, of *c.* 1810. In the attics, a large room to the S. contains a fireplace with a reeded and moulded surround, a pulvinated reeded frieze, a cornice and shaped mantelshelf all of *c.* 1810–20. The N. building contains a Carron cast-iron hobgrate decorated with the Prince of Wales badge etc. In the centre block, timber construction partly visible in the W. wall comprises a cambered tie-beam with straight studs above and below, with some brick-on-edge infilling. The N. room in the front block has a N. wall of fragmentary timber-framing and brickwork, with a 19th-century sash window. Parts of the original structural timber exposed include, in the W. wall, a tie-beam some 1½ ft. above floor level which has a large oak brace morticed into it toward the N.W. end where there is a post with enlarged head into which the N. wall-plate fits. Above the wall-plate, the N. wall is of timber and brickwork of the early 19th century.

(75) HOUSE, Nos. 73, 75, is probably the 'new erected messuage' sold on 23 October 1730 by John Riley, bricklayer, and his wife to John Riley of New Malton (YCA, E.93, f. 57). By 1763 the property belonged to Matthew Smith, yeoman, but had been let in succession

to two whitesmiths, John Simpson and Thomas Cave (1715–79), the latter the father of the engraver William Cave (1751–1812) and grandfather of Henry Cave the artist; Simpson was the master to whom Thomas Cave had been apprenticed in 1729 (YCA, E.94, f. 50; T. P. Cooper, *The Caves of York* (1934)). The house was later occupied by a succession of private tenants, including *c.* 1820–5 Captain John Beckwith, until 1830, after which it became a grocer's shop (Borthwick Inst., Rate Books of St. Martin-cum-Gregory).

The street front, three-storeyed with attics above and three bays wide, is in good brickwork with finer red brick dressings. At first and second floors are brick bands of oversailing courses and, above, a deep moulded cornice. The ground floor, converted to a shop in the 19th century, has to the W., in the position of the original main entrance, a doorway to No. 75; the doorcase is of the 18th century. The windows at first and second floors are set at irregular intervals. The rear elevation is two-storeyed, the attic being in the roof space; there is a small two-storeyed addition against it.

Inside, the ground floor is divided into a shop, occupying the front and back rooms of the former house, and the approach to the flat above, consisting of the original entrance passage and central staircase. At the S. end of the E. side wall of the passage is an archway to the staircase with a timber casing with panelled pilaster-like features on the responds, a moulded semicircular arch with sunk panels on the soffit, and very deep key-block. In the staircase hall are two doorways, one with an early 19th-century architrave with reeded jambs etc., and plain square blocks in the corners, the other with an original moulded architrave and door with six fielded panels; a doorway to the cellar, below the upper flight of the staircase, has a similar moulded architrave and door.

The *Staircase* has, to the lower stage, a cut string, treads with rectangular cheeks and long overlaps, two turned balusters (Fig. 17i) to each tread, swept moulded handrail, and turned newel. From the first floor it has a closed string with deep mouldings and plain newels varying in size; the upper flights are steeper and the handrail is not swept up to the newels.

The main N.W. room on the first floor is panelled and has three sash windows; the moulded cornice breaks and returns forward over the fireplace and window openings; above the door is applied Adamesque decoration, which is probably the frieze reused from a small late 18th-century fireplace. The floors of the attics are of gypsum plaster.

(76) HOUSE, Nos. 74, 76, is of mid 18th-century origin though the only evidence of this is the brickwork and band of the upper storeys of the street front. Early in the 19th century, the building was divided into two separate dwellings, at first of equal size since the rate assessment was £5 on each moiety in 1822. Further alterations probably took place then, for from 1823 onwards the assessment on No. 74, occupied by the owner, Harman Richardson, butter and bacon factor, was £6, while that of No. 76, sublet to a succession of tenants, was £4 10*s*. By this time a shop had doubtless been formed in the ground floor of No. 74. The building was altered again in the late 19th or early 20th century.

The street front is of three storeys. The first-floor windows of No. 74 retain original arches of single rubbed bricks and red brick dressings; all the other windows were more or less altered *c.* 1900, and the shop fronts are also of this period. The rear elevation, of four storeys, has original brickwork to the level of twenty courses above second-floor windows. No. 74 appears to have been heightened in brick early in the 19th century and No. 76 added to in the second half of that century. A two-storey wing of Regency date adjoins No. 74. Inside, all fittings, such as reeded cornices and simple door architraves with simple paterae to the angles, are of Regency date.

The *Staircase* has a mahogany handrail swept up to each terminal newel, square-section mahogany balusters, and shaped ends to the risers.

(77) HOUSE, No. 77, was built probably in *c.* 1790. It is of three storeys with attics. A shop was inserted during the mid 19th century, at which time the N. and W. elevations were faced with stucco and the windows reglazed. Until 1827 the house was occupied by the owner, Mr. Carrack, but it was afterwards let to Robert Carr, a druggist (Rate Books; Directories).

The street front has broad bands and continuous window sills at first and second-floor levels, the stucco facing of the zones equal to the full height of the windows being rusticated. The rustication continues across the blind central windows. At the wallhead is a simple modillion cornice, typical of the late 18th century. On the ground floor, to the E. of the shop front, an entrance doorway has an original doorcase, tall and slender, with narrow pilasters and console brackets supporting a cornice with a frieze enriched with triple flutings; it has a rectangular fanlight.

The W. elevation to Trinity Lane has bands carried round from those of the main front; the arrangement of windows, open and blocked, is also repeated.

Inside, the ground floor has been converted to a shop, and the partition between the entrance passage to the E. and the adjoining room has been removed. The round-headed archway to the small stair hall, formerly from the passage, has simple pilaster-responds with moulded caps. The *Staircase*, which is directly opposite the entrance, has slender turned softwood balusters, two to a tread, simple shaped cheek pieces to the risers, and a thin moulded handrail swept up to each angle. It is top-lit.

The *Saloon* at the front on the first floor extends the whole width of the house. This and the rooms on the second floor contain a number of simple original fittings.

(78) HOUSE, No. 83 (Plate 46), built probably in the second quarter of the 18th century, is mostly original, but the front was altered early in the 19th century to include a bow-window and doorway on the ground floor. A lean-to extension at the back is modern.

Each of the three elevations is of red brick, in no regular

bond, but that to Micklegate is now painted. Returning three-course bands mark the first-floor and attic-floor levels. The street front has a corbelled and dentilled cornice. The bow-window is segmental on plan and the doorway beside it is flanked by fluted half-columns and surmounted by a semi-circular fanlight and timber hood supported on fern-leaf brackets. The original window openings have elliptical arches turned in a single ring of brick headers and with blind tympana above the rectangular sash frames. The E. gable end, which has brick kneelers, five courses deep, and a brick parapet, contains a round-headed recessed panel flanked by two small windows, the whole of Venetian-window form.

Inside, each floor has essentially the same plan, of two rooms with chimney and staircase between; the ground floor is further sub-divided by a passage from the front door to the staircase and rear room. The *Staircase*, rising from ground floor to attic, has the treads housed into a single central square newel post. The rooms contain original and early 19th-century fittings.

(79) HOUSES, Nos. 85, 87, 89 (Plate 174), form a timber-framed range containing three separate tenements under one roof, parallel with the street. Structural evidence shows that the division into three is part of the original design; each tenement is of two bays. Although not marked on Speed's map of 1610, the building is probably of late mediaeval date. The double-jettied front (Fig. 55), embattled bressumers and crown-post roof (Fig. 13e) are features of the late 15th or early 16th century. A manifest addition at the back, itself of late 16th or early 17th-century date, is further evidence supporting this dating, and it seems probable that the range was built for letting as 'rents' along the street frontage of the precinct of Holy Trinity Priory at a date well before the Dissolution. Its survival comparatively unaltered, may perhaps be due, as often elsewhere, to use for the butcher's trade. So far as documentation survives it shows occupation by butchers, of No. 85 probably in the mid 18th century and certainly by 1838, when William Stodart Stoker was assessed on the house and slaughter-house; of No. 87 from 1708 by George Chapman, butcher, and his successors in business; and of No. 89 by William Pearson, butcher, in 1777–96. (Deeds in the hands of the Ings Property Trust; Rate Books; Directories.) The freeholds were independently owned, but a fee-farm rent payable out of No. 87 to John Tempest esq. in 1807 may have been a survival from post-dissolution grants of former Priory property. In recent restorations rendering has been removed to expose the framing, and sash windows have been replaced by casements. The ground floor contains shops with modern fronts. Pantiles have given place to plain tiles throughout.

Each of the three tenements originally consisted of a single room to each floor, probably subdivided by curtains or flimsy partitions. No evidence for the ground-floor layout survives. At the E. end of the original rear wall of No. 87 the disposition of pegholes in the first-floor timber-framing, allowing for an opening 2 ft. 2 in. wide, suggests a contemporary timber-framed wing or staircase annexe, refurbished in the 18th century and mostly replaced or enlarged in the 19th century. A late 16th-century or early 17th-century timber-framed wing at the rear of No. 89 stands on the site of such an earlier annexe. A stair inside the first-floor room led up to the second-floor room, open to the roof. All fenestration, probably with oriel windows, was on the street front, except possibly in No. 87. All chimney breasts are of a later date and there is no evidence for the original heating arrangements. Only No. 87 has been fully surveyed, but the timber-framing of No. 89 has been recorded during restoration.

Fig. 55. (79) Nos. 85–89 Micklegate. Jetty framing.

On the street front (Plate 174), in No. 87 the late Victorian shop front replaces a bow window which existed in 1886 (G. Benson and J. England Jefferson, *Picturesque York* (1886), Pl. 2). To E. of it is the shop doorway, adjacent to a further doorway to a through passage beneath a rectangular fanlight. Inside, a dog-leg staircase opens off the through passage. The first-floor landing is to the S. of the main timber-framed house and may represent the original timber-framed stair annexe, since fragments of framing survive in the S. wall. The room at the back is a later addition. The original room, now subdivided, is entered through an 18th-century round-headed doorway. A short straight stair against its E. wall leads upwards; studding pegged to the underside of the horizontal beam to the W. indicates that this stair follows the original disposition. The second-floor ceiling at tie beam level hides the roof construction, apart from curved braces from the main posts to the cambered tie beam; the single room of two bays was originally open to the roof. In the course of restoration only the original central post and one stud in each bay were found to survive: the present framing is based on the evidence of mortices and pegholes. Although cut back, sufficient evidence remained of the embattling on the bressumer. A central N.–S. beam, 10 in. by 9 in., carries the N. post and is tenoned into the S. post and has two E.–W. beams, 11 in. by 9 in., off-set from one another, tenoned into it on each side. Each of the four main divisions contains six N.–S. joists, about 8 in. by 6 in.

To No. 89 a timber-framed back range of late 16th or early 17th-century date has been added at right angles. Its roof is of four bays. A curved brace between the eaves-plate and the central post has been removed to make way for a doorway from the second floor of the front block into the attic, where four of the trusses are visible. The first truss 'A' is only 1¾ ft. from the S. wall of the main block. The third truss, 'C', of similar type, has a cambered tie beam, 12 in. by 7 in., instead of a horizontal one; the trusses 'B' and 'D' have tie beams like 'A'. The third bay is largely occupied by brick chimney-breasts of various dates from the early 18th century onwards. The bays are about 6¼ ft., with three pairs of coupled rafters, without collars, to each bay. On the first floor of the front range, a gap of 2 ft. 2 in. between stud and post, indicated by the mortices and pegholes, at the E. end of the S. wall may mark a doorway from the annexe stair, as in No. 87.

(80) Bathurst House, No. 86, one of the finest houses in Micklegate, was built in the early years of the 18th century (Figs. 56–7). The site was that which, *c.* 1230,

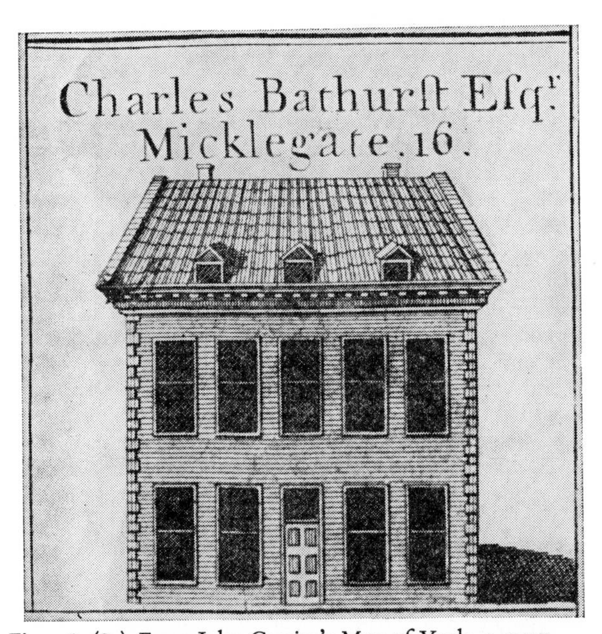

Fig. 56. (80) From John Cossins's Map of York, *c.* 1727.

had belonged to Agnes, first wife of Nicholas de Bugthorp, and was later granted by him to maintain a canon for ever in the Priory of Healaugh Park, under the description 'between the lane next St. Gregory's church and the house of William son of Agnes, and from Myclegate to North-street' (*YASRS*, xcii for 1935 (1936), 155–6, 161). Drake (280) describes it as 'Charles Bathurst's house newly built at Gregory [Barker] Lane end', and it appears on John Cossins's plan of *c.* 1727, when it belonged to Charles Bathurst, then High Sheriff of Yorkshire. That the house is earlier than this is proved by the survival of rainwater pipes bearing the Bathurst crest with heads on which are the letters C.F.B., evidently for Charles Bathurst, father of the High Sheriff, and his wife Frances, both of whom died in the first six months of 1724 (Davies, 167). The younger Bathurst never married and died in 1743, after which the house was occupied for a time by the Hon. Abstrupus Danby (Borthwick Inst., Rate Books of St. Martin-cum-Gregory; *cf.* Davies, 139). The house was regarded as of considerable importance, since it was individually marked, not only upon Cossins's plan, but also on John Haynes's prospect of York of 1731. By 1752 the property had passed to Henry Masterman senior, who advertised it to let as '4 rooms below and 5 chambers above, with 5 good garrets, a kitchen, washhouse, laundry, large cellars, garden, 2 coach-houses, stable for 9 horses' (*York Courant*, 9 Oct. 1753).

Originally of two storeys with attics, the latter lit by three dormer windows as shown in the view on Cossins's plan (Fig. 56), the house was heightened to a full third storey *c.* 1820–5 (probably early in the occupancy of Mrs. Lucy Willey, who appears in the Rate Books from 1818 to 1839; in 1823 the assessment was raised from £16 to £17). At the same period various interior fittings were replaced in Regency style. Later extensions and alterations were made to the service wing at the rear, some of the work being done in 1873 for the North-Eastern Railway, which owned the premises in 1872–9 and converted them into offices for the District Goods Manager, Southern Division (Title Deeds; information from British Transport Historical Records, York).

The property, after the death of Henry Masterman senior in 1769, descended through his son, Henry Masterman junior, to Henrietta Masterman who married Sir Mark Masterman Sykes. After passing through several occupations, including that of William Cadday (d. 1806), Sheriff of York in 1797–8 (Skaife ms.), the house was sold by the Sykes for £1350 in 1813 to Mrs. Elizabeth Richardson (YCA, e.96, f. 214), and soon afterwards resold to Mrs. Lucy Willey, who occupied the house herself until 1838. It was later the home of the Misses Sandys (1838–49), of William Frederick Rawdon (1850–5), and of Caleb Williams, surgeon, who died in 1871. After the period of occupation as railway offices it again became a surgeon's residence from 1879 to 1909; from 1911 to 1921 it was the Central Hotel and thereafter for almost 40 years the York Y.W.C.A. It is now the property of York University (Title Deeds; Rate Books; Directories).

The street front (Fig. 57) is in good quality red Flemish-bonded brickwork, the upper storey being of slightly larger bricks. When the upper storey was added, the original rain-

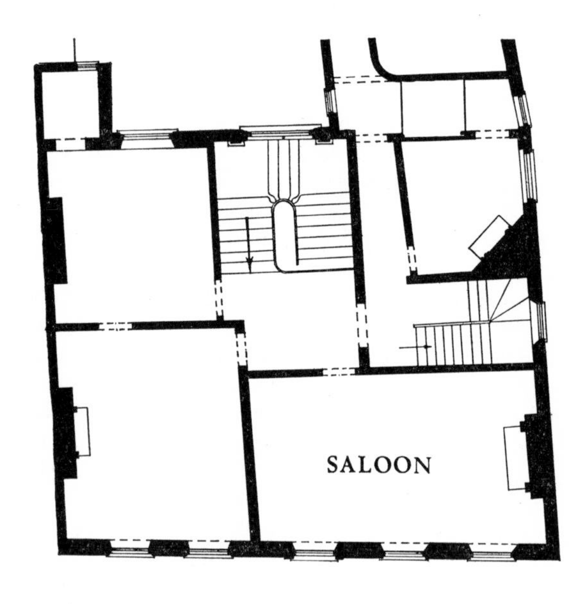

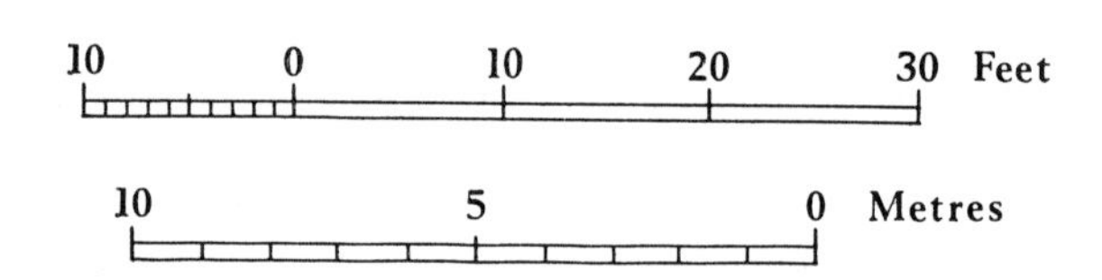

Fig. 57. (80) Bathurst House, No. 86 Micklegate.

water heads and cornice were re-employed and matching lengths of fall pipes were inserted. Most of the architectural features are plainly shown in Plate 175. The Roman Doric surround to the entrance is of the second half of the 18th century, but the glazed doors are modern. In the windows, narrow glazing bars replace the originals, which presumably were thicker. Above the entrance, the narrow central bay breaks forward some 3 in., the upper part being slightly widened at the head of the central window. To E. and W. are large lead rainwater heads (Plate 81) decorated with cherubs' heads and initials, 'B' above, 'C' and 'F' below; the fall pipes of square section have brackets enriched with the Bathurst crest.

The E. side elevation to Barker Lane has a flat-topped gable, in part with ashlar coping. The good quality red brickwork of the front is returned about the S.E. angle and the first-floor band is continuous. A side entrance has a semicircular gauged brick arch and painted stone imposts. There is evidence in the brickwork of changes in the fenestration.

The rear elevation, three storeys high, has a small closet wing to the W. and the service wing to the E. A brick band again occurs at first-floor level, and at the wall head is an early 19th-century cornice with simple paired brackets. The most noticeable feature is the large staircase window with a semicircular head turned in gauged bricks. Inside, the entrance hall, with simple plaster ceiling and cornice, has inserted doorways with pedimented doorcases of *c.* 1840 in the E. and W. walls. To N. is a round-headed archway with wood panelled responds and soffit and a fluted key-block; it opens to the stair hall.

On the ground floor (Fig. 58), most of the fittings in the S.E. room are in the Regency style. The window casing has linear decoration; the N. doorcase has a reeded surround with paterae at the angles though the door itself is early 18th-century. The W. doorway and fireplace are Victorian. The *Staircase* (Plates 82, 84) has an open string, swept oak handrail terminating at the foot above a fluted newel post, and turned balusters (Fig. 17h) mostly three to a tread. The balusters, with pierced twisted shafts alternating with two fluted shafts, are further elaborated with pedestals which, in York, are unusual for the multiplicity of mouldings. The oak dado has fielded panels. The asymmetry of the surround to the large staircase window to fit the rise on the turn of the stair is also unusual (Plate 90). The ceiling over the staircase itself has a heavily moulded and enriched plaster cornice with a cove above rising to a square panel with quatrefoil centrepiece; that over the first-floor landing has an enriched geometrical design with the Bathurst crest in the middle. This same landing has a round-headed archway with flanking panelled pilasters and a key-block in the E. wall and, in the other walls, doorways with simple architraves and doors of six ovolo-moulded fielded panels.

The *Saloon* windows and the N. doorway have reeded or fluted architraves with paterae at the angles all in Regency style, but the doorway contains a reused early 18th-century door of eight fielded panels. The fireplace is modern. The ceiling is decorated in Adam style, probably in embossed paper, with a round panel in the middle surrounded by delicate swags of husks and florets. The N.E. room contains some early 18th-century panelling on the E. wall and in the S.E. angle the fireplace survives from the original fittings.

(81) MICKLEGATE HOUSE, Nos. 88, 90 (Plate 177; Figs. 58, 59) the most important Georgian residence S.W. of the Ouse, was built for John Bourchier (1710–59) of Beningbrough as his town house, and was finished by 1752. It is said to have been designed by John Carr of York (G. Benson in *Architectural Review*, II (1897), 111), and though there is no proof of this, it resembles other houses known to have been designed by him. The house passed to Bourchier's widow who died in 1796, and was then leased to James Walker and, from 1811, to Joshua Crompton (1755–1832), who bought the freehold in 1815 and in whose family the house remained until 1896. When the house was for sale in 1815 it was described as comprising, on the *Ground Floor*, Entrance Hall, Dining Room, Parlour, large Kitchen, Back Kitchen, Pantry, Housekeeper's Room, Butler's Pantry, Water Closet; *First Floor*, two elegant adjoining Drawing Rooms, two good Lodging Rooms one with attached Dressing Room, the other with a Light Closet; *Second Floor*, six good Lodging Rooms, one Dressing Room; *Attics*, four good Servants' Rooms, and a Light Closet; Vaulted Cellars and a Laundry; good Garden, Coach-house, Stables for 11 Horses, 'a large Reservoir Well supplies the house with Water' (*York Courant*, 3 April). After 1896 the house became business premises and much of the panelling was removed and sold; a fireplace from the best bedroom is now in the Treasurer's House. Though now only a skeleton of its former self, it remains the best house of its date in York, at least in regard to the main staircase and the plaster ceiling over it.

The main front (Plates 177, 178) is built in good red brick with stone dressings above a stone-faced basement and has not suffered from any alteration; the doorway at the E. end which departs from the strict symmetry of the elevation is part of the original design. The rainwater pipes have been renewed and the wrought-iron railings have been restored. A pair of gates matching the railings is in the Yorkshire Museum.

The back (Plate 55) is built of buff-pink brick with pale red brick dressings. The storeys are divided by projecting brick bands. Some of the windows still have original sashes, others have replacements with thinner glazing bars. A lead rainwater

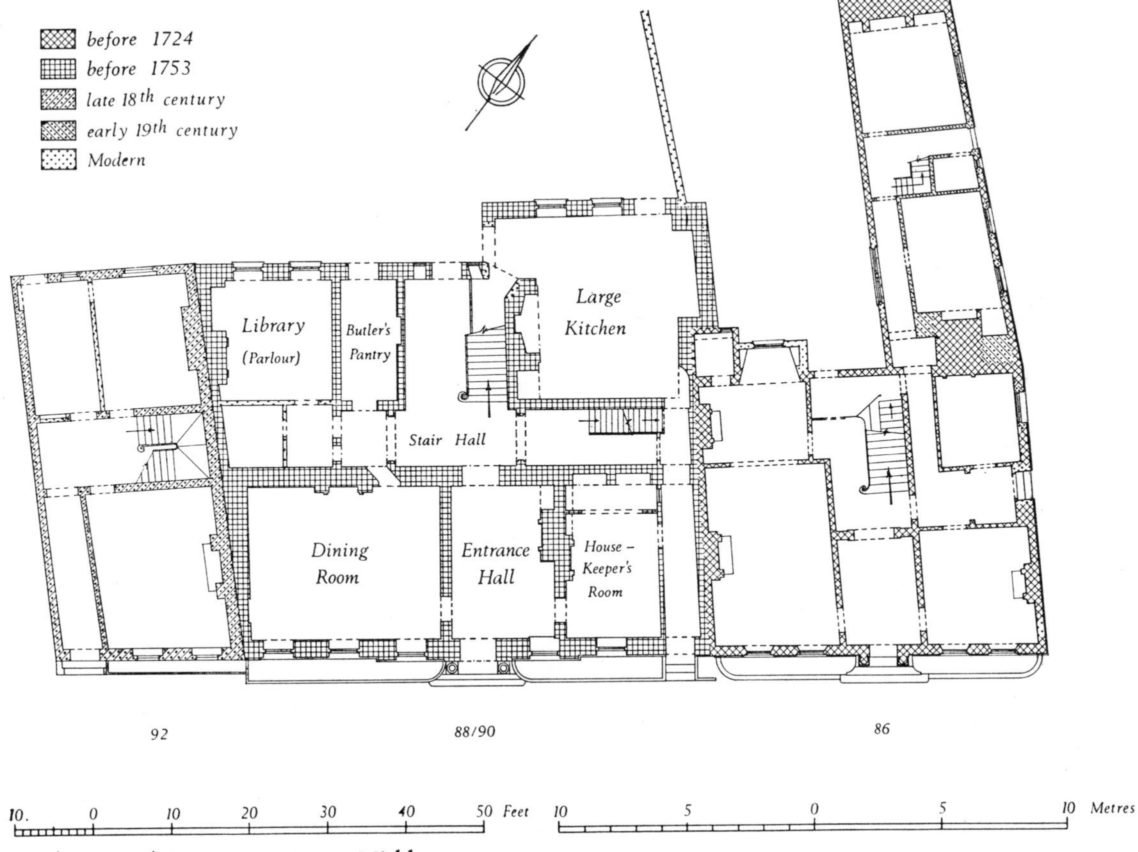

Fig. 58. (80, 81, 83) Nos. 86, 88, 90, 92 Micklegate.

head (Plate 81) bears the initials of John and Mildred Bourchier with the date 1752 and each holdfast to the downpipe bears the Bourchier crest, a man's head wearing a ducal coronet and a long tasselled cap hanging forward.

Inside, the hall has a panelled plaster ceiling, and a fireplace with a marble surround (Plate 73); to N. an archway leads to the stair hall (Plate 179). To the W., the *Dining Room* is lined with simple panelling under an enriched cornice, and has a fireplace with a marble surround (Plate 74). Behind the dining room and under the Best Bedroom was the *Library*, also lined with panelling and having an original fireplace surround. The windows contain two painted glass panels (Plate 181): a spaniel, 'Dick', sitting on a cushion, and a greyhound, 'Rover', both signed 'W. Peckitt Pinxit 1756'. Peckitt's account book (York City Art Galley, Box. D.3) has the entry: '1756 February—For John Boutcher Esq. a setting Dog £1. 1. 0; For Miss Boutcher a lap Dog £1. 11. 6.' The E. front room was the *Housekeeper's Room* and has a series of fitted drawers and cupboards in a recess. Behind this room the *Servants' Staircase* has cut strings and turned balusters with square knops similar to the servants' staircase in No. 55 Micklegate. To the N.E. below the Retiring Room, the *Kitchen* has a large segmental-headed fireplace which contained a steel grate (Plate 92) with ovens, two movable trivets with swivelling circular hobs, and a swing arm for a kettle. Above is an elaborate mechanical spit, worked by a smoke jack in the main flue, transmitting a drive through bevel gears to a horizontal, square shaft. (*Range and spit removed.*) The main staircase (Plate 182) rises to the first floor only; it has broad cantilevered steps and enriched balusters (Plate 84; Fig. 180) and is lit by a round-headed window flanked by pilasters (Plate 90).

The *Cellars* are covered by brick vaults except under the kitchen where there is a large room with a fireplace and casement windows.

On the first floor the stair hall has an enriched modillioned cornice and a ceiling in rococo style (Plate 180), resembling one at Fairfax House, Castlegate. To the N. and S. are medallions containing busts in relief, one representing Shakespeare. Over the landing is an extension of the ceiling with an oblong panel.

In the front of the house are two intercommunicating rooms. The larger, formerly called the *Reception Room* (31 ft. by 19 ft.), has been stripped of most of its fittings. The smaller *Drawing Room* (25 ft. by 19 ft.) has also been stripped, the carved mantlepiece having been removed to Esholt Hall, but the rococo ceiling remains. It has a centre piece with a spaniel barking at a water bird, perhaps a heron, framed in asymmetrical scrolls and foliage, with four heads in panels at the angles (Plate 77), perhaps representing the seasons. In the room to N.E., once called the *Retiring Room* (23 ft. by 19½ ft.), is a plain fireplace of white marble. In the former *Best Bedroom* the walls are plainly panelled. The modern fireplace in the W. wall takes the place of that removed to the Treasurer's House (Plate 71).

The second floor is reached from the servants' staircase, and above are attics where parts of massive oak roof trusses are exposed. The joints are notched and pegged, and bear position-marks in Roman numerals. There are four intermediate trusses visible, spaced about 10 ft. apart.

(82) HOUSE, Nos. 91, 93, consists of a two-storey brick range along the street and a lower range of one storey and semi-attic running S.E., forming an L-shaped plan. It is uncertain whether both ranges are of the same date; the heights are different but the brickwork appears similar. Some features, such as the tumbled gable to the back range and the sawtooth corbelled eaves on the front, indicate a late 17th or early 18th-century date. The whole property belonged to Phillip Tate senior, merchant tailor, by 1741, and 5 years later was sold to another of the same trade, John Monckton or Mountain, whose widow Jane lived in the house until 1776. It was then sold to Thomas Kilby, a brewer, who let it to a succession of tenants (YCA, E.94, ff. 198, 200v.; Rate Books of Holy Trinity). The first evidence of a division into two tenements is in 1830, when Mary Collins, dressmaker, seems to have occupied the smaller shop, No. 91; the larger shop, kept by Ellen and later by William Gregg, was a Post Office in 1851, when No. 91 was occupied by Robert Nutbrown, a gardener. The space enclosed by the two early ranges was filled in the 19th century, and there is a long 20th-century extension to the S.E.

The street front is shown in the elevation opposite p. 69. The interior has been considerably altered and the ground floor contains two shops separated by a through passage. No original features survive, apart from axial beams on both floors of the front range.

(83) HOUSE, No. 92, a good example of the smaller town house (Fig. 58), was built *c.* 1798 (when the rating assessment was increased), probably by the firm of John Carr; the staircase balusters are like those of the Black Swan, Coney Street, and No. 18 Blake Street, the latter being by Peter Atkinson senior. A two-storey range at the back, possibly of the first half of the 18th century, was modernised in the 19th century, when a first floor was added to the vestibule joining it to the main building. In 1948 wrought-iron scrollwork was added to the side of the entrance doorway (York Civic Trust, *Report* 1948–9).

Robert Fairfax of Steeton, a Captain in the Royal Navy, who bought the site, became M.P. for York in 1713, and Lord Mayor in 1715, when he gave a fine brass chandelier to St. Martin's Church, Micklegate. In 1725 he died, aged 60, and was buried at Newton Kyme; his wife, who died in 1735, was buried at St. Mary Bishophill Senior (Davies, 141–4). The house was sold, in 1805, by John Fairfax of Newton Kyme, grandson of Robert, to Mary Coates of York, widow, who had been the occupier since 1800 (YCA, E.96, f. 39; Rate Books). Later owners were Thomas Backhouse, seedsman, from 1817 until his death in 1845, and his

MICKLEGATE, looking N.E. from Priory Street.

(38) BISHOPHILL. Bishophill House, Nos. 11, 13. Ceiling of Saloon, *c.* 1765.

(38) BISHOPHILL. Bishophill House, Nos. 11, 13. Details of Saloon ceiling, *c.* 1765.

(47) House, No. 40. *c.* 1750.

(46) Houses, Nos. 32, 36. 1747.
BLOSSOM STREET.

(43) BLOSSOM STREET. Windmill Hotel, Nos. 14, 16. Late 17th century and later.

SOUTH PARADE. 1825–8.

BLOSSOM STREET, from W.

(71, 73) MICKLEGATE. Houses, No. 67 and Nos. 69, 71. *c.* 1600, refronted.

Houses, Nos. 104(*l*)–82(*r*).

Houses, Nos. 126–116.
THE MOUNT

(52) HOLGATE ROAD. Holgate House, No. 163, from N.E., *c.* 1770 and later.

Ground to second floor.

Newels and balusters.

Second floor to attics.

(54) MICKLEGATE. House, Nos. 4, 6. Staircase, early 18th century.

MICKLEGATE BAR, from Blossom Street, by Moses Griffith, *c.* 1785 (York City Art Gallery).

MICKLEGATE, looking towards the Bar, by Thomas Rowlandson, 1800 (York City Art Gallery).

(55) MICKLEGATE. Houses, Nos. 3, 5 and Queen's Hotel. *c.* 1720.

(55) MICKLEGATE. Queen's Hotel, Nos. 7, 9. Fireplace in Saloon, early 18th century.

House No. 5. Ceiling and cornice above former stair well.

Queen's Hotel. Saloon from S.W.

(55) MICKLEGATE. House, No. 5 and Queen's Hotel. Early 18th century.

Doorway in N.E. room.

Fireplace in N.E. room.

Entrance doorway to Saloon.

(55) MICKLEGATE. Queen's Hotel. First floor. Early 18th century.

Wall.

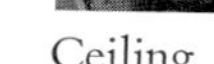

Ceiling.

59) MICKLEGATE. House, No. 17. Painted room decoration, late 19th century.

(65) MICKLEGATE. House, Nos. 53, 55. First-floor landing, *c.* 1755.

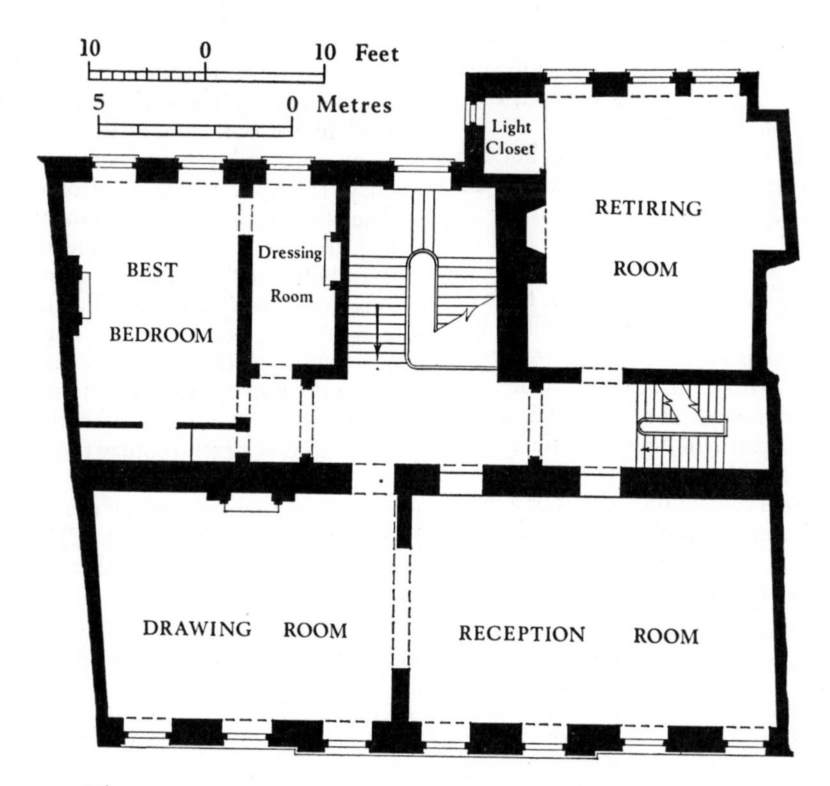

1st Floor Plan

Fig. 59. (81) Micklegate House, Nos. 88, 90 Micklegate.

N

brother James Backhouse (1794–1869), nurseryman and seedsman, who lived here until he moved to Holgate House (52) in 1859.

The main front (Plate 53) is in red brick, with stone dressings, and has a slated roof. The windows have deep flat arches and stone sills, and the modillioned and dentilled cornice overlaps the adjacent Micklegate House (Plate 78); to E. is a moulded rainwater head with a fall-pipe secured by holdfasts decorated with opposed fleurs-de-lis. The iron railings are original.

The entrance leads to a through passage off which the staircase is placed transversely between front and back rooms. The front dining room is typical of a York house of the period. It has a moulded skirting and dado rail, applied plaster mouldings to the walls, and a dentilled plaster cornice, with transverse fluting to frieze. The fireplace, though correct in style, is said to be a reproduction. The staircase has a swept moulded pinewood handrail, spiralling over a newel at the bottom; cantilevered treads with plain cheeks and moulded edges; slender, turned balusters, three to a tread, with an urn feature under a square knop; and on the inner side, a moulded dado rail and skirting. Some of the upper rooms have original cornices and fireplace surrounds.

(84) The Falcon Inn, No. 94, was one of the two great hostelries of Micklegate (Drake, 280), and its site included the present No. 96. The inn was ruined by the coming of the railway to York and the site was redeveloped in 1842–3, when the rates first show the present division into two houses (Borthwick Inst., Rate Books of Holy Trinity, Micklegate).

No part of the ancient buildings survives, but the carved sign of the Falcon (Plate 92) is attached to the present front. It is probably of the late 18th or early 19th century.

(85) House, No. 95, of three storeys, on plan forms a long narrow rectangle and is of two separate builds: the front part, timber-framed, is probably of the 16th century; the extension at the back, in brick, is of the late 17th century, as are the chimney and adjacent staircase at the junction of the two parts. In Victorian times, the ground-floor room to Micklegate was converted to a shop and the upper storeys cement rendered. For a half-century from 1800 this was the barber's shop of James Mackerill; since the mid 19th century the house has belonged to the owners of No. 97.

At the front the simple shop front and the windows are of the 19th century, but the original jetty of the first floor survives. The back is built of thin bricks typical of late 17th-century work in York; it has a projecting band at second-floor level.

Inside, the timber studding of the side walls is visible on all floors. A fireplace to the rear has a late 18th-century cast-iron grate, by Carron, of duck's-nest design, enriched with foliage, with vesica-shaped panels containing cherubs playing a drum and a flute. The staircase has shaped flat balusters (Fig. 17d). At the centre of the back part is a roof truss with curved upper crucks.

(86) House, No. 98, was built *c.* 1770–80. At the time it belonged to Denis Chaloner, a cooper, who sold it in 1785 to George Smithson, innholder of the adjacent Nag's Head (88) (YCA, E.95, f. 28v.). It formed part of the public house until after 1850, but was remodelled about 1828, when the assessment was raised (Rate Books) A modern wing has been added at the back.

The street front, of three storeys under a slated roof, has a modern shop front to ground floor, two upper storeys of painted, Flemish-bonded brickwork with a stone band at first floor, and a modillioned cornice. At first floor is an inserted bay window, of the second half of the 19th century. At the back a small wing projects at the W. side. Inside, the staircase is original and has a swept moulded handrail, closed string, and slender turned balusters with square knops.

(87) Houses, Nos. 99, 101, 103, 105, 107, 109, comprise a two-storey range of seven timber-framed tenements with a frontage to Micklegate of nearly 100 ft., and to the rear of Nos. 107 and 109 three buildings of the 15th, 17th and 19th centuries, known collectively as No. 111 (Figs. 60, 61). The front range, Nos. 99–109, was built in the mid or late 14th century, probably before 1369, the N.E. house, No. 99, standing against the now-vanished 13th-century gateway to Holy Trinity Priory (*see* p. 12, Fig. 27). These tenements were combined and divided in various combinations at different periods. Nos. 99, 101 and 103 had attics inserted in the 17th century, and No. 103 was refronted in brick in the middle of the 18th century by a member of the Greenup family. In 1774 Nos. 105, 107 and 109 were 'newly and uniformly fronted' with a three-storeyed brick elevation by Thomas Peart (YCA, E.96, f. 47). No. 111 consists of three buildings, here described as (A), (B) and (C) (*see* plan). (A) was built, probably *c.* 1806, in the angle between the two parts of No. 113 (93), and incorporates the S.W. corner of No. 109. (B) at the S.E. end of the No. 111 complex, is a 15th-century timber-framed building; from its position far back from the street frontage, it may have formed some minor element in the conventual layout of Holy Trinity Priory. (C) is also timber-framed and is a late 16th or early 17th-century house. Its position, linking (B) to the back of Nos. 107 and 109, fits the pattern noted in No. 89 Micklegate (*see* p. 82) of buildings being erected within the grounds of the Priory after it had come into lay possession. The planned use of a semi-attic in No. 111(C) is interesting, and the constructional details of the 14th-century range are worthy of note.

Nos. 99, 101 now form a single shop, but were originally two separate tenements. The front retains its original jetty, but

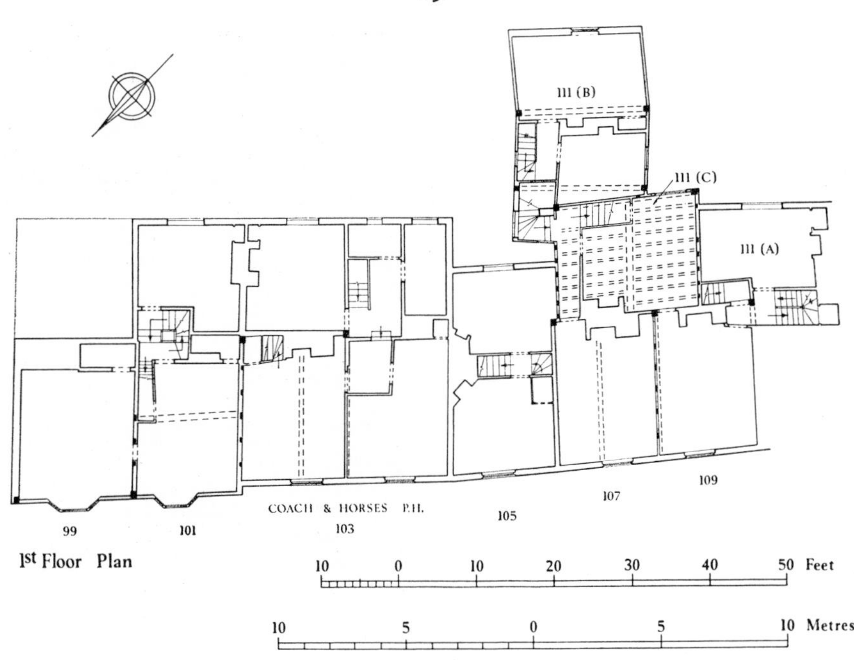

Fig. 60. (87) Nos. 99–111 Micklegate.

is otherwise remodelled. In 1958 the back wall was rebuilt in brick and the gabled E. end cloaked by a brick skin (Plate 49).

The ground floor retains traces of timber-framing, and the W. room (No. 101) has an 18th-century plaster cornice. The central through passage and staircase are of 19th-century date. On the first floor the E. room has had its inserted attic floor removed, and much of the timber-framing is visible (Plate 49). The roof-trusses are of crown-post construction with archaic features (*see* p. lxxii and Fig. 13a). At the back of No. 101 is a 19th-century brick addition.

No. 103, The Coach and Horses, comprises two of the original tenements and has a 17th-century block added at the rear. The front wall was rebuilt in brick (now roughcast) in the mid 18th century and retains the first-floor sash windows of that date. The back was also rebuilt in brick in the 18th century; it has been considerably altered.

The ground floor has been extensively modernised internally, but on the first floor some of the timber framing is visible, and the E. room has a late 18th-century hob grate. In the attic the central and E. roof-trusses are visible, the W. one being mostly masked by wattle and daub. In the E. truss the crown post tapers evenly from top to bottom (Plate 50), but in the central truss it has the usual enlarged head (Plate 49). Both trusses are of the same type as in Nos. 99 and 101.

Nos. 105, 107, 109 present a unified front elevation of brick, of three storeys under a roof covered with modern sheeting. They have a band at second floor and a modillioned and dentilled cornice; most of the openings are Victorian. Between No. 107 and 109 is a rainwater head inscribed 'TP 1774'. At the back only a mid 18th-century extension behind No. 105 is visible; it was originally three storeys, but has been reduced to two. The lower storeys contain many timbers of the original frame as well as late 18th and 19th-century hob grates. The staircase to No. 105 is of mid 19th-century date.

No. 111(A) has a S. elevation of three storeys in brick with pantiled roof and brick bands between the storeys. The base of the S. wall is of early 18th-century date with a contemporary window. All the internal fittings are of the early 19th century.

No. 111(B) is of two storeys with pantiled roof. The external timber-framing, apart from main posts, has been replaced by brickwork. The S. and W. walls are of late 17th or early 18th-century brick, the S. end having a tumbled gable. The ground-floor room has a chimney-breast contemporary with the brick walls, and panelling of both 17th and 18th-century dates. The first floor is reached by a late 18th-century staircase. Both rooms have early 18th-century fireplace surrounds with 19th-century hob grates. The 15th-century roof has two trusses of crown-post construction (*see* p. lxxii, Plate 49, and Fig. 62).

No. 111(C), of two storeys and semi-attic, is timber-framed with brick patchings and pantiled roof. The gabled S. end is of plastered brick, and the W. wall was rebuilt in the 19th century. The first floor is approached by a 19th-century staircase in an external addition. Originally the entrance must have been from No. 107 or 109, and the opening to the staircase a window (*see* p. lxxviii). The original framing survives, almost complete, on the E., S. and W. walls, and in the N. wall is a fireplace with a 19th-century hob grate. The semi-attic has 3 ft.-high walls, and the central roof-truss has sole-pieces carrying kerb-principals (*see* p. lxxiv, Fig. 63). It is lit by a window in the S. gable. *Nos. 105–111 demolished in 1961.*

(88) THE NAG'S HEAD, No. 100, consists of a three-storey unit of three bays, with its end to Micklegate and jettied, erected *c.* 1530; and at the back, behind a big

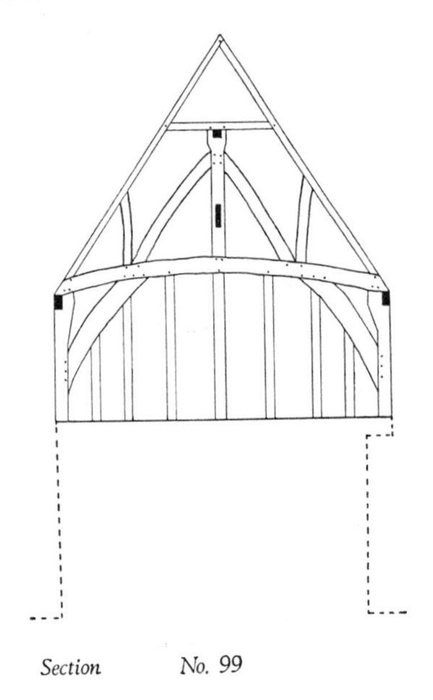

Fig. 62. (87) No. 111B Micklegate. Crown-post truss.

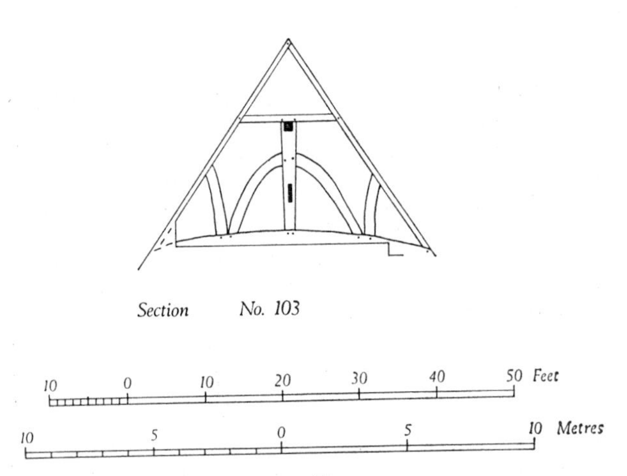

Fig. 61. (87) Nos. 99, 103 Micklegate.

brick chimney, a narrower wing originally of one storey only and perhaps built in the 14th century (*see* p. lxxii). In the 18th century the street front was rebuilt, and an attic formed in the main house. New staircases were put in an annexe, which had filled in the angle between the two old blocks in the late 17th or early 18th century. At an uncertain date a floor was inserted in the rear building. In the mid 18th century, the property belonged to Isabella Stockdale, who left it to two sisters, Jane and Sarah Coghill, who by 1777 were married to John Cawthery and William Powell of Leeds. In 1815–23 the premises were sold by William Stockdale Powell, only son and heir of William and Sarah, to trustees for John Fryers Kilby, brewer (YCA, E.96, f. 246v.; E.97,

Fig. 63. (87) No. 111C Micklegate. Sole-piece truss.

f. 167v.), when the property included several buildings adjoining the yard and extending N. to Toft Green, formerly a malt kiln and stables. The messuage was 'used as a Public House called the Nag's Head, now in the occupation of John Smith Aledraper', formerly of Robert Mortimer aledraper and afterwards of Francis Lumb. From 1785 until after 1850 the Nag's Head also included No. 98 (86).

The narrow front with gable to the street is of 18th-century brick. A modern ground-floor window cuts into a projecting brick band. The first floor is lit by a handsome segmental bow window of *c.* 1810, having three sash lights under an entablature with dentilled cornice and fluted frieze. Internally, in the front part of the house some of the original main timber posts and beams are exposed and, on the second floor, some of the studwork of the side walls crossed by long straight struts. The roof is constructed of common rafters carried on side purlins supported by curved struts from the tie beams; there may be collars above the ceiling.

In the back range, main posts are exposed. One truss has a boldly cambered tie beam carrying a king-post and two raking struts but the upper part is concealed. A second truss has had the tie beam and braces under it all cut away. None of the original fireplaces remain; the existing staircase is partly modern and partly of *c.* 1840.

(89) HOUSE, Nos. 102, 104, is of late 17th-century date, but in the early 18th century the main front was refenestrated. The premises were divided into two in 1812, and bay windows were inserted in the Micklegate front *c.* 1850. The property, with other houses adjacent (*see* Nos. 100, 106, (88, 90)) belonged in the mid-18th century to Isabella Stockdale. In 1817 William Stockdale Powell sold Nos. 102, 104 to Francis Calvert, cordwainer, who had been the occupant from 1812. At that time the house was divided, the rates being apportioned on assessments of £3 10*s.* 0*d.* for No. 102, and £4 10*s.* 0*d.* for No. 104. Calvert himself occupied No. 102, and let No. 104 to a series of private residents until 1834, when he made this his own residence, while No. 102 became the business premises of Calvert & Son (YCA, E.97, f. 35v.; Borthwick Inst., Holy Trinity Rate Books; Directories).

The house is of three storeys and built of brick. It runs parallel to Micklegate with a relatively long frontage. The elevation to the street has at first floor two three-sided bay windows added *c.* 1850, and between them a sash window with narrow rubbed brick flat arch; above these windows a brick band may have been removed. At second floor are three windows with narrow flat arches; all windows have red brick reveals. Above is a heavy modillioned cornice returned at each end (18th century). The roof, with prominent equilateral gables (late 17th century), is covered with Welsh slates.

At the back are two projecting wings, and between them the wall of the main range is faced with 18th-century brickwork in which is a blocked round-headed window. The W. wing (Plate 54), of narrow red brick, is of two storeys and attics; at first-floor level is a moulded brick string. In the N. gable the first floor is lit by an early 19th-century sash window, in the blocking of a late 17th-century opening with a label above a flat arch. Above is a tumbled gable having a window with a brick pediment above a flat arch, and a blocked œil-de-bœuf above (Plate 51).

Some of the interior fittings are of the 18th century. The upper part of the staircase in No. 102 has a closed string, simple square newels with attached half balusters, and closely set turned balusters with square knops, comparable with the altar rails at Holy Trinity, Goodramgate.

(90) HOUSE, Nos. 106, 108, of the late 18th century, is undistinguished apart from a good entrance. It was probably built in 1778, as the rating assessment was raised in 1779 from £6 to £22. This was another property of W. S. Powell (*see* (88, 89)), let to private residents until 1822; from then until 1849 it was a girls' boarding school kept by the Misses Harriet Palmer and Elizabeth Ellis. Later in the 19th century a shop front was inserted, and in recent years a warehouse was added behind the ground floor.

The house is of three storeys and has a Victorian shop window, to E. of which the original entrance remains: Roman Doric columns support a pedimental hood containing a semicircular fanlight. The upper storeys are in brickwork in Flemish bond with a continuous band joining the first-floor window sills. There are two sash windows to each upper floor below a modillioned and dentilled cornice at the eaves. Internally, the upper part of the staircase, placed transversely between front and back rooms, has slender turned balusters, two to each tread, and a slender mahogany veneered handrail. *Demolished 1967, but original entrance doorway reinstated in new building.*

(91) HOUSE, No. 110, built in the early 18th century, has been so drastically remodelled that only the staircase and roof remain unaltered. About 1778, when the adjacent house (Nos. 106, 108) was built, the staircase windows were blocked and the house modernised; most fittings are of the late 18th or early 19th century. Like the adjacent houses (88–90) this formed part of the Stockdale property and was sold by W. S. Powell in 1815 (YCA, E.96, f. 250v.). In 1820 it was resold to the occupier, Thomas Mawson, a combmaker (E.97, f. 110).

The house is of three storeys and the front is stuccoed above modern shop fronts; each of the upper floors has two windows in original openings, but with modern casements. At the eaves is a dentilled and modillioned cornice, carried through from Nos. 106, 108.

The plan of the house is similar to that of No. 106 but has the fireplaces placed diagonally across the corners of the rooms. The staircase which now starts at the first floor has square newels, heavy moulded handrail and closed string with boldly overhanging capping which carries turned balusters, some of an early bulbous form and others of rather later design with and without square knops.

(92) HOUSE, No. 112, until recently the Red Lion public house, was built in the 16th century as a timber-framed structure with a gable to Micklegate, possibly jettied. In the 18th century it was refronted with brick and the roof hipped back; *c.* 1860–70 the front was re-fenestrated, with a large shop window to ground floor, a bay window to first floor and one sash replacing two above. This was an ancient property of the Vicars Choral of York Minster and the leases show that it was partly rebuilt between 1745 and 1748 (York Minster Library, Sub-Chanter's Book 3, 75).

The house is of three storeys, built in two bays with a chimney against the back wall. Some of the timber framing is exposed inside, with curved braces rising from the posts to rails and wall-plates.

The original roof is almost intact. The truss to N. has principal rafters and a straight tie-beam; curved struts from tie to rafters support the purlins. The central truss is carried on great posts, their heads not enlarged, supporting a markedly cambered tie-beam; curved braces from each post support the tie-beam. The principals are connected by a collar; from the tie curved braces, morticed into the collar, support purlins under the principals. The S. truss vanished when the roof was hipped, but a good series of common rafters remains, all with collars.

(93) HOUSE, No. 113, was built about 1740; it was of three storeys and comprised a front block and a wing projecting at the back. About 1811, a fourth storey was added to the front block, the street front was re-modelled, and the upper part of the back wing was altered. The back range was extended *c.* 1820–30, and a pleasant S. front to the garden created; a little later a small block was added at the S.E. corner. In 1736, when occupied by Mrs. Jane Palmes, this house was bought by John Theakston, whitesmith (YCA, E.93, f. 87), who sold it in the next year to William Greene, gent., owner of the adjacent 'large messuage' on W., the former No. 115 (E.93, f. 103). It was probably the 'large new built sash'd house in occupation of Roger Meynel esq.' advertised by Mr. William Green in 1742 (*York Courant*, 23 Feb.), with a coach-house and two stables for six and two horses respectively. A succession of tenants occupied the house during the later 18th and early 19th centuries, among them Thomas Jennings, esq., from 1779 until 1821. During his occupancy, in 1812, the rating assessment was advanced from £5 to £9, and this presumably indicates that the extensive alterations and additions to the house had just been done. Apart from general reassessment of the parish, no further change took place until after 1850, when the owner and occupier was the Rt. Rev. Dr. John Briggs.

The front elevation, of four storeys, in pale-coloured stock brick with red brick dressings above a stone plinth, is of the 18th century, with alterations of *c.* 1811. The entrance, tall and narrow, has an early 19th-century doorcase flanked by reeded columns with formalised leaf caps, under a flat entablature. To E. are two sash windows and to W. a modern window. At first floor are two large segmental bow windows (*c.* 1811), each having a moulded wooden base with brickwork above, three sashes, reeded pilaster jambs, plain roundel paterae to the caps, a fluted frieze, and simple cornice. The second floor has two sash windows, and in the top storey are two windows the same width as those below but less than half the height.

The front block has one front room beside the entrance hall on the ground floor and a saloon extending the full width of the house on the first floor; the back wing contains the staircase and two rooms beyond. The extension provided one large room on each of the main floors. Nearly all the fittings are of the 19th century but the doorway to the saloon has the original enriched architrave and cornice to the landing, and towards the room a 19th-century architrave under the original cornice (Plate 66). The 18th-century stairs up to the second floor remain (Plate 86), with cut string and turned balusters (Fig. 18j). There are also some 18th-century doors, of three fielded panels (Plate 66). *Demolished 1960.*

(94) HOUSE, No. 114, was built in the second half of the 17th century, with two storeys and cellars. A third storey was added in the mid 19th century and during the same century many alterations were made throughout. The freehold of the property belonged in the 18th century to Ann Thorp, who married John Heron of Sculcoates, and to their descendants, but the house was leased to Edward Bedingfield, esq. (1730–1802), a son of Sir Henry Bedingfield, 3rd baronet. Edward settled at York about the time of his marriage in 1754 to Mary, daughter of Sir John Swinburne of Capheaton. Of the ten children born to Edward and Mary between 1754 and 1771, while this was their home, only one had issue, Anne, born here on 21 March 1758: she married Thomas Waterton of Walton Hall near Wakefield and became the mother of Charles Waterton the celebrated naturalist. Edward died here in 1802 (*York Courant*, 17 May), and his widow moved to No. 109 The Mount (50), letting this house to Richard Hansom (1746–1818), carpenter, who soon afterwards bought the freehold for £450 (YCA, E.96, f. 6v.). Hansom's grandson Joseph Aloysius Hansom (1803–82), architect and inventor of the hansom cab, is stated to have been born here on 26 October 1803, and the property continued in the family until late in the 19th century (Borthwick Inst., Rate Books of Holy Trinity, Micklegate; Directories). It included a garden extending back to Toft Green, where the former stable was converted by the

Hansoms into a joinery workshop. By 1881 the premises were occupied by a provision merchant.

The street front (Plate 53) is of red brick above a modern shop front; the larger bricks of the top storey show its later date. The back has a 19th-century gable above 17th-century brickwork of the lower storeys.

The ground floor has been converted to a shop. The *Staircase* (Plate 84) rises against the E. wall, and has turned oak balusters like those of the altar rails at Holy Trinity, Goodramgate, dated 1675. Between the lower floors, the staircase is lit by a two-light window with original oak mullion and high transom. On the first floor, the larger of the two front rooms is lined with fielded panelling (Fig. 16f) with dado rail and cornice; it has an early 19th-century marble surround to the fireplace. A room at the back retains an 18th-century moulded fireplace surround.

(95) HOUSE, Nos. 118, 120, was built *c.* 1742, by Robert Bower, who had bought the site for £170 on 28 August 1741 (YCA, E.93, f. 131) from Samuel Waud. The new house was certainly built before March 1750/1, when the tenants of the Vicars Choral in No. 116 were complaining that the 'new house of Mr. Bowers . . . blinds some of their lights' (York Minster Library, Sub-Chanter's Book 2, 489). Bower, who already owned the adjacent No. 122 (96), may have lived here for a time (*York Journal*, 3 Feb. 1747), but in 1757 he sold the property for £900 to Matthew Chitty St. Quintin (YCA, E.94, f. 12v.), who had been the occupier at least from 1754 (Miss I. Pressly in York Georgian Society, *Report* 1948–9). St. Quintin died in 1785, leaving the house to his nephew Sir William St. Quintin, 5th baronet, who advertised it for sale as 'the very neat convenient dwelling house, with stable and coach house and garden adjoining, paintings and pictures' (*York Courant*, 23 Aug.). The purchaser was Stephen Atkinson of Knaresborough, who lived here for a short time but in 1790 sold the property for £830 to William Taylor (E.95, f. 100v.), from whose widow it was acquired in 1806 by George Peacock, printer and proprietor of the *York Courant*, Lord Mayor in 1810 and in 1820, who made it his home until his death in 1836. For many years the house continued to be a private residence, but by 1872 it was occupied by Edward Sherwood, a warehouseman, and the present shopfront is of the late 19th century. In 1893 the freehold was bought by Edward Williamson, a dyer, who later pulled down the coach-house and stable on Toft Green and built a factory there (Mrs. E. Wilson in York Georgian Society, *Report* 1968, 31–5).

During the later 18th century, probably after 1763 (when an act of Parliament provided that houses fronting a main street should lead the water from roof to ground in proper fall-pipes), a parapet was built above the dentilled string of the top storey, and the present door-case (Plate 63) added to the entrance. At Peacock's death in 1836 the house was advertised as comprising 'a drawing room and dining-room, three lodging rooms, with a dressing room attached to each; two servants' sleeping rooms and four small attic rooms; and certain convenient kitchens, a butler's pantry, larder and good cellars' (*Yorks. Gazette*, 16 Jan.) (Fig. 64). In the early 19th century, the second floor S. bedroom and its dressing room were combined, various fireplaces modified, and the overmantel in the panelled room to S. enriched with decorative motifs, perhaps replacing a painting. The back range was probably added by 1836, and all windows, except one in the attics and a stair light, given new glazing bars. In 1948–9 the house was restored by J. Stuart Syme, becoming the York Georgian Society's headquarters until 1967. The house is now again in private ownership and in 1968 the ground floor was skilfully restored to its original character.

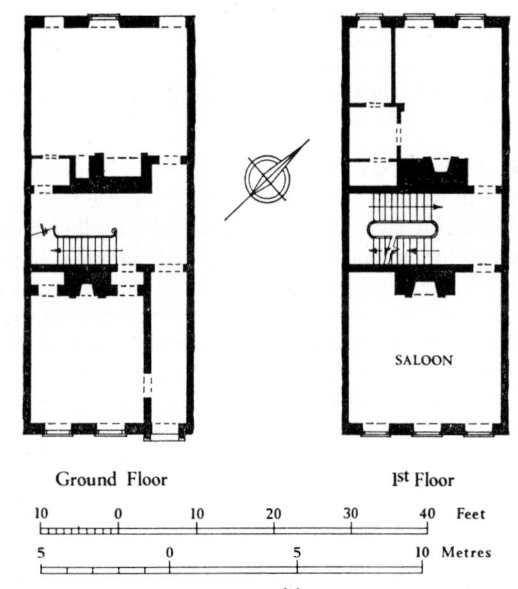

Fig. 64. (95) Nos. 118, 120 Micklegate.

The front (Plate 184) is in brickwork with fine red brick dressings, projecting brick bands, and brick cornice. The added parapet has piers to the angles. The back wall (Plate 184) is in pale brick, with fine red brick dressings; it rises to a brick gable above a brick-on-edge dentilled cornice. At the N.E. angle is a lead fall-pipe with holdfasts enriched with fleurs-de-lis.

The front ground-floor room, though so long used as a shop, retains remarkably good early Georgian fittings: moulded skirting, chair rail, panels with moulded raised surrounds, and a very deep dentilled cornice. In the N.W. wall, the central fireplace (Plate 72) has an overmantel with a seascape by Adam

Willaerts (b. Utrecht 1577, d. 1666) under an enriched entablature with an eagle and swags on a central panel and broken pediment above. To either side of the fireplace are matching doorcases and doors, one opening to the stair hall, one to a cupboard. The *Staircase* (Plate 87) rises from cellar to attics, in the middle of the house between front and back rooms. It has cantilevered treads, heavy turned newels and turned balusters three to a tread, with square knops and every third stem twisted. Under each landing is a moulded ceiling, and, in the cellar, a floor of Sicilian marble.

On the first floor the *Saloon* at the front of the house is a splendid example of early Georgian interior decoration; it has a plain dado between skirting and chair rail, moulded and fielded panelling (Fig. 16c) above with panelled shutters to the windows and a modillion cornice. The elaborate fireplace (Plate 183), in the N.W. wall, has in the overmantel a later insertion consisting of a panel with Classical figures in bas relief, probably by the firm of Wolstenholme, within a border of vine guilloche. The iron grate is a reproduction of Adam type by Carron. At the back of the house are a bedroom and dressing room panelled throughout, the former with moulded skirting, dado rail and cornice. In the S.E. wall is an original Georgian marble fireplace (Plate 73), with frieze, cornice and overmantel above; the frieze, of carved pine, has a naturalistic foliage arabesque, and the overmantel a raised eared surround. The rooms on the second floor retain original fireplaces and other fittings but are not panelled.

(96) House, Nos. 122, 124, 126, was originally two houses. In the late 17th century, the E. house was built; it still retains two bolection-moulded panelled rooms at first floor, a flight of the original staircase at attic level, and the original roof. Early in the 18th century, the house to W. was rebuilt and the two merged into one, the lower three flights of the staircase, with the hall and landing, being rebuilt in the latest fashion. These alterations were probably carried out at the time of the marriage of Robert Bower, mercer (1705–77), to Tabitha Burdett of Sleights in 1738 (Northallerton N.R. Registry of Deeds, A 487. 594; C 74.32). Bower occupied the house until his death, and his widow lived there until 1784, after which it passed through the hands of several private residents until it was sold in 1833 (*Yorks. Gazette*, 10 Aug.). From 1834 the occupant was John Hopps, surgeon (d. 1850), who was succeeded by William Drinkall, a grocer; by 1876 the property had been redivided and was in the hands of Edward Hill, grocer, and Mr. Cutforth, shoemaker (Rate Books; Directories). In 1756 railings in front of the house were 'maliciously broke and destroyed' (*York Courant*, 24 Aug.).

The *Front Elevation*, of three storeys, in brick, is of two separate builds; a straight joint, seen above the 19th-century shop front, divides them. No. 122 (to E.) has, at first floor, two sash windows with flat single gauged brick arches; the upper half of the second floor has been considerably rebuilt, with a new parapet. No. 126 has, at first floor, two tall sash windows with flat arches and stone sills and, at second floor, a projecting band, and two casements with red brick dressings and modern concrete arches. The *Rear Elevation* is also of two separate builds: that to E. (Plate 184) is of rather narrow red brick with projecting bands and a gable. The house to W., considerably set back, is of three storeys in pinkish-white brick. It is roofed in two spans parallel to the street.

The ground floor, of L-shaped plan, has three rooms and central entrance passage, to which the staircase is set at right angles. The entrance passage has an enriched plaster cornice; leading off to the stair hall is a moulded arch supported on fluted pilasters. In the stair hall the moulded ceiling, of simple geometrical pattern, remains intact. In the S.W. room, now a shop, the E. and W. walls retain panelling and a modillioned cornice.

The *Staircase* rises from ground floor to attics in four flights; the first three flights, with open string, are mid 18th-century (Plates 84, 87; Fig. 18m), the fourth and landing balustrade 17th-century (Plate 87). The later staircase has rail, string, balusters and treads all of softwood. The last flight and top landing have a moulded rail, square newels with moulded caps, a closed string and bulbous balusters, set fairly far apart. On the first floor the two rooms in the earlier part to the E. are lined with panelling (Fig. 16b) and one has a simple early 18th-century fireplace (Plate 74) and cupboards. Above, the original 17th-century oak roof timbers remain.

(97) House, Nos. 128, 130, 132, is on a site owned in 1739 by the Widow Mountain and occupied by James Thompson (YCA, E.93, f. 118); it passed to Henry Jubb, an eminent medical practitioner (1720–92), who probably built the present house about the time he became Sheriff in 1754–5; he was Lord Mayor in 1773 (Davies, 140). By 1798 the property belonged to his only child Dorothy and her husband, Major William Thompson, who sold it in 1801 to Robert Swann, the banker (d. 1858). Swann occupied the house until 1833 and left it to his son John, who immediately sold it to William Cook, joiner; it was sold to Frederick John Day, veterinary surgeon, in 1863, and to John Henry Shouksmith in 1883. For a period about 1850 the house was leased to the Ordnance Survey as their York headquarters (Title Deeds; Directories). The property, originally two messuages, became united when the present house was built, but by 1863 had been divided again.

The street front, three-storeyed in stuccoed brick, has string-courses above and below the top storey, and a parapet. Above a modern shop front each storey has four sash windows, with flat arches and stuccoed key-blocks. The roof is covered with modern pantiles. The back has small projecting wings to E. and W. (Plate 55). There are brick bands continued around the wings, at first and second-floor levels.

Internally, some of the rooms retain original cornices and doors; part of the staircase is made up of old materials including turned and twisted balusters of *c.* 1730–40.

(98) House, Nos. 134, 136, occupies the site of old houses sold by Jeremiah Ridsdale, baker, to Thomas Brown, esq., of Middlethorpe (YCA, E.93, f. 118). Brown built the present house in 1740, when a Roman statue was found while digging the foundations (*York Courant*, 8 and 22 April). In 1759 Brown sold to Charles Radcliffe, esq. of Heath the house and offices 'as the same have been lately new built by the said Thomas Brown ...' in the occupation of the Hon. Marmaduke Langdale (later 5th Lord Langdale, d. 1778). Radcliffe died in 1768, leaving the house to his daughters Frances and Elizabeth. Between 1775 and 1790 this was the town house of the widowed Lady Goodricke and her son Sir Henry, the 6th baronet (d. 1802), the owner of Trinity Gardens on the other side of Micklegate. They were followed by members of the family of Duffin, occupants until 1851, (Davies, 133; Rate Books; Directories). In 1863 the freehold passed from descendants of the Radcliffes to Henry Crummack, surgeon and apothecary, who left it to his wife; in 1883 the property was sold to Samuel Richard Brown Franks. The red brick front was added during the period when the building was well-known as Franks' Hotel.

The front was refaced with cherry-red brick *c.* 1900; openings remain in their original positions at first and second floors. Above is a modillioned cornice, almost certainly reused, and an early 19th-century waterhead, with fluted bowl and moulded cornice, and lead pipe with opposed fleurs-de-lis on the holdfasts. The roof is hipped. The W. side, nearly all cased with the same modern brick, has a dormer with pedimental head, probably original. The original three separate square or oblong stacks of the main chimney are now conjoined by a later head. The back (Plate 55) is of flecked red brick with projecting bands and double gable. Only the first-floor windows retain original sashes.

The entrance hall at the W. side of the house, leads to a central lobby off which the main staircase lies to the back of the house and a secondary staircase (partly removed) to the E. The rooms have been modernised, but that to the front retains a good plaster ceiling. A fireplace surround removed from this room belongs to the York Georgian Society.

From ground to first floor, the main *Staircase* (Plate 87) has turned balusters (Fig. 18l); at the bottom the oak handrail is swept down to a turned and enriched newel (Plate 84). At the half-landing is a large stair light, with moulded round arch supported on Composite-order pilasters; its lower half has been removed. From first floor to attics, the staircase has coarser balusters and square newels (Plate 85).

Most of the rooms on first and second floors have original cornices and one retains a fireplace with heavy moulded stone surround. In the attics the floor is of gypsum plaster; most doors are original, having two large fielded panels. The roof timbers are also original.

(99) House, No. 138, was built as a timber-framed L-shaped structure in the 17th century, remaining evidence being some framing in the back range, a chimney, and the upper part of the staircase. The back walls were rebuilt, or cased in brick, *c.* 1700; window openings and a fireplace surround of this date remain. Soon after 1850 the house was drastically remodelled and refronted. The house was for several generations the property of the Fothergill family, who sometimes lived in it, and it remained a private residence until after 1850 (YCA, E.93, f. 118; Rate Books; Directories).

The front, of stuccoed brick, has a mid 19th-century shop front and four sash windows above with a stucco-dressed band at sill level. The back of the main range, of narrow brick, has a two-course band between the storeys and, above, to W., a gable. To E. a wing projects at right angles, built of the same brick and with the same narrow band as the main range. A chimney has a 17th-century diagonally-set shaft.

On the first floor there is timber framing in the back wing and a fire-place with bolection-moulded surround. The remnant of a 17th-century staircase (now sealed off) has a heavy moulded rail, closed string, square newel with half-baluster attached to one face, and heavy turned bulbous balusters; those on the top landing are somewhat thinner.

(100) House, Nos. 142, 144, 146, was probably the ancient house of the Waller family and may incorporate remains of the new house called 'le read-brick house' of Thomas Waller (d. 1609), mentioned early in the 17th century (Davies, 138). The main campaign of building may have been due to Robert Waller, Sheriff in 1674–5, Lord Mayor 1684, and M.P. 1690, who died in 1698. In 1720 Ann Waller, widow of Matthew Waller, gent., sold the freehold to Nathaniel Wilson, a merchant (YCA, E.93, f. 6); the house was then occupied by Thomas Selby, esq., still in residence in 1746 (*York Journal*, 10 Feb. 1747). Subsequent occupiers included members of the families of Wintringham, Lawson and Corneille (Davies, 139). Major Bartholomew Corneille was the owner in 1774–7 (Rate Books; YCA, E.94, f. 197v.), and his widow continued to live in the house until 1806. For many years thereafter the property remained in the hands of William Gage, esq., and his daughters, Miss Margaret Gage and Mrs. George Anne, until 1843. It was probably Mr. Gage, *c.* 1810, who rebuilt the upper part of the Micklegate front; the rear dates mostly from the late 17th century.

The front elevation, comprising a range of three houses, now contains shop windows at ground floor; above are sash windows with flush frames, timber sills and flat arches of gauged rubbed brick; and on the second floor, small sash windows over those below. From the level of the first-floor window arches, the building has been heightened, in coarser brickwork. The back (Plate 54), covered at ground floor by a modern industrial building, had originally two late 17th-century gabled wings with a narrow space between, which was later filled by a brick three-storeyed structure with a gable, probably in the early 18th century. The wings have limestone ashlar quoins and shaped kneelers; a three-course brick band at the base of each gable has been matched in the central structure. At first floor, the E. gable has a modern window set in a 17th-century opening with a three-centred arched head; the central window of the W. wing has been partly concealed but part of its arch remains. In the central section the arch of an old window remains, probably of the 18th century. The inside has been much altered and most of the fittings are of the first half of the 19th century.

MOUNT, THE—*See* BLOSSOM STREET p. 62, and pp. 127–8.

MOUNT EPHRAIM, like Cambridge Street and Oxford Street, was laid out in 1846 and completed by 1851 as terrace housing mainly occupied by railway employees (*see* p. 128).

MOUNT PARADE, laid out in 1823, seems to have been the earliest example in York of a new type of development, the suburban road planned for genteel residences. The little terrace faces S.W., with a series of small front gardens towards the roadway. Building proceeded slowly, and only nine houses were occupied by 1830: five by gentry, one by a coach-guard and one by a stone and marble mason as his private house. Some houses were not finished until *c.* 1840. Cumberland House, on the S.W. side of the Parade, was built *c.* 1834 (*see* p. 128).

MOUNT TERRACE was developed on a building lease of 1824 as a terrace of five houses with a larger house at the end towards the Holgate Road, built by 1827 and occupied by 1828. Like Mount Parade, which it adjoins, this was a genteel development of the suburbs (*see* p. 128).

NORTH STREET. This name, already Nordstreta by *c.* 1090, was applied from the 13th century to the whole of the street running N. from Micklegate near to the W. bank of the Ouse and then turning W. on a line parallel to Micklegate. This latter part of the street has long been known as Tanner Row, from a quarter of tanneries which lay along it and in the area between it and the city wall. In the 14th century and later the tanners formed a large proportion of those parishioners of All Saints' North Street rated to subsidies or identifiable in other ways. At a later date the street had some association with the building trades: bricklayers, joiners, masons, and sand and timber merchants. A few persons of higher standing had houses there, and Peter Atkinson senior, the architect, built his own house (No. 26, now demolished) on a site he acquired in 1776. The important yard of William Stead (1752–1834), stone and marble mason, was at Nos. 36, 38, from 1802 onwards.

(101) Houses, Nos. 6, 10 (Fig. 65), with the yard and subsidiary buildings, formed in the 18th century a single property in a number of tenancies. The freehold belonged to John Playter, a coal merchant, who sold it in 1788 to John Dodsworth, ironmonger, and John Dodsworth, brewer, for £850 (YCA, E.95, f. 67). The premises later passed to Thomas Cattley, who in 1819 conveyed them to Caleb Fletcher and Christopher Scarr, wholesale grocers (E.97, f. 93v.). The yard was still used in part as a coal-yard, but there was also a small garden containing a garden house, and the main house was said to have been occupied by John Playter, John Dodsworth brewer, and Thomas Cattley successively as a private residence. This refers to No. 10, built in the first half of the 18th century, along with some of the subsidiary buildings towards the river, probably warehouses; these also included a timber-framed building of the late 16th century with a roof carried on sole-pieces and with diminishing principals (*see* Sectional Preface, p. lxxiv). Though fragments of reused timber found during demolition might suggest that No. 10 was on the site of an earlier timber-framed house, it was No. 6 that appeared to have been the original messuage, built in the late 16th century. This was originally of three storeys and attics, jettied to North Street and to the narrow lane on the S.; its top storey was removed in the 18th century, leaving part of a 17th-century staircase in the front attic. Along the lane stretched a rear wing, in part of late 16th-century build, but extended in the late 17th century, when the range was cased in brick. An important staircase was built in this range and the plaster ceiling and part of the main stair light are of *c.* 1700. In the first half of the 18th century, when No. 10 was built, the whole front to North Street was redesigned as a single unit to include No. 6, and faced in brick. Early in the 19th century new sashes with thin glazing bars were inserted in most windows, and a shop front in the ground floor of No. 10. Before 1850 (OS 1852) the central passageway between Nos. 6 and 10 had been widened to S. to form the present carriageway to the yard.

The front to North Street is in Flemish bond and has a brick plinth with weathered offset. The shop front in No. 10 takes the place of two original windows, of which the flat arches remain. On the first floor are three tall sash windows with stone sills and flush frames in each half of the front, and a blind recess placed centrally. The N. window retains its original heavy glazing bars. The coved lath-and-plaster cornice to No. 10 is original; to S., over the central bay and No. 6 is a late 18th-century timber cornice with modillions. The roofs are covered partly with pantiles, partly with plain tiles. The back of the main range is of brickwork in random bond. The S. end of the front range has been underpinned with brickwork of the early 19th century, but the framed jetty

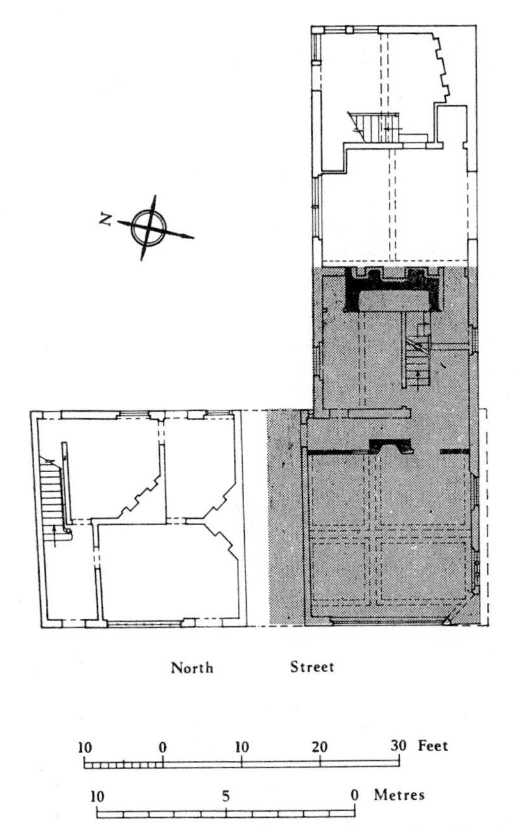

Fig. 65. (101) Nos. 6, 10 North Street. Area of earliest house stippled.

survives on the first floor. The S. elevation of the back wing, towards the lane, has a brick casing of the late 17th century throughout. Above the ground floor is evidence of a former moulded brick and tile string-course, which survives on the E. end. There is a brick plinth cut by a wide modern opening beneath two windows (early 18th-century). The staircase window consists of two late 17th-century mullioned and transomed windows placed one above the other with inserted glazing bars of the late 18th century.

There are cellars built of stone and brick beneath the whole building, with timber ceilings. In No. 6 the ground floor front room, though converted to a shop, retains an important decorated plaster ceiling (Plate 186) of the first half of the 17th century shortened to N. by the formation of the carriageway. In the back range some of the rooms have stop-chamfered beams carrying chamfered ceiling joists of the 17th century. The late 18th-century staircase has turned balusters with square knops and a swept rail, without newels. Above is a plaster ceiling with a shaped panel containing a wreath surrounded by leaves and flowers within an oval (*c.* 1700). When the staircase in this position was formed in the 17th century, a main chimney stack was removed, but the 16th-century breast still remains in the cellars.

Above the front range the *Attics* are reached by a late stair formed within the remains of the original staircase, of which a moulded rail and two oak newels with attached half-balusters survive. This staircase evidently led to the upper storey, taken down when the house was transformed and refronted in the early 18th century. The roof is of the early 18th century, of softwood, with large tie-beams (10 in. by 8 in.) and principals and collars (all 12 in. by 3 in.). Over the back range is a loft containing two original 17th-century oak trusses.

In No. 10 the front room on the *Ground Floor*, now a shop, is lined with bolection-moulded fielded panels, under a heavy cornice. In the E. wall is a door of three fielded panels, with original L-hinges. In the S.E. corner is a fireplace with Adam-esque surround (Plate 73) and a Carron hob-grate (Plate 73) decorated with trumpets, entwined ribbons and Prince of Wales feathers, probably for George IV when Prince Regent (1810–20). The *Staircase* has square newels, a heavy handrail and turned balusters (Fig. 17f), two to each tread; every third baluster has a twisted shaft. On the *First Floor* the main front room has painted pine panelling with bolection mouldings and raised panels. *Demolished 1963; the 17th-century plaster ceiling from No. 6 re-erected in the King's Manor for the University of York.*

(102) House, No. 19 (Plate 187), originally of the early 18th century, consists of a main block towards the street and a gabled wing behind, possibly but not certainly a somewhat later addition. Internally the building has been much altered, especially in recent years, and the original ground-floor plan completely destroyed by the formation of a carriageway to S. and an inserted shop front. At first floor one front room has original panelling. The staircase is of the late 18th century.

On the E. front four sash windows to the upper storey have stone sills which appear to have been lowered about 1 ft. or more; the window heads extend to the timber eaves cornice. The back wing is of red brick (Plate 187). All the openings in the S. elevation have elliptical arched heads formed of a single course of headers. The W. gable is mostly in stretcher bond but with occasional courses of headers; the gable parapet is of good tumbled brickwork. The N. side, completely stuccoed, was formerly a party wall with an adjoining building now demolished, of which parts of a stone chimney survive.

Internally few old features survive. The late 18th-century staircase has turned balusters with square knops, moulded and swept handrail, and square newels; from first floor to attic the design is simplified. In the attic is a short length of reused 17th-century balustrading (Plate 187). A first-floor room has original panelling with bolection mouldings, surmounted by a plaster cornice.

(103) House, No. 26, was presumably designed by Peter Atkinson senior (1725–1805), as he bought the site for £150 in 1776 and later occupied the present building (YCA, E.94, f. 182; E.97, f. 137). In the present century the ground floor was structurally altered to make a through carriageway; steel girders support the upper floors which comprise front and back rooms with a staircase between them.

The front, in Flemish-bonded red bricks of high quality, has gauged brick arches to all windows. At ground floor only the angles remain, the carriage opening occupying most of the frontage. At first floor are two sash windows with flush moulded frames and a stone band at sill level, now flush but probably originally projecting. The second floor has two almost square sash windows. Above is a wooden cornice with shallow modillions and dentilling, returned at both ends.

On the first floor is a very good late 18th-century fireplace surround in Adam style, with carved pine applied decoration, of honeysuckle motifs, to the frieze. The marble slip remains *in situ*, but the surrounding wooden moulding is badly damaged and the firegrate has been removed. Both rooms retain their cornice and skirting. The *Staircase*, in poor condition, has slender turned balusters and a slender mahogany veneered handrail. Over the stair well was originally a large glazed rectangular fanlight above a coved ceiling, now blocked. *Demolished 1966.*

(104) CHURCH COTTAGES, No. 31 North Street and Nos. 1, 2 All Saints' Lane (formerly Church Lane), are owned by the Church now and probably ever since their construction in the late 15th century (Plate 185; Fig. 66). They face the N. side of All Saints', North Street. Evidence for the period of construction is the S.E. corner post, with embattled cresting and rose-like paterae. The timber framing and roof structure with

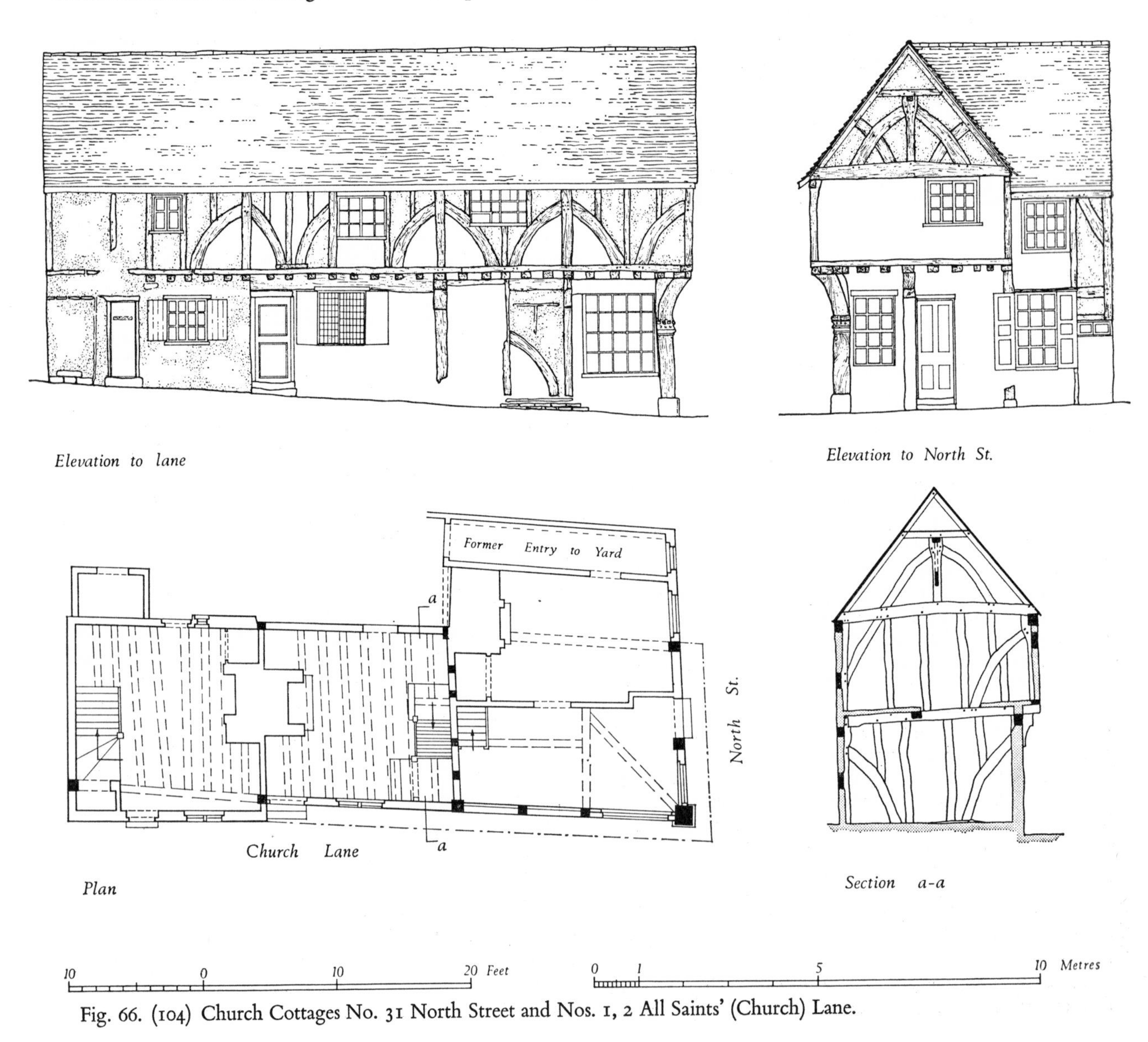

Fig. 66. (104) Church Cottages No. 31 North Street and Nos. 1, 2 All Saints' (Church) Lane.

common rafters, crown-posts and collar-purlins, are typical of York buildings of that period (Fig. 13c).

The building comprises a two-storey range of six unequal bays, and a small wing at the E. end. It formed three dwellings each occupying two bays, that at S. end having a small hall in the projecting wing. The upper floor is jettied along the S. side and across the E. gable end. No original openings remain intact, as windows and doors have been replaced at various times. However, it is possible to conjecture a reconstruction of the first-floor windows to S., where a series of pegs occurs in a horizontal line on some of the studs and braces, midway between the wall-plate and bressumer; they serve no structural purpose now, but originally must have carried a horizontal timber, or a series of brackets, and suggest shallow oriel windows of timber construction, with sills pegged to the main timbers. In the second bay from E., a small 18th-century oriel window exists, probably replacing a similar one. The N. elevation is largely rebuilt in brick and the W. gable is partly covered by a later structure. Internally there are exposed joists to the ground-floor rooms and a dragon-beam to the S.E. angle. All three dwellings have fireplaces inserted in the 18th century, and no evidence remains of any earlier form of heating. The central and W. cottages have simple steep staircases of ladder-like appearance and of uncertain date, replacing original ones which must have been similar, as the arrangement of the joists to accommodate them remains unaltered. The upper rooms of the central and W. cottages are open to the roof.

(105) Houses, Nos. 46, 48, 50, include buildings of two main periods. No. 46, possibly late 17th-century, is two-storeyed and in stuccoed brick, with a pantiled roof. A central carriageway has been driven through it. Nos. 48 and 50, to N., are a pair of late 18th-century cottages. Adjoining the rear of No. 50 is a pair of three-storey houses of *c.* 1840, in brick with one course of headers to four of stretchers. All internal fittings have been removed. *Demolished 1965.*

(106) House, No. 62, was built probably shortly before 1760. In 1761, when sold by William and Ann Peckitt (parents of the glass-painter) to William Fentiman, it was described as 'all those two houses . . . now divided into several tenements' (YCA, E.94, f. 36v.); this and the existence of two staircases indicate that it was designed from the beginning as two dwellings, in one of which the Peckitts had been living. The house underwent extensive refurbishing, including renewal of the roof, in the first half of the 19th century, probably 1833. A deed of 1834 refers to 'all those four dwelling-houses or tenements . . . adjoining to and behind the same . . .', but no reference is made to these in a deed of 1832, nor any earlier deeds; the scar of these dwellings, with a chimney breast, is visible to the rear. Later the S. room on the ground floor was converted into a warehouse, and the rear buildings were demolished.

The elevations are of brick in an irregular stretcher bond, and the roof is of plain tiles. Inside, the house is in two separate tenements, described as (A) and (B), to N. and S. respectively.

(A) On the ground floor, a central doorway leads into a passage. The staircase, approached through an archway at the end of the entrance passage, has turned and moulded balusters with square knops, a closed string, and moulded handrail; this last is discontinuous at the turns, where it is swept up to a turned and moulded newel.

On the first floor, the W. room contains a fireplace with a wooden moulded surround with a dentilled cornice shelf; the early 19th-century hob-grate is decorated with oval medallions containing seated figures and with dolphins. On each side of the fireplace is a cupboard with a door of six fielded panels. On the second floor, the partition between the W. and E. rooms is of planks 10 in. wide with a central rail 4½ in. deep, chamfered on both arrises.

(B) The ground floor, though converted, retains the staircase. This has a closed string, moulded and turned balusters with square knops, a simple handrail and plain square newels. On the first floor, the W. room contains a fireplace with painted stone surround which has a cambered head with a central key-block carved with a scallop-shell. Above the fireplace is a plain frieze with shaped ends and moulded cornice shelf. Owing to the collapse of the ceiling of the second floor, the roof over the whole house is visible; it is of *c.* 1830–40 and divided into three bays by two king-post trusses. The king-posts, enlarged at the foot to carry raking struts supporting the purlins and at the top for the abutment of the principals, carry a plank ridge; there is one purlin on each side, slightly offset in adjacent bays, halved through the principal and fastened on the other side. *Demolished 1957.*

(107) House, No. 64 (Fig. 67), was perhaps the 'new built house for sale' in 1772 (*York Courant*, 2 June), but probably part of an earlier structure remains, some of the brickwork of the W. and N. walls being of the 17th century.

The street front (Plate 188), in Flemish-bond brickwork, has a brick plinth and a timber cornice with modillions and dentils. The entrance has a doorcase, a good example of the period, with slender fluted side pilasters with simple caps and bases, carrying fluted brackets supporting a pediment; the door, with semicircular glazed fanlight above, has six sunk panels with applied raised moulding. The spandrels over the fanlight and the tympanum contained applied composition ornament, but this is either decayed or in part broken away. Two windows retain their shutters. An original lead rainwater head with part of the down-pipe and three holdfasts decorated with fleurs-de-lis survive. The hipped roof is covered with pantiles.

A warehouse, built against the W. side, has been demolished, exposing housings for joists at first and second floors. The brickwork is of three types, narrow (4 courses rising 9½ in.) and mostly in stretcher bond to ground floor, modern to second floor, and the rest, though in stretcher bond, very similar to that of the S. elevation to the street. At the back, the N. elevation is extensively defaced by modern openings,

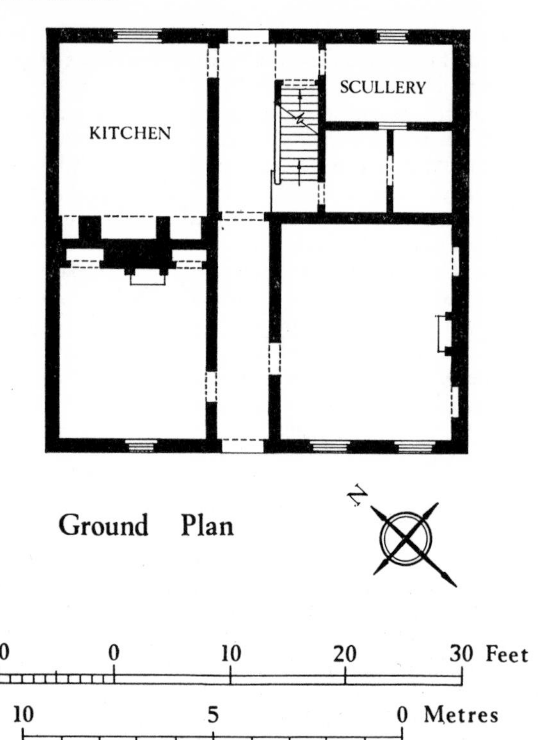

Fig. 67. (107) No. 64 North Street.

and the upper walling is in larger brickwork than the remainder. It retains a number of original openings.

Inside, on the ground floor is a central entrance hall with a stair hall beyond and entertaining rooms to each side; to N.W. is a large kitchen and to N.E. a scullery and larders. There are small cellars. The *Staircase* has a closed string, turned newels and balusters and a moulded handrail. *Demolished 1957. Late 18th-century fireplace surrounds from the ground floor reset, one at 8 Water End, Clifton, York; one at Skelton Hall, York.*

NUNNERY LANE, formerly Baggergate, probably the road of the 'badgers' or hucksters, was a suburban lane between the city moat and a series of gardens and orchards. No extensive housing development took place until *c.* 1829. *See* p. 128.

OXFORD STREET was laid out in 1846 as terrace housing for railway employees; demolished in 1962. *See* p. 128.

PARK STREET, a side street off The Mount, was begun in 1836 with No. 7, built by Thomas Rayson senior, one of York's leading building contractors, as his own residence. Nos. 9 and 11 soon followed, but it was not until after 1847 that the rest of the street was developed. Like Mount Parade and South Parade, this was laid out as a single-sided street, with gardens opposite to each house on the S.W. side. *See* p. 128.

QUEEN STREET, formerly a branch of Thief Lane, had been named by 1830, but those houses which have survived later street-widening were mostly built in the 1840s, after the coming of the railway. *See* p. 129.

RAILWAY STREET, at first named Hudson Street, in honour of George Hudson the railway promoter, was being laid out in 1846 as part of the approaches to the (old) Railway Station but no houses were built until after 1851. (Renamed George Hudson Street 1971).

ROUGIER STREET was laid out in 1841 as part of the new approaches to the (old) Railway Station and was mostly built by 1843. It was named after the family of Rougier, proprietors of a large comb-making factory and of much of the area. Almost all the houses were demolished in 1961. *See* p. 129.

ST. MARTIN'S LANE, a narrow lane, was *Littlegate* in the 13th century; in 1282 it contained twenty tofts owned by sixteen proprietors. Later it provided some accommodation for light industry and warehousing.

(108) WAREHOUSE, No. 12, was built in the late 17th century, apparently as a warehouse, on an L-shaped plan, with a substantial main range and a smaller wing at the S.W. corner, both having interconnecting vaulted basements with external entry only. That the building was not originally a dwelling is proved by the absence of original fireplaces, the very small number of original windows, and the lack of any evidence of former partitions. The floors, with their substantial joists, appear to be later replacements. A small wing at the S.E. corner is an early addition.

The E. elevation, where not concealed by stucco, is of stretcher-bond brickwork with every sixth course in headers and with a band at first-floor window-sill level. To the first floor and the attics are original windows with elliptical arched heads of one course of headers. The gable end has tumbled brickwork; to the S. the lowest part of the gable parapet and the kneeler courses are incorporated in the E. wall of the loftier S.E. addition. The S. elevation of the S.E. addition is mostly masked by modern additions; the gable parapet is in tumbled brickwork, obscured by later stucco. In the W. wall, on the first floor is an early 18th-century window with sliding sashes with thick glazing bars. In the S. wall of the main range the central entrance to the basement is original, the ground-floor entrances are not. Centrally on the first floor is a late 17th or early 18th-century window with pegged frame.

ST. PAUL'S SQUARE was not laid out until *c.* 1855. No. 46 (*see* p. 129) is an earlier house adjacent to Holgate Road.

SKELDERGATE, mentioned in the 12th century, was the main dockside street of York, running parallel to the W. bank of the Ouse below the Bridge. On its eastern side were the Queen's Staith and wharves, with the Crane at the S. end, together with warehouses. On the inland (W.) side were the houses of prosperous merchants. At the Crane was the quay (The Old Wharf) at which all foreign merchandise was landed and stored in bond; and until the late 19th century this was the port of York at which seagoing vessels docked.

In 1282 the Husgable Roll shows that there were sixty-eight tofts in Skeldergate, rather more than half the number in Micklegate, to which it ranked second among the streets of York to S.W. of the river. That this large and important thoroughfare should take its name from the shield-makers, as suggested by the forms of the place-name, seems almost inconceivable, apart from the fact that there is no evidence of any kind that shield-makers, or any related craftsmen, ever worked here. It appears more probable that the name is a corruption from ON *skelde*, a shelf, and refers to the position of the street on the strand of the Ouse, beneath the low cliff, terraced in Roman times, above which stood Bishophill; alternatively, the name may be derived from a person, Skjoldr, by analogy with Skelderskew in Guisborough (EPNS, *North Riding*, 149). *See also* p. 129.

(109) HOUSE, No. 16, early 19th-century, was originally a dwelling, but in the latter part of the century it was converted to a malt house by removal of the first floor, roof and partitions, leaving only the outer walls. A heating chamber was inserted, with the malting floor at the level of the original first-floor window sills. At ground level a passageway runs on all four sides, between the outer walls and the furnace. The building is of brick with slate roof. The street front, of two storeys and five bays wide, was originally symmetrical. The central entrance is a later 19th-century replacement; the windows have flush moulded frames, segmental arches and red brick dressings. Above the oversailing courses which formed the eaves cornice the brickwork has been heightened by four courses. The malting floor, of glazed, perforated tiles, approximately 12 in. square by 2½ in. thick, is supported on angle irons. *Demolished 1965*.

(110) WAREHOUSE, Nos. 18, 19, has been formed from two houses, both of the early 19th century though of separate builds. Only the fronts now survive. They are of three storeys, of brick in Flemish bond and with slate roofs. No. 18 has a first-floor sash window with flat-arched head in rubbed brick; the second-floor window has a sliding sash. There is a timber cornice with simple mouldings. No. 19 has two windows with sliding sashes, stone sills, and stone lintels engraved to imitate voussoirs and keystone. There are projecting timber eaves supported on coupled brackets. *Demolished 1965*.

(111) HOUSES, Nos. 21, 22, were built as a pair in *c.* 1740. In the late 19th century, the whole of the ground floor of No. 21 was removed to make a carriageway through to commercial premises behind. It is doubtful whether any window in either house retains its original frame and glazing, and three original window openings on the first floor of the Skeldergate front of No. 21 have been blocked.

The front to Skeldergate, of two storeys and six bays in width, has been greatly altered. The back elevation towards the river, of three storeys and in good quality Flemish-bonded brickwork, has a band at first-floor level and a secondary band at the level of the sills of the first-floor windows. The entrance has a semicircular head, and the window beside it a segmental head. To the first floor are three windows with flat arches of fine rubbed brick, four courses deep, with saw cuts simulating joints. To the second floor are three smaller sash windows with flat arches of rubbed brick, three courses deep.

Inside No. 22, the *Staircase* has a moulded closed string, turned balusters of uniform height with square knops, and rectangular newels; on the landings the newel posts are rather narrow but extend across the handrails of both flights (1¾ in. by 5¼ in.); at the top of each flight, the upper moulding of the handrail is swept up to the newel. Exposed in the attics is the front principal and collar of a collar-beam roof truss supporting two purlins.

(112) HOUSE, No. 48, probably of very early 18th-century origin, was considerably modified by the addition of a storey later in the century; the entrance is also of the later date. Further alterations belong to the Regency period, when the staircase was reconstructed.

The street elevation (Plate 189), of four bays, is in stock brick with good red brick dressings. There is a handsome late 18th-century entrance in the second bay from the E. (Plate 63). A change of brickwork occurs at sill level on the second floor. The late 18th-century moulded cornice is of timber. The W. part of the rear elevation projects some 8 ft. To the W. of an outbuilding, on the ground floor is a large sash window with a flat arch in rubbed brick; on the first floor are two sash windows with segmental arches of brick headers and ashlar sills. There is a lead rainwater head and fall-pipe; opposed fleurs-de-lis decorate the holdfasts.

Inside, the ground-floor rooms to the rear are in derelict condition. The N.W. room, with a simple 18th-century plaster cornice, has two windows with simple architraves and, in the S. wall, a doorway with moulded architrave and a dentil cornice above a swept frieze. The *Staircase* has a closed string, slender balusters (Fig. 18s) and mahogany handrail. Up to the half-landing above the first floor it is of *c.* 1820; thence it is of early 18th-century date with substantial turned balusters (Fig. 17c).

On the first floor, the N.E. room has in the S.E. corner an angle fireplace with moulded surround and frieze and with a cast-iron firegrate enriched with urns and foliage typical of the late 18th century. The N.W. room, redecorated in the Regency style, has a reeded plaster cornice with floriate

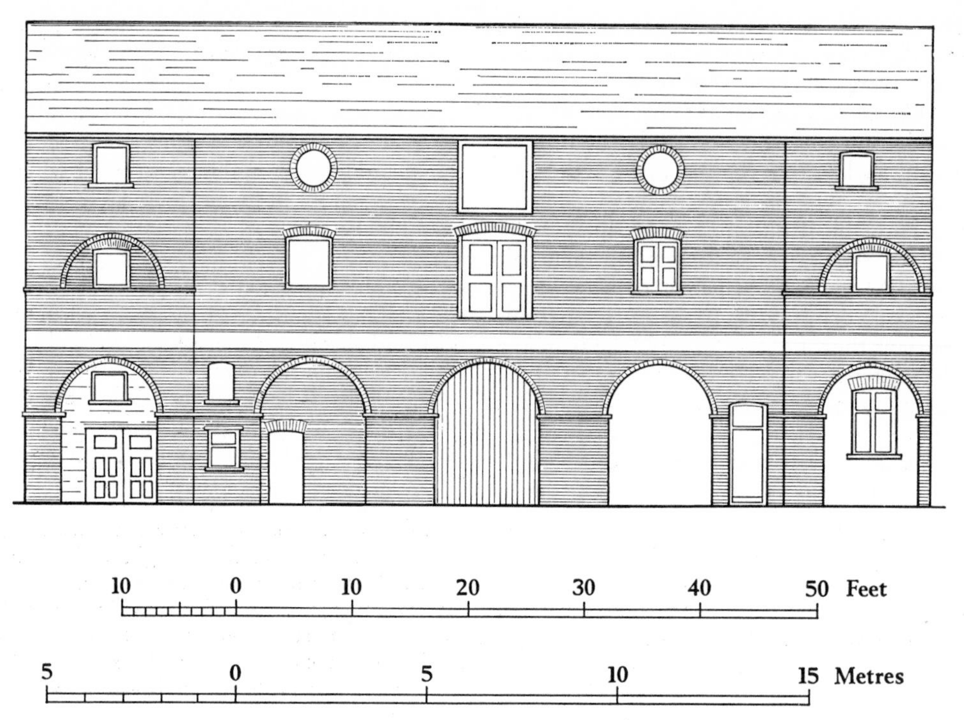

Fig. 68. (114) Sawmill No. 52 Skeldergate.

paterae at the angles. The window and a doorway to the landing have reeded architraves with square angles and paterae; and in the S. wall the white marble fireplace has reeded jambs with incised circular paterae. *Demolished 1964.*

(113) HOUSE, No. 51, was built in the late 18th century; it is of moderate size with an entrance passage running through the house at one side and the staircase opening off the passage, between the front and back rooms. It has been extended for use as offices.

The street front (Plate 189) is built in Flemish bond with a plain string at first-floor level and a narrower band joining the sills of the first-floor windows. Over the windows are flat arches of gauged rubbed brick. The entrance has a simple late 18th-century doorcase with flanking pilasters and an open pedimental hood supported on brackets; over the door is a semicircular fanlight. At the wall head is a timber eaves cornice enriched with dentils and modillions. Inside, on the ground floor the entrance passage is spanned by simple arches at the entries to the stair hall from both front and back. The *Staircase* has slender turned balusters of stained softwood and a swept handrail; there are no newels. In the first-floor N. room is a modillion cornice.

(114) SAW-MILL, No. 52, was built for John Henry Cattley; the foundation stone was laid in May 1839 (*Yorkshireman*, 18 May). The walls are of brick and ashlar and the roof is covered with slates and asbestos.

The street front (Plate 189; Fig. 68) is symmetrical and of five bays, those at the ends projecting slightly. The five ground-floor round-headed arches were originally open. Most of the rear elevation is obscured by later additions. Near the middle are two semicircular arched openings and to the E. is a doorway with a segmental arch of brick headers. On the first floor are two casements, a wide central opening and a doorway further E., all with segmental brick arches. The openings to the second floor have been blocked with brickwork. The hipped roof has seven king-post trusses of heavy scantling, with oblique struts and a ridge-beam; the main joints are reinforced with iron straps.

(115) HOUSE, No. 53, incorporates some remains of an 18th-century building which belonged to John and Dennis Peacock, timber merchants. In its present form it dates from *c.* 1840, when it was acquired by J. H. Cattley, owner of the adjoining saw-mill, No. 52 (114), with which it is roughly contemporary. The ground floor has been altered, but the first floor remains as planned and is unusual in having two small internal rooms or large cupboards. The wing behind retains parts of the red brick walls of the earlier building.

The street front is of white brick in Flemish bond. On each floor are two windows, those on the ground floor being offset to balance the doorway; this last has a later doorcase. The windows of the top floor are shorter than those below. The

Saloon.

Main stair hall.

(66) MICKLEGATE. Garforth House, No. 54. Ceilings, *c.* 1757.

(65) MICKLEGATE. House, Nos. 53, 55. *c.* 1755.

(65) MICKLEGATE. House, Nos. 53, 55. Staircase, *c.* 1755.

(66) MICKLEGATE. Garforth House, No. 54. *c.* 1757.

68, 70) MICKLEGATE. Houses, Nos. 57, 59 and No. 61. 1783 and *c.* 1785.

(79) MICKLEGATE. Houses, Nos. 85, 87, 89. *c.* 1500 and later.

(80) MICKLEGATE. Bathurst House, No. 86. Early 18th century.

(72) MICKLEGATE. House, No. 68. Staircase, *c.* 1650.

(81) MICKLEGATE. Micklegate House, Nos. 88, 90. *c.* 1750.

(81) MICKLEGATE. Micklegate House, Nos. 88, 90. Main entrance, *c.* 1750.

(81) MICKLEGATE. Micklegate House, Nos. 88, 90. Entrance to stair hall, *c.* 1750.

STAIR HALL CEILINGS

(81) Micklegate House, Nos. 88, 90. *c.* 1750.

MICKLEGATE

(65) House, Nos. 53, 55. *c.* 1755.

'Dick', a lap dog.

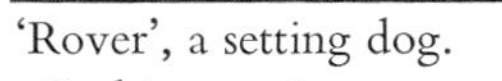
'Rover', a setting dog.

(81) MICKLEGATE. Micklegate House, Nos. 88, 90. Glass in Library, by William Peckitt, 1756.

(81) MICKLEGATE. Micklegate House, Nos. 88, 90. Main staircase, *c.* 1750.

(95) MICKLEGATE. House, Nos. 118, 120. Fireplace in Saloon, *c.* 1742 and later.

(95) House, Nos. 118, 120. *c.* 1742.
MICKLEGATE

(95, 96) Houses, Nos. 118, 120 and Nos. 122, 126. Backs. *c.* 1742 and late 17th century.

openings have flat arches of rubbed brick, plastered soffits and jambs, and stone sills. The wooden eaves cornice is supported on shaped brackets. The roofs are of slate. In the rear elevation, the two windows lighting the stair have round heads and stone sills. In the rear wing, the S.E. wall, which is of four storeys, incorporates older brickwork up to about 12 ft., which is not bonded in to the brickwork of the later house. The S.W. wall is also of older red brick in the lower part and has a central vertical strip of red brick up to about two-thirds of the wall height, which may be an earlier chimney incorporated in the later wall.

Inside, the ground floor has an entrance passage leading through to the stair hall, which is demarcated by two arches across the passage. In the side walls are two corresponding arched recesses enclosing doorways to the front room and to the adjoining premises. In the stair hall the arched opening to the passage is matched by an adjoining arched recess, which may also have been a doorway to the front room. The main *Staircase* has turned balusters (Fig. 18t) and an open string with shaped ends to each tread. The service staircase in the rear wing has square balusters and newels. The large front room on the first floor has an elaborate plaster cornice and a white marble fireplace surround. The windows have splayed jambs and moulded architraves. In the ceiling is a gas fitting suspended from a large circular feature of radiating acanthus leaves. The rooms on the second floor have fireplaces with very simple surrounds, excepting one in the rear wing which has a timber surround with sunk panels and a mantel shelf with ribbed edge.

(116) HOUSE, No. 55, now divided into three flats, was built in the late 18th century. It has been greatly reduced in size since 1850 by the removal of a block of buildings, possibly service quarters, from the back. Little of architectural merit remains since rebuilding and alterations following upon war damage in 1942. The house was in a single occupation until after the middle of the 19th century, the last private occupier, from *c.* 1825, being John Kirlew, a merchant. By 1879 it was 'two messuages formerly in one' occupied by the owner William Parker, an engineer and millwright, and Joseph Burill (Title deeds, Ings Property Co.; Directories).

The street front, rebuilt or cased, has coarse brickwork to the ground floor and cement rendering to first floor. The W. half of the rear elevation is cement rendered, presumably covering the rough brickwork left when the back range was demolished; the windows of this elevation are modern. Inside, the only feature of note is the *Staircase*, with turned balusters with square knops, a closed string, turned newels and a swept handrail.

(117) HOUSE, No. 56 (Plate 189), is a large town house built in the second half of the 18th century, probably to designs by John Carr. The site belonged before 1760 to George Pawson, a York merchant who moved to London and leased the 'house, garden and cellars' to tenants until 1769, when he sold the freehold to Ralph Dodsworth, merchant, Lord Mayor of York in 1792. Dodsworth was Sheriff in 1777–8, and since it was customary for the sheriffs to entertain in their own houses, it seems likely that the present building dates from about that time. After Dodsworth's death in 1796 the house was let to Thomas Smith the younger, a York merchant, who bought the freehold in 1807 for £2,615, including the cellars, two other houses and a riverside warehouse. In 1825 Smith conveyed No. 56 alone to William Cooper for £900, and on his death in 1841 it was inherited by his son Henry Cooper, who was the occupant, with the 'garden, warehouses and wine-vaults in Skeldergate' where he carried on business (Title deeds, Ings Property Co.; Rate Books of St. Mary Bishophill Senior, at St. Clement's; Directories). The house was greatly disfigured in *c.* 1925 by driving through it a carriageway to give access to a yard at the rear, for which important rooms have been destroyed and a service wing has been demolished.

The street front is shown in the accompanying drawing (Fig. 69) with the two destroyed windows reinstated, and in Plates 189 and 63; it is of brick with ashlar dressings. The ashlar plinth which projects well forward formerly carried iron railings. The windows have fine brick splayed arches, five courses deep, the voussoirs being single large gauged bricks with saw-cuts simulating joints. The central window on the first floor has a moulded and eared architrave with a pulvinated frieze. The flanking windows have moulded cornices above the brick arches. The pedimented cornice is of timber. At the apex and on either end of the pediment are stone-capped plinths, those at the ends carrying urns. The W. side elevation is in good quality brickwork, with the first-floor band returned from the front elevation. On the first floor, to the S., is a Venetian window which originally lit the staircase. At the back, the S., all the brickwork excepting that of the W. extremity and the uppermost part and pediment, is irregular as a result of the demolition of a great bow window and the domestic wing. A lunette window remains in the pediment but all the other openings are modern.

Inside, the ground-floor plan originally comprised two reception rooms divided by a central entrance passage leading to a stair hall at the rear with the staircase leading off at right angles. The E. room retains most of its original fittings; it has a moulded skirting and dado, the latter enriched with wave motif, and simple panels above formed by stucco mouldings, the panel over the late 19th-century fireplace in the E. wall having a central urn and swag enrichment at the head; the plaster cornice is enriched and the ceiling is delicately patterned with scrolls, roundels, etc. (Plate 61).

The former stair hall has been greatly altered by the blocking or destruction of the arched openings in all four walls. The surviving pilaster-responds to the archways have sunk panels modelled with beribboned husks and simple caps enriched with acanthus foliage; there is a modillion cornice throughout. The original main staircase has been destroyed, and a modern staircase, incorporating some of the original balusters and part of the handrail has been inserted in the entrance passage. A late

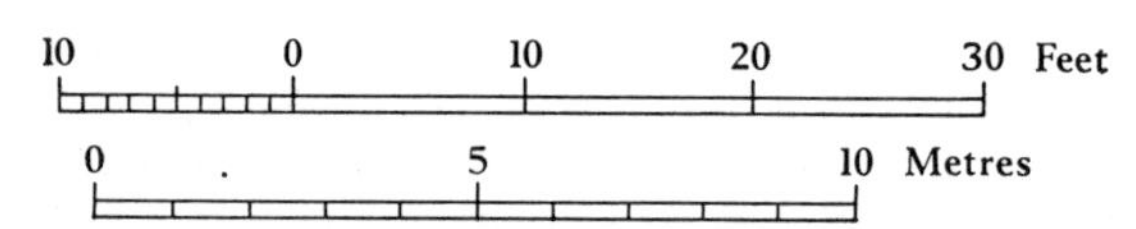

Fig. 69. (117) No. 56 Skeldergate.

18th-century fireplace surround (Plate 73) has been removed.

The first floor has been greatly altered by conversion into flats and insertion of partition walls. The E. room remains intact, with original fittings; at the S. end is a recess with enriched soffit and flanking Tuscan pilasters, and quarter pilasters. A fireplace in the E. wall has marble slips, moulded surround, flanking pilasters with sunk panels decorated with ribbon bows and corn husks, and fluted consoles supporting a cornice shelf; the unusual frieze has six raised elongated octagons with linking bars and containing carved flowers with swirling petals. The cast-iron grate is of *c.* 1835.

(118) THE PLUMBERS' ARMS, No. 61 (Plate 46; Fig. 70), of two storeys and attic, consisted for the most part of a timber-framed house of *c.* 1575, which was jettied on the N.E. front to the street and had two large projecting brick chimney stacks on the N.W. side. In *c.* 1600 a three-storeyed annexe (a on plan), jettied on both upper floors, was added in the N. corner. At the same time the space between the original brick stacks was enclosed to form a closet on each floor (b). The wainscotting throughout the house may well have been inserted then or shortly after. In the early 18th century a staircase was added on the S.E. (c); this structure had become enveloped in 19th-century brick additions. During the late 18th century a large, two-storey bay window (d) was added on the S.W. end of the original house. The premises did not become a public house until after 1850 (Benson, III, 166; Directories). The building was

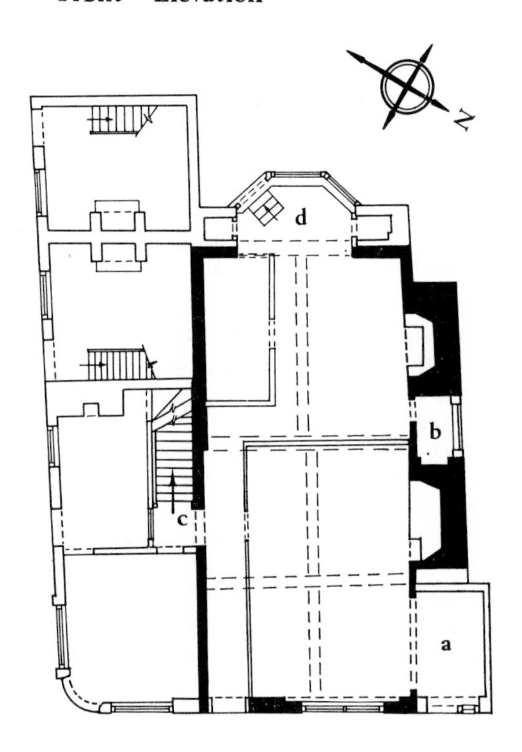

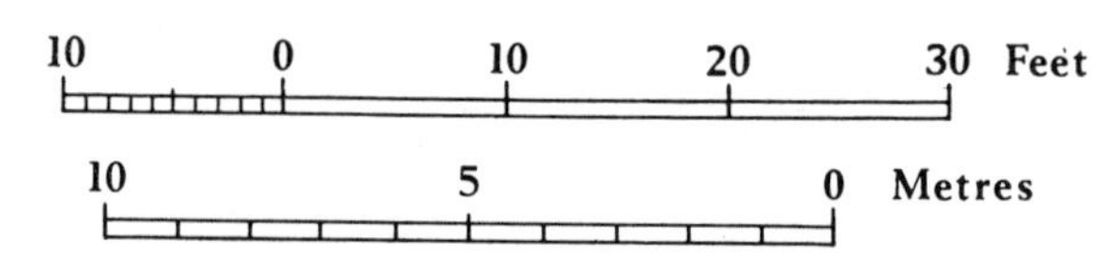

Fig. 70. (118) Plumbers' Arms, No. 61 Skeldergate.

demolished in 1964, when valuable dating evidence came to light.

The original front to Skeldergate was rendered over some of the timber framing. Fixed to the butt ends of the joists of the jettied floor was a fascia board carved with Renaissance egg-and-dart ornament. The fascia boards of both jetties of the N. annexe were copies of the foregoing. All the openings in the front were of the early 19th century and later, though the modern casement window projecting on timber brackets on the second floor of the annexe possibly replaced an earlier window of similar design. The N.W. side of the building was mostly covered by adjoining late 19th-century premises. The S.E. side was masked by a 19th-century brick addition of two storeys, replacing an older addition. The back (S.W.) elevation was mostly rebuilt in the late 18th century when the bay window was added, though the original tie-beam was retained in the rebuilt gable.

The building technique of the timber framing and the roof construction of the original house were revealed during demolition. The close-set studding had infilling of bricks set on edge and keyed to the framing by mortar, the studs being grooved to form the key (Fig. 71). The simple common-rafter roof had single side purlins and arched collars at intervals along it; the rafters fitted into notches cut in the wall plates.

Inside, the ground floor was divided into two rooms from the first by a timber-framed partition wall, but the latter had been replaced later in brick. Each room had in the N.W. wall an original brick fireplace concealed by a modern one. The front room, with access to the annexe, had moulded ceiling beams intersecting at the centre. The S.E. partition consisted of panelling of *c.* 1600, brought forward from the outer wall to leave a passageway behind. The rear room had been sub-divided on the S.E. with panelling reused as partitioning. The early 18th-century staircase in the S.E. addition had square-section newel posts with attached half balusters, finely-turned oak balusters (Fig. 17g), a closed string and moulded handrail.

Fig. 71. (118) Plumbers' Arms. Framing at junction of main building and annexe.

The first floor originally comprised one large room heated by two fireplaces in the N.W. wall. When the house was wainscotted a partition wall of timber studding was inserted under the south-western transverse beam, creating two rooms, each with one fireplace. The front room, which had square-sectioned transverse and longitudinal ceiling beams intersecting at the centre, was panelled throughout. The annexe room entered from it was also panelled throughout, and behind the panelling were found shop-made timber-framed windows (*see* p. lxxviii) of two, three and ten lights, that of ten lights in the N.W. wall being transomed; all had been blocked. The rear main room was also fully panelled. Removal of the panelling and the 18th-century fireplaces in both rooms revealed the original 16th-century brick fireplaces; they had four-centred arches with hollow chamfers continued down the jambs to simple run-out stops. Between the two chimney-stacks a small closet had been formed and the original timber-framed wall cut through for access to it; the closet contained a two-light timber-mullioned window. A similar addition had been made below on the ground floor. The room on the second floor of the annexe had timber-framed walls (Plate 48) with two and three-light mullioned windows, all shop-made as before. *Demolished 1964.*

SOUTH PARADE was a terrace of twenty houses to be built by the York Commercial Building Company, consisting of forty subscribers. A foundation stone was laid in 1824, and by the end of the next year the first three houses were assessed to the rates. By 1827 Nos. 1–11, 13–15 and 17 were inhabited, and all were completed by the end of 1828. No. 16 was the home of Thomas Rayson, senior (1764–1836), from 1828, and it is probable that he was the contractor for the series; at No. 17 the first occupier, from 1827 to 1833, was Peter Atkinson (II), architect, very likely the designer of the terrace. The parade was designed as a venture in relatively high-class housing for genteel occupiers, on a larger scale than that provided by Mount Parade (*q.v.*).

SWANN STREET was built as small-scale terrace housing in 1828–9; it was demolished by 1967. *See* p. 129.

TADCASTER ROAD is a continuation of The Mount and Mount Vale, leading to Dringhouses. *See also* pp. 116–18 and 129.

(119) THE WHITE HOUSE, Nos. 238, 240 (Plate 196), was built in the 17th century; it is called 'New Inn' in an engraving by John Haynes, dated 1731, when it was still a simple oblong on plan. In the later 18th century additions were made to N. and W., and the whole house remodelled inside and out. An angle, formed by the added rooms, was filled by a one-storey extension in the 19th century, when various offices were erected against the W. side. For a long time this was called the Gallows House, being opposite 'Tyburn', the York gibbet on the Knavesmire. In the 19th century it became a starch factory; then it was used by the York and Ainsty Hunt as kennels, and later became licensed again, as the White House Inn. Eventually the proprietors, the Station Hotel, allowed the licence to lapse (J. W. Knowles, York Notes MS., 1890, 72–3) and the premises became a laundry, later divided into two houses.

The front elevation, plastered and colour-washed, with a low stone plinth of flagstones set on edge, is of brick, part refaced in the 18th century. To N., at ground floor, is a sash window with shutters, each with three fielded panels. The doorway, to N., has a round-headed opening with a radial fan-light, fluted pilasters, and a plain frieze with triglyphs, supporting a moulded and dentilled pediment. There is a plain doorway further S. The fenestration is irregular with hung-sash windows and projecting bay windows. At the eaves is a cornice with dentils and modillions, and there are two fluted urn-shaped water heads (late 18th-century).

On plan the original part of the house comprises a straight range of three rooms, all with fireplaces, and an entrance hall. The two rooms to S. share a common chimney with fireplaces back to back and an entrance lobby at the side of the chimney. The N. room is separated from the others by the entrance hall and has a chimney projecting boldly on the N. end wall. The original open fireplaces, with openings spanned by chamfered oak beams, have been partly filled in to receive later firegrates. The staircase (Plate 86) in the entrance hall has a moulded rail and string, and turned balusters; it is of the early 18th century. The first-floor fireplaces have surrounds of the late 18th and early 19th centuries. There is some 17th-century panelling in one of the bedrooms but it is reset. *Demolished 1955.*

TANNER ROW with TOFT GREEN formerly constituted a part of North Street and was a back row at the northern end of the tofts along Micklegate. To some extent this street has always constituted a service road for Micklegate properties, with warehouses, coach-houses, stables and other subsidiary buildings linked to main burgages on the Micklegate fronts of the tofts. However, many of the back plots had been severed at an early date, and this allowed part of Tanner Row to be developed as an independent street. As its name denotes, many of these properties belonged to tanners, and this area was for long almost exclusively devoted to the leather industry.

Toft Green, though a name now applied to a street, was in mediaeval times a public open space next to The King's Toft, a royal holding on which stood the King's Houses and the royal chapel of St. Mary Magdalene. Soon after the Conquest the houses and chapel were in the keeping of the Maleshoures family, who held by serjeanty of finding benches for the County Court of Yorkshire and wax for the county seal. This implies that the King's House was then the administrative centre for the whole county, and this may have been so before the Conquest. These arrangements had become obsolete by the 13th century, and in 1227 the site and chapel were granted by Henry III to the Dominican Friars. The friars were later permitted to extend their boundaries, but when they wished to take in the whole Green in 1307 it was found upon inquisition that Toft Green, 17 perches long by 11 perches from the highway to the city ditch, was the only place in York large enough for the assembly of show of arms, the only spot for erecting military engines of defence, a common market from time immemorial, and the place of duel for trial by combat. This last function was evidently a survival from a period when the Shire Court was held here.

Toft Green was also the place where the properties of the guilds taking part in the Pageant Plays were kept, and became alternatively called Pageant Green. A Friday cattle-market continued to be held here until after the end of the Middle Ages, and a Wednesday swine-market had only been discontinued 'very lately' when Hargrove wrote in 1818. By that time, however, most of the area had been taken up by the New House of Correction, a prison for the City and Ainsty, finished in 1814. In 1839 this disappeared when the whole area was acquired by the York & North Midland Railway for the (old) Station. *See also* pp. 128, 133.

(120) HOUSE, No. 1, with No. 39 North Street, of two storeys has walls of brick and plastered timber framing with tiled roof. It was built in the late 15th century as a Wealden house having an open hall with a two-storeyed block on the E. side, jettied to both N. and E.; there was another jettied block on the W. side of the hall, but not communicating with it. Probably this represented part of a similar two-block Wealden house (*cf. Med. Arch.* VI–VII (1962–3), 216). This W. block was rebuilt or refronted in brick in the 18th century (G. Benson and J. England Jefferson, *Picturesque York* (1886), pl. 7). It fell into disuse towards the end of the 19th century and by 1929 had been demolished. In the early 17th century the hall and E. block were divided into two tenements: the hall was divided into two storeys, the new floor being jettied out to match the flanking jetties, and a central chimney-breast and two staircases

inserted. In the early 19th century further internal alterations were made. The roof was completely renewed probably when the W. block was demolished, and the building, after lying derelict for some time, has recently been renovated. It is of great interest as one of the most northerly recorded examples of the Wealden House.

The *North Elevation* to Tanner Row is plastered above chamfered rubble base-courses which carry a wooden sill-plate. Centrally there is a 19th-century doorway, and to W. a horizontal sliding-sash window of the early to mid 18th century. At the E. end is the corner post with moulded and carved cap (Plate 48) above which it curves out to carry the end of the dragon-beam. The upper storey is plastered with a jetty, the E. half original, the W. of the early 17th century.

The *East Elevation* to North Street, mostly rendered in cement, retains the chamfered rubble base-courses as far as the first doorway. On these rests the sill-plate. Next to the corner post is an early 19th-century bow window, and to S. of this the original doorway to the house, altered in the early 19th century. To S. of this is a second doorway, entirely of early 19th-century date. The plastered upper storey is jettied with later windows. The *South Elevation*, now cloaked by the adjoining house, has been rebuilt in brick.

The *West Elevation*, timber-framed with plastered gable, was originally the partition between the hall and the W. range. The plain framing is exposed and shows at the N. end the twin corner-post construction required by the Wealden design (Plate 48). Pegholes and one surviving curved strut indicate that the roof was probably of the standard York crown-post type (*cf.* Fig. 13c). The existence of the W. range is shown by grooves and mortices in the W. face of the main timbers.

On the *Ground Floor* the E. room has the timber-framing exposed. In its N. wall is an original opening (now blocked) for either a window or an open shop-front (*see* p. lxxviii). In the E. wall are the two jamb-posts of the original doorway; part of the chamfer is visible on the S. post. The rebuilt W. wall has an early 19th-century grate inserted into the 17th-century chimney-breast. To S., behind a later brick partition is a staircase, probably of the 17th century, with wooden octagonal newel post. The W. room, which formed the lower part of the hall, has no original features visible, but in the S.E. corner is a second staircase also with an octagonal newel post.

On the *First Floor* the E. room retains the original framing in the N. wall; the layout suggests two small oriel windows, similar to those in Church Cottages, North Street (104). The W. room represents the upper half of the open hall, but the timber-framed N. wall is of 17th-century date except for the corner posts at each end. These belong to the jettied blocks flanking the hall, and each carries a small solid spandrel-piece to support the eaves across the front of the hall (Plate 48). The E. and W. walls also retain the original timber-framing. Near the N. end of each is the corner post of the original front wall of the hall. These posts carry the original N. wall-plate and under the floorboards, the corresponding middle rail. There is an intruded chimney-breast in the middle of the E. wall.

(121) House, No. 7 (Plate 190), goes back to the late 17th or early 18th century, but the only early feature is the upper part of the staircase which has bulbous balusters characteristic of that period. The front is of the later 18th century, as are also some internal features, such as the doors to the front rooms. It might, therefore, appear that the house is of late 18th-century date with parts of a reused staircase, but from the irregularity of the plan it seems more likely that an earlier house was refronted. There is nothing to indicate that the building was originally timber-framed, or that it had any relationship to the Old Rectory (No. 7A, (122)), adjoining on the W. In the S. wing some work was done in the first half of the 19th century; the back door has mouldings of *c.* 1840, and the lower part of the staircase has square balusters, which may be of the same date.

The *Front Elevation*, of rather poor quality red brick in Flemish bond, has at first-floor level a projecting band three courses deep and at the eaves, a dentilled wooden cornice. On each floor are two sash windows with flush frames and arches of headers; those at ground floor have almost flat arches and those at first floor low elliptical ones. The 19th-century door has four panels and a simple pilaster-type surround. To the extreme W. is the door to No. 7A with a plain surround. The *South Elevation*, faced with stucco, has a Yorkshire sliding-sash window at ground floor and a hung-sash window at first floor. The E. end of the front range is also stuccoed.

Internally, the doors are simple, with two or four panels, sunk or fielded, with ovolo-moulded framing. All the fireplaces have plain, unmoulded surrounds. The staircase to the first floor is early 19th-century, with square-section balusters, turned newels at top and bottom, and a swept, moulded rail. The upper part, to the attic, has turned bulbous balusters and a square newel with ball finial.

(122) The Old Rectory, No. 7A (Plate 190), bears on modern plaster at the back the dates 1498–1937; the latter is the date of a restoration undertaken by the Rev. P. Shaw when Rector of the adjacent All Saints' Church. No evidence for the date 1498 has been discovered, and the structure is probably of *c.* 1600 or later. The building is timber-framed in three bays; it originally had no chimney and there is no evidence of any internal partitions. This suggests that it may not have been built as a house, but because of its situation near the river frontage it would have been suitable for a warehouse. The house has been considerably altered by the insertion, probably in the late 17th century, of a central chimney-stack with back-to-back fireplaces at ground floor, and partition walls in both storeys. A spendid oak staircase of *c.* 1640[1] (Plate 82), said to have come from Alne Hall, was inserted in 1937.

[1] The date is suggested by the extremely close resemblance to the dated work of 1641–2 at New Parks, Shipton, by Thomas Ventris of York, carver.

The elevation to Tanner Row is gabled, with jetties at first-floor and at eaves level; the ground floor has been rebuilt in brick with hung-sash windows; the upper part is stuccoed and also has later windows. On the W. side the framing is exposed; the windows, two blocked below and two renewed above, occupy original window positions.

Internally much of the original framing remains exposed in the E. and W. walls together with the chamfered beams and the joists which carry the upper floor (Plate 190); between the main cross-beams there are spine beams carrying the joists except in the N. bay where the joists run longitudinally to form the jetty. In the attic the floor is of gypsum carried on exposed timbers. The roof is carried on simple trusses with purlins framed to the principal rafters (Fig. 13l).

(123) HOUSE, No. 16 (Plate 46), is largely of *c.* 1820 but contains elements of an earlier structure, probably erected in the late 18th century and represented by lower features of whitish-red brick and the ashlar stone front.

In the early 19th century Sir John Simpson, corn factor, lived here, giving his name to Simpson's Yard at the back, later incorporated in Botterill's Horse Repository. He was Sheriff in 1826–7; Alderman 1834 and re-elected 1835; Lord Mayor 1836 and knighted that year, fined for Lord Mayor in 1847 and again in 1853; he died in 1854, aged 58. His firm, L. & J. Simpson, corn millers, suffered a great fire in 1820 (*York Gazette & Herald*, 14, 28 Oct.), and it is not certain whether the house was affected, but about this time a third storey and attics, in a large dark-coloured brick, were added. The front porch was formed, the entrance hall remodelled, and flues added to the earlier chimney stacks. Fireplaces with good surrounds and Carron grates, although in the later build, would conform better with the earlier work of the late 18th century and may have been moved from the lower storeys at the remodelling.

The three-storeyed *Front Elevation* has a gabled finish. Ground and first floors are in ashlar with some stucco, and part of the second floor has ashlar facing to a former gable some 10–15 ft. below the present gable. At ground floor stucco covers the plinth and lower wall up to a band at window-sill level. To W. is the entrance, of Regency date, with reeded pilasters to the opening and a simple entablature; the door, set back in the thickness of the wall, has six fielded panels, a segmental fanlight, reeded pilasters with lions' masks on the caps, and small flanking lights, blocked with shutters. Two large windows to E. have segmental heads and raised key-blocks. At first floor are three sash windows with flat 'arches', each of a single block, and a plain band at sill level; above is a moulded string-course. Set within a central round-headed recess at second floor is a sash window with plain ashlar sill and, on either side, a flush-framed sash window with plain sill and flat brick arch. The ashlar carries up over the central window, indicating that the building was originally of two storeys with attics, lit by a window in the gable. The *Rear Elevation*, mainly of whitish-red brick but with the top floor and attics of a later, larger brick, has a projecting plinth, three bricks deep, and above it a stone sill band. There are two doorways, a two-storey bay window and hung-sash windows under flat arched heads.

The house is divided by a central passageway between the front office and two back rooms, leading from the entrance hall on the W. to the staircase on the E. The fittings to the ground floor are all of *c.* 1820 or later. On the upper floors some of the rooms have been modernised but some retain their fittings of *c.* 1820. The third-floor fireplaces have iron grates by Carron of late 18th-century style.

The *Staircase* is all of one date: the two flights to the first floor have cantilevered stone treads, stone slabs at the half-landing and landing, and a moulded pinewood rail, under which an iron strap holds the tops of slender cast-iron balusters, copied from turned wooden prototypes with a square knop; at the bottom, the rail spirals over an iron newel of the same form as the balusters; the rest of the staircase has softwood treads on a cut string, a rail as before, and turned wooden balusters like the others but sturdier, incorporating some of *c.* 1830 which have a series of annulets at knop level. The main feature of the staircase is the extreme slenderness of its members; there are similar staircases at the Black Swan, Coney Street, and 18 Blake Street, both of the late 18th century. The skirting of the first three flights is different from that of the others, which is common to the upper floors.

(124) THE UNICORN, No. 17, was originally a small 18th-century house, two-storeyed with an L-shaped plan comprising a range, one room wide, along the street and a wing extending to S. By 1804 it had already become The Unicorn public house, occupied by William Dale, and was owned by John Kilby, brewer of Tanner Row (YCA, E.96, ff. 30–1). In the first half of the 19th century a three-storeyed wing was built behind, filling in the angle between the former ranges and projecting further S. The original disposition of rooms has been lost in later alterations, and the interior is now substantially of *c.* 1840–50, or later.

The N. front was completely altered in the 19th century, though the upper windows could be in the original positions. The ground floor has windows and doors framed by timber pilasters and entablature of *c.* 1840–50. At the eaves is a timber cornice.

The whole of the ground-floor front range of the original building is occupied by the bar; simple moulded door architraves and the bar counter with panelled front are all of the mid 19th century.

TRINITY LANE already existed, under its present name, before the middle of the 16th century. It does not seem to have been of much importance, in spite of being the main link between Micklegate and Bishophill. In the 17th century it contained some light industry, as represented by the soap-boiling factory built by Nicholas Towers and long known as Towers' Folly (now No. 29). Through the 18th and 19th centuries its inhabitants included a few individuals of standing; among them was William Walton, Roman Catholic Bishop and Vicar Apostolic, who lived from 1774 until his death in 1780 in No. 25, now demolished (Rate Books; information from Rev. Fr. Hugh Aveling).

(125) JACOB'S WELL, a house of late mediaeval date, consisted of an open hall with a cross-wing to E.; it is impossible to determine whether there was also one to W. Another house of the same date now forming the S. half of the cross-wing, was originally a separate tenement. This common mediaeval form, of an open hall on the ground floor, is unusual in York. The hall was horizontally sub-divided by the insertion of a floor possibly *c.* 1600. By 1564 both tenements were in the possession of Isabel Warde, who conveyed them in 1566 to the parishioners of Holy Trinity as feoffees (Deeds in possession of the Feoffees). Isabel Warde, who died in 1569, was the last prioress of Clementhorpe Nunnery. In 1651 the property was described as belonging to 'the dissolved Priory in Micklegate' and in 1815 £130 was spent on it (MS book, Rectory, 45, 46). By 1822 it had become the Jacob's Well public house but in 1904–5, becoming unlicensed, it was purchased by the Rector and underwent extensive restoration to the designs and at the expense of Walter Harvey Brook; it is now used as a parish room for Holy Trinity Church.

The form of the original house presents some interesting features. The open hall was probably only one bay in length, the upper part of the W. wall being original; on the ground floor the hall was open to the room in the cross-wing to E. The entrance was on the N. side of the cross-wing and adjoining this now restored doorway are two original windows, each of three lights, in excellent condition. The ground-floor walls of the cross-wing, except to the N., are of stone and may be original or may have survived from an even earlier structure; otherwise, the whole structure was timber-framed, the cross-wing having first-floor jetties to the N. and E. The property now includes a two-storeyed brick-built section to N. of the cross-wing, and a third storey to the cross-wing, now inaccessible except by trap-door; either or both of these additions may be of 1815. Nothing survives of any cross-wing to W. of the hall, the modern rectory of Holy Trinity being built against that end. The very thorough restoration of 1905 included much replacement of old or lost timbers by new ones; the addition of a staircase to N. of the hall and a large bay window on the S. side; and restoration of the adjoining part of the cross-wing in the same style. A most important feature of the house is the pair of carved brackets supporting the canopy over the E. door; they came from the Old Wheatsheaf inn, Davygate, belonging to the Bishop of Durham (Cooper MSS. 4, 13), and are now the only example of their kind surviving in York.

The stonework of the E. wing (Plate 191) is confined to the ground floor on the E., S. and W. elevations except on the W. side where it extends to the top of the first floor at the S. end, but in this upper part the bonding is rougher and the work is probably not original. At the S. end of the E. front, where the jetty has been lowered, the stonework gives place to brick. Timber framing with widely spaced studs is exposed on the first floor of the E. front to Trinity Street and in the N. end of the cross-wing but here the later addition covers much of it; the ground-floor doorway and original windows and the jetty (Plate 191) are visible internally. The jetties of N. and E. elevations meet with a diagonal dragon-beam. The hall was wholly timber-framed but on the N. side only the main posts and rail remain and on the S. the lower part has been rebuilt in stone with a modern bay window.

In the N. wall of the cross-wing the doorway has been restored and a canopy over it removed; the adjoining windows retain the original timber mullions of diamond section, but the glazing is modern. On the E. front the timbering at the S. end is not original and the doorway is modern but has late mediaeval carved brackets (Plate 191), from the Old Wheatsheaf, supporting the canopy. They are cusped with figures carved on the points of the cusps and the spandrels are decorated with an eagle, a Tudor rose, and conventional leaves and flowers. On the W. side of the cross-wing two 17th-century windows to the first floor have chamfered brick jambs and mullions. In the N. wall of the hall a modern window follows the design of the original ones.

Internally, at the S. end of the cross-wing, is a substantial chimney-breast at the E. side of which is a small room which is said to have contained a staircase; it is separated from the main room by a stone wall. The E. end of the hall appears always to have been open to the ground floor of the cross-wing. The hall is now divided into two storeys with an upper floor carried on a chamfered beam and chamfered joists. The roof over the hall consists of paired rafters and collars supported by side purlins and a central collar-purlin carried at the W. end by a crown-post truss (Fig. 13b).

(126) HOUSE, Nos. 2, 4, was originally a late 16th-century timber-framed structure of at least three bays, jettied to the street. In the early 19th century, the jettied front was replaced by a flush brick elevation, probably on the line of the first-floor jetty; the back bay was rebuilt and extended in Victorian times.

The S.W. front (Plate 188), is of stuccoed brick, with flush-framed sash windows to each floor. The surround of the doorway to No. 4 (actually in the adjoining No. 6) has reeded pilasters carrying enriched brackets.

Some of the original framing is visible and on the second floor studs in the S.E. wall are all grooved for infilling (*cf.* Fig. 71) and numbered, and on an adjacent main post is a carpenter's mark (Fig. 11b). *Demolished 1956.*

(127) House, No. 6 (Plate 188), timber-framed and probably of the early 17th century, may have been jettied along the street front, but the front was rebuilt in brick, probably with the fronts of Nos. 2 and 4, in the early 19th century. The house consisted of at least two bays roofed parallel with the street; a central chimney breast was inserted in the 17th century. Enough of the roof structure survived to show that it had a slightly cambered tie-beam, with side purlins clasped between a collar and rafters. In the gable the studs (5 in. by 3 in.) are grooved for infilling and the rafters are pegged at the apex, with no ridge (Fig. 13k). *Demolished 1956.*

(128) House, No. 27, is the survivor of a pair built *c.* 1735 for Messrs Dawson and Hillary, wine merchants. The site was bought by Richard Dawson between October 1733 and January 1734/5 (YCA, E.93, ff. 75, 79, 80) and included a warehouse and cellars. By 1736 the house had been built (Drake, 265), and in 1740 Dawson advertised his property as 'a large convenient well-built House, with a garden and very large Cellars' (*York Courant*, 6 May). In 1746 he mortgaged the 'messuage and warehouse, counting house and cellars joining and under' (YCA, E.93, f. 183). Later in the 18th century the house was taken by William Tuke and his wife Esther to be a girls' boarding school under the auspices of the Society of Friends. The school opened on 1 January 1785 and continued until 1812, when it was taken over as a private venture, but closed in 1814 (H. W. Sturge and T. Clark, *The Mount School, York* (1931), 1–31). No doubt because of its extensive cellarage the building was again used by wine merchants and later became the Trinity House public house. In the early 19th century, narrow window openings in various places were replaced by broad ones. The house to E. was demolished before 1950.

The *Front Elevation*, of pale pinkish brick, has a plain plinth and, above the ground floor, a projecting band. A magnesian limestone cornice supports a brick parapet with stone coping. The main doorway has a round arch of rubbed brick with stone key-block and moulded stone imposts and a radial fanlight with very heavy glazing bars. The plan comprises a central entrance and stair hall with one room to each side and a wing projecting N. and W. All fittings have been removed from the ground floor. The staircase rises in three flights and has a moulded handrail swept upwards at each landing, a cut string and moulded newels and balusters. In a room on the first floor is a fine early 18th-century fireplace (Plate 70), with pine enrichment. A room to N.W., entered through an arched doorway with moulded architrave and panelled reveals, is panelled in two heights with moulded dado rail and skirting and fielded panels. The fireplace, set obliquely in the S.E. angle, has a moulded stone surround, with a heavy entablature and key-block. *Demolished 1961; enriched fireplace from S.W. room presented to the York Civic Trust and in 1969 reused in Nos. 17, 19 Aldwark, York.*

(129) House, No. 29 (Plate 188), basically a large 17th-century structure, four storeys high, has undergone such extensive alteration that its original form and use cannot be determined from internal evidence, though it would seem likely to have served some industrial purpose. The site seems to be that of a building, with large chimney-stacks and Dutch gables, seen on several of the early views of York and identified as Towers' Folly by Lodge (etching, 'The Ancient and Loyall Citty of York', no. 10). This implies that it was the soap-boiling factory built *c.* 1638 by Nicholas Towers (d. 1657), Sheriff at the time of his death (Skaife MS.). Its subsequent history is still obscure until *c.* 1800 when it became the horn and shell comb factory of John Nutt, a use which continued until after 1851 (Borthwick Inst., Rate Books of Holy Trinity Micklegate; Directories). In the next few years it was occupied by the Rev. Henry Vaughan Palmer whose daughter Henrietta, born here on 13 January 1856, was later an authoress under the pseudonym 'John Strange Winter' (T. P. Cooper, *Literary Associations of the City of York*, 7). It later housed St. Stephen's Orphanage for many years.

Of the first building there now survive the outer wall, much altered and with no original windows, but including the large chimneys on the S. side; internally one ground-floor fireplace at the W. end, and the basement. Later alterations include late 17th-century windows, since blocked or altered, and a very fine early 18th-century fireplace, now on the top floor. In the early 19th century the whole was completely remodelled and the central part heightened. Most of the existing windows and much of the fittings and decorations are of this date, though there has been subsequent alteration and subdivision of the rooms.

VICTOR STREET—*see* BISHOPHILL

ACOMB

Acomb lies on the W. side of York, beyond the hamlet of Holgate; it remained a rural parish in the Wapentake of Ainsty until its incorporation into the city in 1937. It was an ancient possession of the Church of St. Peter, York, from before the Conquest, though nearly one-eighth of its land belonged to the king until it was given by Henry I to the York Hospital. In 1222 the lordship of the manor was granted by Archbishop Walter de Gray in perpetuity to the Treasurer of York Minster, and the advowson of Acomb church was added 5 years later (C. T. Clay ed., *York Minster Fasti*, i, YASRS, CXXIII for 1957, 1958, 23). The manor of Acomb included the township of Holgate, though that formed part of the city parish of St. Mary Bishophill Junior, and also scattered parts of Clifton on the opposite side of the

Ouse. It was a populous area in the 14th century, for there were over seventy households in Acomb and Holgate in 1379 (*The Returns . . . of the Poll Tax*, YAS, 1882, 299). Acomb continued to be held by the Treasurers until the abolition of their office in 1547, when the manor passed to the Crown. In 1623 it formed part of an exchange whereby it reverted to the Archbishops, who leased it to tenants, for a long period members of the Barlow family, to whom it was sold in 1855.

Acomb consisted mainly of open arable fields and common waste until inclosure in 1776. The only houses were those in the village, mainly in Front Street but a few beside the Green and near the church. The Court Rolls refer to a number of houses as new built, lately rebuilt or to be rebuilt, in *c.* 1668 (roll of 1728), 1691–3 (Mon. 149), 1758, 1760, 1761, 1774 (two), 1783, and 1784 (YCA, deposited Acomb manorial records, and typed transcripts; H. Richardson, *A History of Acomb*, YPS, 1963; *Court Rolls of the Manor of Acomb*, YASRS, CXXXI (for 1968–9), 200, 248, 251, 252). *See also* pp. 123, 125, 130.

FRONT STREET is the main street of Acomb, running from E. to W., always with the main concentration of houses, occupying tofts regularly laid out on both sides.

(130) HOUSE, No. 2, was built in the second half of the 18th century; since then, fireplaces have been modernised and modern annexes erected to S. of the main block and at the S.E. corner. In the late 18th and early 19th centuries, the Pickford family lived here. Mrs. Pickford was a Radcliff and her son Joseph took that name; he was created a baronet in 1812 (G. Benson, 'Notes on Acomb', AASRP, XXXVIII, 93).

The house is nearly square on plan; the entrance hall between the front rooms gives access to a wide stair hall between the back rooms. The front is symmetrically designed, with a central doorway flanked by plain pilasters and shaped brackets supporting a cornice. On each side is a large three-sided bay window and at first floor are similar bays on either side of a sash window with a flat rubbed brick arch. Between the floors is a projecting brick band. Above, and running round the bays, is a dentilled cornice, and at each end is an oblong chimney-breast. At the back there is a new block to E. and a 19th-century addition projecting in the centre.

The *Cellar* has three divisions, each with a barrel vault; that to S.W. has an open fireplace between segmental-headed recesses, and some candle recesses. The *Staircase* has a moulded rail, a large turned newel at the bottom, with moulded cap and base and turned balusters with a square feature.

On the first floor two rooms retain 18th-century fireplaces and in one of the front rooms the original ceiling cornice is carried round the bay window showing the bay windows are part of the original design.

(131) VINE TREE HOUSE, No. 3, was built in the late 18th century incorporating some walling of an earlier 18th-century building. It became for a time the Black Swan public house. Most internal features date from the later 19th century, when a two-storey annexe was built at the N.W. angle.

The *South Elevation*, of yellowish red brick, has to E. a shallow curved bay window with two sashes, and a dentilled cornice. To W. is a doorway with plain pilasters and frieze, and a moulded canopy carried on modillions (19th-century). At first floor is a similar bay window and, over the doorway, a sash window with flat arch. On the E. side it is apparent that the wall to the top of the ground floor is of older brick, and that the first floor and the S. front are later. On the N. side is a large gable, with large sash windows on ground and first floors, and a small casement light to the attics.

Four rooms retain early 19th-century fireplaces, two of them being of marble. *Demolished 1965.*

(132) HOUSE, No. 4, of one and two storeys, with brick walls, was originally an early 16th-century cottage; it was extended in the 18th century, and has since been largely rebuilt and remodelled. The exterior, stucco-covered and with modern openings, is undistinguished.

A room to W., rebuilt on the old foundations, has a chamfered beam from an old house in Skeldergate. In the roof of a room to E. is a 16th-century moulded ceiling beam, said to belong to the original cottage. The room is panelled from floor to ceiling with oak wainscotting of the early 17th century said to have been brought from a house in Colliergate. The remainder of the house has 18th-century doors, with four fielded panels, and some reused beams.

(133) HOUSE, No. 6 (Plate 192), was built in the first half of the 18th century as an oblong block with a one-storey annexe to S., on which a second storey was built in the later 18th century. Another addition was erected against the W. gable and the openings in the N. front were modernised in the 19th century.

At ground floor, in a room to N.E., is a fireplace with an early 18th-century eared and moulded surround, beneath a pediment formed by two volutes and decorated with a shell and festoons. Above, added in the late 18th century, is a mantel, consisting of two fluted pilasters supporting a plain shelf, and a frieze decorated with a mask between two oval paterae. A room to N.E., at first floor, has a fireplace with panelled jambs, segmental head with enriched key-block, and oval paterae at the angles. In a room to S.E. is a late 18th-century fireplace with wooden surround and moulded mantel. The mid 18th-century *Staircase* has a heavy moulded rail, no string, and turned balusters.

(134) HOUSE, No. 8 (Plate 192), was built in the early 18th century; it was altered in the late 18th century, when the front bay was added, and has been modernised.

In the N.E. room at ground floor, is a late 18th-century fireplace (originally at first floor) having reeded jambs bound with garlands of ivy, reeded lintel, and central panel decorated with Venus and Cupid (Plate 76). On the first floor, the fireplace in a room to S.E. has a late 18th-century basket grate with urns and festoons on the jambs. In another room is a late 18th-century fireplace with reeded jambs and lintel, and oval paterae at the angles.

(135) HOUSE, No. 9, was built in the middle of the 18th century; a two-storey extension was made to N. in the 19th century, and in modern times an annexe has been added to the N.W. corner and two shops built against the S. front.

The street front (Plate 135), partly obscured by later shops, has at first floor two windows, each with a rubbed brick flat arch. The E. side is symmetrically designed with a three-course band between the stages. The central doorway has a moulded cornice supported by shaped consoles (19th-century), and a window on either side; there are two similar windows at first floor. Both elevations have a modillioned cornice at the eaves and a dormer in the hipped roof.

The ground floor has been modernised but retains an 18th-century archway to the stair hall, having an enriched architrave and fluted pilasters with enriched caps. The *Staircase* has a moulded rail and string, heavy turned newels and turned balusters with a square knop. It is perhaps earlier than others of the same form in Acomb, having a marked string which the others omit (*c.* 1750). On the first floor are two simple 18th-century fireplaces. *Demolished 1965.*

(136) THE WHITE HOUSE, No. 12, was built in two parts in the first half of the 18th century, that to W. being perhaps the earlier; in the second half of that century a W. wing and staircase were added and a servants' stair inserted in the older part. The whole was remodelled in the early 19th century, when various windows were blocked, bay windows added, and the roofs remade in places, the S. wall of the W. range being raised to make the roof line correspond with that to E.

The elevations are irregular; there is a projecting band between the storeys and a small early 19th-century cornice at the eaves. The entrance doorway has reeded pilasters and shaped brackets flanking a rectangular fanlight and carrying an entablature with reeded frieze.

Internally, the cellars are covered by early 18th-century vaults. The main staircase, of the second half of the 18th century, has balusters alternately turned and twisted; the servants' staircase, with turned balusters, is of the same date. Several rooms have fireplaces of the 18th and 19th centuries, all of simple design.

Outbuildings etc. (i) To S.E. of the house is a block of buildings of four different dates, including a stable, brewhouse and dovecote of the early 18th century, and a coachhouse and cottage of the early 19th century. In the stables are two loose-boxes, entered under round-headed archways with moulded architraves, supported on a column and pilasters, with plain timber shafts and moulded caps. Under the dovecote, is a high room containing a large open fireplace with a big copper at its side. (ii) In the garden, to S., is a small brick summer-house, square and with hipped pantile roof. The door is of early 19th-century Gothic type. (iii) Many stones used in the garden rockeries have deeply cut 13th-century mouldings, and perhaps came from St. Mary's Abbey. *House and outbuildings demolished 1960.*

(137) MANOR FARM, No. 14, which formed a single building with No. 14a (138), was erected in the late 15th or early 16th century, perhaps the house of the 'farmer' of the manor before the Reformation. It later became the Manor house but in the late 19th century it was divided into two cottages, and new windows, doors and staircases inserted. It is the only timber-framed building remaining in Acomb, and the oldest house.

None of the framing is visible on the exterior, but some is exposed internally and is remarkable for its massive size. Ceiling beams on the ground floor are all cased but good framing is visible on the upper floor, in general consisting of heavy posts carrying wall-plates and supporting large tie-beams. In the cellar is a large chamfered beam and oak joists. In the roof, the principal rafters are about 1 ft. 6 in. square, and chamfered, and there are collar-beams and queen-posts.

(138) HOUSE, No. 14a, comprises a central portion built in the late 15th or early 16th century, which was originally a wing of the Manor Farm (137), rooms to N. not much later in date and having similar framing, and to S. a brick kitchen of the late 17th or early 18th century. At the end of the 18th century the earlier walls were largely recased in brick, and the section connecting this wing to monument (137) may have been removed at this time. The scullery was added in the 19th century and most of the windows were renewed in the same century.

Internally some of the heavy timber posts and beams are exposed. Repairs to the N. wall have shown that it consists of studs and infilling of old narrow bricks. On the upper floor are a number of 18th-century doors showing two panels on one side and vertical planks on the other.

(139) HOUSE, No. 17, was built in the middle of the 18th century on an L-shaped plan against No. 19; its S. front was altered at the end of the 18th century. Most of the fireplaces were inserted at the beginning of the 19th century, when a large bay window was built, extending the dining room to N., and a passage constructed, giving access to a new kitchen to N.W.

The S. front, plastered and with rusticated quoins, has at ground floor two large sash windows and a doorway having fluted engaged pillars with moulded bases and foliated caps, supporting a simple entablature. Between the stages is a band, and at first floor are three windows similar to those below. The early 19th-century cornice has coupled gutter brackets and the roof is of large Westmorland slates.

Internally, 18th-century fittings include panelled doors, window shutters and dado panels under the windows, and the main staircase which has a moulded rail, no string, a square newel at the top and a cluster of four balusters at the bottom; the balusters are turned, with a square knop, alternate ones being twisted. *Demolished 1966.*

(140) LABURNUM HOUSE, No. 19, was built in the early 18th century, but in the middle of that century a new extension was built to E. (No. 17, (139) *above*), and this part deteriorated. In the 19th century, it lost most of its old features.

The S. front, originally symmetrical and with a band between the storeys, has a central doorway with fluted pilasters having moulded caps, a light of four panes between fluted friezes, a moulded cornice above, and door with six fielded panels (late 18th-century). The windows all had sliding sashes under segmental arched heads, but some have been replaced by hung sashes. The only original fittings remaining are doors of two or three fielded panels with angle hinges. *Demolished 1966.*

(141) THE LODGE, No. 21 (Plate 193), built in the late 17th or early 18th century on an L-shaped plan, with a side entry leading to the kitchen in the N.W. wing, was originally of two storeys. In the late 18th century a third storey was added to both ranges; two bay windows were inserted in the S. front, each extending over two floors; other openings were altered, the whole was plastered and, internally, rooms were altered. In the 19th century and modern times, sashes were rehung, some windows altered, and annexes built in the angle at the back.

The S. front has a doorway with plain pilasters, a moulded canopy, and door with six fielded panels and a fanlight above (late 18th-century); it is symmetrically designed except that to W. is a small sash window with flat arch and key-block, representing the original entrance to the kitchen wing. Bay windows flank the central doorway and the second floor has five narrow windows with late sashes. At the eaves is a cornice with modillions and dentils, the E. half being slightly deeper than that to the W. At both ends of the elevation are late 18th-century waterheads; the pipes have two fleurs-de-lis at each junction, a feature common in York at the end of the century. On the other elevations there is a brick band of two courses between the floors.

Internally the only original feature remaining is the staircase, with closed string, square newels and turned balusters (Plates 84, 87). In most of the rooms skirtings, cornices, and doors are of the late 18th century, contemporary with the addition of the third storey, and fireplaces have simple marble surrounds of the 19th century. In the kitchen are cupboards made up of reused early 17th-century panelling.

(142) ACOMB HOUSE, No. 23, built in the first half of the 18th century, consisted of a symmetrically planned block, with a small scullery block projecting to N. In the later 18th century, the rooms to E., above and below, and the dining room were altered, a servants' staircase inserted, and attics added. In the early 19th century, the dining room was extended to N. by the addition of a large bay window, to W. of which was added an annexe and, to E., a porch with alcove above; the ground floor of the front bay was also remodelled. Gent's *Ripon* (1733, part 2, p. 4) mentions that Acomb is 'grac'd by a handsome antient seat of the Family of the Blanshards', thought to have preceded the present house. Successive inhabitants of the house since the late 18th century have been George Lloyd; Foljame; Ralph Creyke, M.P. for York; T. B. Whytehead and Major Lindberg (Benson in *AASRP*, XXXVIII, 93).

At the centre of the symmetrically designed *South Front* (Plate 193) is a projecting bay with chamfered angles, the lower part forming a porch. On either side of the bay are windows with rubbed brick heads and key-blocks, and above a moulded cornice the second floor has small casement windows without key-blocks. The brick of the attic stage differs from that below; as the bay is nowhere bonded-in below this storey, it must have been added at the same time as the heavy cornice which carries round it. The *East Side* is gabled, with two chimney-stacks; it has two late sash windows at ground floor; at first floor is an original window with early sashes and above is a round-headed stair light with rubbed brick arch and key-block, and square imposts, between two small blocked windows. On the *West Side* two chimney-breasts project. At ground floor, the original *North Elevation* is partly obscured by a porch with staircase alcove above, and a large bay window, all of the early 19th century. The staircase alcove is lit by a round-headed window with original glazing bars, reset. At first floor of the main block are four windows with rubbed brick arches, key-blocks, and early 18th-century ovolo-moulded sashes; and, above, a broad band and five attic casement windows as on the S. front.

The *Entrance Hall* has a moulded cornice, dado rail and skirting and doorways with moulded architraves, pulvinated friezes and dentilled cornices. The staircase hall is entered by a large round archway having fluted pilasters with moulded caps and bases, moulded architrave, large key-block, and jambs with plain panels on the inner faces. The front room to E., refitted in the late 18th century, has moulded cornice, dado rail and skirting, and heavy moulded architraves and shutters to the windows; the fireplace has a plain moulded surround with fluted key-block, a stone or plaster moulded and dentilled mantel, and an overmantel consisting of a simple moulded eared panel. On either side is a cupboard doorway with moulded architrave, large fielded panel above, and door of six fielded panels. In the S.W. corner is an early 19th-century built-in cupboard with fluted surround and plain panels. A front room to W., wainscotted from floor to ceiling in two heights, with deal bolection-moulded panelling, has an entablature with heavy modillioned cornice, moulded dado rail and skirting. The fireplace (Plate 194) has a light brown marble surround and panelled overmantel. Doorways to the hall and to cupboards have enriched architrave, enriched pulvinated frieze and cornice (Plate 67). The windows have heavy architraves enriched with egg-and-dart (early 18th-century). The *Dining Room*, to N.W., of late 18th-century character, with moulded cornice, dado rail and skirting, has in the W. wall a brown veined marble fireplace with enriched eared surround, a frieze of Adamesque character with central urn, small urns of flowers on either side, and slender festoons, and a cornice with key-blocks and dentils. The iron fireplace, itself of the same period, has swags on either side, coupled with an urn to right and musical instruments to left, and above are two medallions, each with a cherub, swags and festoons.

The *Main Staircase* (Plates 84, 195) has balusters alternately turned and twisted. From the half-landing, a round-headed archway, having fluted pilasters with moulded caps and bases, moulded head and key-block, leads to an alcove above the back porch; over it is a glazed cupola light. Inserted in rooms to S.E. and rebuilt at first floor the *Servants' Staircase* has a heavy moulded rail and plain string and turned balusters with a square knop (*c.* 1740–50).

On the first floor are four principal rooms all with moulded skirting, dado rail and cornice. The *Saloon* to S.W. has enriched plaster panels, and heavy enriched modillioned cornice. The fireplace (Plate 72) has a brown veined marble surround and elaborate overmantel. The doorways each have an enriched moulded architrave, and entablature with enriched dentilled cornice and pulvinated frieze, below a large enriched panel; the doors have six fielded panels. The later 18th-century ceiling in shallow relief is painted in pale blue and pink, and at each end is an oblong panel (Plate 77). A bedroom to N.W. has a fireplace with simple moulded stone surround of the early 18th century, a dentilled cornice added later in the same century and an overmantel of an eared panel. Two bedrooms to E. are reached through a small plaster-vaulted lobby. Each room has a fireplace with bolection-moulded brown veined marble surround, a moulded and dentilled mantelshelf above (18th-century), and an overmantel of a simple moulded eared panel; on either side is a doorway with moulded architrave, large fielded panel above, and doors with six fielded panels.

(143) HOUSE, No. 25, was built in the early 18th century and modified in the second half of that century; the front was remodelled in the mid 19th century.

The symmetrically designed front has a central doorway with fluted pilasters, moulded caps and moulded canopy; and, on either side, a window with segmental rubbed brick arch and late sashes. There are three similar windows at first floor.

On the ground floor are fireplaces with panelled stone surrounds and shell decoration; doors have three or five fielded panels. The staircase has a moulded rail, no string, and turned balusters alternately straight and spiral; it is lit by a round-headed window. On the upper floor is a fireplace with enriched moulded surround and panelled overmantel; other fireplaces are simpler but one retains an old iron grate with scrolled and foliated arch.

(144) HOUSE, No. 27, was built in the second half of the 18th century with a through passage giving access to a front room, a central transverse staircase and a back room; an extension was erected at the N.W. corner later in that century. In the 19th century two small additions were made to N. and a bay window inserted in the S. front.

The S. front has a doorway with plain pilasters and a pedimental canopy supported on brackets, and a large Victorian bay window. Between the stages is a three-course brick band; the upper floor has three segmental-headed windows with 19th-century sashes. Above is a modillioned cornice, and a hipped Mansard roof with a late dormer. There are houses against the E. and W. sides. To N., on the E. side, is a gable with large segmental-headed sash windows to both floors and later windows to the attics.

Internally, all doors have six fielded panels and original furniture. The front room has a fireplace with moulded architrave, jambs with swags hanging from medallions, a frieze with oval paterae, and a moulded cornice; on either side is a round-headed recess with moulded imposts and moulded architrave to the arch. Archways similarly treated divide the entrance passage to give a spatially separate stair hall. The *Staircase*, rising to the attics, has a moulded rail, no string, a turned newel and turned balusters with a square knop. On the first floor, the front room has a fireplace like that below, but the original grate has a scrolled and foliated round iron arch, with an inner white marble surround.

(145) TUSCAN HOUSE, No. 31 (Plate 47), was built in the early 18th century and remodelled in the second half of that century, when a new main front was erected, and a new staircase and stair light inserted. In the early 19th century various annexes were added.

Straight joints on either side show that the S. front is not original. At ground floor, which has a stone plinth, is a doorway with moulded architrave, and dentilled cornice supported on slender, shaped brackets, on either side of a frieze of triglyphs and moulded paterae; the door has six fielded panels. To E. is a three-sided bay window with sashes, and to W. a sash window with rubbed brick flat arch; there are similar windows at first floor. Each floor has a stone string, which carries straight across and acts as a sill to the windows. There is a semicircular light to the attics. The E. wall, has a two-course brick band between the stages and a wooden modillioned cornice.

In the entrance hall the window W. of the entrance has a seat formed of fielded panels. The staircase, at the back of the hall, has a moulded rail, no string, and turned balusters with a square knop and alternately twisted and plain stems; the plain balusters have all been removed, except those on the landing. At the bottom, a newel reproduces the form of one of the twisted balusters, but in much heavier form. To E. of the staircase a round-headed doorway with moulded architrave and key-block, and fielded panels to the spandrels and jambs, leads to a passage to the rear. The front room has a fireplace with fluted pilasters, and a moulded cornice (late 18th-century); the bay window has a dado of plain panelling, and hung shutters. In the kitchen behind, the fireplace has fluted pilasters with round medallions at the head and fluted panels on the lintel, on either side of a similar medallion (early 19th-century). In the cellar below are some reused oak joists and an ovolo-moulded beam, of the 17th century.

Upstairs, in the main front room the fireplace has a moulded architrave between fluted pilasters with moulded caps and bases, the caps having round paterae, and a cornice. Doors throughout the house have fielded panels, some with angle hinges and original locks. *Demolished 1965.*

(146) HOUSE, No. 51, built in the second half of the 18th century, has been modernised.

The symmetrically designed front has a central doorway with fluted pilasters, lintel of a shaped fluted panel, and moulded cornice (early 19th-century), between segmental-headed windows, that to W. enlarged. Above a brick band of three courses is a segmental-headed recess, between similar windows; in the roof is a dormer with low pitched gable and modillioned cornice.

There are six-panelled doors, of 18th-century type, throughout the house. On the ground floor, a room to N.W. has exposed oak joists slightly chamfered. The *Staircase* has a moulded rail, plain tapered newels with moulded caps and bases, a plain string, and turned balusters. At first floor, in a room to E., is a fireplace with moulded architrave and cornice.

(147) HOUSE, No. 67, has been very drastically renovated and, although perhaps of the original form, retains nothing of note but two 18th-century doors, each with six fielded panels. To E. of the doorway in the S. wall is set a white limestone slab, discovered during the restoration and inscribed in good lettering 'Elizabeth 168[4] Stafford'.

(148) HOUSE, No. 103, was built in the 18th century, and a range was added to N.W. later in the same century.

The symmetrically planned front has a central doorway with plain pilasters, having moulded caps and bases, and a moulded canopy; and, on either side, a sash window with shutters. Above is a brick band of three courses and, at first floor, two similar windows, with a segmental-headed recess between, below a two-course band. At the eaves there is a brick dentilled cornice. The W. gable has two bands and a small light to each floor, and the E. gable similar bands, with a further one at eaves level, and a small light to the attics.

Internally, the ground floor is modernised, except for a fireplace in the N. range, with plain stone surround and wooden dentilled cornice, and a chamfered beam in the same room. At first floor are several 18th-century doors, with two or four fielded panels.

(149) WHITE ROW, consists of six cottages which are wholly modernised, their present character being of 1920–30, but there is no doubt that they represent five houses built soon after 1691, the homes of five copyholders whose names are added at the end of the Court Roll of the Manor for 1693 but not mentioned in the Roll for 1691 (Acomb Manorial Records on deposit in the York City Library). The measurements of the present rooms agree with those in the document mentioned below, and during modernisation chamfered oak bressumers were removed from the fireplaces, and oak joists were seen when the ceilings were reinforced. A proposal to erect the houses (in the Manorial Records, undated, but on palaeographical evidence of the late 17th century) gives a plan and details (Plate 6). They were to be built in a row 186 ft. long, 13 ft. 6 in. wide, and with side walls 8 ft. high. The identification of the compartments is interesting, for only the five rooms marked 'A' were to have a habitable loft above, reached by a ladder. Otherwise, each house has a kitchen ('B') and a stable open to the roof ('C').

The walls are of brick covered with roughcast but such brickwork as is exposed is all of the mid 19th century or later. The roof is covered with pantiles. Three wings have been added at the back. The original five dwellings, having been converted to six, now have more doors and windows than the original plan provided and no original features remain.

The building of five dwellings in one row must have represented an advanced design for rustic housing, despite the fact that rows of houses existed in the city of York from *c.* 1320. Each house was of 'long-house' type, with the entrance through a doorway into the stable, whence a doorway at the side of the chimney led into the kitchen, with the parlour, chambered over, at the far end. This plan is typical of small 17th-century farm houses in many parts of the west and northwest of England, especially Cumberland and Westmorland, but the compartments here described as stables were elsewhere more commonly cow-byres.

GALE LANE runs S. from the W. end of the village and was the way to the former Common Moor of Acomb called Acomb Knole. It was not built up beyond the junction with Front Street.

(150) HOUSE, No. 1, was built in the 18th century; at the end of the century, a chimney-breast and fireplace were added in a room to N. In the 19th century most of the present openings were formed and an annexe built to S.

The front is symmetrically designed with a central four-panelled door and a stepped brick cornice at the eaves. On the N. side a two-course band is carried across at eaves level. The windows are fitted with sliding sashes. The room to N. has a late 18th-century fireplace with plain wooden surround, moulded architrave and dentilled cornice. There are some doors with fielded panels.

THE GREEN lies behind the backs of the built-up plots on the N. side of Front Street, and to S. of the old road to Knapton. It is triangular, the apex lying to E. on the York Road, the W. side having formerly faced the open lands of the Chapel Field. The only early buildings were along the N.E. side near the Church.

(151) HOUSE, Nos. 17, 19, was built in the middle of the 18th century, an annexe being added to the W. end of the N. front a little later. In modern times, a large extension was built on the S.E. side.

The almost symmetrical front has a central doorway and open porch, with two round fluted pillars and corresponding pilasters with simple caps, supporting a moulded entablature with small central medallions, and a lead roof; the door has six fielded panels. On either side are large windows with flat brick arches and 19th-century sashes; there are similar windows at first floor, but over the doorway the original window has been blocked and a modern one inserted. To E. is a further bay, with a modern door and a window in original openings. At the eaves is a brick dentilled cornice.

The entrance hall has a moulded cornice, and a round-headed archway to the staircase has plain jambs and moulded imposts. The staircase has a moulded rail, low string, oblong newel and turned balusters with square knops (*c.* 1740). In the room to W., doors on either side of the fireplace, and to the hall, have six fielded panels. A first-floor fireplace has reeded jambs and lintel, oval angle pieces with paterae, an enriched mantel, and enriched architrave round the inner stone surround (late 18th-century). Some of the rooms have good doors with fielded panels.

(152) DANEBURY HOUSE, was built in the late 18th century. The front was remodelled in the 19th century and there have been further alterations to convert the house into three tenements. To W. is a 19th-century annexe (No. 3).

The plastered front was originally symmetrically designed, as shown by the remaining five second-floor windows. Below are 19th-century bay windows, three to each floor; there are doorways between the bays. There are three chimney-stacks. The back has various late windows and a round-headed stair light.

There are doors with six fielded panels throughout the house. On the ground floor is a fireplace with moulded architrave, dentilled cornice, and stone inner surround; on the first floor are five fireplaces, each with plain wooden surround, moulded architrave and mantel; all are of the 18th century.

A *Stable* range, to N.E., has rusticated stone quoins. Further N. are the remains of a ha-ha and terraced *Gardens.*

DRINGHOUSES

Dringhouses lies to S.W. of York along the main Tadcaster Road: it was a manor given by Archbishop Walter de Gray to his brother and it then descended in the family of Gray of Rotherfield and later to the Deincourts and the Lovells. Ecclesiastically the township belonged mainly to Holy Trinity, Micklegate, but in part to St. Mary Bishophill Senior (earlier to St. Clement) and to Acomb. It forms a long street village stretching along the main road for nearly a mile beyond the ancient boundary of York. Until enclosure in 1835 there were open arable fields cultivated in strips (*see* p. 2) and areas of common waste including part of the Knavesmire. Enclosure from Dringhouses Moor was taking place as early as 1742–6 (YCA, E.93, ff. 150, 183). Dringhouses was incorporated into the city in 1937.

TADCASTER ROAD was the only built-up street of Dringhouses before 1850.

(153) HOUSE, Nos. 13, 15, 17, is of two main builds; No. 13, the oldest part, was originally a low 17th-century cottage of one storey and attics, and Nos. 15 and 17 were erected against it in the early 18th century; later in that century No. 13 was given a new front and an upper storey. Various additions were made in the 19th century. Although added later, the front of No. 13, is similar to the remainder of the front elevation, from which it is separated by a straight joint.

The ground floor has hung-sash windows with segmental arches, and plain four-panelled doors. Between the stages is a four-course brick band; at first floor are three square-headed sash windows, and above is a dentilled brick cornice. The older part of the heightened N. gable is of tumbled brickwork; in this end are blocked 17th-century two-light windows. The S. gable is also tumbled but has modern openings. The back is partly masked by additions.

In No. 13, at ground floor, is a plain 19th-century fireplace and, in the bedroom, a fireplace with plain surround and basket grate with urn features on the jambs. The sitting room in No. 15 has a stop-chamfered beam and a 19th-century fireplace with plain surround. In a bedroom to N. is an 18th-century fireplace. In the sitting room of No. 17 is a stop-chamfered ceiling beam, and a stop-chamfered bressumer over a large fireplace. In both Nos. 15 and 17 the stair landing has a moulded rail, chamfered newel, and slender square balusters of the early 19th century.

(154) HOUSE, No. 23, was built in the 17th century, if not earlier, as a broad N.–S. range, on plan comprising three rooms with an internal chimney between the central and S. rooms. So early a date can be assigned because the house is wide and the beams have a disposition common to buildings of the earlier 17th century. No part of the original house is visible externally, as it was cased in brick in the 18th century. In the early 19th century a two-storey annexe was added to S.E. and a staircase built; *c.* 1850 the dining room was remodelled, and later a further extension was made to E. The whole has been modernised and offices added to the back.

A brick band of three courses, now cut back, divided the storeys of the front elevation. The windows have hung sashes and those on the ground floor are fitted with panelled shutters. The back of the main block, of red brick, colour-washed, has a two-course band between the stages. At ground floor is a modern French window and, at first floor, a sliding-sash window and a large early 19th-century window lighting the stairs. The roof, of shallow pitch, projects boldly.

On the *Ground Floor* the drawing room to N. has been

enlarged by the incorporation of a small room to E. The central room has a cased ceiling beam, and a stone fireplace, which has fielded panelled jambs and a key-block with shell decoration between two shaped and fielded panels on the lintel (late 18th-century). In the dining room to S. are two cased ceiling beams, and windows and fireplaces of mid 19th-century date. The chimney-breast between the two last rooms, over 9 ft. broad, must be earlier than the 18th century. A room to E. has a moulded 18th-century fireplace surround. At *First Floor* most of the doors have six panels and ogee-moulded frames (early 19th-century); the fireplaces, in general, have plain stone surrounds and basket grates (late 18th or early 19th-century). In one room the fireplace is flanked by fielded panelling of the 18th century. The main *Staircase* is of the early 19th century and has a moulded mahogany rail, no string, slender square balusters, a large turned newel at the bottom, and shaped spandrels to the treads.

(155) CROSS KEYS HOTEL, No. 32, was built in the early 18th century, but an inn has existed on this site since *c.* 1250. Later in the 18th century bay windows were inserted, and a large central segmental-headed carriageway was blocked; buildings to W. were erected in the 19th century, and the whole modernised in 1900.

The symmetrically designed *Front Elevation* has, at ground floor, cloaked by a porch, a large segmental-headed archway, now blocked, with stone imposts and a modern doorway in the blocking. On either side, to each floor, are large inserted three-sided bay windows, in which red brick contrasts with the earlier yellow-red brick of the remainder of the elevation. Between the storeys is a short section of an original three-course brick band, and between the bay windows, at first floor, a large blocked segmental-headed window, with a smaller sash window in the filling. The top of the wall, the cornice and the roof are all modern. The N. wall, with a band at eaves level, has been heightened. The S. wall has two large bay windows extending over both storeys; all features otherwise are modern. The W. elevation is modernised.

Little old work remains at ground floor. The *Staircase*, of the first half of the 18th century, has a moulded rail and string, square newels with half-balusters attached, and turned balusters with a square knop; it has been much restored. At *First Floor* is a fireplace with panelled jambs, each with a star form on the entablature; on the lintel shaped panels flank a key-block with shell decoration (late 18th-century).

(156) HOUSES, Nos. 33, 35, were built in the late 18th century, and a parallel two-storeyed block was added to E., in the 19th century.

The street front has, at each end of the ground floor, a segmental-headed sash window with shutters of two plain panels, and centrally a pair of doorways with flat arches, containing doors each with four plain and two glazed panels. Between the storeys is a brick band of four courses and above are three windows, similar to those at ground floor. A 19th-century barn has been built against the N. end. The S. end, only 4½ in. thick and formerly abutting the chapel built in 1747, has a modern window at ground floor. In the sitting room of No. 33 is a fireplace, originally open, with a segmental arch at the head and a moulded cornice. No. 35 has several 18th-century doors.

(157) HOUSE, No. 34, was built in the 18th century, but largely remodelled in the early 19th century; much of the gable ends, and all the W. front are of the later date, as are most of the internal features. In modern times, a room at the S.W. corner has been extended.

The street front has at ground floor an original small sash window to N., two larger windows of the early 19th-century, and to S., the entrance with a Roman Doric doorcase also of the 19th century. At first floor are two 18th-century sash windows and, to S., a shallow curved bow window with sashes framed by fluted uprights and a moulded cornice (early 19th-century). Under the roof are coupled gutter brackets and a moulded cornice (early 19th-century).

In the entrance hall is a doorway with reeded architrave having round paterae at the angles. The staircase has a moulded mahogany rail, no string, a turned newel with moulded cap and base, and slender square balusters. The stair light has a round head with moulded architrave. There are some original doors, with six fielded panels; other doors, also with six panels belong to the remodelling.

(158) MANOR HOUSE, No. 41, though standing on an old site, appears to have nothing earlier than the early 18th century. The central section is of this date, the N. cross range of the early 19th century, and the S. cross range modern.

Only the central part of the W. front has not been cased in new brick, and it has a brick band of three courses between the storeys. At ground floor are three sash windows with plain shutters, and a modern door; at first floor are three similar windows. All other elevations are now cased in new brick, except the sides of the N. range which are of early 19th-century brick; all openings are of this date or modern.

Internally there are some cased beams which, together with the W. wall, suggest an early 18th-century date. At ground floor are doors with six fielded panels and, in the entrance hall, a later 18th-century stone open fireplace. The balusters of the staircase are turned and have fluted stems with square knops, and an umbrella-shaped decoration (early 18th-century); they came from No. 46 Bootham, as did the early 17th-century panelling in a room to N.; in the same room is an 18th-century cupboard with two fielded panelled doors.

(159) HOUSE, No. 60, was built in the late 18th or early 19th century, but drastically altered in the late 19th century, to which date most of the exterior belongs; a straight joint in the S. wall shows that the earlier house was then extended to W. Internally there are many doors with six fielded panels. In a bedroom is an early 19th-century fireplace with moulded jambs and lintel, and round paterae at the angles.

(160, 161) MANOR FARM, Nos. 64, 66, built in the early

18th century, was divided into two in the late 18th or early 19th century, when a staircase was inserted in the S. part and a chimney-breast, with obliquely placed fireplace, in the N. part. Most of the present openings are modern.

The whole house is stuccoed. Its front elevation has a large sash window on either side of a broad modern door, with another door to S. Above a band at first floor are four similar windows; there is a cornice of diagonally set bricks. The N. gable is plain, except for a two-course band between the floors, which stops short of the W. side; the S. gable, with no band, has five modern openings. The W. elevation has one old sash window with very small panes; other windows are modern.

No. 64, has a simple plan: the entrance hall leads through a round-arched opening to the staircase behind, with front and back room to N., both with a diagonally placed fireplace. The early 18th-century *Staircase* has a moulded rail rising to a square newel with a pendant at the base, a moulded string and turned balusters with a square knop. No. 66 has an entrance hall with a later staircase contrived in the S.E. corner of the building. The staircase has a moulded rail, plain string and square balusters (early 19th-century). At *First Floor*, in a bedroom to N.E. (in No. 64) is a diagonally placed fireplace with plain surround, beneath a cupboard with 'butterfly' hinges (18th-century); the modernised room to N.W. has a similar cupboard over the fireplace. On the first floor is an 18th-century fireplace with moulded surround, pulvinated frieze, and moulded mantelshelf. In both houses there are old fitted cupboards.

MIDDLETHORPE

Middlethorpe is a hamlet to S.E. of the Knavesmire; it was a small lordship within the manor of Dringhouses and belonged in the Middle Ages to Byland Abbey. It was leased by the abbey to Reginald Beasley, and in 1558 William Edrington and Edward Beasley obtained the reversion from the Crown (*CPR*, 1557–8, 390). The early settlement of Bustardthorpe, lying between York and Middlethorpe, had disappeared by the 16th century and its area was added to that of Middlethorpe proper. In 1385 Richard II appointed Master Robert Patryngton and John Heyndale, masons, to build a stone cross in Middlethorpe on the spot where Sir Ralph Stafford had been killed by the king's half-brother, Sir John Holand (*CPR*, 1385–9, 13). The cross stood beside the Bishopthorpe Road about one-third of a mile N. of the houses of Middlethorpe.

(162) MIDDLETHORPE GRANGE, in Sim Balk Lane, was built in the late 17th century. It is a two-storeyed house of brick covered with roughcast, roofed with pantiles. It was already called Middlethorpe Grange on White's map of 1785 but in 1818 (Greenwood) appears as Middlethorpe Lodge. By 1831 (Cooper's map) it was 'Grange' again and the name became fixed when Alderman James Meek in 1836 built a new house called Middlethorpe Lodge in Dringhouses (*Yorkshireman*, 21 June). The original house was L-shaped, of two storeys with a single-storey wing at the back. Additions were made in the second half of the 18th century. In 1847 Archbishop Musgrave took a great interest in this farm (C. E. W. Brayley, *Annals of Bishopthorpe*, 13), and the house was occupied by George Bennett, bailiff to successive archbishops in the later 19th century, the period of the main remodelling.

The main S. front towards the fields is roughcast and has a two-course brick band and an eaves cornice of diagonally placed bricks. In general the exterior was remodelled in the second half of the 19th century and few traces of the original work survive.

The original *Staircase* (Plate 83) has a simple rail, shaped plank balusters and square newels, all of oak but painted; an extension to the loft is in softwood (late 18th-century). Some of the doors are of 18th-century date. The main loft has an 18th-century floor but the roof itself is relatively modern.

Fig. 72. (163) Middlethorpe Hall. Sketch by Samuel Buck (BM, Lansdowne MS. 914, f. 30), *c.* 1740.

(163) MIDDLETHORPE HALL was built *c.* 1699–1701 (Plate 198) by Thomas Barlow of Leeds who bought the property in 1698 (Borthwick Inst., List R. Dru., 1–22). An entry in Ralph Thoresby's diary, on 17 September 1702, states: 'Received a visit from Mr Barlow of Middlethorp near York, which very curious house he built after the Italian mode . . .' (*Diary of Ralph Thoresby*, 1830, I, 399) (Fig. 72, BM, Lansdowne MS. 914, f. 30). It is probable, however, that Sir Henry Thompson, who died in 1692, inhabited a house on this site with which the dovecote and various outbuildings, of late 17th-century character, were associated. Side wings were added to the main S. front in the mid 18th

(104) NORTH STREET. Church Cottages, No. 31 and Nos. 1, 2 All Saints Lane. Late 15th century.

(101) NORTH STREET. House, No. 6. Ceiling, early 17th century.

Back from S.W., early 18th century.

Street front, early 18th century.

Reused balustrading in attic, 17th century.

102) NORTH STREET. House, No. 19.

107) NORTH STREET. No. 64. *c.* 1770.

(126, 127) TRINITY LANE. Nos. 2, 4 and No. 6. *c.* 1600 and later.

(129) TRINITY LANE. No. 29. *c.* 1638 and later.

(114) Saw-mill. No. 52. 1839.

(112) No. 48. Early 18th century and later.

(113) No. 51. Late 18th century.

SKELDERGATE

(117) No. 56. Late 18th century.

(121, 122) House, No. 7 and The Old Rectory, No. 7A. *c.* 1600 and later.

(122) The Old Rectory. First floor. *c.* 1600.

TANNER ROW

Street front.

Interior showing jetty.

Canopy bracket, from Old Wheatsheaf Inn, Davygate.

(125) TRINITY LANE. Jacob's Well. Late 15th century.

(135) No. 9. Mid 18th century and later.

(133, 134) Nos. 6 and 8. Early 18th century and later.
ACOMB. Front Street.

(141) The Lodge, No. 21. *c.* 1700.

(142) Acomb House, No. 23. Early 18th century.
ACOMB. Front Street.

(142) ACOMB. Acomb House, No. 23 Front Street. Fireplace in S.W. room, ground floor, early 18th century.

(142) ACOMB. Acomb House, No. 23 Front Street. Main staircase, early 18th century.

(119) TADCASTER ROAD. The White House, Nos. 238, 240 (*Photo. Evelyn Collection*). 17th century and later.

MILL MOUNT GRAMMAR SCHOOL. 1850.

Main block from N.W. *c.* 1700.

E. wing from S. Mid 18th century.
(163) MIDDLETHORPE. Middlethorpe Hall.

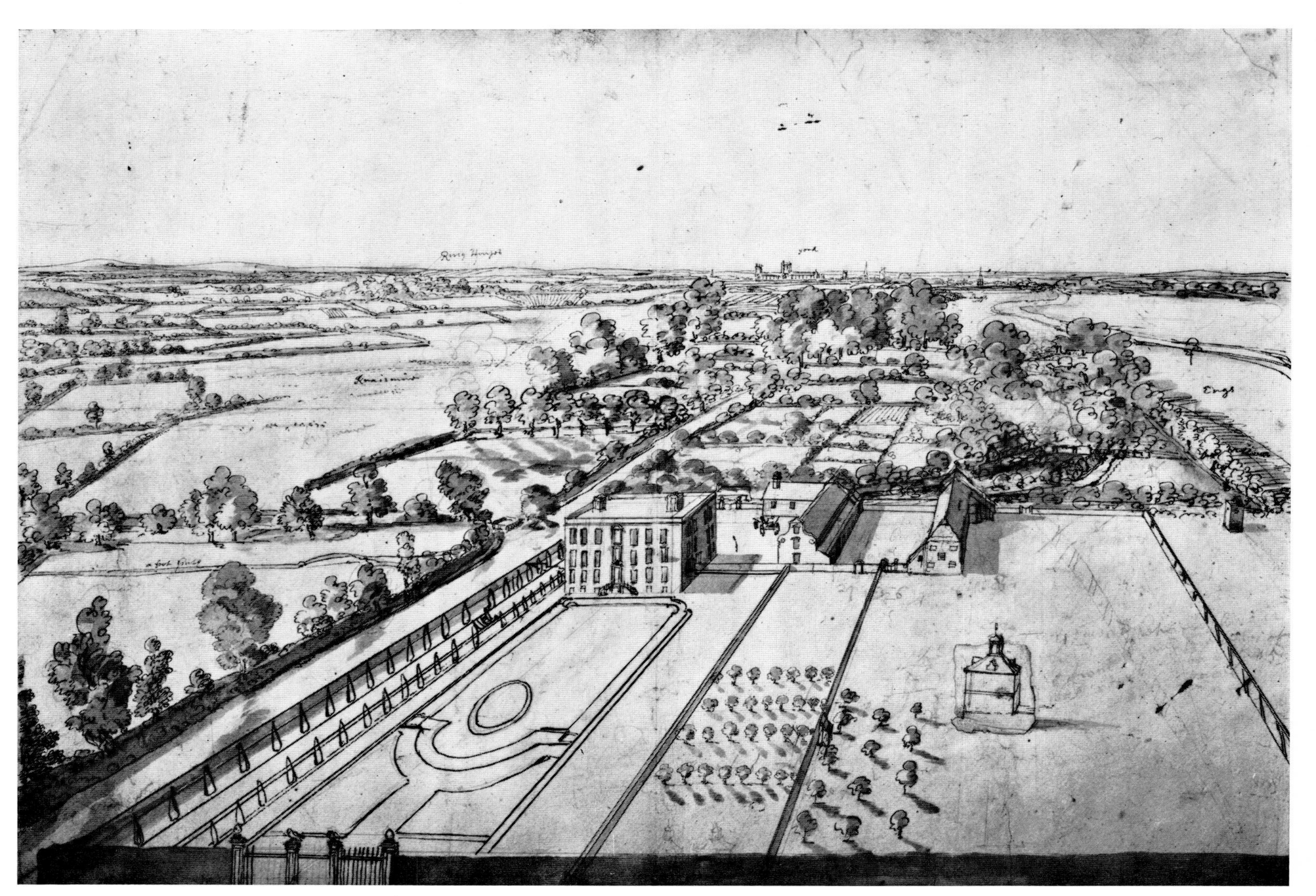

MIDDLETHORPE HALL AND YORK from S., by Francis Place, *c.* 1705 (York City Art Gallery).

(163) MIDDLETHORPE. Middlethorpe Hall, from S. *c.* 1700 and later.

(163) MIDDLETHORPE. Middlethorpe Hall. Fireplace in Dining Room, early 18th century.

N.E. room, first floor. *c.* 1700.

Dining Room. *c.* 1700.
163) MIDDLETHORPE. Middlethorpe Hall.

Ballroom ceiling. Early 19th century.

Main staircase. *c.* 1700.

Dining Room wall. *c.* 1700.

(163) MIDDLETHORPE. Middlethorpe Hall. Interiors.

Entrance Hall. *c.* 1700.

Ballroom. Early 19th century.

Stair Hall. *c.* 1700 and later.

Drawing Room. *c.* 1700 and later.

(163) MIDDLETHORPE. Middlethorpe Hall. Doorways.

(164) MIDDLETHORPE. Middlethorpe Manor. *c.* 1700 and later.

THE MOUNT. Elm Bank. Stair hall. *c.* 1898

THE MOUNT. Elm Bank. Overmantel by George Walton. *c.* 1898

century by Francis Barlow (1690–1771), High Sheriff of Yorkshire in 1735. At the beginning of the 19th century, there were important alterations: the W. wing was enlarged to make a ballroom; the entrance hall and staircase hall received new ceilings; doorways and open porches were added to the N. and S. entrances; a room to W. of the staircase, at first floor, was remodelled; various fireplaces were added; the main roof was heightened; and the balustraded parapet was removed perhaps at this time (Fig. 73). The S.W. prospect of York by John Haynes (1731) shows the Hall with a crowning balustrade. An architect's drawing at the Hall, entitled 'Elevation of the N. Front, Middlethorpe' (early 19th-century), shows the late portico and also, against the E. wing and near to the main block, a large portico since removed, with similar pillars and a Doric frieze. In front of the house is shown a red brick wall with two iron gateways and to E. of them a carriage entrance to the stable yard, as at present; this suggests that the gateways too are of the early 19th century. In modern times a partition has been inserted in the entrance hall; bathrooms fitted in the N.E. corner of each floor; the plain walls of various rooms enriched by applied moulded strips, giving the effect of panelling; and a kitchen range built to N. of the E. wing.

The house remained in possession of the Barlow family until the first half of the 19th century. Lady Mary Wortley Montagu resided there in 1713–15. By 1838 it was held by the Rev. Edward Trafford Leigh, husband of Frances, daughter of John Barlow, and in 1875 by her second husband, M. A. E. Wilkinson; by 1893 their son, Col. G. A. E. Wilkinson, was owner, retaining possession until his death. During most of the 19th century the Hall was let: in 1823 to Lady Mary Stourton, in 1875 to Miss Ann Marion Johnson who kept it as a boarding school, and after 1912 to L. C. Paget. In 1836 arms were granted to John Bower, described as 'of Middlethorpe Hall and Broxholm, co. York' (Harleian Soc., LXVII (1916), 41).

The symmetrically designed *North Elevation* (Plate 197) to the road is of fine brick in Flemish bond, tuck-pointed, with limestone ashlar dressings and rusticated quoins. Although an important feature on the S. front, the side wings are here unobtrusive. Below a moulded stone plinth are basement windows of two square-headed lights with chamfered reveals and mullions, and plain surrounds. The ground floor has a central doorway, with eared and bolection-moulded surround of *c.* 1700 cloaked by an early 19th-century stone pedimented porch (Plate 62). On either side of the doorway are three sash windows, each having a moulded stone architrave with simple key-block, and a moulded sill. Above, and also between first and second floors, is a flat stone band with hollow chamfered edge. At first floor are windows similar to those below, but the central one has a more elaborate eared architrave and moulded cornice. The second-floor windows are similar but

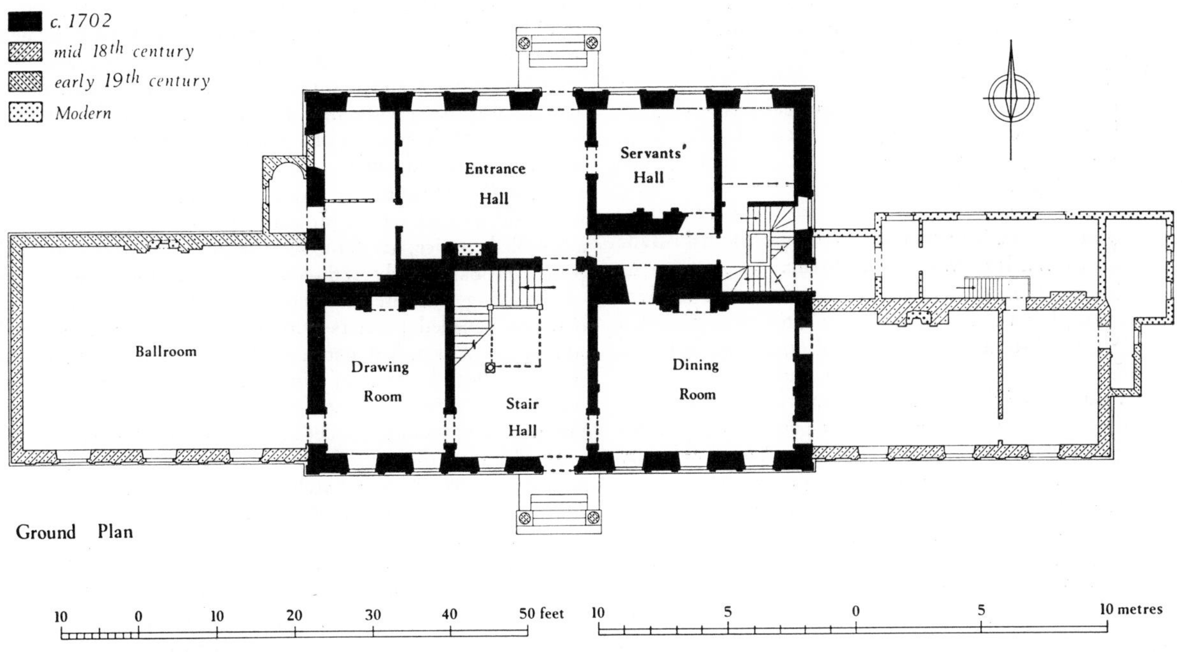

Fig. 73. (163) Middlethorpe Hall.

smaller; the W. end has been rebuilt and a window here is blocked. All the windows have been refitted with 19th-century sashes. Above a simple, bold cornice, which projects over the central window, is a centre-piece of three oblong panels with pilasters between; the centre panel is surmounted by an eagle *displayed*, in limestone, for Barlow.

The *East* and *West Elevations* are each in four bays uniform with the N. front but partly obscured by the wings and by added chimneys; a number of the windows are blocked. In the W. wall of the ballroom the different periods of construction are clearly marked by a change in the size of the bricks.

On the *South Elevation* (Plate 199), the main front of the house, the principal block is visually supported by the side wings, to make a symmetrical façade. The central block has a similar elevation to that to N., but there are slight differences: some of the basement windows have lost their mullions; the porch has pillars with stylised leaves on the caps, moulded bases, and an entablature without pediment; the crowning feature has a festoon and swags in each panel, and the eagle is in better condition. The side wings are each of three bays and of one storey with basement. The E. wing (Plate 197) has, in a plinth with square stone capping, three basement windows each with segmental arch and hung sashes. The bays are divided by stone Composite pilasters and between them are windows, much taller than those in the main block, each having a stone surround with square head and key-block, and early 19th-century sashes. The pilasters support a heavy stone entablature, above which is a parapet with balustrading over each window. The W. range is similar, but the basement windows have sliding sashes and the hung sashes above have exceptionally narrow bronzed glazing bars. The roof, raised behind the balustrade and hipped to W., is of Westmorland slate.

The *Entrance Hall*, divided by the insertion of a modern partition wall, has a coved ceiling with attenuated dentilled cornice (early 19th-century) and a floor of stone flags. In E. and W. walls are doors, each having an enriched bolection-moulded surround (Plate 203), and formerly with a laurel-leaf pulvinated frieze, and an enriched cornice (early 18th-century); the panelled doors are replacements. A large chimney breast in the S. wall has a black marble fireplace with bolection-moulded surround and black marble hearth.

The stair hall, to S., with a floor of black and white marble squares, has in the N. and S. walls large doorways (Plate 203) with early 18th-century doorcases, and later doors with eight fielded panels. The *Main Staircase* (Plate 202), of oak, has a heavy moulded rail, no string, and square panelled newels; the rail rises at each newel, with a spray of foliage in the spandrel; the balusters are fluted and enriched and stand on steps with carved panelled ends (Plate 83). Under the second turning point of the staircase is a handsome Corinthian column. On top of the S.E. newel on the first-floor landing is scratched 'IB 1764 & S B' (for John and Samuel Barlow). The first-floor landing has four doors, each with a bolection-moulded surround and enriched dentilled cornice, and a coved ceiling with a central moulded oval containing a delicate centrepiece set in shell ornament and with radiating features (early 19th-century).

The *Drawing Room* is panelled in two heights with plain panels with grained moulded frame, possibly of deal, and has an enriched cornice, moulded dado and skirting. The fireplace has a moulded and enriched architrave, pilasters with long swags hanging from lions' masks, a fluted and enriched cornice (second half of the 18th century), and an overmantel consisting of a large panel. In recesses on either side are book-cases with enriched cornices and fluted sides. Doorways in the E. and W. walls have enriched surrounds (early 18th-century) and later eight-panel doors (Plate 203). The windows have fielded panelled shutters. The *Dining Room* (Plate 201) is lined with painted oak panelling with enriched cornice, dado rail and skirting. The fireplace, with green marble surround, is flanked by fluted Ionic pilasters above which the cornice is increased to a full entablature (Plate 200). In the E. and W. walls are corresponding centrepieces (Plate 202) of round-headed panels flanked by pilasters. In the corners of the room the cornice is also increased to a full entablature above fielded panels; in the intervening spaces are taller bolection-moulded panels. The doorways have carved architraves (Plate 66).

The *Servants' Staircase* to E. (Plate 83; Fig. 17b) rises from basement to attics. The servants' hall, wainscotted from floor to ceiling in two heights of bolection-moulded oak panelling, with panelled pilasters between each pair of panels, has a moulded cornice, dado rail and skirting, and later fielded panelled shutters to the windows. In the W. wall is a doorway with bolection-moulded surround and eight-panelled door. The fireplace has a white marble bolection-moulded surround and overmantel consisting of a pilaster on either side of a large bolection-moulded panel.

In the *East Wing* the E. room, wainscotted in two heights with fielded panels set in a projecting moulded frame, and with a dentilled cornice, has in the N. wall three bolection-moulded panels, probably early 18th-century reused. In the W. room of the wing, the fireplace has a moulded eared architrave between panelled pilasters with fluted consoles at the head supporting a dentilled cornice; above is a scrolled broken pediment with festoons of laurel leaves enclosing a block on which is a cartouche with eagle supporters, and a scroll inscribed 'Hic posuisse gaudet'. Above this again is a pediment originally over the main S. doorway. To E. are bookshelves having a projecting round arch with console key-block above and, on either side, fluted pilasters and consoles.

The *West Wing* contains a large *Ballroom* with an enriched Adamesque cornice, enriched dado rail and moulded skirting; modern mouldings have been added to give a panelled effect. The ceiling (Plate 202) has an oval of interlacing laurel leaves round a fluted circle with festoons and swags round the outside, and a foliated centrepiece. In the N. wall is a white marble fireplace with panelled pilasters on either side, over which are medallions of musical instruments; the head has a central block bearing martial accoutrements and a moulded cornice. The doorways (Plate 203) have doors with six Regency-type panels. The windows have reeded surrounds. A small closet to N. has a large round-headed niche in the N. wall, with moulded imposts and fluted pilasters on either side (early 19th-century). In general, the *Cellars* are barrel vaulted, that under the dining room having a fireplace with plain pilasters, panelled lintel and key-block, and a cornice supported by fluted consoles (19th-century).

On the *First Floor* a room over the Dining Room is wainscotted in two heights with bolection-moulded panelling, and has a moulded cornice, dado rail and skirting (early 18th-century). In the N. wall is a fireplace with a veined white marble inner surround, having an enriched egg-and-dart architrave, jambs on either side with a volute at the head and a swag beneath; a frieze with elaborate foliated enrichment and, at either end, a medallion; a fluted and enriched cornice; a veined white marble hearth (second half of 18th century) and a grate of the same date. To E. is a dressing room with similar panelling and a plain early 19th-century fireplace. The room to N. of the main staircase, 'The Blue Room', (Plate 201), wainscotted in two heights with bolection-moulded panelling, has a moulded cornice, dado rail and skirting; between each bay of panelling is a panelled pilaster. There are two doors each with a bolection-moulded surround. Set diagonally in the S.W. corner, beneath an earlier bolection-moulded panel between two fielded ones, is a fireplace with an enriched and moulded architrave round a brown veined marble inner surround; and a plain frieze beneath a moulded and dentilled cornice (late 18th-century). The N.W. room has been modernised but retains an early 18th-century fireplace with bold bolection-moulded brown veined marble surround, and a late 18th-century basket grate. The S.W. room, was refitted in the early 19th century.

On the *Second Floor*, fireplaces are mostly of the late 18th and early 19th centuries. To W. is a room with moulded cornice, dado rail and skirting (early 18th-century), but probably modern panelling, formed by applied moulding; beneath a large bolection-moulded panel (early 18th-century), is a fireplace with moulded jambs and lintel and square angle pieces, and a plain mantel (early 19th-century). The door has six fielded panels set in a boldly moulded frame, and an original brass lock and handles (early 18th-century); most doors in the house are of the same type and likewise most windows had fielded panelled shutters fitted later in the century when the glazing bars were altered. Two bedrooms have late 18th-century fireplace surrounds. A large modernised room in the S.W. corner has a fireplace with enriched architrave (late 18th-century), enriched pulvinated frieze, and moulded enriched cornice (early 18th-century). The staircase to the loft, with slender square balusters, is of the early 19th century. The roof has been remodelled to eliminate a central valley and now covers the moulded and panelled lower parts of the chimneys. The trusses are elaborately constructed with queen-posts and cross bracing.

Outbuildings: N.E. of the house is a stableyard with entrances to N. and W. flanked by gate piers of brick with stone bases and crowning entablatures; those to the N. are surmounted by ball finials, those to W. by eagles, the Barlow crest. Flanking the N. entrance are two-storey buildings which originally had open archways in three sides of the lower storey; some of the arches are now blocked. These buildings are of the early 18th century but incorporate some earlier brickwork in the lower courses of the N. walls. The W. gateway is flanked by small single-storey buildings. A gateway to S. is plain and partly rebuilt. Two-storey buildings to the E. are, in part, of *c.* 1700 remodelled in the 19th century. Further E. is a former coach-house with a semicircular arch over the entrance, and beyond is an open yard approached through a gateway with brick piers surmounted by the remains of stone entablatures, uniform with the gateways to the stableyard.

Dovecote, S.E. of house, of *c.* 1700, is square on plan; it has brick walls and a late hipped tiled roof. The inside is lined with nesting boxes now cut into by an inserted floor and later windows. *Garden*, enclosed by 18th-century brick walls, has two piers with 18th-century lead urns. *Ha-ha*, S. of house, is revetted with brickwork. Two *Urns*, S.E. of house, of artificial stone, are decorated with allegorical figures and have handles comprising volutes and birds' heads (perhaps late 18th or early 19th-century).

(164) MIDDLETHORPE MANOR was the capital messuage of a subordinate lordship within the Manor of Dringhouses. In the Middle Ages this lordship belonged to Byland Abbey and after the Dissolution the Waller family became Lords, Thomas Waller in 1566 making an agreement with the City about right of pasture on Knavesmire. Early in the 17th century it was owned by William Brearey, merchant, Sheriff in 1598–9 and Alderman 1609–10 (d. 1637), whose will of 9 August 1637 bequeathed: 'To my son Christopher Brearey my new Hall I built at Middlethorpe'; the foundations of his hall may be those found in the higher land to E. of the present house. The central block of the present house was built in the late 17th or early 18th century, a date suggested by the narrow brick in Flemish bond in the cellar and the oak timbers of narrow, deep section in the roof; the staircase and hall panelling and fittings are also of this date, but are said to have been inserted by Col. Bryan Fairfax, who lived there in 1927. In the second half of the 18th century a block was added to E.; *c.* 1840–50 another large extension was added to W.; and at the turn of the century, a long N.–S. range was built against the W. end of the N. front. In modern times a staircase has been added at the S.W. angle, a dining room has been built on the N. side, various offices added, and the roof of the central block remade.

When Middlethorpe Hall (163) was built by the Barlow family early in the 18th century, the existence of two major houses in the manor led to difficulties and about 1735 (Drake, 382) the lordship of the manor was in dispute between Francis Barlow and Dr. Brearey. The last Brearey lord, Christopher, died in 1826 and his estates were broken up. In 1893 the Hon. E. W. Lascelles was Lord of the Manor; his daughter married Capt. H. D. Brocklehurst, who became owner by 1901; H. E. Preston resided there in 1908; Brigadier General the Hon. O. U. G. A. Lumley in 1917; and by 1927, Col. Bryan C. Fairfax was Lord of the Manor.

All the elevations are stuccoed above a shallow stone plinth and have a timber cornice at the eaves. The *South Elevation*

(Plate 204) consists of a central block recessed between lateral ones. The E. block has two sash windows with stone sills at each floor, one being a dummy. The central block has a central modern porch with a large sash window, probably 19th-century, to each side, and at first floor a small sash window between two large ones each of three lights; the eaves cornice is modern and replaces a parapet. The W. block has a modern twin-light sash window which does not range with the others. The *East Elevation* of the E. block has, to S., a large semi-circular bay, with a French window between two modern sash windows, at ground floor. The *North Elevation* of the E. block, stepping forward a little, has the same disposition as that to S., also with a painted dummy window. Built against the lower part of the central block are a modern dining room and offices; above are sash windows and a round-headed stair window. The W. block, in the same plane, has five windows to ground floor and three above. Built against the W. end of the W. block, the N.–S. range has in its E. wall four segmental-headed windows to each floor. At the W. end of the house, recessed behind the W. block, is a modern two-storeyed addition, and further S. a modern scullery built up on early 18th-century brickwork.

The large *Entrance Hall*, comprising the whole of the S. part of the central block, has been formed out of an original hall with a small room on either side by removal of the intervening walls. It is lined with plain pine panelling in two heights, of *c.* 1700, under a modern cornice. In the N. wall is a fireplace (Plate 75) with decorated timber surround and brown veined marble slip. In the W. part is a fireplace with an enriched pinewood architrave.

In the E. block the S. room has a moulded enriched cornice, moulded dado rail and skirting, and a fireplace (Plate 75) with Corinthian pilasters, decorated frieze and veined white marble slip (*c.* 1770).

The *Main Staircase*, of *c.* 1700, has a moulded string and a heavy moulded rail swept up to square newels over solid spandrel pieces, all of pinewood; the heavy turned balusters are of sycamore. Against two newels at ground floor are elaborate foliated volutes (*c.* 1700).

Cellars, beneath the central block, are built chiefly of 2 in. brick in Flemish bond, with timber floors above. To N.E. is a small wine cellar, with an 18th-century brick barrel vault.

On the *First Floor* the main landing has a panelled wood dado mostly of *c.* 1700 and three-panelled doors of the same date. In the E. block the rooms open off a small central hall lit by a cupola in the ceiling. In the central and E. block some of the rooms have 18th-century fireplaces and ceiling cornices. In the W. block the fittings are of the mid 19th century.

Garden features include: (1) to E. of the house an elaborate *Ha-ha*, with brick revetment on the inner side (late 18th-century); (2) to S.E. of the house two plain *Urns*, of magnesian limestone; (3) to S.E., in a rockery formed in a coppice, various *Moulded Stones*.

TOPOGRAPHICAL LIST OF HOUSES BUILT BETWEEN 1800 AND 1850

The following is a complete list of houses now, or recently, surviving built between 1800 and 1850. Buildings embodying older work are not included. A definite terminus is given by the 1/1056 OS map for which the survey was completed in 1851; this may be the only evidence for a pre-1850 date; but most of the houses can be dated more closely either stylistically or documentarily as noted. All houses have brick walls and Welsh slate roofs unless otherwise stated. Windows usually have flat arches and hung sashes, and external doorways have a timber surround in the form of moulded or panelled pilasters supporting a simple entablature. The gutters are often on paired wooden brackets. Staircases usually have moulded rails, cut strings and square balusters; the fireplaces have wooden or marble surrounds, and the grates are generally by Carron of Falkirk. Drawings mentioned as by Peter Atkinson II and J. B. and W. Atkinson are in the archives of Messrs. Brierley, Leckenby & Keighley (1967). Where a house is described as of 1½ bays, this means a room with a passage at the side of it.

STREET	NUMBER	DATE	OTHER DETAILS
Acomb Park	–	*c.* 1850	5 bays × 2 storeys. Gentleman's residence.
Acomb Road	11	*c.* 1840	3 bays × 2 storeys. Symmetrical. Gutter on coupled brackets. Windows with flat arches of stone or cement. Good railings (Plate 80).
	65–73 (odd)		Called Severus Place in 1853. Newly erected houses *YG* 23/2/1828, 9/8/1828, 2/3/1829.
	65	*c.* 1828/9	3 bays × 3 storeys and basement. Symmetrical. Porch with free-standing pillars. Windows with flat arches of stone or cement (Plan, Fig. 19).
	67	*c.* 1828/9	2 bays × 2 storeys and basement. Gutter on brackets. Doorway with brackets under entablature. Windows with flat cement arches.
	69, 71, 73	*c.* 1828/9	Terrace. Each house with 2 bays × 2 storeys and basement. Gutters with coupled brackets. Doorways with moulded jambs and lintels and round paterae at angles.
Albion Street	8, 9, 10	1815/20	Up by 1823 (Baines Directory (1823), II, 70–2). Cottage and tenements erected and street projected in 1815 (YCA, E. 96 ff. 243v, 249v); not laid out until 1816, when described as 'new made' (E. 97 f. 19v) but not named. Houses erected *c.* 1815 by John Taylor, cabinet maker, and H. Headley, flax merchant. Terrace. Each 1½ bays × 2 storeys. Doorways with reeded surrounds and rectangular fanlights. Windows with red brick flat arches (Plan, Fig. 19).
Askham Lane (Acomb)	Westfield House	*c.* 1850	3 bays × 2 storeys, double width. Symmetrical. Projecting gabled centre. Hung-sash windows with plate glass.
Bishopgate Street			Not shown clearly on Robert Cooper's map (1832) but relatively early in style. Building land for sale *YG* 19/2/1842.
	4	*c.* 1830/5	1½ bays × 2 storeys; low. Gutter on simple brackets; doorway with recessed panels to jambs and round paterae on blocks at top, supporting moulded lintel; door with six fielded panels.
	5	*c.* 1830/5	2 bays × 2 storeys and basement. Gutter on coupled brackets. Doorway with reeded pilasters and raised panels on frieze. Door with six panels and raised mouldings. Windows with plain brick arches.
	6	*c.* 1830/5	3 bays × 2 storeys and basement. Symmetrical. Coupled gutter brackets. Porch with free-standing fluted columns supporting plain entablature; false two-leaf doors with marginal panes.
	7	*c.* 1830	1½ bays × 2 storeys and basement; low. Gutter on coupled brackets. Doorway with fluted pilasters, simple entablature and fanlight with marginal panes. Windows with red brick flat arches.
	26, 27	1850	1½ bays × 3 storeys. Only these two are on the 1852 OS map. Gutter on shaped brackets. Doorway with recessed panels on jambs and entablature with shaped brackets. Three-sided bay windows to ground floor.
Blossom Street	5, 7, The Punch Bowl (Public House)	*c.* 1835	3-storey building, nearly symmetrical; central carriageway separates two shops from Punch Bowl. Ground floor stuccoed and rusticated. Five windows above with plain brick arches.
	16	rebuilt 1822	*See* (43) stage 3.

STREET	NUMBER	DATE	OTHER DETAILS
Blossom Street (continued)	21	1814, 1820, 1846	1 bay × 3 storeys. June 1814 bought by the Bar Convent (Archives 7 B 3(6)). Richard and Henry Hansom rebuilt the front elevation for £174 in 1820 (7 B 4(1)). 1846 rebuilt by G. T. Andrews for £745. John Lakin, bricklayer; Noah Akeroyd, mason, provided an 'extra Bardilla chimney peice'; Wilson, joiner; John Henry Cattley roofed it; Richard Knowlson did plastering (7 B 13).
	27	c. 1840	3 bays × 3 storeys.
	29	c. 1820	1½ bays × 3 storeys.
	31, 33	c. 1850	2 bays × 3 storeys.
	38	c. 1820	2 bays × 2 storeys. *Demolished 1965.*
	39 (Lion and Lamb)	1828	3 bays × 3 storeys. Symmetrical. Assessment raised in 1828 from £6 5*s*. to £8.
	43	1823/9	3 bays × 3 storeys and basement. Forms the end of South Parade.
Bridge Street			The first shops in New Bridge Street (Briggate) appear in the 1816 Directory, but only four were open by 1818; the street as a whole was built up by 1822 (Baines).
	1	c. 1815/20	1 bay × 4 storeys to Bridge Street and 3 bays × 4 storeys to Queen's Staith. Cornice with square modillions. Windows with rubbed red brick flat arches, stair window with marginal panes. Probably by Peter Atkinson II.
	2	1842	1 bay × 4 storeys and very lofty. Designs by J. B. and W. Atkinson, March 1842, for Mark Rooke; contractors: Thomas Jackson, bricklayer; Richard Knowlson, plasterer; and John Shaftoe, mason. Cornice on shaped brackets.
	3, 4	c. 1830	2 bays × 3 storeys. Gutters on coupled brackets. Windows with brick flat arches of common bricks.
	5 (Public House)	c. 1840/50	1 bay × 3 storeys. Stuccoed.
	6	c. 1815/20	2 bays × 3 storeys. Windows with red rubbed brick flat arches.
	9	c. 1815/20	1 bay × 3 storeys. Façade of fine red brick. Dentilled cornice on modillions of 18th-century type.
	11, 12	c. 1815/20	2 bays × 4 storeys to Bridge Street; 3 bays × 4 storeys to North Street. Gutters on coupled brackets. Windows with red rubbed brick flat arches. *Demolished.*
	13, 14, 15, 16	c. 1815/20	4 bays × 3 storeys and attics. Façade of fine red brick. Cornice on square modillions.
Cambridge Street	2–34 (even)	1846	Terrace houses consisting of blocks with varying detail, all 1½ bays × 2 storeys, some with basements. Erected for railway employees and catering for different categories of workers. Probably by G. T. Andrews. Good railings. *Demolished.*
Cherry Hill (*but see also* Clementhorpe)		c. 1830	Only one entry for Cherry Hill in the 1830 Directory. Mentioned in White's Directory (1838) II, 721.
Clementhorpe (called St. Clement's Place 1850)	7–11	c. 1845	Three houses for sale already occupied *YG* 11/3/1848. 1½ bays × 2 storeys each. Coupled gutter brackets. Doorways with unusual volute motif.
	12, 13, 14 and 1, Cherry Hill	1823	Mentioned in Baines Directory (1823) II, 70. 1½ bays × 2 storeys. Doorways with reeded attached columns and panelled entablatures. Windows have flat cement arches (Plan, Fig. 19). *See also* Brunton Knight, 668.
Cygnet Street (formerly Union Street)	11–15, 22–6	1846	Terrace—each house 1½ bays × 2 storeys. Doorways with plain pilasters and entablatures. Windows with flat arches of plain brick (Plan, Fig. 19). *Nos. 22–6 demolished.*
Dale Street	34–46	1823/8	In Pigot's Directory of 1828. Erected between 1823 and 1830 (Brunton Knight, 668). Seven dwellinghouses for sale *YG* 2/9/1848. Terrace of one build. Each house 1½ bays × 2 storeys.
Dove Street	39 houses	1827/30	Each 1½ bays × 2 storeys; third storey added to Nos. 15, 16, 18. To be built in 1827 (*YH* 13/10/1827); in 1829 Joseph Shouksmith, plumber and glazier, and his wife Hannah were erecting houses 'in the parish of Bishophill the Younger in the new road to Baggergate Lane' (YCA, E.98, f. 83). Mentioned in Hargrove's *Guide to the City of York* (1838). Fine variety of doorcases; window arches of plain brick or cement with key blocks (Plate 58). *Largely demolished.*

STREET	NUMBER	DATE	OTHER DETAILS
EAGLE STREET	13 houses	c. 1845	Railway development. Each 1 bay × 2 storeys. Paired doorways. Window arches of plain brick. *Demolished.*
FRONT STREET (Acomb)	29	c. 1840/50	3 bays × 2 storeys. Walls stuccoed. Roof pantiled. Gutter on widely spaced brackets. Doorway with consoles. Sliding-sash windows.
	52	c. 1850	Pleasant symmetrical design, 3 bays × 2 storeys. Large bricks. Gable over projecting centre with doorway and round-headed window within round arched recess. Windows on either side, sliding-sash with flat arches of plain brick.
	53	c. 1840/50	Probably built as two houses, 1½ bays + 1½ bays × 2 storeys. Large bricks. Coupled gutter brackets. Doorway with plain pilasters and entablatures; oblong fanlight. Each house had three-sided bay windows to each floor in wood.
	55	c. 1850	Sandwiched in between 53 and 57. 1½ bays × 2 storeys. Large brick. Roof pantiles.
	56, 58	c. 1850	1½ bays + 1½ bays × 2 storeys. Large brick. Hipped roof. Doorways on outer sides. Hung-sash windows.
	57, 59	c. 1850	Each 1½ bays × 2 storeys. Gutter on simple brackets. Plain doorways. Windows hung sash with slightly segmental arches of plain bricks.
	77	c. 1840/50	3 bays × 2 storeys. Hipped roof; pantiles. Three-sided bay windows on one side and small broad hung sash on the other. Two upper windows with slightly segmental arches of plain headers.
	79	c. 1850	2 bays × 2 storeys. Large bricks. Shop front and matching doorways all with large consoles. Two plate-glass hung-sash windows to first floor.
	87	c. 1850	3 bays × 2 storeys. Symmetrical. Stuccoed brick. Roof of pantiles. Gutter on coupled brackets. Doorway in centre with plain surround. Hung-sash windows, plate glass.
	95	c. 1840/50	3 bays + 2 bays × 2 storeys. Pantiled roof. Simple gutter. Plain doorways. Sliding-sash windows.
	105, 107	c. 1850	3 bays × 2 storeys. Symmetrical design. Large buff bricks. Pantiled roof. Paired doorways at centre. Sliding-sash windows.
THE GREEN (Acomb)	3, 5	c. 1800	Two blocks: (a) 2 bays × 2 storeys. Pantile roof. Brick cornice. On either side a three-sided bay window to ground and first floors, each with plate-glass hung sashes. (b) Higher. 1½ bays × 2 storeys. Same bay windows.
	16, 18	c. 1830	1½ + 1½ bays × 2 storeys. Coupled gutter brackets. Paired doorways with moulded jambs and reeded panels on frieze. Plate-glass hung-sash windows.
	25	c. 1830	1½ + 2 bays × 2 storeys. Stuccoed walls. Pantiled roof. Hung-sash windows, with small panes.
	29, 31, 33	c. 1830	Tall; each 1½ bays × 2 storeys and attics. Pantiled roofs. Openings nearly all modernised.
	Sun Inn	c. 1850	3 bays × 2 storeys. Stucco on large bricks. Coupled gutter brackets. Openings modern. Hung-sash windows.
	56	c. 1830	1½ bays × 2 storeys. House and stable. Coupled gutter brackets. Good doorway with moulded jambs and lintel and moulded round paterae at angles; fanlight with marginal panes. Broad hung-sash windows with small panes.
	58	c. 1820/30	Earlier than last which is built over its end wall. 2 bays × 2 storeys. Blocked doorway. Sliding-sash windows.
GREEN LANE (Acomb)	9	c. 1840	1 bay × 2 storeys; lofty.
	11, 13	c. 1810/20	1½ bays × 2 storeys each. Fireplaces with Adamesque details illogically arranged.
HOLGATE ROAD (formerly Holgate Lane)			1823 building land for sale and building progressing. (*YG* 19/7/1823). In Baines Directory 1823. 1828 York Villas being erected (*YG* 9/8/1828). Building land for sale 1842 (*YG* 22/1/1842, 19/2/1842).
	2, 4	c. 1840/50	2 bays × 3 storeys. Large bricks; hung-sash windows with heavy flat arches and plate glass.
	6, 8, 10, 12, 16	c. 1840/50	2 and 3 bays × 2 storeys.
	18, 20	c. 1830	Each house 1½ bays × 2 storeys. Good doorway to No. 20.
	22 (Holgate Villa)	c. 1840	3 bays × 2 storeys. White brick, three-sided bay windows. Iron balusters to staircase. Probably by G. T. Andrews. *Demolished.*
	26	c. 1840	1 bay to Holgate Road, 3 bays to Lowther Terrace × 2 storeys.

STREET	NUMBER	DATE	OTHER DETAILS
Holgate Road (continued)	28–50 (even)	c. 1845/50	1½ bays+2 bays × 2 storeys, brick and cement flat window arches. 36, 38 with modern shopfront. Some good doorcases. Some windows with marginal panes.
	45 (St. Catherine's Hospital)	1834/5	Designed by G. T. Andrews in 1833 for Messrs. Simpson and completed 22 August 1835 (YCL City Archives K.82 and M.17A, M.17B). *Demolished.*
	52	c. 1845/50	1 bay × 2 storeys.
	54, 56, 58, 60	c. 1845/50	1½ bays × 3 storeys. Called Blenheim Place on stone band. Built in two blocks of two houses. Windows with marginal panes.
	62	c. 1845/50	3 bays × 3 lofty storeys; windows with red brick flat arches.
	63, 65	c. 1830	On Robert Cooper's map, 1831/2. 4 bays × 3 storeys. Doorways with attached fluted columns. Bay windows polygonal on ground floor, segmental on first floor. Simple railings.
	64	c. 1845/50	2 bays × 3 storeys. Good cornice.
	66 (The Crystal Palace)	c. 1845/50	3 bays × 2 storeys. Symmetrical. Cement arches with key-blocks. Set back with yard in front and probably originally built as a Public House.
	70, 70A, 72	1846/7	Each 2 bays × 3 storeys. All of one build; 70A and 72 have paired doorways with Ionic columns on outer sides. One house 'nearly new . . . belongs to Mr. Shafto, a builder' 1850 (Letter YCL YL/Gray letters, No. 3).
	71, 73	c. 1830	On Robert Cooper's map 1831/2. 4 bays × 3 storeys; paired doorways; curved bay windows to ground floor. Narrow red rubbed brick flat arches to other windows.
	74, 76	c. 1845/50	4 bays × 3 storeys and basement. Flat stone arches to upper windows. Doorways and contemporary three-sided bays have modillioned entablatures.
	75	c. 1830	On Robert Cooper's map, 1831/2. 1½/2 bays × 2 storeys and basement. Good doorway.
	77	c. 1828/30	On Robert Cooper's map, 1831/2. 2 bays × 3 storeys. Good porch and curved bay to ground floor.
	78, 80	c. 1845/50	2 bays × 3 storeys. Generally similar to 74, 76. Stone window heads. 80 retains shutters.
	82, 84, 86	c. 1845/50	As last, but with heavy cornice on shaped modillions. Each 2 bays × 3 storeys.
	88, 90	c. 1845/50	4 bays × 2 storeys and attics. Plain doorways and three-sided bay windows.
	92	1850/1	3 bays × 2 storeys. Symmetrical. Vicarage to St. Paul's, designed by J. B. and W. Atkinson. Porch with engaged fluted columns and an entablature with triglyphs. Three-sided bay windows. Good railings.
	96–118 (even) (Holgate Terrace)	1846/51	Brunton Knight, 668. Each 2 bays × 3 storeys. Good doorways. Three-sided bay windows. Noteworthy cast-iron railing standards.
	120, 122	c. 1840	4 bays × 3 storeys. Symmetrical pair. Doorways in recessed side wings.
	124 (Holgate Hill Hotel, formerly The Poplars)	c. 1850	3 bays × 2 storeys. Symmetrical. Italianate villa.
	126	c. 1835/40	126, 130 on Bishophill Junior Tithe Map 1847 but not on Robert Cooper's map of 1831/2. 3 bays × 2 storeys. Symmetrical. Central doorway with plain pilasters and entablatures, and fanlight with marginal panes, and original lamp. Windows with flat brick arches and on ground floor marginal panes.
	130	c. 1835	3 bays × 2 storeys. Symmetrical. Hipped roof. Small porch with gable; outer doorway has radial fanlight and door with six fielded panels. Windows with cement flat arches.
	171	c. 1830	Appears to be on Robert Cooper's map 1831/2. 1½ bays × 2 storeys.
Lowther Terrace		1846	Generally 1½ bays × 2 storeys. Included Providence Place and Denton Terrace, shown on the 1852 OS map. *Demolished 1967.*
Micklegate	28	1810	2 bays × 3 storeys. Built by Ambrose and Robert Gray, bricklayers.
	39, 41	1835	3 bays × 2 storeys. Symmetrical. Being built in 1835 (*YG* 23/5/1835) and designed by J. B. and W. Atkinson for Mr. Varvill. No. 39 was the office of J. B. and W. Atkinson between 1837 and 1851 (Plate 79).

STREET	NUMBER	DATE	OTHER DETAILS
MICKLEGATE (continued)	40	*c.* 1840/50	2 bays × 4 storeys. Windows with plain brick arches. Moulded cornice on shaped brackets.
	50, 50A		
	63	*c.* 1840	2 bays × 3 storeys. Windows with rubbed brick heads.
	78, 80, 82, 84	1821/2	6 bays × 3 storeys. Three houses by Peter Atkinson II. Light buff brick, hung sash windows with red rubbed brick flat arches. Land to be cleared March 1821. Bones found November 1821. Building Lease granted 24 April 1822 (G. Benson, *Notes on the Parish of St. Martin-cum-Gregory*, I; *Gentleman's Magazine* (1821), pt. 2, 557; *YG* 27/10/1821, 24/11/1821). (Plates 60, 65, 68; Plan, Fig. 19.)
	94, 96 (The Falcon Inn)	1842/3	4 bays × 4 storeys. Two houses first mentioned 1842/3 having replaced the Falcon, one of the important inns of York. Cornice on shaped brackets. Doorways with large consoles. Windows with plain brick flat arches. *See* (84) p. 88.
MILL MOUNT HOUSE (now Mill Mount School)		1850	Partly 2 storeys, partly 3 storeys and basement. Important house built by J. B. and W. Atkinson for Charles Heneage Elsey. Plan dated March 1850 and later. Parcel of land for sale for building 1844 (*YG* 24/2/1844). (Plan *see* p. xciv, Fig. 19.) Stock brick. Stuccoed entrance porch. Large windows with flat arches of plain brick (Plate 196).
THE MOUNT	69, 71, 73, 75	*c.* 1830	Large terrace, each 2 bays × 3 storeys, cellars and attics. First rated 1832/3 but 69 not rated until 1834 and perhaps the last to be built. Good porticos and windows with rubbed brick flat arches.
	77, 79 (previously Park Place)	1831/2	3 bays (front) + 5 bays (side) × 4 storeys. Designed by Peter Atkinson II for Alderman William Dunsley; signed detail of November 1831 in Brierley drawings. Mr. Bayliff of Kendal made chimney pieces for it in 1832, including one of Italian marble for drawing room, one in black Kendal marble for dining room (Gunnis, 42.) (Plates 59, 79.)
	89 (St. Stephen's Children's Home)	*c.* 1810/35	Detached house of two storeys. Stucco rendering conceals a complicated building history.
	92, 94	1821	3 bays × 3 storeys, cellars and attics. Erected by Joseph Bullock, brickmaker (YCA E.97, f. 162). Two newly built houses for sale *YG* 22/9/1821. Symmetrical design. Buff-red brick. Cornice with square modillions. Paired doorways with fluted attached columns and entablature with triglyphs. Two curved bays, upper windows with rubbed brick flat arches (Plates 60, 65, 76, 79, 89; Plan, Fig. 19).
	100, 102, 104	1807/8	3 bays × 3 storeys, cellars and attics. Building in 1807 and leased by 1808 (Title Deeds). Roman vault found 17 August 1807 (Hargrove, II, 506). For sale *YC* 19/8/1812. Brick but centre house stuccoed. Moulded and dentilled cornice. Doorways with fluted pilasters and tall slender foliated consoles. Good curved bays. Back kitchen added by 1813 (Plates 59, 65, 79; Plan, Fig. 19).
	116, 118	*c.* 1840	2 bays × 3 storeys each. On map of 1847 by Thomas Holliday (Plate 159; Plan, Fig. 19).
	117	1833/4	3 bays × partly 2 and partly 3 storeys. First rated 1835/6, and probably finished by 29 April 1834 when sold to Thomas Pickersgill. Gothic (Plan, Fig. 19).
	119	1833	3 bays × 2 storeys, basement and attics. First rated 1836/7 (Plate 58).
	120	1842/3	2 bays × 3 storeys. Mentioned as built in deeds of No. 122 (21 April 1843). On map of 1847 by Thomas Holliday.
	121	1833	Grecian villa. 3 bays × 2 storeys and attics. 1833 painted on ridge beam. Not rated until 1834/5 (Plan, Fig. 19).
	122	1848/9	Plot vacant on Tithe Map of 1847. Stone façade. Same style as 124, 128 The Mount (Plate 80).
	123	1833	Villa. 3 bays × 2 storeys, cellar and attics.
	124	1843	2/3 bays × 3 storeys. Similar to 122, 128 The Mount.
	125	1833	Villa. 3 bays × 2 storeys and attics. Similar to 123, 127 The Mount. First rated 1839/40.
	126	*c.* 1840	2 bays × 3 storeys, cellar and loft.
	127	1833	Villa. 3 bays × 2 storeys and attics. First rated 1838/40. Like 123 The Mount. Conservatory (Plates 76, 89).

STREET	NUMBER	DATE	OTHER DETAILS
THE MOUNT (continued)	128	*c.* 1840	2/3 bays × 3 storeys, basement and attics. Like 122, 124 The Mount. Stucco front.
	129	1839/40	Originally 3 bays × 2 storeys, but very much altered.
	130, 132	*c.* 1830	Two of an original group of three. Originally 6 bays × 3 storeys, attics and basements (Plates 60, 65, 80, 91).
	131	1833	Villa. 3 bays × 2 storeys and basement. First rated 1833/4.
	136–44 (even)	1824	Terrace. Lease for building 1824. Each house in general 1½ bays × 3 storeys, basement and attics (Plates 58, 64, 76, 79, 91; Plan, Fig. 19).
MOUNT EPHRAIM		1846/51	Railway housing like Cambridge Street
	3		Newly built for sale *YG* 13/5/1848.
	7–21 (odd)		1½ bays × 2 storeys. *Demolished 1963.*
	44, 46		1 bay × 3 storeys. *Demolished 1963.*
MOUNT PARADE			Mount Parade (Plate 59) to be built *YG* 19/7/1823. Houses for sale *YG* 15/3/1828. In 1828 Nos. 1, 2, 4, 5, 6, 7 occupied (Pigot's Directory).
	1–7	1823/8	2 bays × 2 storeys.
	8, 9, 10	1829/30	2 bays × 2 storeys. All separate but of the same shape. Variant porticos—that of No. 8 reused and of 18th-century type.
	11	*c.* 1830	2 bays × 2 storeys. Red brick façade.
	16, 17, 18	*c.* 1840	2 bays × 2 storeys. Two newly erected houses for sale *YG* 8/7/1848.
	19	*c.* 1830	2 bays × 2 storeys. Exceptional façade and good details (Plate 47).
	20, Cumberland House	after 1834	3 bays × 2 storeys. Symmetrical. On plot of land advertised as suitable for building *YG* 4/1/1834 (Plan, Fig. 19).
MOUNT TERRACE (off Holgate Road)		1827/8	New houses for sale *YG* 24/3/1827, 15/3/1828, 11/1/1834, 26/7/1834. Nos. 1 to 5 in Pigot's Directory 1828; No. 6 already built in 1827.
	1, 3, 4, 5	1827/8	2 bays × 2 storeys and basement.
	6	by 1827	Villa. 3 bays × 2 storeys and basement. Symmetrical. John Hargrove, printer, lived here in 1838 (Directory). (Plan, Fig. 19.)
MOUNT VALE (Newington Place)			Roman material found in 1823 (*Gentleman's Magazine*, 1823, pt. 1, 633; YPS *Report* 1825; Hargrove, *A Guide to the City of York* (1838), 35).
	147	1827	First rated 1828. 2 bays × 3 storeys and attics (Plate 59; Plan, Fig. 19).
	149, 151	1823	First rated 1824. 2 bays × 3 storeys and attics.
	159 (Herdsman's Cottage)	*c.* 1840	*Cottage orné.* Cruciform × 1 storey and attics. Steep roofs with elaborate barge-boards to gables.
	206–12 (even)	*c.* 1840	Each 1½ or 2 bays × 3 storeys. Individual houses form a terrace. These four parcels of land were sold in 1837 by the Corporation (YCL, Council Minutes 1).
	214	*c.* 1840	2 bays × 2 storeys.
	218	*c.* 1820	3 bays × 2 storeys. Symmetrical. Two-storey bay windows flanking porch with reeded columns.
	220	*c.* 1840	3 bays × 2 storeys. Symmetrical.
	222, 224	1820/40	Built as one unit, each 1½ bays × 2 storeys.
	226, 228	*c.* 1830/40	Each 1½ bays × 2 storeys.
NORTH STREET	Behind 17 Bridge Street	1838	Warehouse for Messrs. Varvill by J. B. and W. Atkinson. *Demolished.*
NUNNERY LANE			Twelve houses by 1830 (York Directory). New-built house in Nunnery Lane at entrance to Dale Street *YG* 4/4/1829.
	17A, 21	*c.* 1830	3 bays × 2 storeys. Windows with cement arches and key-blocks.
	23	*c.* 1820	1 bay × 2 storeys, low. Gutter on coupled brackets.
	51, 55, 57	*c.* 1830	Terrace. 1½ and 2 bays × 2 storeys. Gutters on coupled brackets. Windows with cement flat arches. Shown on Robert Cooper's map, 1831/2.
OXFORD STREET	1–41	1846/50	Houses for railway employees. Each 1½ bays × 2 storeys. Built in twelve groups. *Demolished 1961/2.*
PARK STREET			Building land for sale *YG* 6/3/1847. All up by 1850 when first rated.
	7	1836	3 bays + 1 bay added × 3 storeys. Built in 1836 by Thomas Rayson, contractor, for himself. See lithograph. *c.* 1840, YCL, Evelyn Collection.
	9, 11	*c.* 1835/40	Each 2 bays × 3 storeys and basement.
	13, 15	*c.* 1847/50	Each 2 bays × 3 storeys and basement. Open passageway between doorways.

STREET	NUMBER	DATE	OTHER DETAILS
PARK STREET (continued)	17	c. 1847–50	3 bays × 3 storeys. Symmetrical. Moulded stone architraves to windows of lower storeys.
	19	c. 1847/50	Villa. 3 bays × 3 storeys and basement. Ground floor window with narrow side lights. Bay window opposite. Stone sills joined as bands.
	21	c. 1847/50	Villa. 3 bays × 2 storeys and basement. Symmetrical. Moulded stone architraves to windows. Sills joined to form band.
QUEEN'S STAITH	Varvill's Warehouse (Ebor Electrical Co.)	1849	Designs by J. B. and W. Atkinson dated October 1849.
QUEEN STREET	11–16	1840/50	Each 1½ bays × 2 storeys and basement. Doorways with plain surrounds. Windows with slightly segmental arches of ordinary brick.
	17–20	c. 1835	First rated 1836. Each 1½ bays × 2 storeys. Gutters on coupled brackets. Doorways with big consoles. Windows as last.
RAILWAY STREET		1846	Being formed in 1846 but no houses in it on the OS map of 1852. (YCL, Council Minutes III.)
ROSARY TERRACE	1–10	1843/6	1½ bays × 2 storeys. Very plain railway housing. 7–10 with scullery annexes. *Demolished.*
ROSEMARY TERRACE Skeldergate		1823	In Baines Directory (1823) II, 70/2. *Demolished before 1940.*
ROUGIER STREET	1–16	1842/3	Built in pairs, forming a long terrace. Each 1½ bays × 3 storeys. *1–14 demolished 1961.*
ST. CLEMENT'S PLACE (*See* CLEMENTHORPE)			
ST. PAUL'S SQUARE	46	c. 1835/40	Two blocks, one 4 bays × 2 storeys, the other 1 bay × 3 storeys. Fine windows with marginal panes (Plate 91).
SKELDERGATE	31	1843	1 bay × 3 storeys. Designs by J. B. and W. Atkinson for 4 dwelling houses of which only two were built. Houses being built by Mr. Atkinson *YG* 10/6/1843. Three-sided bay windows.
	53	c. 1840/50	White brick building, 2 bays × 3 storeys. Windows with flat arches of rubbed brick. Staircase with turned balusters.
SOUTH PARADE	3–20	1825/8	A large important terrace, probably built by Thomas Rayson senior, bricklayer, who was himself living in No. 16 from 1828 to 1836 (YCA, E.98, f. 135v. and Rate Books). Being built by forty subscribers in 1825 (Illustrations to Drake, Hudson MS., YCL, f. 159 dorso). No. 4 for sale *YG* 31/10/1829. No. 13 available 6 months earlier, *YG* 11/4/1829 (Plate 157).
SWANN STREET		1828/9	Newly built house for sale *YG* 23/3/1828. Nine newly built houses for sale *YG* 31/10/1829. *Demolished by 1967.*
TADCASTER ROAD	300	1833	Villa. 3 bays × 2 storeys, cellars and attics. *Demolished 1960.*
	302	1833	Villa. 3 bays × 2 storeys, cellars and attics, with additions. Erected by Mr. Eshelby, builder, of Dringhouses. Roman tile tomb found in excavations (Hargrove, *A Guide to the City of York* (1838), 50). (Plates 68, 80, 91.)
	304	1833	Villa. 3 bays × 2 storeys and basement.
	306	1833	Villa, 3 bays × 2 storeys and basement. Very complete and of good quality. (Plates 58, 64, 79, 89; Plan, Fig. 19.)
TADCASTER ROAD DRINGHOUSES	34	c. 1830	Detached house. 3 bays × 2 storeys.
	60	c. 1850	3 bays × 2 storeys. Coupled brackets. Doorway with big consoles. Three-sided bay windows.
	72, 74	c. 1850/60	1½+1½ bays × 2 storeys. Openings modern.
	76, 78	c. 1850	Each 1½ bays × 2 storeys. Pantiled roof. Simple doorway. No. 78 stuccoed and with mock timber framing.
	80–88	c. 1850	Each cottage 1½ bays × 2 storeys. Low. Pantiled roof. All openings modern. Stucco on brick and mock timber framing. One-storey annexes at back.
	159 (Ashfield)	c. 1850	4 bays × 2 storeys (garden front). Gentleman's residence.
TANNER ROW	12, 14	c. 1840	Part of a terrace. Each 1½ bays × 3 storeys. Shown on Y & NM Railway plan 1840. *Demolished 1960.*

STREET	NUMBER	DATE	OTHER DETAILS
TANNER ROW (continued)	39 (The Grapes)	*c.* 1845/50	2 bays × 4 storeys. Heavy moulded and dentilled cornice. Windows with plain brick flat arches.
	43	*c.* 1845/50	4 bays × 3 storeys and basement. Lofty. Light buff-red brick. Cornice on shaped brackets. Doorway with recessed panels, bold cornice on modillions and oblong fanlight. Entry with round-headed archway with stone surround. Windows with plain brick flat arches.
	56	*c.* 1850	4 bays × 3 storeys. Large brick. Gutter on brackets.
TOFT GREEN	Toft Green Chambers	*c.* 1845	5 bays × 3 storeys. Symmetrically designed pair. Buff-red brick; cornice on shaped brackets; continuous stone sill to first floor. Doorways in end bays, with plain stone surrounds. Windows with rubbed brick flat arches, flanking central blind recesses.
TRINITY LANE	43, 45, 47	1846	Terrace, each 1½ bays × 3 storeys. Designed for Michael Varvill by J. B. and W. Atkinson. Contractors, William Coulson, John Jackson and John Armstrong. *Demolished.*
VICTOR STREET (formerly St. Mary's Row)		1811	Terrace. First two houses erected by Thomas Rayson, and other plots up for sale, in 1811 (YCA, E.96, f. 169). This was apparently the earliest development laid out as a terrace in York S.W. of the Ouse. *Demolished 1960.*
YORK ROAD, ACOMB	60	*c.* 1840/50	Italianate villa. 3 bays × 2 lofty storeys. Projecting centre with gable. Very bold oversailing eaves on coupled brackets. Stuccoed. Door with six fielded panels. Windows with marginal panes to hung sashes, and shutters. Called 'The Cottage' on 1853 OS map.

(E.A.G.)

CHRONOLOGICAL LIST OF HOUSES BUILT BETWEEN 1800 AND 1850

c. 1800	3, 5 The Green, Acomb.
1807/8	100, 102, 104 The Mount.
1810	28 Micklegate (part of the Adelphi Hotel).
c. 1810/20	11, 13 Green Lane, Acomb.
1810/35	St. Stephen's, 89 The Mount.
1811	St. Mary's Row, Bishophill, later Victor Street.
1815/20	8, 9, 10 Albion Street; 1, 6, 9, 11–16 Bridge Street.
c. 1820	29, 38 Blossom Street; 218 Mount Vale; 23 Nunnery Lane.
c. 1820/30	58 The Green, Acomb.
c. 1820/40	222, 224 Mount Vale.
1821	92, 94 The Mount.
1821/2	78, 80, 82, 84 Micklegate.
1822	16 Blossom Street.
1823	Rosemary Terrace, Skeldergate; 1 Cherry Hill; 12–14 Clementhorpe; building started in Holgate Road; 149, 151 Mount Vale.
1823/8	34–46 Dale Street.
1823/30	43 Blossom Street; 1–10 Mount Parade.
1824	136, 138, 140, 142, 144 The Mount.
1825/8	South Parade.
By 1827	6 Mount Terrace.
1827	147 Mount Vale.
1827/8	1, 3, 4, 5 Mount Terrace.
1827/30	Dove Street.
1828	39 Blossom Street; York Villas, Holgate Road.
1828/9	Swann Street; 65, 67, 69, 71, 73 Acomb Road (Severus Place).
c. 1828/30	77 Holgate Road.
1829	House at entrance to Dale Street, Nunnery Lane.
1829/30	8, 9, 10 Mount Parade.
c. 1830	7 Bishopgate Street; 3, 4 Bridge Street; 16, 18, 25, 29, 31, 33, 56 The Green, Acomb; 18, 20, 63, 65, 71, 73, 75, 171 Holgate Road; 69, 71, 73, 75, 130, 132 The Mount; 11, 19 Mount Parade; 17A, 21, 51, 55, 57 Nunnery Lane; 34 Tadcaster Road, Dringhouses.
1830/35	4, 5, 6 Bishopgate Street.
c. 1830/40	226, 228 Mount Vale.
1831/2	77, 79 The Mount.
1833	117, 119, 121, 123, 125, 127, 131 The Mount; 300, 302, 304, 306 Tadcaster Road.
After 1834	Cumberland House, 20 Mount Parade.
1834/5	St. Catherine's Hospital, Holgate Road.
1835	39, 41 Micklegate.
c. 1835	5, 7, The Punch Bowl Public House Blossom Street; 130 Holgate Road; 17, 18, 19, 20 Queen Street.
1835/40	126 Holgate Road; 9, 11 Park Street; 46 St. Paul's Square.
1836	7 Park Street.
1838	Warehouse, North Street.
1839/40	129 The Mount.
c. 1840	11 Acomb Road; 27 Blossom Street; 9 Green Lane, Acomb; 22, 26, 120, 122 Holgate Road; 50, 50A, 63, Micklegate; 116, 118, 126, 128 The Mount; 16, 17, 18 Mount Parade; 159, 206, 208, 210, 212, 214, 220 Mount Vale; 12, 14 Tanner Row.
c. 1840/50	5 Bridge Street; 29, 53, 77, 95 Front Street, Acomb; 2, 4, 6, 8, 10, 12, 16 Holgate Road; 40 Micklegate; 11–16 Queen Street; 53 Skeldergate; 60 York Road, Acomb.
1842	2 Bridge Street; new houses in Holgate Road.
1842/3	94, 96 Micklegate; 120 The Mount; 1–17 Rougier Street.
1843	124 The Mount; 31 Skeldergate.
1843/6	Rosary Terrace.
c. 1845	7–11 Clementhorpe; Eagle Street; Toft Green Chambers.
c. 1845/50	28–90 (even), Holgate Road; 39, 43 Tanner Row.
1846	21 Blossom Street; Cambridge Street; Cygnet Street; Lowther Terrace; Railway Street (Hudson Street) being laid out; 43, 45, 47 Trinity Lane.
1846/50	Mount Ephraim; Oxford Street; Holgate Terrace.
1847/50	13, 15, 17, 19, 21 Park Street.
1848/9	122 The Mount.
1849	Warehouse, Queen's Staith.
1850	26, 27 Bishopgate Street; Mill Mount School.
1850/1	St. Paul's Vicarage, 92 Holgate Road.
c. 1850	Acomb Park; Westfield House, Askham Lane; 31, 33 Blossom Street; 52, 55–59, 79, 87, 105, 107 Front Street, Acomb; Sun Inn, The Green, Acomb; Holgate Hill Hotel; 60, 72, 74, 76, 78, 80–88, 159 (Ashfield) Tadcaster Road, Dringhouses; 56 Tanner Row.

ARMORIAL INDEX

OF ROYAL AND OTHER HERALDRY

This list contains blazons of Royal Arms before 1850 and of other arms before 1700. The blazons are given as they appear on the monuments surveyed and are not necessarily the generally accepted versions.
The suffixes 'a' and 'b' denote the first and second column of the page.

ROYAL ARMS

GENERAL ARMORIAL

UNIDENTIFIED COATS

1. Quarterly: (1) *gules, a fess or* (? Beauchamp); (2) *gules, a fret or* (? Audley), (3) *gules, a bend or* (? Folliott or Hastings); (4) *argent, on two bars gules six bezants three and three* (? Martyn). (4), p. 8b.
2. [Blank] impaling *gules, a chevron argent.* (4), p. 8b.
3. *Gules, on a chevron between three unicorns passant argent three mullets sable* impaled by YORKE. (6), p. 19a.

GLOSSARY

OF THE MEANING ATTACHED TO THE TECHNICAL TERMS USED IN THE INVENTORY

ABACUS—The uppermost member of a capital.

ABUTMENT—The solid lateral support of an arch.

ACANTHUS—A plant represented in stylised form in Classical and Renaissance ornament, in particular in the capitals of the Corinthian and Composite Orders.

ACHIEVEMENT—In heraldry, the shield accompanied by the appropriate external ornaments, helm, crest, mantling, supporters, etc. In the plural the term is also applied to the insignia of honour carried at the funerals and suspended over the monuments of important personages, comprising helmet and crest, shield, tabard, sword, gauntlets and spurs, banners and pennons. (*See also* HATCHMENT.)

ACROTERIA—In Classical architecture, blocks on the apex and lower ends of a pediment, often carved with honeysuckle or palmette ornament.

ALTAR TOMB—A modern term for a tomb of stone or marble resembling, but not used as, an altar.

ANNULET—In architecture, a small flat fillet encircling a column or shaft.

ANTHEMION—Honeysuckle or palmette ornament in Classical architecture.

APSE—A semicircular or polygonal recess, semi-domed or vaulted, in or projecting from a building.

ARABESQUE—A highly stylised fret-ornament in low relief, common in Moorish architecture, found in 16th and 17th-century work in England.

ARCADE—A range of arches carried on piers or columns. *Blind arcade*, a series of arches, sometimes interlaced, carried on shafts or pilasters against a solid wall.

ARCH—The following are some of the most usual forms:

Equilateral—A pointed arch struck with radii equal to the span.

Flat or straight—Having the soffit horizontal.

Four-centred, depressed, Tudor—A pointed arch of four arcs, the two outer and lower arcs struck from centres on the springing line and the two inner and upper arcs from centres below the springing line. Sometimes the two upper arcs are replaced by straight lines.

Lancet—A pointed arch struck with radii greater than the span.

Ogee—A pointed arch of four or more arcs, the two uppermost being reversed, *i.e.*, convex instead of concave to the base line.

Pointed or two-centred—Two arcs struck from centres on the springing line, and meeting at the apex with a point.

Relieving—An arch, generally of rough construction, placed in the wall above the true arch or head of an opening, to relieve it of most of the superincumbent weight.

Segmental—A single arc struck from a centre below the springing line.

Segmental-pointed—A pointed arch, struck from two centres below the springing line.

Skew—Spanning between responds not diametrically opposite.

Squinch—*See* SQUINCH.

Stilted—An arch with its springing line raised above the level of the imposts.

Three-centred, elliptical—Formed with three arcs, the middle or uppermost struck from a centre below the springing line.

ARCHITRAVE—The lowest member of an entablature (*q.v.*); often adapted as a moulded enrichment returned round the jambs and head of a doorway or window opening.

ARCHIVOLT—In Classical architecture, the moulding round an arch.

ARRIS—The sharp edge formed by the meeting of two surfaces.

ASHLAR—Masonry wrought to an even face and square edges.

ASHLARING—In carpentry, a series of short vertical timbers in the lower angle of a pitched roof.

ASTRAGAL—A small semicircular moulding or bead.

ATTIC—A low storey above an entablature or cornice; also, a storey wholly or partly within the roof.

ATTIC BASE—A moulded column-base with a profile comprising two *torus* (convex) mouldings divided by a *scotia* (concave) between two fillets. In Romano-British examples the fillets are often omitted.

AUMBRY—Cupboard in a church for housing the sacred vessels.

BACK HOUSE—A residential building behind the main part of a house.

BALL FLOWER—In architecture, a decoration, peculiar to the first quarter of the 14th century, consisting of a globular flower of three petals enclosing a small ball.

BAND OR PLAT BAND—A flat projecting horizontal strip of masonry or brickwork across the face of a building, as distinct from a moulded string.

BARGE BOARD—A timber plank, often carved, fixed to the edge of a gabled roof at a short distance from the face of the wall, to protect projecting timbers.

BAROQUE—A style of architecture and decoration emerging in the 17th century which uses the repertory of classical forms with great freedom to emphasise the unity and pictorial character of its effects. The term is also applied to sculpture and painting of a comparable character.

BARREL VAULTING—*See* VAULTING.

BATTLEMENT—In fortification, the alternating merlons and embrasures on the parapet or breastwork of a rampart walk; hence *Battlemented* (or *Embattled*, a usage for the decorative adaptation of the feature).

BAY—The main divisions of a building or feature, on plan or in elevation, defined by recurring structural members, as in an arcade, a fenestrated façade or a timber frame.

BEADING—Small round moulding.

BILLET—In architecture, an ornament used in the 11th and 12th centuries consisting of short attached cylinders or rectangles with spaces between.

BOLECTION MOULDING—A bold moulding of double curvature raised above the general plane of the framework of a door, fireplace or panelling.

BOND—*See* BRICKWORK.

BOSS—A projecting square or round ornament, covering the intersections of the ribs in a vault, panelled ceiling or roof, etc.

BOTTOM RAIL—The lowest horizontal timber of a door, partition, window sash, etc.

BOWTELL—A round moulding.

BRACE—In timber framing and timber roof construction, subsidiary timber rising obliquely from a major vertical member to support a major horizontal member (in contradistinction to a STRUT, *q.v.*). *Arch-brace*, when curved. *Wind-brace*, a subsidiary timber placed diagonally between the principals and purlins of a roof to increase resistance to wind-pressure.

BRATTISHING—Ornamental cresting on the top of a screen, cornice, etc.

BRESSUMER—A spanning beam forming the direct support of a wall or timber-framing above it.

BRICKS and BRICKWORK

Cutter—A brick of very fine quality used for arches, quoins, etc., and capable of being cut.

Header—A brick laid so that the end appears on the wall face.

Stretcher—A brick laid so that the side appears on the wall face.

English Bond—A method of laying bricks so that alternate courses appear as all headers and all stretchers on the wall face.

English Garden Wall Bond—In which three courses of stretchers to one course of headers appear on the wall face.

Facing Bond—In which expensive bricks concealing a core of common brickwork are laid, usually in English Garden Wall bond (*q.v.*).

Flemish Bond—In which alternate headers and stretchers in each course appear on the wall face.

Stretching Bond—In which only stretchers appear on the wall face.

BROACH STOP—A half-pyramidal stop against a chamfer, to effect the change from chamfer to right angle.

BUTTRESSES—Projecting masonry or brickwork support to a wall.

Angle—Two meeting, or nearly meeting, at right-angles at a corner.

Clasping—Returned to encase an angle.

Diagonal—Projecting diagonally at a corner.

Lateral—At a corner of a building and axial with one wall.

CABLE MOULDING—A moulding carved in the form of a rope or cable.

CANONS (of a bell)—The metal loops by which a bell is hung.

CANOPY—A projection or hood over a door, window, etc.; the covering over a tomb or niche.

CARINATED—Having an angular profile.

CARTOUCHE—In Renaissance ornament, a tablet imitating a scroll with ends rolled up, used ornamentally or bearing an inscription or arms.

CARYATID—Sculptured female figure used as column or support.

CASEMENT—1. A wide hollow moulding in a window jamb, etc.; 2. the hinged part of a window which opens sideways.

CAVETTO—A hollow moulding, in profile a quarter-circle.

CENTRING—Temporary wooden framework used to support an arch or vault during construction.

CHALICE—The name used in the Inventory to distinguish the pre-Reformation type of Communion cup with a shallow bowl from the post-Reformation cup with a deeper bowl.

CHAMFER—The small plane formed when a sharp edge or arris is cut away, usually at an angle of 45°; *hollow chamfer*, when the plane is concave; *sunk chamfer* when it is recessed.

CHANTRY CHAPEL—A chapel built for the purposes of a chantry (*i.e.* a foundation, usually supporting a priest, for the celebration of masses for the souls of the founder and such others as he may direct).

CHEVRON—In architecture, a decorative form resembling an inverted V and often used in a consecutive series.

CINQUEFOIL—*See* FOIL.

CLEARSTOREY or CLERESTORY—An upper stage, pierced by windows, in the main walls of a church or domestic building.

CLEAT—Projecting block of wood.

COFFERS—Sunk panels in ceilings, vaults, domes and arch-soffits.

COLLAR BEAM—In a roof, a horizontal beam framed to and serving to tie together a pair of rafters at some distance above wall-plate level.

COLLAR PURLIN—*See* PURLIN.

CONSOLE—A bracket with a compound-curved outline.

COPED SLAB—A slab of which the upper face is ridged down the middle, and sometimes hipped at each end.

CORBEL—A projecting stone or piece of timber for the support of a superincumbent weight. *Corbel-table*—A row of corbels, usually carved, and supporting a projection.

CORNICE—A crowning projection. In Classical architecture, the crowning or upper portion of the entablature.

CORONA—The square projection with vertical face and wide soffit in the upper part of a Classical cornice.

COVER PATEN—A cover to a Communion cup, used as a paten.

CREST—A device worn upon a helm or helmet. *Cresting*, an ornamental finish along the top of a screen, etc.

CROCKETS—Carved projections spaced, usually at regular intervals, along the vertical or sloping sides of spires, canopies, pinnacles, hood-moulds, etc.

CROP-MARK—Trace of a levelled or buried feature revealed on the land surface by differential growth of crops, especially after drought.

CROSS—In simplest form, with plain arms at right-angles, as for St. George. Other forms include: *Crosslet*—with a smaller arm crossing each main arm; *Flory*—with the arms ending in fleurs-de-lis; *Formy*—with the arms widening from the centre and square at the ends; *Potent*—with a small transverse arm on the end of each main arm; *Saltire* or *St. Andrew's*—X-shaped; *Tau* or *Anthony*—T-shaped.

CROWN POST—A vertical post standing centrally on a tie-beam to give direct support to a collar purlin, and often also with four-way struts to the nearest rafter-couple and to the collar purlin.

CRUCK TRUSS—*See under* ROOFS.

CUSHION CAPITAL—A capital cut from a cube with its lower angles rounded off to adapt it to a circular shaft.

CUSP—A pointed projection from the soffit of an arch, formed by two arcs of smaller radius. The foils in Gothic windows, arches, panels, etc., are formed by *cusping* and *sub-cusping*, often ornamented at the ends (*cusp-points*) with carving.

CYMA—A moulding with a wave-like outline consisting of two contrary curves.

DADO—The separate protective or decorative treatment applied to the lower parts of wall-surfaces to a height, normally, of 3 ft. to 4 ft. *Dado rail*, the moulding or capping at the top of the dado.

DENTILS—The small rectangular tooth-like blocks used decoratively in Classical cornices.

DIAPER—All-over decoration of surfaces with squares, diamonds, or other patterns.

DIE—The part of a pedestal between the base and the cornice.

DOG-LEG STAIRCASE—*See* STAIRCASE.

DOG-TOOTH ORNAMENT—A typical 13th-century carved enrichment consisting of a series of pyramidal flowers of four petals; often used to enrich hollow mouldings.

DORMER WINDOW—A vertical window on the slope of a roof and having a roof of its own.

DOUBLE-OGEE MOULDING—*See* OGEE.

DRAGON BEAM—A ceiling beam on the diagonal into which are housed the ends of the joists that form jetties on two adjacent fronts of a building.

DRAWBAR—A stout timber for securing a door; it fits in a long tunnel in one jamb whence it can be pulled, across the door, to engage in a socket in the opposite jamb.

DRESSINGS—The building materials specially chosen or treated defining or emphasising the architectural features of an elevation.

DRIP STONE—*See* HOOD MOULD.

EMBATTLED—*See* BATTLEMENT.

EMBRASURES—The openings in an embattled parapet, or the recesses for windows, doorways, etc.

ENTABLATURE—In Classical and Renaissance architecture, the part of an order above the column, the full entablature comprising *architrave*, *frieze*, and *cornice*; often used alone, in whole or in part, as a horizontal architectural feature.

ENTASIS—The convexity or swell on a vertical or near vertical line or surface, to correct optical illusion of concavity.

EXTRADOS—The outer curve of the voussoirs of an arch.

FANLIGHT—Glazed opening immediately over, and integrated within the framing of, a doorway.

FASCIA—A plain or moulded facing board.

FILLET—In mouldings, a plain narrow band between, or adjacent to, more complex mouldings.

FINIAL—A stylised ornament at the top of a pinnacle, gable, canopy, etc.

FOIL (*trefoil, quatrefoil, cinquefoil, multifoil*, etc.)—The shape defined by the curves formed by cusping.

FOLIATED (of a capital, corbel, etc.)—Carved with leaf ornament.

FOUR-CENTRED ARCH—*See* ARCH.

FRIEZE—The middle zone in an *entablature*, between the *architrave* and the *cornice*; generally any band of ornament or colour immediately below a cornice.

GADROONING—Decorative enrichment comprising a series of convex ridges, the converse of fluting, forming an ornamental edge or band.

GARDEROBE—Wardrobe. Antiquarian usage applies it to a latrine or privy chamber.

GAUGING—In brickwork, cutting and rubbing bricks to a particular shape. Specially made soft bricks are used for the purpose.

GESSO—A mixture of whiting and size, spread on stone or wood as a ground for painting.

GRAFFITO—Scratched inscription or design.

GRID-IRON TRACERY—*See* TRACERY.

GRISAILLE—Painting in shades of grey.

GROINING, GROINED VAULT—*See* VAULTING.

GROUND SILL—The lowest horizontal timber on which a wall or partition is erected.

GUILLOCHE—A geometrical ornament consisting of two or more undulating bands intertwining to form a series of circles.

GUTTAE—Small stud-like projections under the triglyphs and mutules of the Doric entablature.

GYPSUM—Hydrated sulphate of lime ($CaSO_4 + 2H_2O$), a comparatively soft mineral found in Yorkshire along the W. side of the Vale of York. On rehydration after heating it will set hard.

HALF-HIPPED ROOF—*See* HIPPED ROOF.

HALL—In a mediaeval house, the principal room, often open to the roof.

HAMMER-BEAMS—Horizontal brackets of a timber roof projecting at wall-plate level (as if a tie-beam with the middle part cut away); they are braced and help to diminish lateral pressure by reducing the roof span. Sometimes there is a second and even a third upper tier of these brackets. *Hammer-post*, a stout vertical post rising from the forward end of a hammer-beam to support a plate or purlin or the principal rafter above.

HATCHMENT—In modern usage, the large square or lozenge-shaped framed painting displaying the armorial bearings of a deceased person. It was first hung outside his house and then laid up in the church.

HIPPED ROOF—A roof with sloping instead of vertical ends. *Half-hipped*, a roof with ends partly vertical, and partly sloping.

HOG-BACK—A type of late Saxon stone grave-cover shaped with a curved ridge forming a 'hog-back'.

HOOD MOULD or LABEL—A projecting moulding on the wall face above an opening or feature; it may follow the form of the arch or head of the same or be square in outline.

IMPOST—The projection, often moulded, at the springing of an arch, upon which the arch appears to rest.

INDENT—The sinking in a slab for a monumental brass.

INTAGLIO—A cutting or engraving into any substance for decorative effect; thus the pattern is within the surface of the material.

INTRADOS—The inner curve of an arch.

JACK RAFTER—A shortened rafter, *e.g.* running from hip to eaves or from ridge to valley.

JAMBS—The sides of an archway, doorway, window, or other opening.

JETTY—The projection of an upper storey of a building beyond the plane of a lower storey.

JOGGLING—The method of cutting the adjoining faces of the voussoirs of an arch with rebated, zigzagged or wavy surfaces to provide a better key or lodgement.

KERB ROOF—*See* ROOFS, *Kerb*.

KEYSTONE—The middle voussoir in an arch.

KING-POST—A vertical post extending from a tie-beam or a collar-beam to the apex of a roof, and supporting a ridge-piece.

KNEELER—In a parapeted gable, the stone or block built well into the wall to resist the sliding tendency of the coping.

LACING COURSE—In masonry or brickwork, a bonding course binding the wall-facing together or to the wall core.

LANCET—A tall, narrow window with a pointed head, typical of the 13th century. *See also* ARCH.

LINENFOLD PANELLING—Wainscot ornamented with stylised representation of folded linen.

LINTEL—The horizontal beam or stone bridging an opening.

LOCKER—A small cupboard formed in a wall. *See also* AUMBRY.

LOOP—A small narrow light, often unglazed.

LOUVRE—A lantern-like structure surmounting the roof of a hall or other building, with openings for ventilation or the escape of smoke; the openings are usually crossed by sloping slats (*louvre boards*) to exclude rain. Louvre boards are also used in the windows of church belfries, instead of glazing, to allow the bells to be heard.

MANDORLA—An aureole or glory in the form of a pointed oval surround. *See also* VESICA PISCIS.

MANNERIST—A use of the repertory of revived antique forms in an arbitrary way.

MANSARD—*See under* ROOFS.

MASK STOP—*See under* STOPS.

METOPES—The panels, often carved, filling the spaces between the triglyphs in the Doric entablature.

MIDDLE RAIL—A horizontal rail between ground sill and wall-plate in a timber-framed wall.

MILL RIND—The iron axle fitting to the centre of a millstone.

MISERICORDE—A bracket, often elaborately carved, on the under side of the hinged seat of a choir stall. When the seat is turned up the bracket comes into position to support the occupant during long periods of standing.

MITRE—The junction of mouldings or strips meeting at an angle; in joinery the joint is commonly on the line of the mitre. A junction in which the joint is straight and not at the angle of the mitre is termed a *mason's mitre*.

MODILLIONS—Brackets under the cornice in a Classical entablature.

MORTICE—A socket cut in a piece of wood, usually to receive the end, the *tenon*, of another piece.

MULLION—An upright of timber, stone or brick dividing an opening into lights.

MUNTIN—In panelling, an intermediate vertical timber between panels and butting into or stopping against the rails.

MUTULES—Shallow blocks under the corona of the cornice in a Classical entablature.

NAIL-HEAD—Small architectural enrichment of pyramidal form, used extensively in 12th-century work.

NECKING or NECK MOULDING—The narrow moulding round the lower extremity of a capital.

NEWEL—The central post in a circular or winding staircase; also the principal post at each angle of a dog-legged or well staircase.

OFFSET—A ledge formed by the set-back of a wall.

OGEE—A compound curve of two parts, one convex, the other concave. A *double-ogee* moulding is formed by two ogee mouldings meeting at their convex ends.

OGEE-BAR STOP—*See* STOP.

ORDERS (of arches)—Receding concentric rings of voussoirs.

ORDERS OF ARCHITECTURE—In Classical or Renaissance architecture, the five systems of columnar architecture, known as Tuscan, Doric, Ionic, Corinthian, and Composite. *Colossal Order*, one in which the columns or pilasters embrace more than one storey of the building.

ORIEL WINDOW—A projecting bay window carried upon corbels or brackets.

OVERSAILING COURSE—A brick or stone course projecting beyond the course below it.

OVOLO MOULDING—A Classical moulding forming a quarter round or semi-ellipse in section.

PALIMPSEST—1. Of a brass: reused by engraving the back of an older engraved plate. 2. Of a wall-painting: superimposed on an earlier painting.

PALL—A cloth covering a hearse.

PALLADIAN or VENETIAN WINDOW—A three-light window, with a tall round-headed middle light and shorter lights on either side, the side lights with flanking pilasters and small entablatures forming the imposts to the arch over the centre light.

PATEN—A dish for holding the Bread at the celebration of Holy Communion.

PATERA, -AE—A square or circular flat ornament applied to a frieze, moulding or cornice; in Gothic work it commonly takes the form of a four-lobed leaf or flower.

PEDIMENT—A low-pitched gable used in Classical and Renaissance architecture above a portico, at the end of a building, or above doorways, windows, niches, etc.; sometimes the gable angle is omitted, forming a *broken pediment*, or the horizontal members are omitted, forming an *open pediment*. A curved gable form is sometimes used in this way.

PELICAN IN PIETY—A pelican shown, according to the mediaeval legend, feeding her young upon the drops of blood she pecks from her breast.

PELLET ORNAMENT—An enrichment consisting of balls or flat discs.

PILASTER—A shallow pier of rectangular section attached to a wall.

PISCINA—A basin for washing the sacred vessels and provided with a drain, generally set in or against the wall to the S. of the altar, but sometimes sunk in the pavement.

PLAT BAND—*See* BAND.

PLINTH—The projecting base of a wall, generally chamfered or moulded at the top.

PODIUM—In Classical architecture, a basis, usually solid, supporting a temple or other superstructure.

POPPY HEAD—The ornament at the heads of bench-standards or desks in churches; generally carved with foliage and flowers, and somewhat resembling a fleur-de-lis.

PORTICO—A covered entrance to a building, colonnaded, either constituting the whole front of the building or forming an important feature.

PRINCIPALS—In a roof of double-frame construction, the main as opposed to the common rafters.

PULVINATED FRIEZE—In Classical and Renaissance architecture, a frieze having a convex or bulging section.

PURLIN—*Collar purlin*, a beam running longitudinally immediately beneath the collars joining pairs of common rafters. *Side purlin*, a horizontal longitudinal member giving intermediate support to the common rafters.

QUARRY—In glazing, small panes of glass, generally diamond-shaped or square set diagonally.

QUEEN-POSTS—In a roof truss, pair of vertical posts equidistant from the centre line of the roof. *See also under* ROOFS.

QUOINS—The dressed stones at the angle of a building, or distinctive brickwork in this position.

RAFTERS—Inclined timbers supporting a roof-covering. *See also under* ROOFS.

RAIL—A horizontal member in the framing of a door, screen, or panel.

REAR ARCH—The arch on the inside of a wall spanning a doorway or window-opening.

REBATE—A continuous rectangular notch.

REEDING—The converse of fluting, *i.e.* with convex not concave moulding.

REELS—Ornament resembling a line of bobbins, used in Classical architecture.

RELIQUARY—A small box or other receptacle for relics.

REREDOS—A screen of stone or wood at the back of an altar, usually enriched.

RESPONDS—The half-columns or piers at the ends of an arcade or at each side of a single arch.

RETICULATED TRACERY—*See under* TRACERY.

REVEAL—The internal side surface of a recess, especially of a doorway or window opening.

RIDGE (or RIG) AND FURROW—Remains of old cultivations.

RISER—The vertical piece connecting two treads in a flight of stairs.

ROCOCO—The latest (18th-century) phase of Baroque, especially in Northern Europe, in which effects of elegance and vivacity are obtained by the use of a decorative repertory further removed from antique architectural forms than the earlier phases and often asymmetrically disposed.

ROLL MOULDING or BOWTELL—A continuous prominent convex moulding.

ROOD (*Rood beam, Rood screen, Rood loft*)—A cross or crucifix. The *Great Rood* was set up at the E. end of the nave with accompanying figures of St. Mary and St. John; it was generally carved in wood, and fixed on the loft or head of the rood screen, or on a special beam (the *Rood beam*) reaching from wall to wall. Sometimes the rood was merely painted on the wall above the chancel-arch or on a closed wood partition or tympanum in the upper half of the arch. The *Rood screen* is the open screen spanning the E. end of the nave, shutting off the chancel; in the 15th century a narrow gallery was often constructed above the cornice to carry the rood and other images and candles, and it was also used as a music gallery. The loft was approached by a staircase (and occasionally by more than one), either of wood or built in the wall, wherever most convenient, and, when the loft was carried right across an aisled building, the intervening walls of the nave were often pierced with narrow archways. Many of the roods were destroyed at the Reformation, and their final removal, with the loft, was ordered in 1561.

ROOFS

Collar-beam—a principal-rafter roof (*q.v.*) with collar-beams connecting the principals.

Cruck (or *Crutch*)—with principals springing from below the level of the wall-plate. The timbers are usually curved.

Hammer-beam—hammer-beams (*q.v.*) instead of tie-beams, braced from a level below the wall-plates, form the basis of construction.

Crown-post—a trussed-rafter roof with a central post (crown-post) standing on a tie-beam and carrying a centre purlin supporting the collars.

Kerb—A double-framed roof in which the principal rafters rise only to the collar, which carries purlins known as kerbs, above which the upper part of the roof is commonly but not necessarily of lower pitch (when it is termed a Mansard roof). *Kerb principal*—A principal rafter rising only from wall-plate to collar.

King-post—in which a central post (king-post) standing on the tie-beam or collar-beam of a truss directly supports the ridge.

Mansard—characterised in exterior appearance by two pitches, the lower steeper than the upper (see *Kerb* roof above).

Principal-rafters—with rafters of greater scantling than the common rafters framed to form trusses at regular intervals along the roof; normally called by the name of the connecting member used in the truss, tie-beam or collar-beam. Post-mediaeval roofs of this kind often have queen-posts.

Queen-post—with two vertical or nearly vertical posts (queen-posts) standing towards either end of the tie-beam of a truss and supporting the collar-beam or the principal rafters.

Scissor-truss—as Trussed-rafter (*q.v.*), but with crossed braces instead of, or as well as, collars.

Tie-beam—a principal-rafter roof with a simple triangulation of a horizontal beam linking the lower ends of the pairs of principals to prevent their spread.

Trussed-rafter—in which each pair of common rafters (all the timbers in the slopes being common rafters of uniform size) is connected by a collar beam, which is often braced. At intervals, pairs of rafters may be tenoned into a tie-beam.

Wagon—a trussed-rafter roof with curved braces forming semi-circular arches springing from wall-plate level.

RUBBLE—Walling of rough unsquared stones or flints. *Coursed Rubble*, rubble walling with the stones or flints very roughly dressed and levelled up in courses; in *Regular Coursed Rubble* the stones or flints are laid in distinct courses, being kept to a uniform height in each course.

RUSTICATION—Primarily, masonry in which only the margins of the stones are worked, also used for any masonry where the joints are emphasised by mouldings, grooves, etc.; rusticated columns are those in which the shafts are interrupted by square blocks of stone or broad projecting bands. *Rupilation*—Masonry faced to resemble a waterworn rock surface.

SACRISTY—A room generally in immediate connection with a church, in which the holy vessels and other valuables are kept.

SARCOPHAGUS—A stone coffin, usually inscribed and often embellished with sculptures, intended to be viewed above ground or in a tomb chamber.

SCALLOPED CAPITAL—A development of the cushion capital (*q.v.*) in which the single cushion is elaborated into a series of truncated ones.

SCUTCHEON or ESCUTCHEON—A metal plate pierced for the spindle of a handle or for a keyhole.

SEDILIA (sing. *Sedile*, a seat)—In a church or chapel, the seats, usually incorporated in the wall or screen south of the altar, used by the ministers during the Mass.

SHAFT—A slender column. *Shafted jambs*, reveals of a wall opening elaborated with one or more shafts, either engaged or detached.

SIDE-PURLIN—*See* PURLIN.

SILL—The lower horizontal member of a window or door-frame; the stone, tile or wood base below a window or door-frame, usually with a weathered surface projecting beyond the wall face to throw off water. In timber-framed walls, the lower horizontal member into which the studs are tenoned.

SOFFIT—The under side of an arch, staircase, lintel, cornice, canopy, etc.

SOFFIT CUSPS—Cusps springing from the flat soffit of an arch, and not from its chamfered sides or edges.

SOLE-PIECE—In a timber-framed building a short horizontal timber forming the junction between a wall-post and a prinicipal rafter, where there is no tie-beam.

SPANDREL—The more or less triangular space between an angle and a contained curve.

SPLAY—A sloping face making an angle of more than a right angle with another face, as in internal window jambs, etc.

SPRINGING LINE—The level at which an arch springs from its supports.

SPUR—Carved tongue, foliage or grotesque filling each spandrel between a circular base and a square or polygonal plinth.

SQUINCH—An arch thrown across the angle between two walls to support a superstructure, such as the base of a stone spire.

SQUINT—A piercing through a wall to allow a view of an altar from places whence it could otherwise not be seen.

STAGES—The divisions (*e.g.* of a tower) marked by horizontal string-courses.

STAIRCASE—A *close-string* staircase is one having a raking member into which the treads and risers are housed. *An open-string* staircase has the raking member cut to the shape of the treads and risers. A *dog-leg* staircase has adjoining flights running in opposite directions with a common newel. A *well staircase* has stairs rising round a central opening more or less as wide as it is long.

STANCHIONS—The upright iron bars in a screen, window, etc.

STILE—The vertical members of a frame into which are tenoned the ends of the rails or horizontal pieces.

STOPS—Blocks terminating mouldings or chamfers in stone or wood; stones at the ends of labels, string-courses, etc., against which the mouldings finish, frequently carved to represent shields, foliage, human or grotesque masks; also, plain or decorative, used at the ends of a moulding or a chamfer to form the transition thence to the square. *Ogee-bar*, a form of stop in which the chamfer is barred by a bowtel, beyond which an ogee moulding forms the transition to the square.

STOUP—A receptacle, normally by the doorway of a church, to contain holy water; those remaining are usually in the form of a deeply-dished stone set in a niche or on a pillar.

STRAIGHT JOINT—A vertical joint in a wall usually signifying different phases of building.

STRAPWORK—Decoration comprising carved or painted interlacing and intersecting bands, much used in Elizabethan and Jacobean work.

STRING, STRING COURSE—A projecting moulded band across a wall. *See also* STAIRCASE.

STRUT—In timber framing and roof-construction a subsidiary oblique timber rising from a horizontal member to give support to a vertical post or to a rafter (in contradistinction to BRACE, *q.v.*).

STUDS—The common posts or uprights in timber-framed walls.

SWAG—Decorative representation of a festoon of cloth or flowers and fruit suspended from both ends.

TESSERA, -AE—A small cube of stone, glass, or marble, used in mosaic.

TIE-BEAM—The horizontal transverse beam in a roof, tying together the feet of pairs of rafters to counteract thrust.

TOOLING—Dressing or finishing a masonry surface with an axe or other tool, usually in parallel lines. *Diagonal*—in England, often characteristic of Norman masonry. A change from diagonal tooling to vertical has been noted at Wells Cathedral *c.* 1210 (*Arch. Jour.* LXXXV). *Hammer-dressed or Nigged*—hewn with a pick or pointed hammer, instead of a chisel.

TORUS—In Classical architecture, a convex moulding, generally a semicircle in section.

TOUCH—A soft black marble, quarried near Tournai and used in monumental art.

TRABEATION—The use of horizontal beams in building construction; descriptive in the Inventory of conspicuous cased ceiling-beams.

TRACERY—The ornamental work in the head of a window, screen, panel, etc., formed by the curving and interlacing of bars of stone or wood, grouped together, generally over two or more lights or bays. *Grid-iron*, a form of late Perpendicular tracery in which mullions are crossed by transoms, giving the effect of a grille. *Reticulated*, formed by repetitive curvilinear quatrefoils or trefoils.

TRANSOM—An intermediate horizontal bar of stone or wood across a window-opening. The horizontal member of a door-frame beneath a fanlight.

TREAD—The horizontal platform of a step or stair.

TRELLIS—Latticework of light wood or metal bars.

TRIGLYPHS—Blocks, with vertical channels, placed at intervals along the frieze of the Doric entablature.

TRUSS—A number of timbers framed together to bridge a space, to be self-supporting, and to carry other timbers. The trusses of a roof are generally named after a particular feature in their construction, e.g. *King-post, Queen-post*; see under ROOFS.

TYMPANUM—The triangle in the face of a pediment or the semicircle in the head of an arch.

VAULTING—An arched ceiling or roof of stone or brick, sometimes imitated in wood or plaster. *Barrel vault*, a tunnel vault unbroken in its length by cross vaults. *Groined vault* (or *cross vault*), resulting from the intersection of simple vaulting surfaces. *Ribbed vault*, with a framework of arches carrying the covering of the spaces between them. One bay of vaulting divided into four quarters or compartments is termed *quadripartite*.

VENETIAN WINDOW—*See* PALLADIAN WINDOW.

VERGE—The slightly projecting edge of a roof-covering along the sloping gable-end of a roof.

VESICA PISCIS—A pointed oval setting, usually for the figure of Christ enthroned or the Virgin, familiar in mediaeval representations. *Also* MANDORLA (*q.v.*).

VOLUTE—An ornament in the form of a spiral, *e.g.* in the Ionic capital.

VOUSSOIRS—The wedge-shaped stones forming an arch.

WAGON ROOF—*See under* ROOFS.

WAINSCOT—Wood panelling. Oak imported for this purpose from the Baltic was also so called.

WALL-PLATE—A timber laid lengthwise at the wall top to receive the ends of the roof rafters and other joists. In timber-framing, the studs are also tenoned into it.

WAVE MOULDING—A compound moulding formed by a convex curve between two concave curves.

WEATHERBOARDING—Horizontal, overlapping planks nailed to the uprights of timber-framed buildings to keep out the weather. The boards are generally wedge-shaped in section, the upper edge being the thinner.

WEATHERING (to sills, tops of buttresses, etc.—) A sloping surface for casting off water.

WELL STAIRCASE—*See* STAIRCASES.
WIMPLE—Scarf covering chin and throat.
WIND BRACE—*See* BRACE.
WINDOW—*Casement*, with hinged glazed panels. *Sash* (or *hung-sash*), glazed panels sliding vertically. *Sliding-sash*, with glazed panels sliding horizontally (also *Yorkshire sash*).

YORKSHIRE SASH—*See* WINDOW—*Sliding-sash*.

INDEX

Numbers in brackets refer to the serial numbers of the monuments.
The letters 'a' and 'b' denote left and right-hand columns respectively.

Dd 147344 K 24